T0226904

Trace Elements in Human
and Animal Nutrition—Fifth Edition
Volume 1

Trace Elements in Human and Animal Nutrition—Fifth Edition

Volume 1

Edited by

WALTER MERTZ

U. S. Department of Agriculture
Agricultural Research Service
Beltsville Human Nutrition Research Center
Beltsville, Maryland

Editions 1–4 were prepared by
the late Dr. Eric J. Underwood

ACADEMIC PRESS, INC.
Harcourt Brace Jovanovich, Publishers

San Diego   New York   Berkeley   Boston
London   Sydney   Tokyo   Toronto

ACADEMIC PRESS, INC.
San Diego, California 92101

United Kingdom Edition published by
ACADEMIC PRESS LIMITED
24-28 Oval Road, London NW1 7DX

Library of Congress Cataloging-in-Publication Data

(Revised for vol. 1)

Trace elements in human and animal nutrition.

  Rev. ed. of: Trace elements in human and animal
nutrition / Eric J. Underwood. 4th ed. c1977.
  Includes bibliographies and indexes.
  1. Trace elements in nutrition. 2. Trace elements
in animal nutrition. I. Mertz, Walter, Date.
II. Underwood, Eric J. (Eric John), Date. Trace
elements in human and animal nutrition. [DNLM: 1. Nutrition. 2. Trace Elements. QU 130 T75825]
I. Mertz, Walter, Date. II. Underwood, Eric J.
(Eric John), Date. Trace elements in human and
animal nutrition.
QP534.T723 1986   599′.013   85-20106
ISBN 0-12-491251-6 (v. 1 : alk. paper)

PRINTED IN THE UNITED STATES OF AMERICA
89 90 91 92 93   9 8 7 6 5 4 3
Transferred to digital printing 2005

Contents

Contributors   ix

Preface   xi

1. Introduction
Eric J. Underwood and Walter Mertz

    I. The Nature of Trace Elements   1
   II. Discovery of Trace Elements   5
  III. Mode of Action of Trace Elements   9
  IV. Trace Element Needs and Tolerances   11
   V. Trace Element Deficiencies   14
      References   17

2. Methods of Trace Element Research
J. Cecil Smith, Jr.

    I. Nature's Experiments   21
   II. Early Attempts Using Purified Diets to Study Trace
       Elements   23
  III. Scientific Era: Development of Adequate Purified Diets   24
  IV. The "New" Trace Elements   26
   V. Advances in Analytical Methodology   26
  VI. Factors to Consider in the Design of Human Studies   40
 VII. Suggested Future Research   49
      References   51

3. Quality Assurance for Trace Element Analysis
Wayne R. Wolf

    I. Introduction   57
   II. Accuracy-Based Measurement   59

III. Sampling   62
IV. Generating Valid Analytical Data   64
V. Development of Analytical Methodology   64
VI. Validation of Analytical Methodology   72
VII. Quality Control of Analytical Methodology   74
VIII. Data Handling and Evaluation   74
IX. Conclusion   76
References   77

4. Iron

Eugene R. Morris

I. Iron in Animal Tissues and Fluids   79
II. Iron Metabolism   91
III. Iron Deficiency   108
IV. Iron Requirements   116
V. Sources of Iron   119
VI. Iron Toxicity   125
References   126

5. Cobalt

Richard M. Smith

I. Cobalt as an Essential Trace Element   143
II. Cobalt in Animal Tissues and Fluids   148
III. Cobalt Metabolism   150
IV. Cobalt in Ruminant Nutrition   152
V. Cobalt in the Nutrition of Humans and Other
Nonruminants   172
VI. Cobalt Toxicity   174
VII. Mode of Action of Cobalt in Vitamin B_{12} 175
References   176

6. Manganese

Lucille S. Hurley and Carl L. Keen

I. Manganese in Animal Tissues and Fluids   185
II. Manganese Metabolism   191
III. Biochemistry of Manganese   195
IV. Manganese Deficiency and Relation to Biochemical
Function   197
V. Manganese Requirements   207
VI. Sources of Manganese   211
VII. Manganese Toxicity   213
References   215

7. Chromium

Richard A. Anderson

I.	Chromium in Animal Tissues and Fluids	225
II.	Chromium Metabolism	229
III.	Chromium Deficiency and Functions	232
IV.	Chromium Sources and Requirements	238
V.	Chromium Toxicity	240
	References	240

8. Nickel

Forrest H. Nielsen

I.	Nickel in Animal Tissues and Fluids	245
II.	Nickel Metabolism	248
III.	Nickel Deficiency and Functions	251
IV.	Nickel Requirements and Sources	259
V.	Nickel Toxicity	262
	References	266

9. Vanadium

Forrest H. Nielsen

I.	Vanadium in Animal Tissues and Fluids	275
II.	Vanadium Metabolism	277
III.	Vanadium Deficiency and Functions	279
IV.	Vanadium Requirements and Sources	291
V.	Vanadium Toxicity	292
	References	294

10. Copper

George K. Davis and Walter Mertz

I.	Copper in Animal Tissues and Fluids	301
II.	Copper Metabolism	320
III.	Copper Deficiency and Functions	326
IV.	Copper Requirement	337
V.	Copper in Human Health and Nutrition	342
VI.	Copper Toxicity	347
	References	350

11. Fluorine

K. A. V. R. Krishnamachari

I.	Fluoride Toxicity in Humans	365
II.	Fluoride Toxicosis in Cattle	387

III.	Studies on Experimental Animals	395
IV.	Fluoride Metabolism	398
V.	Fluoride and Dental Caries	402
VI.	Natural Distribution of Fluorides	405
	References	407

12. Mercury

Thomas W. Clarkson

I.	Introduction	417
II.	Mercury in Animal Tissues and Fluids	417
III.	Mercury Metabolism	420
IV.	Sources of Mercury	422
V.	Mercury Toxicity	424
	References	426

13. Molybdenum

Colin F. Mills and George K. Davis

I.	Molybdenum in Animal Tissues	429
II.	Molybdenum in Foods and Feeds	435
III.	Molybdenum Metabolism	437
IV.	Functions of Molybdenum and Effects of Deficiency	439
V.	Responses to Elevated Intakes of Molybdenum	444
	References	457

| *Index* | | 465 |

Contributors

Numbers in parentheses indicate the pages on which the authors' contributions begin.

RICHARD A. ANDERSON (225), U.S. Department of Agriculture, Agricultural Research Service, Beltsville Human Nutrition Research Center, Beltsville, Maryland 20705

THOMAS W. CLARKSON (417), Environmental Health Sciences Center, University of Rochester School of Medicine, Rochester, New York 14642

GEORGE K. DAVIS[1] (301, 429), Distinguished Professor Emeritus, University of Florida, Gainesville, Florida 32601

LUCILLE S. HURLEY (185), Department of Nutrition, University of California, Davis, California 95616

CARL L. KEEN (185), Department of Nutrition, University of California, Davis, California 95616

K. A. V. R. KRISHNAMACHARI (365), National Institute of Nutrition, Hyderabad 500007, India

WALTER MERTZ (1, 301), U.S. Department of Agriculture, Agricultural Research Service, Beltsville Human Nutrition Research Center, Beltsville, Maryland 20705

COLIN F. MILLS (429), The Rowett Research Institute, Bucksburn, Aberdeen AB2 9SB, Scotland

EUGENE R. MORRIS (79), U.S. Department of Agriculture, Agricultural Research Service, Beltsville Human Nutrition Research Center, Beltsville, Maryland 20705

FORREST H. NIELSEN (245, 275), United States Department of Agriculture, Agricultural Research Service, Grand Forks Human Nutrition Research Center, Grand Forks, North Dakota 58202

J. CECIL SMITH, JR. (21), U.S. Department of Agriculture, Agricultural Research Service, Beltsville Human Nutrition Research Center, Beltsville, Maryland 20705

[1]Present address: 2903 S.W. Second Court, Gainesville, Florida 32601.

RICHARD M. SMITH (143), CSIRO Division of Human Nutrition, Adelaide, South Australia 5000, Australia

ERIC J. UNDERWOOD (1)[2]

WAYNE R. WOLF (57), U.S. Department of Agriculture, Agricultural Research Service, Beltsville Human Nutrition Research Center, Beltsville, Maryland 20705

[2]Deceased.

Preface

Eric J. Underwood died from a heart ailment on 18 August 1980, in Perth, Australia. During his long, productive career he saw his field of scientific interest, trace element nutrition and physiology, expand from small, isolated activities into an area of research that has gained almost universal recognition in basic and applied sciences and transcends all barriers of disciplines. He contributed to this development more than any other person of our time, through his own research, which laid the foundation for our understanding of the physiological role of cobalt, through his numerous national and international consulting activities, and through his book *Trace Elements in Human and Animal Nutrition,* which he authored in four editions. The book has assumed an eminent place among publications in trace element nutrition, not only as a source of data but because it offered mature judgment in the interpretation of the diverse results. In his preface to the fourth edition, Underwood stated that "the overall aim of the book has remained, as before, to enable those interested in human and animal nutrition to obtain a balanced and detailed appreciation of the physiological roles of the trace elements. . . ." This goal remains the dominating guideline for the fifth edition.

The invitation of the publisher to assure the continuation of the book was discussed in detail with Dr. Erica Underwood and Dr. C. F. Mills, who had been closely associated with the previous editions. Both agreed that a fifth edition was consistent with Eric Underwood's plans and should be considered. There was little question concerning the format. In our opinion, Eric Underwood was one of the few persons, perhaps the last, whose whole scientific career had paralleled the development of modern trace element research, and who had acquired a direct, comprehensive understanding of new knowledge as it emerged. On this basis he was able to evaluate with authority the progress and problems of the field. If the fifth edition was to preserve Underwood's aim of presenting a balanced account of the field rather than a mere compilation of literature, we considered the efforts of more than one author essential. We recognize that the participation of many authors may affect the cohesiveness of the revised edition, but we believe the risk is minimized by the close past association with Underwood's work and the philosophy of many of our contributors.

The major change in the format of the fifth edition is the presentation of the book in two volumes, necessitated by the rapidly increasing knowledge of metabolism, interactions, and requirements of trace elements. Even with the expansion into two volumes, the authors of the individual chapters had to exercise judgment as to the number of individual publications that could be cited and discussed. No claim can be made for a complete presentation of all trace element research, and no value judgment is implied from citing or omitting individual publications. The guiding principle was to present the minimum of results that would serve as a logical foundation for the description of the present state of knowledge. The inclusion of part of the vast amount of new data published since 1977 was possible only by condensing the discussion of earlier results. We have tried, however, not to disrupt the description of the historical development of our field.

Recent results of research were accommodated by devoting new chapters to the subjects "Methodology of Trace Element Research" and "Quality Assurance for Trace Element Analysis" and by expanding the discussion of lithium and aluminum in separate, new chapters. The first two subjects are of outstanding importance as determinants of future progress. The concern for the quality of analytical data motivated the authors of the individual chapters to review critically and, where necessary, revise analytical data presented in the previous editions. The rapid progress of trace analytical methodology since the mid-1970s has changed what had been accepted as normal for the concentrations of many trace elements in tissues and foods. The new data reflect the present state of the art in trace element analysis, but they may be subject to future revision.

The editor thanks the contributors of this fifth edition for their willingness to devote much time, effort, and judgment to the continuation of Eric Underwood's work. He also apologizes to readers and contributors alike for the delay in the publication of Volume 1, caused by health problems of the designated author of two chapters. The outstanding effort of Mrs. M. Grace Harris as editorial assistant is gratefully acknowledged.

WALTER MERTZ

1

Introduction

ERIC J. UNDERWOOD (Deceased)
and WALTER MERTZ

U. S. Department of Agriculture
Agricultural Research Service
Beltsville Human Nutrition Research Center
Beltsville, Maryland

I. THE NATURE OF TRACE ELEMENTS

Many mineral elements occur in living tissues in such small amounts that the early workers were unable to measure their precise concentrations with the analytical methods then available. They were therefore frequently described as "traces," and the term trace elements arose to describe them. This designation has remained in popular usage even though virtually all the trace elements can now be estimated in biological materials with some accuracy and precision. It is retained here because it is brief and has become hallowed by time.

It is difficult to find a meaningful classification for the trace elements or even to draw a completely satisfactory line of demarcation between those so designated and the so-called major elements. At the present time, less than one-third of the 90 naturally occurring elements are known to be essential for life. These consist of 11 major elements, namely, carbon, hydrogen, oxygen, nitrogen, sulfur, calcium, phosphorus, potassium, sodium, chlorine, and magnesium. If the six noble gases are excluded as unlikely to have a physiological function, 73 elements of the periodic system remain and, because of their low concentration in living matter, are termed the trace elements. A logical, permanent classification of the trace elements, other than by chemical characteristics, is impossible at

1

the present state of our knowledge. The past has repeatedly proved wrong the then-existing arbitrary classifications into "essential" and "toxic" elements, and elements have been shifted from the latter into the former category at an increasing rate. It is evident that the number of elements recognized as essential reflects the state of knowledge at a certain point in time. It is logically wrong to establish a category of "toxic" elements, because toxicity is inherent in all elements and is but a function of concentrations to which living matter is exposed. This was recognized almost 100 years ago by Schulz (49), and subsequently formulated in mathematical form by Bertrand (5). Venchikov (67,68) has expanded this concept and represented the dose response in the form of a curve with two maxima and three zones of action of trace elements. The biological action zone, in which trace elements function as nutrients, is followed by an inactive zone, and, with excessive exposures, by a pharmacotoxicological action zone (Fig. 1A). The intakes or dose levels at which these different phases of action become apparent, and the width of the optimal plateau, vary widely among the trace elements and can be markedly affected by the extent to which various other elements and compounds are present in the animal body and in the diet being consumed.

Venchikov's observations undoubtedly are pertinent to many lower forms of life but do not necessarily describe the response of higher organisms with strong homeostatic defenses. The fact remains that each element produces toxicity at certain concentrations and is compatible with life at a lower range of exposure. Whether this lower range of exposure is only tolerated by the organism or is required to provide a vital function, determines whether an element is essential or

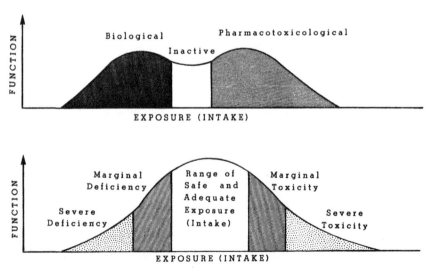

Fig. 1. Dose–response curves. (A) Venchikov's biphasic curve; (B) total biological dose–response curve.

not. Essentiality is established when a further reduction below the range of tolerable levels, better termed "range of safe and adequate intakes," results in a consistent and reproducible impairment of a physiological function. This exposure dependence of any biological action of an essential element, physiological or toxic, can be presented in the "total biological dose-response curve" (38) which, in the ideal case, clearly identifies the span of exposure that is fully adequate to meet the requirements of the organism and, at the same time, is safe from even the slightest signs of toxicity (Fig. 1B). The quantification of the right-hand side of the "total dose–response curve," describing toxic reactions, is easier to accomplish than that of the left-hand side, because of the higher concentrations involved. Our knowledge of that part is therefore more complete than that of the "biological" part of the curve. Indeed, there has always been disagreement during the history of trace element research whether for some elements the left-hand, "biological" side of the curve existed at all (e.g., selenium, chromium, lead, and arsenic).

The above definition of essentiality, relying on the reproducible demonstration of a consistent impairment of a physiological function, is quite liberal. Cotzias (14) has stated the criteria for essentiality with more stringency. He maintained that a trace element can be considered essential if it meets the following criteria: (1) it is present in all healthy tissue of all living things; (2) its concentration from one animal to the next is fairly constant; (3) its withdrawal from the body induces reproducibly the same physiological and structural abnormalities, regardless of the species studied; (4) its addition either reverses or prevents these abnormalities; (5) the abnormalities induced by deficiencies are always accompanied by specific biochemical changes; and (6) these biological changes can be prevented or cured when the deficiency is prevented or cured. Regardless of which criteria are applied, it is obvious that the number of elements recognized as essential depends on the sophistication of experimental procedures and that proof of essentiality is technically easier for those elements that occur in reasonably high concentration than for those with very low requirements and concentrations. Thus, it can be expected that, with further improvement of our experimental techniques, more elements may be demonstrated as essential.

These considerations suggest a logical classification of the 73 trace elements into those with proven essentiality and the rest for which essentiality is not now known. This classification leaves room for the possibility that future research will prove additional elements of the second category as essential. Each of the two categories can be subdivided according to their practical importance under given conditions, for example, local, regional, or national imbalances in the environment, industrial emissions, or specific dietary habits. The practical importance of trace elements is totally unrelated to essentiality; some essential elements may not be of any nutritional concern at all, as is the case with manganese in human nutrition; others, such as selenium, may have the highest regional importance because of deficiency in one area and toxicity in another.

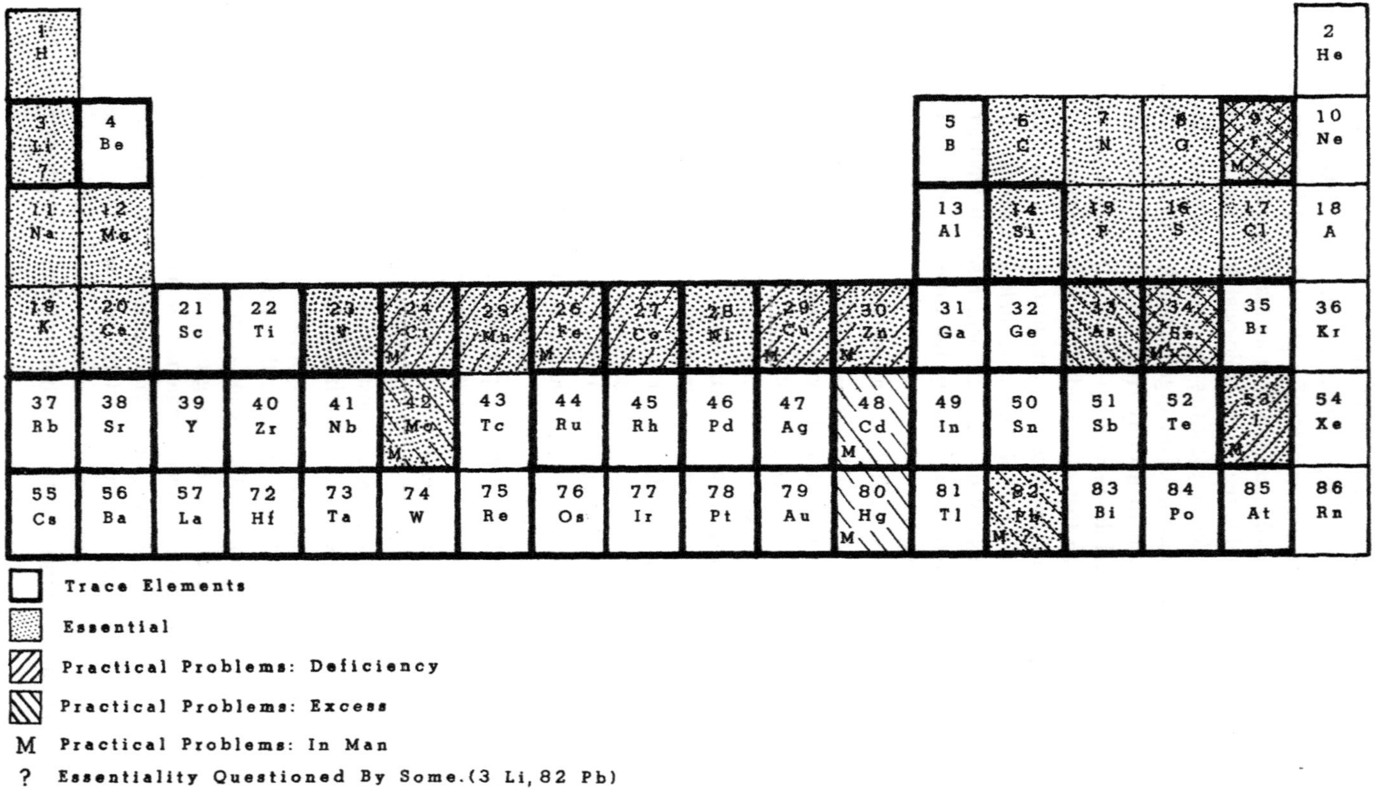

Legend:

- ☐ Trace Elements
- Essential
- Practical Problems: Deficiency
- Practical Problems: Excess
- M Practical Problems: In Man
- ? Essentiality Questioned By Some. (3 Li, 82 Pb)

Fig. 2. List of elements.

The listing of elements as essential is difficult and, depending on the definition of essentiality applied, not unanimous (Fig. 2). There is no disagreement concerning the essentiality of chromium, manganese, iron, cobalt, copper, zinc, selenium, molybdenum, and iodine, although not all present practical nutritional problems to humans. Fluorine is included because of its proven benefits for dental health and its suggested role in maintaining the integrity of bones. We also include the "new trace elements," lithium, silicon, vanadium, nickel, arsenic, and lead because deficiencies have been induced by independent groups of investigators, but we realize that our present knowledge of mode and site of action as well as of human requirements is inadequate. Growth effects of tin have been demonstrated by Schwarz (52) but have not yet been independently confirmed. The essential role of boron for plants is well established but remains to be proven for animal organisms. The list in Fig. 2 is presented as a temporary and arbitrary guideline only. It is temporary, because the results of ongoing and future research may make changes necessary; it is arbitrary, because the weighing of evidence for elements such as lead, lithium, and cadmium involves personal judgment. This uncertainty of classification and its flexibility is an expression of the vigor of modern trace element research.

II. DISCOVERY OF TRACE ELEMENTS

Interest in trace elements in animal physiology began over a century ago with the discovery in living organisms of a number of special compounds which contained various metals not previously suspected to be of biological significance. These included turacin, a red porphyrin pigment occurring in the feathers of certain birds and containing no less than 7% copper (9); hemocyanin, another copper-containing compound found in the blood of snails (22); sycotypin, a zinc-containing blood pigment in Mollusca (37); and a vanadium-containing respiratory compound present in the blood of sea squirts (18). Such discoveries did little to stimulate studies of the possible wider significance of the component elements. These were to come from the early investigations of Bernard (4) and MacMunn (34) on cell respiration and iron and oxidative processes, which pointed the way to later studies of metal–enzyme catalysis and of metalloenzymes and which were to greatly illuminate our understanding of trace element functions within the tissues. This early period also saw (1) the discovery by Raulin (44) of the essentiality of zinc in the nutrition of *Aspergillus niger;* (2) the remarkable observations of the French botanist Chatin (8) on the iodine content of soils, waters, and foods, and the relationship of the occurrence of goiter in humans to a deficiency of environmental iodine; and (3) the demonstration by Frodisch in 1832 that the iron content of the blood of chlorotics is lower than that of healthy individuals, following an earlier discovery that iron is a characteristic constituent of blood (see Fowler, Ref. 21).

During the first quarter of the present century further studies were initiated on iron and iodine in human health and nutrition. These included the isolation by Kendall (28) of a crystalline compound from the thyroid gland containing 65% iodine which he claimed was the active principle and which he named thyroxine, and the successful control of goiter in humans and animals in several goitrous areas by the use of supplemental iodine.

During this period, the advent of emission spectrography permitted the simultaneous estimation of some 20 elements in low concentrations, and the "distributional" phase in trace element research began. Extensive investigations of trace element levels in soils, plants, animal, and human tissues were undertaken at that time (18,58,74) and later, as analytical techniques improved (73). These and other distribution studies referred to in later chapters were responsible for (1) defining the wide limits of concentration of many of the trace elements in foods and tissues; (2) illuminating the significance of such factors as age, location, disease, and industrial contamination in influencing those concentrations; and (3) stimulating studies of the possible physiological significance of several elements previously unsuspected of biological potentiality.

The second quarter of this century was notable for spectacular advances in our knowledge of the nutritional importance of trace elements. These advances came from basic studies with laboratory species aimed at enlarging our understanding of total nutrient needs and from investigations of a number of widely separated naturally occurring nutritional maladies of humans and our domestic livestock. In the 1920s Bertrand of France and McHargue of the United States pioneered a purified diet approach to animal studies with these elements. The diets employed were so deficient in other nutrients, particularly vitamins, that the animals achieved little growth or survived only for short periods, even with the addition of the element under study (6,36). Vitamin research had not then progressed to the point where these vital nutrients could be supplied in pure or semipure form. The Wisconsin school, led by E. B. Hart, initiated a new era when they showed in 1928 that supplementary copper, as well as iron, is necessary for growth and hemoglobin formation in rats fed a milk diet (24). Within a few years the same group, using the special diet and purified diet techniques with great success, first showed that manganese (27,70) and then zinc (61) were dietary essentials for mice and rats. These important findings were soon confirmed and extended to other species. Nearly 20 years elapsed before the purified-diet technique was again successful in identifying further essential trace elements. These were first molybdenum (15,46) in 1953, then selenium (50) in 1957, and chromium (51) in 1959.

During the 1930s a wide range of nutritional disorders of humans and farm stock were found to be caused by deficient or excessive intakes of various trace elements from the natural environment. In 1931, mottled dental enamel in humans was shown by three independent groups to result from the ingestion of excessive amounts of fluoride from the drinking water (10,56,66). In the same year, copper deficiency in grazing cattle was demonstrated in parts of Florida (42) and Holland (54). Two years later, in 1933 and 1935, "alkali disease" and "blind staggers" of stock occurring in parts of the Great Plains region of the United States were established as manifestations of chronic and acute selenium toxicity, respectively (2,48). In 1935, cobalt deficiency was shown to be the cause of two wasting diseases of grazing ruminants occurring in localized areas in South Australia (35) and Western Australia (62). In 1936 and 1937, a dietary deficiency of manganese was found to be responsible for perosis or "slipped tendon" (72) and nutritional chondrodystrophy (33) in poultry. In 1937, enzootic neonatal ataxia in lambs in Australia was shown to be a manifestation of copper deficiency in the ewe during pregnancy (3). Finally, in the following year excessive intakes of molybdenum from the pastures were found to cause the debilitating diarrhea affecting grazing cattle confined to certain areas in England (20).

In the investigations just outlined, attention was initially focused on discovering the cause and devising practical means of prevention and control of acute disorders with well-marked clinical and pathological manifestations of the nutritional abnormalities. It soon became evident that a series of milder maladies involving the trace elements also existed, which had less specific manifestations and affected more animals and greater areas than the acute conditions which prompted the original investigations. The deficiency or toxicity states were often found to be ameliorated or exacerbated (i.e., conditioned) by the extent to which other elements, nutrients, or compounds were present or absent from the environment. The importance of dietary interrelationships of this type was highlighted by the discovery of Dick (16), while investigating chronic copper poisoning in sheep in southeastern Australia in the early 1950s, that a three-way interaction exists between copper, molybdenum, and inorganic sulfate, and that the ability of molybdenum to limit copper retention in the animal can only be expressed in the presence of adequate sulfate. Metabolic interactions among the trace elements were subsequently shown to be of profound nutritional importance and to involve a wide variety of elements. These are considered in the appropriate chapters that follow and have been the subject of many recent reviews (31).

The third quarter of this century has seen many other notable advances in our understanding of the nutritional physiology of the trace elements, besides the great significance of metabolic interactions mentioned previously. These advances have mostly been highly dependent on concurrent developments in analytical techniques, among which atomic absorption, neutron activation, and microelectron probe procedures have been particularly prominent. With such procedures the concentrations and distribution of most elements in the tissues, cells, and even the organelles of the cells can be determined with ever-increasing sensitivity and precision and the metabolic movements and the kinetics of those movements followed with the aid of stable or radioactive isotopes of suitable half-life. At the same time many metalloproteins with enzymatic activity have been discovered, allowing the identification of basic biochemical lesions related causally to the diverse manifestations of deficiency or toxicity in the animal. The elucidation of the role of copper in elastin biosynthesis and its relation to the cardiovascular disorders that arise in copper-deficient animals provide a classical example of progress in the area (see Hill *et al.*, Ref. 26).

The end of that period and the beginning of the last quarter of this century saw a rapid increase in the number of trace elements shown to be essential and a remarkable surge of interest and activity in their significance or potential significance in human health and nutrition. Deficiencies of nine "new" trace elements were described, namely for lithium, fluorine, silicon, vanadium, nickel, arsenic, cadmium, tin, and lead. Although independent confirmation of the findings is still lacking for tin, fluorine, and cadmium, this work has added a whole new dimension to trace element research. An appraisal of these discoveries and their nutritional significance is given in later chapters. Progress in this area was made possible by the use of the plastic isolator technique developed by Smith and Schwarz (55). In this technique the animals are isolated in a system in which plastics are used for all component parts and there is no metal, glass, or rubber. The unit has an airlock to facilitate passage in and out of the so-called trace element-sterile environment and two air filters which remove almost all dust down to a particle size of 0.35 μm. The virtual exclusion of dust is the critical final step in the whole technique, because rats fed the same highly purified crystalline vitamin and amino acid-containing diets and drinking water outside the isolator system may grow adequately and show none of the clinical signs of deficiency that appear in the animals maintained in the isolator. The operation of the system requires much experience and concern not only for contamination control but also for the stability of organic components of the diet fed in the isolator (41).

The availability of laminar-flow facilities has offered a convenient, albeit expensive, alternative to the isolator system, and the application of various purification procedures to dietary ingredients has allowed the production of diets that may contain contaminating elements at concentrations of only 20 ng/g or less (45). Techniques to circumvent the introduction of contaminants that are most difficult to control, namely from sources of protein and mineral salts, have been described and are being used with success. For example, Nielsen has reduced the need for adding mineral salts by using carefully analyzed natural ingredients for his diets (43). Anke (1) introduced the use of ruminant species, which allowed him to substitute chemically pure urea for a significant proportion of trace element-containing proteins.

Perhaps the most significant development in the trace element field during the last quarter of our century was the rapidly increasing recognition that trace elements have a fundamental role in health maintenance and that optimal exposure (within the range of safe and adequate intakes) cannot be taken for granted. The essential function of several elements in defense mechanisms against microbiological, chemical, viral, and oxidative insults became well recognized (7,17), linking some trace elements not only with infectious but also with chronic degenerative (29) and even neoplastic (13) diseases. Epidemiological correlations have suggested increased disease risks from marginal exposures that do not result in any clinical and not always in biochemical signs of deficiency, thus suggesting new criteria for the definition of adequate nutritional status and for national and international recommendations of trace element intake. On the other hand, the possibility that contamination of the air, water supply, and foods with trace elements, arising from modern agricultural and industrial practices and from the increasing motorization and urbanization of sections of the community, may have deleterious effects on the long-term health and welfare of populations has stimulated increasing interest in the concentrations and movements of these elements in the environment and in the maximum permissible intakes by humans (11). Interest in that field of concern has been increasingly concentrated on health risks associated with marginal overexposures, which may affect large segments of the population, although they do not produce clinical signs. The greatly improved techniques for long-term total parenteral nutrition have allowed patients to be kept alive for many years with chemically defined, pure nutrients. This has resulted in inadvertently induced clinical deficiencies for several trace elements. The recognition and therapy of these deficiencies has extended our knowledge of absolute trace element requirements and of deficiency signs, heretofore only extrapolated from animal experiments.

The rest of the twentieth century presents the trace element researcher with the challenge to consolidate the gains that have been made, and to provide the scientific knowledge that is needed to realize a great potential for the improvement of health which now is still based on epidemiological correlations or on animal experiments. The mode of action of the "new trace elements" is incompletely known; of the hundreds of suspected or known interactions of trace elements, only very few can be described quantitatively; and the marginal trace element status, of so much concern to the nutritionist and toxicologist, cannot be readily and reliably diagnosed. If the vigor of trace element research in the first three-quarters of the century can be maintained during the fourth, many of these challenges will be met.

III. MODE OF ACTION OF TRACE ELEMENTS

The only property that the essential trace elements have in common in that they normally occur and function in living tissues in low concentrations. These normal tissue concentrations vary greatly in magnitude and are characteristic for each element. They are usually expressed as parts per million (ppm), micrograms per gram (μg/g), or 10^{-6}, or with some, such as iodine, chromium, and selenium, as parts per billion (ppb), nanograms per gram (ng/g), or 10^{-9}. It should be noted that the magnitude of tissue concentrations or requirements does not reflect nutritional importance or predict the risk of nutritional problems. Certain elements such as bromine and rubidium, for which no essential role is known, occur in animal tissues in concentrations well above those of most of the essential trace elements.

The characteristic concentrations and functional forms of the trace elements must be maintained within narrow limits if the functional and structural integrity of the tissues is to be safeguarded and growth, health, and fertility are to remain unimpaired. Higher organisms possess homeostatic mechanisms that can maintain the concentrations of trace elements at their active sites within narrow physiological limits in spite of over- or underexposure. Such mechanisms include control of intestinal absorption or excretion, the availability of specific stores for individual elements, and the use of "chemical sinks" that can bind in an innocuous form potentially toxic amounts of elements. The degree of homeostatic control varies from one element to another, but continued ingestion of diets, or continued exposure to total environments that are severely deficient, imbalanced, or excessively high in a particular trace element, induces changes in the functioning forms, activities, or concentrations of that element in the body tissues and fluids so that they fall below or rise above the permissible limits. In these circumstances biochemical defects develop, physiological functions are affected, and structural disorders may arise in ways which differ with different elements, with the degree and duration of the dietary deficiency or toxicity, and with the age, sex, and species of the animal involved.

Most, but not all trace elements act primarily as catalysts in enzyme systems in the cells, where they serve a wide range of functions. In this respect their roles range from weak ionic effects to highly specific associations in metalloenzymes. In the metalloenzymes the metal is firmly associated with the protein and there is a fixed number of metal atoms per molecule of protein. These atoms cannot be replaced by any other metal. However, Vallee (64) has shown that cobalt and cadmium can be substituted for the native zinc atoms in several zinc enzymes with some changes in specificity but no inactivation. This worker has stated further that "the key to the specificity and catalytical potential of metalloenzymes seems to lie in the diversity and topological arrangement both of their active site residues and of their metal atoms, all of which interact with their substrates." He has emphasized the importance of spectra as probes of active sites of metalloenzymes, which can reveal geometric and electronic detail pertinent to the functions of such systems (65). His basic studies in this area provide important beginnings to an understanding of the molecular mechanisms involved in the metalloenzymes and the nature of the metal ion specificity in their reactions.

Evidence is accumulating that the protein–metal interactions not only enhance the catalytic activity of enzymes but also may increase the stability of the protein moiety to metabolic turnover. In this connection Harris (23) has shown that copper is a key regulator of lysyl oxidase activity in the aorta of chicks and may be a major determinant of the steady-state levels of the enzyme in that tissue. The tissue levels and activities of some metalloenzymes have now been related to the manifestations of deficiency or toxicity states in the animal, as discussed later. However, most clinical and pathological disorders in animals or humans as a consequence of trace element deficiencies or excesses cannot yet be explained in biochemical or enzymatic terms. There are additional sites of action. Cobalt is the essential metal in vitamin B_{12}, iodine confers hormone activity to the hormonally inactive compound thyronine, and chromium acts as a cofactor for the hormone insulin. There may be roles for certain trace elements essential to maintain structure of nonenzymatic macromolecules. Such a role has been postulated for silicon in collagen (53). The high concentrations of certain trace elements in nucleic acids (69) may play a similar role; the elucidation of their function *in vivo* remains a challenge to trace element researchers.

IV. TRACE ELEMENT NEEDS AND TOLERANCES

Trace element needs and tolerances of animals and humans are expressed either in absolute or dietary terms. The absolute requirement and tolerance for a given element relates to the amounts actually absorbed into the organism, in contrast to the dietary requirement or tolerance that describes the amounts present in the typical daily diet that allow the absorption of the absolute requirement (or absolute tolerance). Because of the different intestinal absorption efficiency for individual elements, the ratio between absolute and dietary requirements varies widely. It approaches unity for elements that are nearly completely absorbed (e.g., iodine), assumes intermediate values (on the average, $1:10$ for iron) and may be less than $1:100$ (e.g., chromium). The absolute requirement can be determined with relative precision but has little practical importance except for total parenteral nutrition, which bypasses the intestinal absorption process. The absolute requirement to maintain a given nutritional status is equal to the sum of the daily losses of the element (via feces, urine, skin and its appendages, milk, menstrual blood, and semen) plus, where appropriate, the amounts needed to maintain existing concentrations of the element in a growing organism.

Practically all recommendations by national and international expert commit-

tees are stated as dietary requirements (12,71). The latter can be expressed as a
certain concentration (or range of concentrations) in a typical national diet that
will meet the trace element requirements if the diet also meets the energy demands of humans or the animal species for which the recommendation was set.
Alternatively, dietary recommendations can be set as the amount (or range of
amounts) of an element to be consumed per day. In contrast to the nutritionally
oriented dietary recommendations or allowances, which do not take into account
nondietary sources of trace element exposure, tolerances usually represent limits
of total exposure from diet, water, and air that should not be exceeded. Dietary
allowances are strongly influenced by the nature of the typical diets consumed,
because numerous interactions among dietary components and different chemical
forms of elements in individual foods determine biological availability. The
multitude of possible dietary interactions is, with few exceptions, not yet quan-
tifiable, and the dietary forms of many trace elements are as yet unidentified.
Therefore, the determination of dietary requirements (and tolerances) and the
subsequent setting of dietary recommendations is much more difficult and beset
with greater uncertainties than that of absolute requirements. For these and other
reasons, it is not surprising that such recommendations for trace element intakes
differ between countries, often to a remarkable degree. An expert committee of
the World Health Organization, for example, was unable to recommend one
dietary zinc intake for all potential users worlwide; instead it arrived at three
recommendations, for populations with diets of low, intermediate, and high
bioavailability of dietary zinc, respectively (71).

Trace element requirements are established by determining in a typical, defined diet that daily intake of the element under question that maintains the
organism in an optimal nutritional status. In the ideal case, sensitive methods can
be applied to measure reliable indicators of status, for example the extent of
tissue iron stores or the levels of iodine-containing thyroid hormones in blood.
Other methods consist of experimental induction of mild, clinical deficiencies
and of determining the amount required to reconstitute and maintain normal
function. Metabolic studies measuring the response of element-specific enzymes
to different levels of intake can give valuable information for some elements, for
example, selenium. Extrapolations from studies in several animal species, combined with the results of national health and nutrition surveys can result in
preliminary requirement estimates. Regardless of how estimates of requirement
are obtained, they are applicable in a strict sense only to the exact dietary
conditions under which they were obtained, because of the pronounced influence
of dietary interactions on biological availability. This important fact must be
taken into consideration when dietary requirements are translated into dietary
recommendations and the latter are applied to public health policies. Dietary
recommendations are formulated to cover the requirements of "practically all

healthy people'' and therefore are set somewhat higher than the estimated average requirement (12). They are not applicable to disease states.

Dietary recommendations can be stated as ranges, rather than as single figures of intake. That presentation takes into account the homeostatic regulation of higher organisms that tends to buffer marginally deficient or marginally excessive intakes by changing the efficiency of absorption and excretion. Furthermore, the statement of a range can take into account the differences of intake that are required to meet the requirement when supplied by typical national diets of different bioavailability. Finally, the range concept reflects the variation of requirements among individuals. The "ranges of safe and adequate intakes" as introduced in the ninth edition of the "Recommended Dietary Allowances" (12) have been defined as adequate to meet the trace element needs of healthy people in the United States. They do not delineate a desirable intake against deficiency on one end and against toxicity on the other; intakes outside of the range are stated to increase the *risk* for deficiency and toxicity, respectively.

Similarly, a series of "safe" dietary levels of potentially toxic trace elements has been established, depending on the extent to which other elements which affect their absorption and retention are present (59). These considerations apply to all the trace elements to varying degrees, but with some elements such as copper they are so important that a particular level of intake of this element can lead to signs either of copper deficiency or of copper toxicity in the animal, depending on the relative intakes of molybdenum and sulfur, or of zinc and iron. The many mineral interactions of this type which exist are discussed in later chapters. However, it is appropriate to mention at this point the experiments of Suttle and Mills (60) on copper toxicity in pigs, which strikingly illustrate the importance of trace element dietary balance in determining the "safe" intake of a particular element. These workers showed that dietary copper levels of 425 and 450 $\mu g/g$ cause severe toxicosis, all signs of which were prevented by simultaneously providing an additional 150 $\mu g/g$ zinc plus 150 $\mu g/g$ iron to the diets.

Estimates of adequacy or safety also vary with the criteria employed. As the amounts of an essential trace element available to the animal becomes insufficient for all the metabolic processes in which it participates, as a result of inadequate intake and depletion of body reserves, certain of these processes fail in the competition for the inadequate supply. The sensitivity of particular metabolic processes to lack of an essential element and the priority of demand exerted by them vary in different species and, within species, with the age and sex of the animal and the rapidity with which the deficiency develops. In the sheep, for example, the processes of pigmentation and keratinization of wool appear to be the first to be affected by low-copper status, so that at certain levels of copper intake no other function involving copper is impaired. Thus, if wool quality is taken as the criterion of adequacy, the copper requirements of sheep are higher than if growth rate and hemoglobin levels are taken as criteria. In this species it has also been shown that the zinc requirements for testicular growth and development and for normal spermatogenesis are significantly higher than those needed for the support of normal live-weight growth and appetite (63). If body growth is taken as the criterion of adequacy, which would be the case with rams destined for slaughter for meat at an early age, the zinc requirements would be lower than for similar animals kept for reproductive purposes. The position is similar with manganese in the nutrition of pigs. Ample evidence is available that the manganese requirements for growth in this species are substantially lower than they are for satisfactory reproductive performance. Recent evidence relating zinc intakes to rate of wound healing and taste and smell acuity also raises important questions on the criteria for adequacy to be employed in assessing the zinc requirements of humans.

In the past, the division of trace elements into an essential and a "toxic" category has tended to confine the nutritionist to concern with dietary recommendations for the first category of elements and the toxicologist to the establishing of tolerances for the second. Lack of communication between the two disciplines sometimes has resulted in difficult situations when unenforceable "zero tolerance levels" were applied to selenium, an essential element of great public health importance. Recent developments in selenium research have demonstrated that deficiency and toxicity can present equal health risks to populations that may not even live far from one another. The concerns of the nutritionist for adequacy of intake and of the toxicologist for tolerance level are not incompatible with each other; both contribute, albeit from different directions, to the construction of the total dose–response curve and the exact definition of the "ranges of safe and adequate intake."

V. TRACE ELEMENT DEFICIENCIES

There is some disagreement among trace element researchers about the use and the meaning of the term deficiency. In accordance with the dual definition in dictionaries either as "a lack" or as an "inadequate supply" of a nutrient, some restrict the term to only the severest forms, while others include the whole spectrum from the mildest to the most severe stages.

Progress in trace element research in the past few decades has shown that "lack of" a nutrient in a diet or organism is not a valid scientific criterion of deficiency. Responsible analytical chemists refrain from reporting a value as "zero," realizing that concentrations well below even modern detection limits may contain millions and billions of atoms in a volume of the analyte. The absolute lack of a nutrient, the "zero concentration," cannot be established with present methods and may escape our capabilities for a long time to come.

Therefore, the second dictionary version, that of an "inadequate supply," remains as a criterion, and a deficiency state can be defined as a state in which the concentration of an essential nutrient at its sites of action is inadequate to main-·tain the nutrient-dependent function at its optimal level.

The number of functions that are accepted as criteria of nutritional status has increased substantially. The two criteria of the past, survival and growth rates, have been extended and complemented by many more, based on determinations of enzyme functions, hormone profiles, metabolism of substrates, and also based on functions related to the immune system and even to cognitive and emotional development. A deficiency is defined as a state in which any one or more of such criteria are consistently and significantly impaired and in which such impairment is preventable by supplementation with physiological amounts of the specific nutrient.

This definition does not take into account the degree of deficiency, nor the severity of the resulting signs and symptoms; yet, the quantitative aspects of deficiency states have eminent practical implications for public health policies. There is almost universal agreement among policy makers that the occurrence of a clinically consequential deficiency in a population calls for intervention by fortification of an appropriate carrier food or by direct supplementation of the groups at risk with the limiting nutrient. The results of such interventions can be evaluated quantitatively and the health benefits are obvious (57). In contrast, marginal deficiencies that do not result in direct clinical consequences are not generally recognized as health risks, and there is disagreement among policy makers and even among scientists as to the need for intervention on a population scale. The controversies surrounding fluoridation of drinking water represent an extreme example (11).

The definition of a deficiency as a state that can manifest itself in a wide-ranging spectrum of metabolic and clinical disturbances calls for a classification of these disturbances into different phases, as a function of the nutrient supply in diet and environment. A definition of four phases has been proposed (39). Phase I, the "initial depletion" phase, is characterized by changes only in the metabolism of the element itself in response to a suboptimal intake, compensating during a certain period of time for the inadequate supply, so that no disturbances of biological structure or function are detectable. Increased intestinal absorption efficiency and/or reduction of excretory losses, desaturation of specific carrier proteins, and gradual diminution of body stores are typical features, but the concentration of the element at its specific sites of action is not affected. This phase has been thoroughly described as a consequence of marginal iron intake, and several routine diagnostic procedures are available. The initial depletion phase may revert back to a normal status with an increasing intake of the element; it may persist throughout a lifetime if the intake remains marginal; or it may lead into phase II with a more severely limited intake. By itself, the initial depletion phase has no adverse consequences for health, but it represents an increased risk for deficiency: the margin of safety is reduced.

Phase II, the "compensated metabolic phase," is characterized by an impairment of certain specific biochemical functions, such as trace element-dependent enzyme activities or receptor affinities.

At some time during the depletion process the homeostatic mechanisms may become inadequate to maintain normal element concentrations at the sites of action, with a resulting impairment of some specific biochemical functions. At first, there are no measurable changes in the level and metabolism of the substrates of these biochemical functions, because other systems, independent of the deficient element, compensate: The "compensated metabolic phase."

Selenium deficiency in experimental animals and humans, in the presence of adequate vitamin E is accompanied consistently by depressed activities of the selenium-specific enzyme, glutathione peroxidase, but not by any indication of peroxidative damage to the organism. During the early phase of chromium deficiency, the resulting defect, insulin resistance, is often compensated by an increase in circulating insulin levels, and no effect on glucose tolerance is observed. Zinc deficiency results in a depression of the zinc enzyme carboxypeptidase A in rat pancreas, but does not impair digestive functions (40).

The compensated metabolic phase represents deficiency, because of the impairment of biological function. Its direct consequences for health are insignificant, but only as long as the various compensatory systems are operative. Any change in dietary or environmental exposure to interacting factors can precipitate a severe deficiency. Superimposing a vitamin E deficiency and feeding high levels of dietary fat to rats in the compensated metabolic phase of selenium deficiency will almost immediately precipitate death. Thus, the state of "compensated metabolic deficiency" is connected with a substantially increased risk for clinical deficiency, a risk that is influenced not only by several known interacting factors but possibly also by some that remain to be identified.

There is disagreement among policy makers concerning the public health importance and risks of that phase. For example, Finland has instituted measures to increase the low selenium levels in its agricultural products whereas no such measures are considered necessary in New Zealand, a country with even lower selenium concentrations in the environment (30,47).

Phase III, the "decompensated metabolic phase" of deficiency, is characterized by the appearance of defects in functions important for health, such as metabolic, immunological, cognitive, emotional, developmental, and those related to work capacity. By definition, the signs and symptoms are not clinical and are detectable only by specialized tests. In addition to the health risks discussed for phase II, the metabolic deviations (e.g., of glucose or lipid metabolism) represent recognized risk factors for clinical disease, such as diabetes and cardiovascular diseases.

Phase IV, the "clinical phase," is characterized by the appearance of disease and, with increasing severity of the deficiency, by death. For many trace elements the clinical phase can be cured by nutritional means; some do not respond to nutritional therapy alone (e.g., severe iodine deficiency goiter).

As stated above, there is no general agreement about the public health implications of the four phases of deficiency. Many countries accept the occurrence of low iron stores in the female and child population (phase I) as indication for fortification programs (12); others reject even the clinical appearance of goiter (phases III and IV) in large population groups as reasons for nutritional intervention (19). Such different attitudes are influenced by national history, by the nutritional situation, and by the prevailing assessment in a country of the risk–benefit ratio of intervention; they are extremely difficult to reconcile.

Yet, it may be possible to agree on some scientific guidelines to assess the relevance of the individual phases of a deficiency for human and animal health, if it is conceded that decisions for or against intervention are also dependent on local features or national background that are not always amenable to scientific treatment.

1. The risk for serious disease increases with the progression of the deficiency from phase I to phase IV.
2. The risk from phases I and II may be considered relatively mild if an adequate balance between the nutrient and its interacting nutritional and environmental factors is ascertained.
3. Diagnosis of phase III, uncompensated impairment of biological functions in an individual, however slight, should be cause for nutritional therapy in that individual. If there is a substantial incidence of phase III deficiency in a population, the risks and benefits of public intervention programs should be discussed.
4. Diagnosis of phase IV in an individual calls for immediate therapeutic measures. A substantial incidence of phase IV deficiency in a population under most conditions should be cause for public intervention efforts.

REFERENCES

1. Anke, M., Grün, M., and Partschefeld, M. (1976). *In* "Trace Substances in Environmental Health" (D. D. Hemphill, ed.), Vol. 10, p. 403. Univ. of Missouri.
2. Beath, O. A., Eppson, H. F., and Gilbert, C. S. (1935). *Wyo. Agric. Exp. Stn. Bull.* **206.**
3. Bennetts, H. W., and Chapman, F. E. (1937). *Aust. Vet. J.* **13,** 138.
4. Bernard, C. (1857). "Leçons sur les effets des substances toxiques et médicamenteuses." Baillière, Paris.
5. Bertrand, G. (1912). *Proc. Int. Congr. Appl. Chem., 8th* **28,** 30.
6. Bertrand, G., and Benson, R. (1922). *C. R. Acad. Sci.* **175,** 289; Bertrand, G., and Nakamura, H. (1928). *C. R. Acad. Sci.* **186,** 480.

7. Chandra, R. K. (1982). *Nutr. Res.* **2,** 721.
8. Chatin, A. (1850–1854). *Nutr. Res.* **2,** 721.
9. Church, A. W. (1869). *Philos. Trans. R. Soc. London Ser. B* **159,** 627.
10. Churchill, H. N. (1931). *Ind. Eng. Chem.* **23,** 996.
11. Committee on Biologic Effects of Atmospheric Pollutants (1971–1977). "Medical and Biological Effects of Environmental Pollutants". Natl. Acad. Sci., Washington, D.C.
12. Committee on Dietary Allowances (1980). "Recommended Dietary Allowances," 9th revised Ed. Natl. Acad. Sci., Washington, D.C.
13. Committee on Diet, Nutrition, and Cancer (1982). "Diet, Nutrition and Cancer". Natl. Acad. Press, Washington, D.C.
14. Cotzias, G. C. (1967). *Trace Subst. Environ. Health Proc. Univ. Mo. Annu. Conf., 1st* p. 5.
15. DeRenzo, E. D., Kaleita, E., Heyther, P., Oleson, J. J., Hutchings, B. L., and Williams, J. H. (1953). *J. Am. Chem. Soc.* **75,** 753; *Arch. Biochem. Biophys.* **45,** 247 (1953).
16. Dick, A. T. (1954). *Aust. J. Agric. Res.* **5,** 511.
17. Diplock, A. T. (1981). *Philos. Trans. R. Soc. London Ser. B* **294,** 105.
18. Dutoit, P., and Zbinden, C. (1929). *C. R. Hebd. Séances Acad. Sci.* **188,** 1628.
19. Deutsche Gesellschaft für Ernährung (1976). *"Ernährungsbericht,"* p. 141. Frankfurt a/M.
20. Ferguson, W. S., Lewis, A. H., and Watson, S. J. (1938). *Nature (London)* **141,** 553; *Jealott's Hill Bull.* No. 1 (1940).
21. Fowler, W. M. (1936). *Ann Med. Hist.* **8,** 168.
22. Harless, E. (1847). *Arch. Anat. Physiol. (Leipzig)* p. 148.
23. Harris, E. D. (1976). *Proc. Natl. Acad. Sci. U.S.A.* **73,** 371.
24. Hart, E. B., Steenbock, H., Waddell, J., and Elvehjem, C. A. (1928). *J. Biol. Chem.* **77,** 797.
25. Henze, M. (1911). *Hoppe-Seyler's Z.; Physiol Chem.* **72,** 494.
26. Hill, C. H., Starcher, B., and Kim, C. (1968). *Fed. Proc., Fed. Am. Soc. Exp. Biol.* **26,** 129.
27. Kemmerer, A. R., Elvehjem, C. A., and Hart, E. B. (1931). *J. Biol. Chem.* **94,** 317.
28. Kendall, E. C. (1919). *J. Biol. Chem.* **39,** 125.
29. Klevay, L. M. (1983). *Biol. Trace Elem. Res.* **5,** 245.
30. Kovistoinen, P., and Huttunen, J. K. (1985). *In* "Trace Element Metabolism in Man and Animals—5." Commonwealth Agric. Bureaux, Farnham Royal, U.K.
31. Levander, O. A., and Cheng, L., eds. (1980). "Micronutrient Interactions: Vitamins, Minerals and Hazardous Elements." *Ann. New York Acad. Sci.* **355.**
32. Liebscher, K., and Smith H. (1968). *Arch. Environ. Health* **17,** 881.
33. Lyons, M., and Insko, W. M. (1937). *Kentucky Agric. Exp. Stn. Bull.* **371.**
34. MacMunn, C. A. (1885). *Philos. Trans. R. Soc. London Ser. B* **177,** 267.
35. Marston, H. R. (1935). *J. Counc. Sci. Ind. Res. (Aust.)* **8,** 111. Lines, E. W. *J. Counc. Sci. Ind. Res. (Aust.)* **8,** 117.
36. McHargue, J. S. (1926). *Am. J. Physiol.* **77,** 245.
37. Mendel, L. B., and Bradley, H. C. (1905). *Am. J. Physiol.* **14,** 313.
38. Mertz, W. (1983). *In* "Health Evaluation of Heavy Metals in Infant Formula and Junior Food" (E. H. F. Schmidt and A. G. Hildebrandt, eds.), pp. 47–56. Springer, Berlin.
39. Mertz, W. (1985). Metabolism and Metabolic Effects of Trace Elements. *In* "Trace Elements of Nutrition of Children." Raven, New York.
40. Mills, C. F., Quarterman, J., Williams, R. B., and Delgarno, A. C. (1967). *Biochem. J.* **102,** 712.
41. Moran, J. K., and Schwarz, K. (1978). *Fed. Proc., Fed. Am. Soc. Exp. Biol.* **37,** 617 (Abstr.).
42. Neal, W. M., Becher, R. B., and Shealy, A. L. (1931). *Science* **74,** 418.
43. Nielsen, F. H., Myron, D. R., Givand, S. H., Zimmerman, T. J., and Ollerich, D. A. (1975). *J. Nutr.* **105,** 1620.
44. Raulin, J. (1869). *Ann. Sci. Nat. Bot. Biol. Veg.* **11,** 93.

45. Reichlmayer-Lais, A. M., and Kirchgessner, M. (1981). Z. Tierphysiol. Tierernaehr. Futter-mittelkd. **46**, 8.
46. Richert, D. A., and Westerfeld, W. W. (1953). J. Biol. Chem. **203**, 915.
47. Robinson, M. F., and Thomson, C. D. (1983). Nutr. Abstr. Rev. Clin. Nutr. Ser. A **53**, 3.
48. Robinson, W. O. (1933). J. Assoc. Off. Agric. Chem. **16**, 423.
49. Schulz, H. (1888). Pflüger's Arch. Ges. Physiol. **42**, 517.
50. Schwarz, K., and Foltz, C. M. (1957). J. Am. Chem. Soc. **79**, 3293.
51. Schwarz, K., and Mertz, W. (1959). Arch. Biochem. Biophys. **85**, 292.
52. Schwarz, K., Milne, D. B., and Vinyard, E. (1970). Biochem. Biophys. Res. Commun. **40**, 22.
53. Schwarz, K. (1978). In "Biochemistry of Silicon and Related Problems" (G. Bendz and I. Lindquist, eds.), p. 207. Plenum, New York.
54. Sjollema, B. (1933). Biochem. Z. **267**, 151.
55. Smith, J. C., and Schwarz, K. (1967). J. Nutr. **93**, 182.
56. Smith, M. C., Lantz, E. M., and Smith, H. V. (1931). Ariz. Agric. Exp. Stn., Tech. Bull. **32**.
57. Stanbury, J. B., Ermans, A. M., Hetzel, B. S., Pretell, E. A., and Querido, A. (1974). WHO Chron. **28**, 220.
58. Stich, S. R. (1957). Biochem. J. **67**, 97.
59. Subcommittee on Mineral Toxicity in Animals (1980). "Mineral Tolerance of Domestic Animals." Natl. Acad. Sci., Washington, D.C.
60. Suttle, N. F., and Mills, C. F. (1966) Br. J. Nutr. **20**, 135, 149.
61. Todd, W. R., Elvehjem, C. A., and Hart, E. B. (1934). Am. J. Physiol. **107**, 146.
62. Underwood, E. J., and Filmer, J. F. (1935). Aust. Vet. J. **11**, 84.
63. Underwood, E. J., and Somers, M. (1969). Aust. J. Agric. Res. **20**, 889.
64. Vallee, B. L. (1971). In "Newer Trace Elements in Nutrition" (W. Mertz and W. E. Cornatzer, eds.), p. 33. Dekker, New York.
65. Vallee, B. L. (1974). In "Trace Element Metabolism in Animals" (W. G. Hoekstra, J. W. Suttie, H. E. Ganther, and W. Mertz, eds.), Vol. 2, p. 5. Univ. Park Press, Baltimore, Maryland.
66. Velu, H. (1931). C. R. Séances Soc. Biol. Ses. Fil. **108**, 750; (1938) **127**, 854.
67. Venchikov, A. I. (1960). Vopr. Pitan. **6**, 3.
68. Venchikov, A. I. (1974). In "Trace Element Metabolism in Animals" (W. G. Hoekstra, J. W. Suttie, H. E. Ganther, and W. Mertz, eds.), Vol. 2, p. 295. Univ. Park Press, Baltimore, Maryland.
69. Wacker, W. E. C., and Vallee, B. L. (1959). J. Biol. Chem. **234**, 3257.
70. Waddell, J., Steenbock, H., and Hart, E. B. (1931). J. Nutr. **4**, 53.
71. WHO Expert Committee. (1973). "Trace Elements in Human Nutrition". World Health Organ. Tech. Rep. Ser. No. 532.
72. Wilgus, H. R., Norris, L. C., and Heuser, G. F. (1936). Science **84**, 252; J. Nutr. **14**, 155 (1937).
73. Workshop on Research Needed to Improve Data on Mineral Content of Human Tissues. (1981). Fed. Proc., Fed. Am. Soc. Exp. Biol. **40**, 2111.
74. Wright, N. C., and Papish, J. (1929). Science **69**, 78.

2

Methods of Trace Element Research

J. CECIL SMITH, Jr.

U.S. Department of Agriculture
Agricultural Research Service
Beltsville Human Nutrition Research Center
Beltsville, Maryland

I. NATURE'S EXPERIMENTS

Historically, certain disease states of humans and domesticated animals have been recognized to occur naturally within the confines of a specific geographic area; however, it was not immediately apparent that lack of trace elements might be involved. Iodine was apparently the first trace element to be associated with a human disease located in specific regions. As early as 1820, the salts of iodine were used therapeutically by the Frenchman Courtois and others. Even before then, the ancient Greeks used burned sponges for the treatment of goiter. During the early 1800s iodine was shown to be present in sponges and seaweeds (13,147). Although administration of the iodine salts resulted in some benefit, the practice fell into disrepute because, in part, toxic symptoms were induced by excessive doses. In addition, the rationale of iodine therapy had not yet been demonstrated. Nevertheless, in mid-nineteenth century the Frenchman Chatin (22) noted the association between some cases of goiter and deficient intakes of iodine. As a botanist he had observed the wide distribution of minute quantities of iodine in soil, drinking water, and foods. However, initially Chatin's pioneering discovery was not readily applied. Apparently this was due, in part, to certain anomalies in his data and to the inability of clinicians to differentiate various types of thyroid diseases caused by excess or deficiency of thyroid activity.

Therefore, several decades passed without his discovery benefiting those suffering from goiter throughout the world. In the 1900s, the goiter–iodine deficiency theory was rediscovered. In the meantime following Chatin's report, Baumann (9), the German chemist, in 1895 provided definitive evidence that iodine was present in the thyroid gland and that the level was decreased in simple or endemic goiter. In addition, in 1915 Kendall (65) crystallized thyroxine, which was shown to contain 60% iodine.

Eighty years after Chatin's demonstration of the association of iodine and the incidence of goiter in specific geographic regions, it became evident that several areas of the world were associated with either deficiencies or toxicities involving other trace elements. The diseases affected not only humans but also their livestock. For example, during 1931 a deficiency of grazing cattle in widely separated parts of the world (Holland and Florida) was reported and corrected by copper administration. In contrast to deficiencies in the soils of certain geographic locations, toxicities of selenium were recognized as being associated with the livestock condition of "alkali disease" (chronic) or "blind staggers" prominent in some regions of the Great Plains of the United States (148).

Another one of nature's experiments led to the recognition of the need for cobalt in ruminant nutriture. During the nineteenth century "wasting disease" or "bush sickness" of sheep and cattle had been recognized in localized areas of the world including New Zealand and Australia. The condition was treated by transferring the sick animals to areas where the disease was not prevalent for varying periods. Scientific work at the end of the nineteenth century in New Zealand established that the condition was not due to infections or toxic substances, and a mineral deficiency or excess was postulated. Indeed, a series of investigations led to the claim by Aston in 1924 (8) that the condition was due to a deficiency of iron, since it could be cured or prevented by dosing with crude salts or ores of iron. However, subsequent studies by Underwood and Filmer (146) in Australia demonstrated that the condition was not caused by iron deficiency but was corrected by administration of minute oral supplements of cobalt alone. The iron compounds which were potent in correcting the bush sickness were contaminated with sufficient cobalt to prevent or correct the deficiency. In the United States, cobalt deficiency was recognized in cattle of the Atlantic Coastal Plain in the early 1940s. A correlation between the cobalt concentration in the forage and the incidence of the deficiency was soon established. Cobalt chloride was added to the mineral mixture of affected cattle in the North Carolina region with a significant improvement.

Geographic regions have subsequently been associated with impaired zinc nutriture in humans (51,104). In this case it appears that the initial observation of the deficiency in Iran in the early 1960s was not simply a result of inadequate zinc intakes due to the food being grown on deficient soil, but was apparently a result of consuming large quantities of naturally occurring chelators, phytate and/or fiber, found in the unleavened whole-wheat bread. Indeed the subjects apparently were ingesting recommended levels of zinc, but an insufficient quantity was bioavailable (134). In addition, the original "zinc-deficient dwarfs" consumed high amounts of clay—a potential binder of trace metals.

The latest example of nature's experiments concerns the observation in China of an endemic disease (Keshan) that was responsive to selenium supplementation (23,66). This condition affected large numbers over a broad region of the northeast to the southwest of China. The geographic region characteristic of Keshan disease was demonstrated by lower selenium levels in foods in specific areas being associated with increased incidence of the condition. The salient features included a cardiomyopathy, arrhythmias, abnormal electrocardiograms, and heart enlargement. The condition occured most frequently in children under 15 years and in women of childbearing age. Earlier investigations demonstrated that selenium was a component of the enzyme glutathione peroxidase (115), an enzyme that degrades potentially toxic lipid peroxides, and thus established a biochemical function for selenium. Before 1979, however, there were no reports of human deficiency.

In addition to the report of Keshan disease in China, during the same year it was demonstrated that a patient receiving total parenteral nutrition (without selenium) for an extended period, responded to selenomethionine (149). This first documented clinical case of selenium deficiency contributed to the establishment of selenium as an essential nutrient.

II. EARLY ATTEMPTS USING PURIFIED DIETS
TO STUDY TRACE ELEMENTS

More than 100 years ago in 1873, Forster (41) of Germany fed purified, salt-poor diets and found that the experimental animals (dogs and pigeons) died within a few weeks. He concluded that his experiments confirmed that certain minerals were essential for life. Although his conclusion was correct, the animals probably died of vitamin deficiencies because the diets used were relatively purified. Indeed, Lunin (81), also of Germany, extended the work of Forster and realized that other factors were missing from the diet but stated that no other explanation was available using the existing knowledge for that time, 1881. He used the ash of milk as a basis for supplying minerals of his purified diets. However, he recognized that milk contained essential factors other than minerals.

Bunge (18), in his textbook of "Physiological and Pathological Chemistry," published in 1902, advanced the thesis that the mineral composition of the body was similar to the mineral content of the milk of a given species, suggesting that milk minerals were present in optimal concentrations and proportions. Several subsequent early investigators, including Osborne and Mendel (99,100), when formulating their "salt mixture IV," used a combination of minerals similar to that found in milk. However, in their classic studies they found that rats died when fed their purified diet unless "protein-free milk" was added. From this observation it was realized that the milk supplied other essential factors (now recognized as vitamins). Other pioneers in the field of developing purified diets in conjunction with animal experiments included Bertrand of France (11,12) and McHargue (86) of the state of Missouri. Their studies were performed during World War I and extended into the early 1920s. Both groups, though widely separated, reported data suggesting a mammalian requirement for growth of the trace elements zinc, copper, and manganese. The experimental animals fed the purified diets also died or showed impaired growth, even when supplemented with the specific trace element being studied. Overcoming this difficulty necessitated adding natural materials to the diets such as yeast, meat extracts, or plant "greens." These products supplied sufficient trace elements to confound the experiments. In addition, they provided other essential dietary factors which had not yet been identified.

Thus, during the 1920–1930 decade, it became evident that the purified diets used for the early trace mineral studies were lacking vitamins. Hence the "greens," yeast, milk, and other natural products were providing the vitamins essential for growth.

III. SCIENTIFIC ERA: DEVELOPMENT OF ADEQUATE PURIFIED DIETS

The second quarter of this century was the period for the discovery of several trace elements essential for mammals. Only two trace elements, iron and iodine, had been shown to be necessary before 1925. During the period of 1925–1930, investigators at the University of Rochester (112) treated experimental hemorrhagic anemia in dogs with a variety of food items. They observed that hemoglobin production rate was not directly related to the iron content of the product and raised the question of the role of other trace minerals. During this same period the Wisconsin group led by E. B. Hart and C. A. Elvehjem made major contributions using diets based on feeding milk, one of nature's products that is naturally deficient in three trace elements: iron, copper, and manganese. They also further developed the use of the purified-diet technique. The group's first major discovery regarding trace elements was the demonstration of the essentiality of copper for growth in 1928 using rats (55) and rabbits (155). This contribution was an unexpected by-product of their work with iron, where they showed that after 4–6 weeks of being fed only milk, young rats failed to grow; anemia developed, followed by death. In treating this condition, they observed anomalies in the efficacy of various natural sources of iron. The supplementation of pure inorganic iron salts failed to prevent the development of anemia. However, if the animals were given a small amount of copper with the iron, the anemia was corrected and the animals recovered. From this observation, Hart *et al.* (55) concluded that copper was essential along with iron.

Within 3 years of this major contribution, the Wisconsin group demonstrated the essentiality of another trace element, manganese. They found that a manganese-deficient diet resulted in infertility of rats and mice. Supplemental manganese reversed the condition. The third trace element shown to be essential in mammalian nutriture by the Wisconsin investigators was zinc. In 1934, Todd *et al.* (142) were apparently the first to formulate a purified diet sufficiently adequate in essential nutrients except zinc. The basic ingredients included casein, egg white, corn starch, and mineral mixture. The vitamins which had been identified at that time were supplied separately each day. They noted that egg white was naturally low in zinc, an observation previously reported by Lutz (82). The casein was prepared by acid precipitation of skim milk, and the corn starch was acid-extracted to remove contaminating zinc. Although some minerals commercially available at that time were of sufficient purity so that they could be used without treatment, a "large number of calcium salts were found to be contaminated to a considerable degree with zinc" (142). Therefore, they prepared calcium phosphate in their laboratory. For several of their experiments they provided each animal with 2 ml of milk daily, which improved the adequacy of the diet, probably by providing additional essential vitamins. The zinc-deficient diet resulted in an impaired growth of nearly 50% compared to the control diet containing zinc at 50 mg/kg diet. In addition, alopecia was noted (with a picture provided) in the animals fed the low-zinc diets. Although all of the then-known vitamins and essential minerals were supplied, they recognized that the purified diet may not have been completely adequate because the zinc-supplemented animals did not grow optimally. Nevertheless other workers, including Newell and McCollum (95) a year earlier, had not been successful in the use of a purified diet to produce a zinc deficiency in rats, although they reported that their diet contained only 0.1 μg of zinc per gram. Todd *et al.* (142) commented that they had no adequate explanation for the difference in results between laboratories. In retrospect, it is probable that the diet fed by Newell and McCollum (95) contained more than 0.1 μg zinc per gram, and/or that the rats obtained additional zinc from environmental sources.

Thus, after numerous attempts by investigators in different parts of the world to develop an adequate purified diet to produce specific trace element deficiencies, the Wisconsin scientists were the first to be successful. Although purified diets have been further refined, including the use of the essential amino acids instead of whole protein, the basic formulation used today remains similar to that reported by those early investigators.

IV. THE "NEW" TRACE ELEMENTS

The discovery of trace element requirements in mammalian systems proceeded much more slowly after the rapid pace during the period of 1925–1935 when copper, manganese, zinc, and cobalt were demonstrated to be essential. During this period numerous trace elements had been shown to be present in animal tissues and milk, although no nutritional role was evident. By spectroscopic and other methods, milk was reported to contain aluminum, boron, copper, iodine, iron, lithium, manganese, rubidium, silicon, strontium, titanium, vanadium, and zinc (161). By 1939, nearly every element known had been found in "living tissues." In the same year, Shohl (125) speculated that the following were probably "not essential for nutrition": aluminum, arsenic, chromium, cesium, lithium, molybdenum, rubidium, strontium, tellurium, titanium, and vanadium. Those of "doubtful value" were cobalt, fluorine, nickel, and zinc.

Nearly 20 years elapsed between 1935 and 1953 without the discovery of any new essential trace elements. In 1953, a physiological role for molybdenum in mammals was presented. Two independent groups (31,111) reported that this trace element was a component of the enzyme xanthine oxidase. Within the same decade, Schwarz and Foltz (119) identified selenium as an essential nutrient, and Mertz and Schwarz (86a) discovered the nutritional importance of chromium as the trace element resulting in impaired glucose tolerance in rats fed diets deficient in that element. At present, recommended dietary intakes for humans have been established for all of the trace elements discovered through 1959, except for cobalt (107).

The "new" trace elements suggested by animal studies in the 1970s as being nutritionally important include molybdenum, vanadium, lithium, fluorine, silicon, nickel, arsenic, cadmium, lead, and tin. Nielsen and Mertz (97) have concluded that cadmium, lead, and tin do not meet the criteria for nutritional essentiality.

V. ADVANCES IN ANALYTICAL METHODOLOGY

In an initial attempt to assess mineral composition of plant and animal materials, investigators during the last century determined the total ash obtained by simple incineration (123). Evidence of the importance of minerals in the physiology of plants and animals was first based on this relatively crude method. It was soon recognized that the ash obtained by high-temperature incineration resulted in analytical losses of minerals. "Newer methods of ashing," using uniform low heat in an electric muffle furnace, gave better results. Although the technique was gross, it did allow comparison of the mineral composition of different tissues. For example, in 1874 Volkmann (154) compared the ash content of the organs and tissues of a human (male) and reported an average of

4.35%, ranging from 22.1% for the skeleton (bone) to 0.70% for skin. The ash content of the mother's milk affected the time required to double the birth weight of an infant, according to a thesis of Bunge (18). From these early observations came the recognition of the need to determine accurately the composition of the inorganic fraction of biological materials including food. Therefore, methods were developed for the analysis of specific minerals including trace elements.

Procedures using micromethods for staining tissue were among the first techniques of qualitatively identifying trace elements in biological samples. For example, the distribution of iron in various organs was observed by the application of hematoxylin to sections of tissue and observation under a microscope. In 1891, in a classic paper entitled "On the Demonstration of the Presence of Iron in Chromatin by Micro Chemical Methods," MacCallum (83) stated, "The discovery of micro-chemical methods for detecting iron in cells has aided me in establishing the generalization that the most important of all elements in the life of every cell is an iron holding compound."

In 1919 Birckner (14), using a turbidimetric method, indicated that it was "possible to determine less than 1 mg of zinc expeditiously and with a fair degree of accuracy . . ." He found zinc widely distributed in human foods including eggs and milk, suggesting an important nutritional role. After meticulous and extensive analysis, Lutz (82), in 1926, used volumetric (and fluorescent) methods to determine the total zinc content of an adult human to be 2.2 g, a value still referred to at present and assumed to be accurate.

Thus, analytical techniques for minerals (macro and trace) analysis progressed from simply obtaining the weight of total ash to developing more specific methods such as turbidimetric, gravimetric, and volumetric procedures. Although these methods were relatively insensitive compared to present techniques, they were a step toward more accurate quantitation. Some of these older methods have been reviewed by Beamish and Westland (10).

Historically, methods of trace element research have focused on the need for more sensitive and accurate methods of analysis. Indeed, the lack of adequate methods for quantitating inorganic components found in minute quantities (micrograms or submicrograms per gram of biological material) resulted in the analyst referring to them as being present in "trace amounts," hence, the term trace elements. Early investigators (124) often referred to them simply as "traces." Although the term has been criticized as not conveying their importance by suggesting that their role is minor, it still prevails.[1]

[1]Davies (27) suggested in 1972 that the words "trace elements" be abandoned and that instead these elements be called "essential biological metals," since they are necessary for the maintenance of health of humans and other mammals. Cotzia and Foradori (25) suggested that the term trace elements be reserved for those elements which occur in tissues at a concentration of 1 μg/g or less. Phipps (103) preferred to call the trace elements "micronutrients," because "some so-called trace elements are actually present in quite significant amounts; it is merely the dietary requirement for them that is low."

As a result of the lack of sufficiently sensitive methods, the toxic aspects of trace elements were often evident before their essentiality was established because the abnormally elevated concentrations were readily quantitated whereas the physiological levels were not detectable.

A major analytical contribution advancing the field was the development and availability of more sophisticated instrumentation, which allowed the accurate quantitation of physiological levels of trace elements instead of the qualitative estimation. For example, during the early part of this century the newly developed technique of emission spectrography gave an important stimulus to trace element research. This method allowed, for the first time, simultaneous determination of more than 20 elements including trace elements of low concentrations, in biological materials including human tissues (148). Immediately, numerous laboratories throughout the world began determining the biological distribution of trace minerals in plants and animals.

The studies of Tipton and Cook in 1963 (141) were the most comprehensive and significant of the early spectrographic investigations of trace elements in human tissue. Underwood (148) has pointed out that although the net effect of these classic studies was a giant step forward, often the data were overextended because the method lacked sensitivity. If a particular element was not detected, there was a tendency to imply that it did not normally exist in that tissue, when in actuality the quantity present was below the detection limit of the technique. Examples were arsenic, iodine, bromine, fluorine, mercury, and germanium. Conversely, some investigators suggested that the presence of a particular concentration of a trace element in an organ or tissue was evidence that it was important and had a function. Although such teleological arguments were not strictly scientific, the analytical presence of trace elements in human and cow's milk did provide impetus for the early workers to concentrate on their nutritional importance and resulted in the discovery of the biological significance of several of the trace elements.

In 1955, Walsh (157), an Australian physicist, developed a new technique which was to become the preferred and most widely employed method for trace element analysis, atomic absorption spectroscopy (AAS). In its simplest form the instrumentation consists of a hollow cathode lamp (light source), a flame atomizer, a grading or prism, and a photodetector. Although AAS was and remains a single-element-at-a-time technique, biological samples could be accurately quantitated for numerous trace elements, including some which were consistently present in all tissues but for which essentiality had not yet been established. The relative low cost and ease of operation, coupled with a high degree of sensitivity for several elements, allowed the field to advance rapidly. The other major advance concerned the development of another multielement technique, neutron activation analysis (NAA). This technique bombards the sample with neutrons, so that the elements present become radioactive and can then be quantitatively detected. The advantage of this multielement technique over the earlier developed emission spectrography method was increased sensitivity.

Although AAS and NAA remain popular techniques for trace mineral analysis, several other methods are available and have been summarized by Wolf (160). These include atomic emission, atomic fluorescence, X-ray fluorescence, and gas chromatography coupled with mass spectrometry.

There has been renewed attention to the use of a relatively new technique using inductively coupled plasma generators as an atomization (heat) source for optical emission spectrometry. A comparison of this method with flame absorption spectroscopy for multielement analysis of diets, feces, and urine gave good agreement for recovery of various elements added including zinc, copper, iron, and manganese (30). Also, the limits of detection for AAS have been lowered with the development of the flameless graphite furnace attachment.

Although no single technique or analytical method is applicable to all trace element analysis, factors to consider when choosing methodology and instrumentation include the following:

1. Whether single-element or multielement analysis is required
2. Type (matrix) of biological samples and potential interferences
3. Number of samples and time required per sample (automated or manual)
4. Sample size required
5. Availability, size, cost, ease of operation, and service of instrumentation
6. Skill, training, and experience of available personnel
7. Whether the instrumentation will be shared with other laboratories and/or used by different personnel

Usually the selection of instrumentation involves compromises. For example, usually multielement analysis provides less sensitivity for individual trace elements compared to single-element analysis.

With the development of more sophisticated and sensitive instrumentation came the challenges of preventing contamination of the sample with exogenous trace elements—those not native to the sample but originating from dust, water, utensils, containers, equipment, and laboratory personnel. For biological samples such as blood and other liquids, storage containers could also be a major source of contamination. In addition to the usual contamination, increasing the level of the analyte to be tested, Thiers (139) advanced the concept of loss of the trace elements by adsorption on the container walls (negative contamination). Another potential problem which may not be recognized is that long-term storage may result in sample dehydration, especially if "frost-free" freezers are used. Dehydration will result in concentration of the sample, resulting in falsely elevated levels of the analyte to be determined. (A technique used to prevent dehydration is to place ice cubes in heat-sealed plastic bags along with the samples.) Other specific details associated with sample collection, sample preparation, and cleaning of glass and plastic ware have been presented elsewhere (94,132,135,160).

A. Animal Experimentation: Approaches to Developing Trace Element Deficiencies

As indicated previously, the essentiality of a specific trace element usually was demonstrated in experimental animals prior to establishing its requirement for humans. Thus, animal models have made a major contribution toward the advancement of trace element research. For example, in the case of zinc, a deficiency was first developed in the rat in 1934 (142), and in swine in 1955 (143), but was not recognized in humans until the early 1960s (104). Likewise, the essentiality of chromium was first demonstrated in rats fed a commercial laboratory ration shown to be deficient in chromium (86a). Similarities between the "white muscle" disease of sheep and Keshan disease, a cardiomyopathy and associated pathology in humans, led Chinese scientists to establish the essentiality of selenium in human nutriture (23,66). Keshan disease has been demonstrated to be associated, in part, with inadequate selenium intake.

Animal models also have been useful in studying the ultratrace elements (96). The models have included monogastric mammals, poultry, and ruminants. A novel approach for studying basic mechanisms of trace element metabolism *in vivo* has been reported by Richards (109,110). This model involves poultry embryos maintained in long-term shell-less culture. Cell culture techniques have also been adapted for studying basic biochemical mechanisms involving trace elements (33,34).

1. Dietary Approaches

The classical methods of establishing that a trace element is essential have centered about producing a deficiency which can be prevented and/or reversed by supplementation with the single trace element being studied (24,88). The key factor in the development of the deficiency is that the test animal ingests from the diet and the environment an amount of trace element that is less than the requirement. Diets formulated for the development of trace element deficiencies can be divided into two major groups: one using natural and another employing purified or semipurified products.

a. Natural Product Diets. Using this approach, a variety of commercially available ingredients (nonpurified) provide the essential dietary nutrients. Each component usually provides several nutrients such as protein, carbohydrate, vitamins and minerals. Schroeder *et al.* (118) was able to avoid the use of complex mineral mixtures (a major source of contaminating trace elements) by using a variety of natural products. In the past, products such as dried skim milk, wheat bran, alfalfa meal, cornmeal, whole-wheat flour, and liver powder have been used. The effectiveness of natural product diets in producing a deficiency is dependent on the individual components being devoid or low in the trace element being studied. In order to find products low in a specific trace element, food-stuffs known to contain high levels are avoided. For example, meats and sea-foods are generally high in zinc, so it would be impossible to formulate a zinc-deficient diet with these ingredients. Likewise whole-egg powder contains high amounts of zinc. In contrast, and curiously, egg white is relatively low in zinc (80) and copper (35); thus it is often used as the source of protein in formulating zinc- or copper-deficient diets. However, egg white can be a major source of selenium and thus cannot be used for preparing selenium-deficient diets. In practice, a series of foodstuffs are initially analyzed to identify classes of prod-ucts that are low in the trace element being studied; then acceptable lots or batches are identified by analysis of several different sources. For example, wheat grown in certain areas of the Dakotas is relatively high in selenium, reflecting the elevated content in the soil. In contrast, wheat grown in low-selenium areas such as New Zealand or Finland is normally low in selenium and could be used for developing deficient diets. When a product low in the trace element has been identified, that batch or lot accumulated and reserved for future studies. However, with natural foodstuffs, storage problems such as ran-cidity may impose limitations on the amount to be stored, especially if they are not stored under refrigeration.

    b. Purified (Semipurified) Diets. Theoretically, diets prepared from pu-rified classes of nutrients would be the lowest in a specific trace element if the latter were excluded from the mineral mixture.[2] A typical purified diet includes amino acids as the protein source, sucrose or glucose as carbohydrates, essential fatty acids, and individual minerals and vitamins. In some cases a ''low-ash'' purified fiber such as cellulose is added to provide bulk and prevent digestive disturbances and diarrhea.

    Semipurified diets typically include an isolated protein such as casein, lac-talbumin, egg white solids, soy, or zein. Frequently, these proteins are major sources of contaminating trace elements and must be treated with a chelating agent to remove the contaminating metals. Although the chelating agent eth-ylenediamine tetraacetate (EDTA) has been used successfully to remove zinc from protein sources to produce a zinc-deficient diet (28), little work has been

---

[2]However, it should not be assumed that highly purified diets, including those whose protein source is amino acids, are more desirable for trace element studies than those using natural products. For example, the nickel content of a dried skim milk diet was 80 ng/g, compared to 190 ng/g for an amino acid-based diet (128). In addition, the cost of purified amino acids makes their routine use in trace element-deficient diets prohibitive at present.

done using other methods of purification. When chelating agents are employed to lower contaminating minerals, caution must be exercised .to assure that residues of the chelate have been removed prior to incorporating the component into the diet. Although some isolated proteins may be sufficiently low in the trace element of interest, different sources or batches may vary widely in their trace mineral concentration. Prior analysis of each dietary component will allow preparation of a deficient diet without purification or extraction.

Theoretically, liquid-type diets could be used for trace element studies, since they can be extracted directly with various chelating agents using appropriate metal-binding agents (after adjusting the pH) in combination with a solvent system such as methylisobutyl ketone. Obviously, traces of the solvent must be removed. However, liquid diets have different physical characteristics compared to commercial laboratory rations, so the results may not be relevant. In addition, they may have no fiber.

Another approach to the development of experimental trace element diets for animals has been the use of exotic foods such as torula yeast (120). Commercially prepared as a source of protein for animal feeds, torula yeast has been demonstrated to be sufficiently low in selenium and has been used as the protein source in selenium-deficient diets. Since it is lacking in sulfur-containing amino acids, methionine is usually added as a supplement. Likewise, chromium-deficient diets have contained torula yeast, although not all batches or sources are sufficiently devoid of chromium. This apparently is due to contamination during preparation. Torula yeast cannot be used as the protein source for copper-deficient diets, since it usually contains too much copper (J. C. Smith, unpublished data); analysis of different lots and sources varied from 3 to 15 μg/g of copper, illustrating the relative high copper content and the variability among sources. One approach to produce zinc deficiency has been to add phytate to a semipurified diet (98). Apparently the phytate binds the zinc, making it unavailable for absorption. Using this approach the deficiency can be enhanced by increasing the calcium content. Indeed, in the first report of parakeratosis in swine, a zinc deficiency condition, the diet contained a relatively high concentration of zinc, 34–44 ppm, but the development of the deficiency was dependent on the combination of high dietary phytate and relatively high calcium in the diet (143). Using another approach, Apgar (6) injected EDTA intramuscularly to produce zinc deficiency expeditiously in pregnant rats.

In addition to protein as a prime source of contamination, another obvious source is the mineral mixture. Although a trace element (or its salt) may not be deliberately added to a mineral mixture, it can be present in significant amounts as an impurity associated with other compounds. For example, a mineral mixture (42) prepared from reagent-grade mineral compounds, but without nickel salts added, contained up to 3 μg/g. If this mixture were used at the 5% level as the mineral source for a diet, the contribution of contaminating nickel would be 150 ng/g—too high to produce a deficiency of nickel in any nonruminant species studied to date. Likewise, contaminating chromium was 1.4 μg/g, a concentration greatly exceeding the requirement for all animals studied. Although several traditional mineral mixtures (46,59,99,100) have not included specific trace elements such as zinc, they should not be used to produce deficient diets because they may lack or have inadequate quantities of other essential trace elements such as chromium, copper, manganese, and selenium. In addition, some mineral mixtures have been responsible for urolithiasis due to abnormal ratios of minerals, especially calcium and phosphorus (113,121,129,130).

Although purification and recrystallization of the minerals and other components, when feasible, is one approach to prevent contamination of the diet with the trace elements being studied, it is usually not practical due to the time and expense required to process the large quantities involved.

In the past, the carbohydrate source of preference has been crystalline sucrose or glucose. Usually, commercial sources of these ingredients are relatively low in trace metal contamination. However, because carbohydrates are the major ingredients, ranging from 60 to 70% of the diet, they have the potential of being major sources of contaminants.

A problem which must be considered an undesirable "side effect" of feeding diets high in purified carbohydrates such as sucrose or glucose is the development of "fatty livers" apparently due to lipogenic properties of simple sugars. The effect of the abnormal lipid deposition, which possible confounding of the trace element metabolism, has not been adequately addressed. It has generally been assumed that complex carbohydrates are more contaminated with trace elements than simple sugars. In the classic study of Todd et al. (142), which first described zinc deficiency using rats fed a purified diet, the investigators extracted the complex carbohydrate source (corn starch) with acid as a method of removing contaminating zinc. However, we have observed that present-day commercial sources of corn starch are as low in copper as glucose, fructose, or sucrose (J. C. Smith, unpublished data). Finally, in regard to the carbohydrate moiety of trace element-deficient diets, it has been demonstrated that the severity of copper deficiency in rats is dependent on not only the level of copper in the diet but also the type of dietary carbohydrate (35–39). Diets containing fructose as the sole source of carbohydrate produced the most severe deficiency, whereas starch-fed animals were least deficient; glucose and sucrose produced an intermediate level of deficiency, although all diets contained the same low level of copper (i.e., 1 ppm). In each case the carbohydrate was provided at a level of 62% of the total diet. Earlier, Hegsted (56) predicted that the dietary type of carbohydrate may interact with trace elements. Indeed, Amine and Hegsted (1) reported that lactose increased iron absorption. Later, Landes (73) concluded from his studies with rats that the type of dietary carbohydrate influenced to some extent the trace mineral (iron, copper, zinc) status.

2. Ultrapure Drinking Water

Although diet is a major source of contamination for experimental animals, the drinking water should be monitored and demonstrated to be free of minerals. Early workers relied on distilling procedures to purify the water. Special glass (i.e., quartz) stills were employed. They were careful not to allow the water to come in contact with metal or rubber (tubing or stoppers). Using these techniques, limited quantities of "triple-distilled" water could be produced. This technique required careful monitoring and was expensive.

Modern methods of providing ultrapure water employ mixed-bed resins to produce "deionized" water having a resistance of up to 18 MΩ (150). Such water is ideal for experimental animals and as an analytical reagent. Water that has been rendered sufficiently pure should not come in contact with potential sources of contamination such as metal or rubber, including the rubber stoppers of the bottles used to provide drinking water to experimental animals. Nonrubber substitutes for stoppers include polyethylene beakers, with a whole punched in the bottom to receive the drinking tube. Previously, though relatively expensive, stoppers fabricated from vinyl or silicon were commercially available. The drinking tubes are another potential source of contamination. They are usually fabricated from glass or stainless steel. Glass tubes have the advantage of being relatively free of contaminating elements and being easier to keep clean than the metal tubes. However, caution must be exercised to prevent injury to hands due to breakage of the glass tubes, especially when initially inserting them into the one-holed stoppers. Obviously, stainless-steel tubes are not appropriate for studies involving chromium, although they are satisfactory for zinc and copper experiments.

Although the majority of animal care facilities have installed automatic watering devices to maintain experimental animals for routine experiments, individual water bottles are recommended for studies involving rodents. Bottles have the advantage of allowing monitoring and measurement of water intake. In addition, the hazard of flooding due to faulty valves is eliminated. If automatic watering devices are used, the trace metal content of the water, at the source of consumption, must be monitored.

3. Ultraclean Environment

Ideally, all animal experiments involving trace elements should be carried out in a controlled environment. Experiments involving ultratrace elements require environments free of contaminating dust and equipment. "White" or "clean" rooms developed to reduce airborne particulates during the fabrication and assembly of aerospace and electronic equipment, can be easily adapted as research laboratories (93).

It was recognized in the 1960s that a controlled-environment system would be required for producing ''new'' trace element deficiencies in laboratory animals. Schroeder *et al.* (118), in an attempt to provide a less contaminated environment, constructed wooden animal quarters ''at the top of a remote hill 500 m high at the end of a mile-long (1.609 km) dirt road, in order to avoid airborne contaminants, especially those from motor vehicle exhausts.'' The materials used in construction of the building were analyzed for the specific trace elements studied. Triple coats of a plastic varnish were used to cover potential sources of contamination. Animal cage racks were constructed of wood, and all nail heads were covered with plastic wood. Shoes were removed prior to entering the animal room. Electrostatic precipitators were used in an effort to remove dust from the incoming air.

The approach of Smith and Schwarz in 1967 (127) was to control the immediate environment in which the animal experiments were conducted. This was of particular importance, since most research laboratories are located in metropolitan areas. Thus an all-plastic controlled environment system was described for small experimental animals (Fig. 1). The system consisted of an air filter

Fig. 1. Metal-free controlled environment system designed for producing trace element deficiencies using experimental animals. From J. C. Smith and K. Schwarz (1967). *J. Nutr.* **93**, 182. Used with permission.

(fiberglass) assembly, a thin-film plastic isolator, with caging and accessories to minimize extraneous trace element contamination. The flexible plastic isolator was a modification of the type used for germ-free studies and had a capacity of 25 rats, in groups of 5 rats per cage. There were no apparent deleterious effects of the isolator system per se since weanling rats fed commercial laboratory ration in the controlled environment grew as well as control animals maintained outside the isolator in metal cages under conventional conditions. In addition, a comparison of the fecal microbial spectrum of rats in the controlled environment with that of conventionally maintained animals indicated that the controlled environment did not result in a major alteration in the gastrointestinal microflora (60). Original plans included a second barrier in the form of a plastic room-sized isolator which would contain several of the smaller ones. However, this precaution proved to be unnecessary. At that time, it was proposed that the controlled-environment technique, coupled with sensitive methods of analysis, would permit definitive experiments to determine the essentiality of specific trace elements.

Since the original report, numerous improvements have been made to control airborne-particulate contamination based on the application of high-efficiency particulate (HEPA) filters to upgrade the quality of air supplied to animal rooms used for trace element studies. These filters have an efficiency of nearly 100% for the removal of dust particles of 0.5 μm or less in the air.

4. Caging

Early in the history of trace element studies it was recognized that metal caging could be a major source of contamination, along with metal food containers and litter or bedding. Studies involving zinc deficiency necessitated avoiding galvanized caging. This was recognized as early as 1926 when Lutz (82), in a study designed to determine the distribution of zinc in rats, cats, and humans, stated, "the rats used as subjects in these experiments were kept in copper cages . . . in order to prevent any possible contamination from galvanized iron cages." In the same year, McHargue (86) reported improvising glass-lined cages for rats fed purified diets deficient in copper, zinc, and/or manganese. These were probably the first efforts to prevent trace element contamination from caging. Stainless-steel cages are a potential source of chromium, nickel, molybdenum, and silicon (128). Animals in contact with litter such as wood chips obviously could contain sufficient quantities of several trace elements from this source. Indeed, in 1963 Schroeder et al. (118) reported that soft wood chips contained relatively high levels of trace elements. Thus plastic caging was fabricated to address the need to eliminate metal and litter (91,127). For chromium studies, metal-free housing units fabricated from rigid clear plastic and polystyrene "egg crate" panels have been reported (102). Unlike plastic cages previously developed for trace element studies, these cages fit into commercially available racks. The all-plastic cages with lids also fit onto laminar-flow stations, permitting the maintenance of animals not only in a metal-free environment but also in a continuous flow of ultraclean air.

B. Human Studies

Historically, the essentiality of most trace elements for humans has first been indicated by producing an overt dietary deficiency in an experimental animal model. On the other hand, toxicity of a specific trace element frequently was detected by poisoning of human or animals due to contamination of food and/or environment. Controlled studies were designed specifically to investigate trace element metabolism using human subjects.

1. Metabolic Balance Technique: Two Approaches

The metabolic balance technique has been routinely employed to determine the requirement and/or bioavailability of specific nutrients, including trace elements. In addition, the technique provides data concerning the total dietary intake, pathways of excretion, amounts and thus percentages excreted, and apparent absorption (137). The two general designs of human studies consist of either maintaining the volunteers in a metabolic ward, as in a hospital, or allowing them to remain in their usual (home) environment. When the special metabolic unit approach is employed, the subjects are usually provided exercising facilities. Frequently, they are restricted to the immediate area of the metabolic unit with continuous recording of food intake and collection of excreta. The meals are formulated by dietitians and specially prepared within the unit.

When the home setting, "free-living" approach is used, the volunteers are trusted to adhere to instructions of the protocol. One form of this approach allows the subject to consume "self-selected" meals prepared at home, usually in conjunction with daily dietary records. After specific instructions and supplying the necessary trace metal-free containers, diet, fecal, and urinary collections can be made, providing data necessary for calculating balance and apparent absorption. Using this approach the Beltsville Human Nutrition Research Center has reported a 1-year study involving 29 adults who remained in their home setting and ate self-selected meals (90). After appropriate training, the subjects kept detailed dietary records which were daily presented to nutritionists for checking. Diet, feces, and urine were collected 1 week out of each of the four seasons to determine the balance for selected nutrients including trace minerals. Thus there were four balance periods of 1 week duration during the year. Nutrient intake was calculated from the daily dietary records and a nutrient data base derived from Agriculture Handbooks No. 8 and No. 456 and nutrient composition information provided by commercial companies and individual research laboratories.

   a. Food. Each subject collected duplicate samples of all food and beverages consumed during the 1 week collection periods. Food was collected daily in polypropylene jars with lids; beverages were collected separately in a similar container. The subjects were instructed not to include any food that was not eaten such as bones, peelings, seeds, or cores. Fruit juices and milk were added to the food container. In the laboratory the food was weighed; a measured quantity of deionized water was then added and the mixture blended in the collection jar using an electric blender that attached directly to the container. (The parts of the blender coming in contact with the food including the blades had been previously checked as a source of contamination for the elements being studied.) Aliquots of the daily food composites for each subject were frozen. At the end of the week, aliquots for all 7 days were blended. A single aliquot of the 7-day food composites was weighed and stored frozen at $-20°C$.

   Duplicate samples were collected of all beverages consumed including water, coffee, tea, soft drinks, and alcohol. Aliquots of daily beverage consumption was stored at $4°C$. The seven daily beverage aliquots were combined at the end of the week, and aliquots of the "master" composite were stored frozen at $-20°C$.

   b. Urine. During the collection week, each subject collected all urine voided in a 24-hr period using polypropylene jars. (For some studies toluene is added to the collection jar to prevent microbial growth.) Each first morning void urine was added to the previous day's collection. The daily collections were taken to the laboratory; the quantity was measured and made to a similar volume (usually 2 liters) with deionized water. An aliquot was removed and stored at $4°C$. To each daily aliquot, 0.5 ml of ultrapure concentrated HCl was added for every 100 ml. Weekly composites were made after the seven daily samples were collected. The weekly composite was stored frozen at $-20°C$.

   Trace element balances were determined for selenium, manganese, zinc, and copper. The balances were calculated, after analysis of the food and beverages, feces, and urine according to the following formula.

$$\text{Balance} = \frac{[\text{intake} - (\text{fecal} + \text{urine excretion})]}{\text{intake}} \times 100$$

   c. Feces. In order to collect for analysis the feces that corresponded to the meals consumed, a gelatin capsule containing 50 mg of brilliant blue dye was ingested with a meal (breakfast) by each subject at the beginning of the collection week. Similarly, at the end of the collection period, a second capsule with the dye was consumed with breakfast. The fecal samples, one defecation per container, were collected in round (1 q) polystyrene containers with lids until the second marker of brilliant blue had passed. The fecal samples were transported daily to the laboratory, where they were weighed and frozen. The fecal material saved for analysis was that collected between the two markers.

2. *Limitations of the Balance Technique*

Traditionally, the balance technique has been used to provide data relative to determining the dietary requirement of specific nutrients including trace elements. However, evidence is accumulating which suggests that prior long-term habitual intake of a trace element being studied may be a major determinant regarding whether an individual is in positive, negative, or equilibrium balance at any particular intake. For example, New Zealand women required only about 24 μg of selenium per day to maintain equilibrium balance (138), whereas women in Maryland required more than twice that amount, 57 μg/day (75). The usual daily selenium intake for the Maryland women approximated 75 μg, compared to about one-third this amount for the New Zealand women. Thus, perhaps adaptation to different habitual intakes over a long period has allowed the New Zealand women to have a lower apparent requirement than those in Maryland. Levander and Morris (75) suggest that throughout their lives the New Zealanders have consumed diets lower in selenium than those eaten by the Marylanders and thus apparently need less dietary selenium to maintain "smaller total body selenium pools." They concluded that selenium requirement, as determined by balance, apparently is, in part, a function of previous habitual intake.

Other studies have indicated that adaptation to different intakes may occur within a relatively short balance period. For example, in a brief preliminary research note in 1973, Buerk *et al.* (17) reported that four volunteers fed a "special diet" providing only 0.6–1.0 mg of zinc per day showed a markedly negative balance for only the first 5–10 days of the experiment; thereafter there was a conservation of the element with a decrease in both urinary and fecal zinc losses. However, complete balance data were not provided, and the design of the experiment involved healing of a standardized wound, which may have affected the results.

When subjects whose self-selected diets provided approximately 12 mg of zinc were given a lower daily intake during a metabolic balance study, they initially showed a negative zinc balance, indicating that the requirement had not been met (2). However, after the subjects consumed the lower zinc intake for a longer period (3 weeks), adaptation—with apparent homeostatic mechanisms—was triggered and near-equilibrium attained. Thus, the initial balance suggested a higher requirement (than the subsequent data) because homeostatic adaptation to the lower intake had not occurred. This example indicates the importance of considering the previous intake of subjects involved in balance studies involving trace elements. In addition, balance studies should be of sufficient duration to allow for homeostatic adaptation. This adaptation may include a change in size of body pools, a conservation of the element with decreased turnover, and/or an alteration in absorption.

Problems associated with balance studies including those involving minerals have been discussed by others (57,137). Obviously, a significant change in body weight would affect the balance. To maintain body weight under controlled conditions, the food intake is adjusted. However, in the study involving 29 adults consuming self-selected diets, food (caloric) intake decreased by an average of 13% during the four 1-week balances compared to the mean for the entire year (67). Apparently, the disturbance and altered daily routine required for complete collection of food, feces, and urine during the balance periods resulted in less food consumed compared to usual intake. In addition, there was a loss of approximately 1 lb of body weight during each collection week. Thus, for free-living individuals consuming self-selected diets the habitual level of intake and selection of food and beverage items may be altered during the intervention of collecting data and samples necessary for metabolic balance studies.

Another possible compromising factor which could affect balance studies of trace elements concerns the form of the food that is offered. For example, for ease of preparation and to assure accurate quantitation and constancy of intake, defined diets have been liquified and the "meals" provided as a drink. Bioavailability and absorption of trace elements from such liquid purified diets, frequently of a composition markedly different from usual foods including foods containing less bulk (fiber), probably are not comparable to conventional meals.

VI. FACTORS TO CONSIDER IN THE DESIGN OF HUMAN STUDIES

Trace element studies involving humans are extremely difficult because of lack of adequate controls and also because of individual differences in metabolism due apparently in part to variations in genetic background. In addition, there are cumulative errors from environmental contamination and analytical methodology. Because of the variability among individuals, in the ideal situation subjects should serve as their own controls. If only two treatments are used (i.e., supplementation or placebo), a crossover design may be ideal if there is little carryover or biological memory. Theoretically, specific trace elements whose major metabolic excretory route is urinary would show less carryover and adjust more rapidly than elements whose metabolism includes endogenous circulation with a major portion eventually excreted via the feces. If there is a carryover, sufficient time should be allowed following the crossover to obtain "steady state." It may be necessary to carry out preliminary experiments to determine the time required to obtain equilibrium for the elements being studied. The experimental design may be influenced by the number of subjects involved, anticipated differences due to individual variation and treatment, as well as resources available. Therefore, statistical expertise should be sought prior to initiation of the study.

In the past, human studies included the frequent use of radioisotopes to provide information on metabolism, including distribution, transport, and excretion of the trace element being studied. In addition, the nonradioactive form of the element could also be quantitated to provide comparative data. However, the use of radioactive materials in healthy subjects is presently considered an unnecessary health risk and is largely limited to diagnostic purposes. On the other hand, "stable" nonradioactive isotopes are being used with increasing frequency in conjunction with balance studies and allow the "tracing" metabolically of specific elements (151,152). The major advantage is elimination of the health risk due to radiation. Quantitative recovery of the trace element is not required because isotopes are chemically identical for analytical purposes, so that incomplete recoveries will equally affect both native analyte and added isotope. However, stable isotopes are not readily available, often expensive, and require highly trained personnel and sophisticated equipment for analyses such as neutron activation or mass spectrometry. Equally important, some stable isotopes of trace elements whose natural abundance is relatively high are poor candidates for use in studies involving humans, since large quantities of the enriched isotope must be added. That is, the stable isotope technique employs enriching or "spiking" the food, extrinsically or intrinsically, with a known quantity of isotope. Therefore, the relative abundance of the two isotopes naturally found is altered, or "diluted," (thus the technique has been coined stable-isotope dilution method) (151). The technique frequently has been used for balance studies involving zinc, selenium, and copper. The least abundant and therefore most desirable stable isotopes of zinc, copper, and selenium are ^{70}Zn (0.62%), ^{65}Cu (30.9%), and ^{74}Se (0.87%), respectively. Unfortunately, the stable-isotope technique cannot be used for those trace elements which are mononuclidic such as fluoride, aluminum, manganese, cobalt, and arsenic.

A. Selection of Subjects

Subjects who volunteer to participate in nutrition studies, especially general and trace element investigations, may not accurately represent the "average" individual. This may be particularly true of volunteers who are recruited for "live-in" metabolic studies where subjects are housed in an institutional setting, requiring major changes in life-style including isolation from social interactions. Individuals volunteering for nutritional studies may have an above-average interest and knowledge regarding fitness and health. In some cases, only males are used in order to eliminate the possible effect of the menstrual cycle and hormonal alterations on the data collected. Obviously, recruiting normal pregnant and lactating women is difficult, especially if isolation in a metabolic unit or hospital ward is required. Thus, of necessity, studies involving pregnant or lactating females frequently use the "free-living" technique. Using this approach, it is extremely important that honesty be one of the prime criteria for subject selection. For example, should a sample be lost, it must be reported by the subject. Likewise, since the subjects are without constant surveillance, they must consume only the foods and beverages provided. Often first-time subjects are unaware of the sacrifices necessary to adhere to protocols involving a long-term commitment as well as the effort necessary. Experimental designs may have built-in checks for determining compliance, but there is no substitution for honesty.

Some subjects may be unable to adhere to the protocol or will not complete the study due to unforseen emergencies. Thus, recruitment should include extra subjects approximating 1 for every 10 entered into the study.

B. Assessment of Trace Element Nutriture in Humans

Identification of inadequate trace element nutriture requires accurate methods of assessment. Ideally, trace element status should not be determined using a single parameter, and biochemical indices are frequently combined with functional tests in an effort to obtain more definitive assessment. The number of indices that are presently available to assess accurately the status of the majority of trace elements is limited. At present, iron has the largest variety of indices for determining status, and impaired iron status can be assessed with a high degree of accuracy. The major indices include hemoglobin, serum iron, serum ferritin, total iron-binding capacity, transferrin saturation, erythrocyte protoporphyrin, and mean corpusular volume.

A novel approach for assessing the iron status of participants of the second National Health and Nutrition Examination Survey (NHANES) used three different models (76). The first model, termed the ferritin model, included transferrin saturation (TS), erythrocyte protoporphyrin (EP), and serum ferritin (which indicates level of stores). The second model used TS, EP, and mean corpusular volume (which can identify severe depletion of iron affecting red blood cells). This model was called the mean corpusular volume model. Two of the three parameters were required to be abnormal in order to determine that the individual had impaired iron status. A third model, the hemoglobin shift, assessed the prevalence of anemia. This technique determined the percentile shift in median hemoglobin levels after excluding subjects with abnormal values for one or more iron status indicators (26,76). Using these models the prevalence of impaired iron status was estimated. It was noted that many of the abnormal values for iron status in older persons may have resulted not from inadequate iron status per se but from inflammatory disease. Thus, assessing iron status in humans has advanced to a stage of sophistication that is yet to be attained for the other trace elements essential in human nutriture.

1. Chromium

The evaluation of the nutritional status of chromium in humans has traditionally included biochemical and oral glucose tolerance tests. The biochemical tests included the analytical determination of the chromium concentration in biological specimens such as serum or plasma, urine, and hair. In the past, due to methodological difficulties and inadequate instrumentation, there have been major discrepancies among laboratories regarding normal ranges of chromium concentrations for tissues (133). Recent improvements in these areas have resulted in general agreement on normal ranges for serum, urine, and hair. In addition, reference materials including serum are being developed to provide aids in quality control (153).

In general, blood (serum) concentrations have been thought not to be an accurate reflection of chromium status, since blood chromium is not in equilibrium with tissue stores (87). However, in a patient receiving long-term total parenteral nutrition who responded to chromium supplementation, blood chromium concentrations were low (0.55 ng/ml) compared to laboratory normals of 4.9–9.5 ng/ml (63). Although the normal values are too high in light of current concentrations obtained with improved instrumentation (133), the relative difference between the chromium-deficient subject and normals would suggest that during severe chromium deficiency blood chromium levels may decrease.

Urinary chromium excretion following a glucose load is a potential index of chromium status. However, because of the very low concentrations (<1 μg per 24 hr), extreme caution must be exercised to prevent contamination. Conventional methods of urine collection, especially under field conditions, are not acceptable. In addition, although methodological improvements for urinary chromium determinations are now available, the techniques require special instrumentation and trained personnel. However, there is presently insufficient evidence to establish that urinary excretion accurately reflects chromium nutriture except perhaps during severe deficiency. The usefulness of urinary chromium measurements in the assessment of status must await definitive experiments involving subjects overtly deficient in chromium due to dietary deficiency. Should this parameter prove helpful in assessing status, there are conflicting data concerning whether 24-hr collections are necessary or if the chromium–creatinine ratio can be used to calculate daily excretion (106). Gurson and Saner (47) indicated that the ratio between urinary chromium and creatinine is related to age, with the ratio highest in newborn infants; for the adult age group a double-blind study indicated no correlation between 24-hr chromium excretion and age (4). The subjects ranged from 21 to 69 years of age.

Another potential parameter is the alteration of serum chromium concentrations following a glucose load. Theoretically, a correlative rise in serum chromium with circulating insulin and/or glucose concentrations would indicate adequate chromium status. However, there is also no general consensus regarding its accuracy, since some studies have indicated an acute rise in serum chromium (44) while others have reported a decrease (5). One suggestion has been to calculate the ratio of serum chromium concentration (1 hr after the glucose load) to the fasting chromium level (79).

Earlier studies used hair as the index of choice because the concentration is high in relation to other tissues, can be sampled readily without special equipment, and can be stored with no refrigeration. On the other hand, the hair's concentration of chromium and other trace elements does not represent the immediate status but rather the long-term nutriture. In addition, contamination may be a problem due to hair treatment. Thus, hair chromium analysis must be interpreted with caution.

Perhaps the most promising approach for assessing parameters suitable for determining status would be to develop relevant animal models and produce a specific chromium deficiency allowing detection of abnormal biochemical and functional tests. Indeed, early animal (rat) studies indicated glucose intolerance of animals fed inadequate dietary chromium (86a,87). Since then, glucose tolerance and chromium nutriture have been reported in other species including mice, squirrel monkeys, and, most importantly, humans. However, numerous other conditions may alter glucose tolerance, so this parameter cannot be used alone. At present, the only critical test for chromium is a definitive response to chromium supplementation under controlled conditions (15).

2. Copper

Copper nutriture in humans is receiving increased attention, in part because of the recommendation of an "estimated safe and adequate" daily dietary intake of 2–3 mg by the Food and Nutrition Board of the U.S. National Academy of Sciences in 1980 (107). Assessment of self-selected intakes by free-living individuals has indicated intakes well below the lower limit. Indeed, a 1-year study reported in 1984 of such individuals reported average intakes of 1.2 mg for 28 adults living in the Washington, D.C. area (101). The health consequences of habitually consuming less than the recommended levels are unknown. However, a link between ischemic heart disease and inadequate copper nutriture has been hypothesized (69–71).

The most frequently employed parameter for assessing copper nutriture has been serum or plasma concentration (136a). Lahey et al. (72) reported that serum copper concentrations were significantly higher in females than in males. In addition, higher copper concentrations have been noted in serum than in plasma, apparently as a result of specific anticoagulants causing fluid shifts from the blood cells (114,135). A diurnal variation in plasma copper also has been demonstrated (58). Other potential complicating factors when serum or plasma are used for assessing status include the hypercupremia associated with pregnancy, oral contraceptive agents (131), infectious diseases, hematological disorders, neoplasms, and neurological, cardiovascular, and rheumatic diseases (85). In contrast, hypocupremia (<80 μg copper per deciliter) has been associated with Wilson's disease, kwashiorkor, marasmus, tropical sprue, macrocytic anemia, malabsorption due to small bowel disease, hyperchromic and hypochromic anemias, and nephrosis (158,159).

A major portion of the copper present in mammalian plasma is associated with the α-globulin, ceruloplasmin. The fraction of total plasma copper associated with ceruloplasmin has been reported to approximate 60% (78), although earlier reports indicated more than 90% (29). At any rate, the serum ceruloplasmin oxidase activity is frequently used for assessing copper nutriture in a clinical setting. However, results from animal studies indicate that ceruloplasmin oxidase activity is not correlated with liver copper concentrations when the animals are severely deficient, because the ceruloplasmin activity decreased to nondetectable levels prior to development of an overt deficiency (35). Liver is considered the prime storage organ for copper, and thus liver copper concentration is usually the parameter of choice for assessing copper status in animals. Obviously, this tissue is not routinely available for assessment of copper nutriture in humans.

Another promising parameter for copper assessment is the superoxide dismutase (SOD) activity of the erythrocytes. This copper metalloenzyme catalyzes the conversion of the superoxide radical to hydrogen peroxide plus oxygen. Animal studies have indicated a strong association between copper status (as determined by liver copper concentration) and liver SOD activity (39). Insufficient data from human studies involving SOD activity of erythrocytes are available to establish its value, but a preliminary report indicates its potential value (108).

Hair copper concentration has also been used as an index of copper nutriture (68). However, a report concerning two children with frank copper deficiency indicated normal hair copper levels and (16) concluded that the "considerable variation in the copper content of hair is probably independent of body stores and may be more closely related to hair production needs. . . ." It appears that hair copper concentrations may not accurately reflect the copper status. Thus, as stated by Hambidge (53), "interpretation of anayltical data on hair copper requires great caution."

Urinary copper is another potential parameter. However, estimates of normal 24-hr urinary excretion indicate a relatively low concentration with wide variations among laboratories. The apparent discrepancy probably results from methodological differences and possible contamination. The low concentration of urinary copper precludes its analysis using conventional flame AAS. More sensitive methodologies using AAS graphite furnace techniques must be developed for urinary copper.

Several studies have indicated abnormal cardiac function as measured by ECG in animals with copper deficiency. It has been suggested that decreased copper status may affect the electrophysical aspects of the intraventricular conducting system of the heart. As a result, abnormal ECGs and premature beats may occur (69,116).

3. Selenium

The traditional techniques for assessment have centered about biochemical indices including selenium concentrations of whole blood, erythrocytes, plasma, and hair (133). Short-term depletion–repletion balance studies have also been used for examining selenium metabolism in humans, but these are not a practical method for assessment.

Studies in human subjects of "low selenium status" suggest a potential usefulness of platelet glutathione peroxidase activity as a functional biochemical index (74). Theoretically, the adequacy of the intake and thus status could be measured by determining this enzyme's activity. However, to determine its validity, methods should be standardized with normal ranges established for specific population groups. In addition, the effect of age, gender, and other variables need to be studied prior to its adoption. However, even if this test, which has become automated in part (62), is demonstrated to reflect selenium status accurately, its widespread use will be hampered by the rather complicated and laborious preparation of sample and apparent loss if the preparation is thawed and refrozen (74).

Although theoretically blood glutathione peroxidase activity may be useful in assessment with less preparation than platelets, Butler et al. (21) reported that "a component of the erythrocyte other than glutathione peroxidase was found to contribute more to the peroxidase activity." This component coeluted with hemoglobin. Earlier, Burke et al. (20) noted that hemoglobin interfered with the assay of guinea pig blood glutathione peroxidase activity and suggested that "similar difficulties may be expected in other species."

There is evidence that 24-hr urine collections may be useful in assessing status, since it has been reported that urinary selenium reflects intake or, more specifically, bioavailable dietary selenium (19,140). However, such samples are difficult to obtain, especially for free-living individuals not institutionalized.

Thus, as with the other trace elements discussed, with the exception of iron, definitive assessment of status requires additional indices which can be determined with relative ease. Ideally, the samples should be capable of being collected under survey conditions. Therefore, until development of such methods, supplementation studies which include measurements of biochemical and/or physiological responses under controlled conditions are the most definitive methods presently available.

4. Zinc

Assessment of zinc status is increasingly requested both on an individual and survey basis and has been reviewed by Solomons (136). For example, assessment of the zinc nutritional status of the U.S. population was attempted using data collected in the second NHANES, in 1976–1980 (77,77a). In addition, because zinc nutriture is important in numerous biochemical functions and altered in several disease conditions, there is a need for establishing indices that can be used for assessment. Serum or plasma zinc is most frequently used as an indicator. Experimental zinc deficiency using animals results in decreased plasma zinc concentrations. Although documented zinc deficiency in humans has usually been associated with decreased serum levels, there have been exceptions. Thus, it must be recognized that plasma zinc may not consistently assess status. Other limitations to the use of serum or plasma are the problems of preventing contamination of the sample. In addition, numerous nonnutritional factors may affect plasma concentrations. For example, there may be a racial difference, since the 1500 blacks aged 3–74 years sampled in the second NHANES study generally tended to have lower mean serum values than did the 11,531 whites of similar age. Although statistically significant, the differences were small: 87.3 ± 14.7 (mean ± SD) µg/dl for the whites (both sexes), compared to 84.6 ± 14.1 for blacks. At present it is unclear whether the difference is attributable to race per se or to other variables including differences in diet. Likewise, several disease conditions may be accompanied by an altered (lower) plasma zinc concentration. These include myocardial infarction (54) and liver diseases (50,52).

In regard to age, serum levels for males are low in childhood, increase significantly during adolescence, peak in young adulthood, and decline after 45 years to values similar to those of the children (77,77a). For females, the change with age appears to be less pronounced than with males. However, it is now well established that physiological changes associated with pregnancy result in a lowering of the serum zinc concentration (61,77,77a). In addition, women who use oral contraceptives have a lower plasma zinc level than nonusers of similar age (49,77,77a,131).

A marked effect of food intake on serum zinc concentration has been demonstrated. Specifically, serum zinc concentrations were significantly higher for subjects who had fasted compared to those who were not required to fast (77,77a). In addition, there is an apparent diurnal variation in plasma zinc, with morning concentrations being higher than afternoon ones (77,77a).

Although serum zinc concentrations are now available from a "reference population," comparisons must take into account time of blood collection, fasting status, gender, and age of subjects. Likewise, for females it is imperative to determine whether they are pregnant or use oral contraceptives.

The use of plasma has been preferred instead of serum by some investigators, because serum may be contaminated with zinc due to hemolysis of the red blood cells. Separation of the serum requires additional contact time between erythrocytes and other cellular fractions during the period required for clot retraction. This prolonged contact between serum and cells may result in transfer of zinc to the serum, since the cells have a higher concentration. Reaming of the receding clot before centrifugation may also result in further hemolysis and/or contamination.

In 1968, Foley et al. (40) reported a higher zinc concentration in serum compared to plasma obtained from the same subjects. They concluded that the difference in zinc concentration between serum and plasma could be attributed, in part, to destruction of the platelets with zinc from the platelets contaminating the serum. Subsequently, Smith et al. (135) also noted higher zinc (and copper) concentrations in serum than in citrated plasma. Kasperek et al. (64) also reported a higher concentration in serum than in plasma, although the difference was less than that reported by Foley et al. (40). Normal values for plasma and serum zinc levels in adults have tended to decrease in the past four decades, although no consistent difference between plasma and serum zinc values is apparent from literature values (135). Makino (84) suggested that there is "a shift of water from erythrocytes to plasma" in citrated blood resulting in a dilution of the plasma and thus a lower zinc concentration. Makino (84) and others (122) have reported no difference between the plasma or serum zinc concentration when heparin was used as the anticoagulant. Thus, the type of anticoagulant can alter the plasma zinc concentration; therefore, published data can be compared only with caution. For a brief discussion of this area and additional references see Makino (84), Shaw et al. (122), and Smith et al. (135, 135a). Although plasma or serum zinc concentration must be reevaluated in regard to being used simply to assess status, it remains the most widely used index (105).

Other biochemical tests include determining the zinc concentration in erythrocytes, platelets, leukocytes, hair, urine, skin, saliva, and buccal cells. Determining the serum activities of several zinc-dependent enzymes or related proteins has also been employed for assessing status. Functional tests have included measuring taste and smell acuity, dark adaptation, neurophysiological function, immune response, in vitro uptake by various cell types in culture, retention and turnover, and zinc tolerance tests. Each test has limitations and in general lacks specificity. Therefore, at present no single test alone can be relied on for definitive assessment of zinc status.

The response to zinc supplementation under controlled conditions remains the most definitive method of determining if a true deficiency exists. However, this method is too cumbersome and time-consuming to be applied to the majority of circumstances involving noninstitutionalized individuals. Thus, the search continues for assessment indices. Determining the concentration of the zinc metalloprotein, metallothionein, in physiological fluids appears promising as an index that may be relatively specific (43). Ideally, serum metallothionein may help assess zinc status in a fashion similar to serum ferritin's contribution to reflecting iron nutriture. However, it may be necessary to develop tests for several indices, combining biochemical and functional parameters as has been the case for assessing iron status.

VII. SUGGESTED FUTURE RESEARCH

Well-designed experiments using animals fed a highly purified diet will continue to provide useful information necessary for the expansion of knowledge. Such an approach may be necessary to reveal the essentiality of additional ultratrace minerals. However, there is the danger that the use of chemically defined diets combined with ultraclean environments may obscure the detection of important nutrient interactions, an area that should receive increased attention. The relevance of interactions of trace elements with other nutrients altering dietary requirements has been established; for example, the vitamin C level in a meal affects the bioavailability and thus the recommended daily dietary intake of iron (92). Sir Fredrick Hopkins recognized the problem when he succinctly stated, "We thought we were feeding our animals proteins, fats, and carbohydrates, but all we were really giving them was carrots and oats" (126). Thus, the practical significance must be considered. This is especially critical when extrapolating results of such animal studies to human nutriture. Formulating practical animal diets using food items consumed by humans may be helpful.

The selection of the most relevant animal model should receive high priority. Although rodents have been most extensively used, they may not always be ideal. For example, rats appear to be poor experimental animals for the study of arsenic, since, unlike other mammals including humans, they concentrate the element in their blood (32). The study of certain conditions such as trace element nutriture during pregnancy and lactation requires relevant animal models for both ethical and scientific reasons. The data provided from studies using nonrodents may be more applicable to humans. Monkeys (45,117) and sheep (7,39a) have been used. Some experiments indicate that swine can serve as a sensitive animal model which may be relevant for identifying indices to assess copper status in humans (137a). The increased use of such models should be encouraged where feasible.

In addition to animal experiments which require rather long-term studies, simple and less time-consuming *in vitro* techniques, ideally capable of being automated, need to be expanded. Measuring the alteration in a metalloenzyme activity as a result of adding increments to saturation of a specific trace element is one example of a method that has been suggested (89) and appears promising. Such a procedure is relatively simple and could be automated for increased efficiency.

Additional *in vitro* functional tests such as the binding of trace elements by specific cell types warrant further development. The use of radioactive trace elements simplified this approach. The expanded use of determining affinities and binding capacities of carrier proteins in plasma may be helpful. However, in many cases, especially for the new trace elements, the carrier proteins have yet to be identified.

Undoubtedly, chemical analyses of selected biological samples will continue to provide data relevant to trace element nutriture. However, the challenge is to identify those tissues which accurately reflect status. In the case of humans, the sample should be available with minimal effort and trauma. As part of the aerospace age, it may be possible to develop noninvasive techniques to measure "whole-body" or specific organ–tissue trace element content without the use of radiation or other health risk methods.

Trace element nutriture may prove to play an important role in the biochemistry of the brain. Theoretically, impaired status of specific trace elements could result in altered metabolism via enzymes involving neurotransmitters and/or brain opiates. Thus, measurement of brain (mental) function and behavior response are potential areas for future exploration. Previous animal studies have suggested that inadequate or toxic levels of specific trace elements may result in impaired brain biochemistry and/or function. For example, offspring born from zinc-deficient rhesus monkeys have been reported to have abnormalities of behavior (117) and memory (48). A few studies have involved humans. As early as 1919, suggestive evidence was obtained regarding a relationship between iron status and mental function (156). Recent studies have associated brain function, as measured by electroencephalogram, with iron status (144,145).

Application of electrophysiology using computerized analyses of EEG and evoked responses promises to allow objective and quantitative measurement of brain function. However, "normal" brain function may be difficult to establish. Nevertheless, reliance on the subjective and qualitative judgments of human observers is reduced. Another advantage is that the technique is noninvasive. The challenge will be causally to relate abnormal status, deficiency, or toxicity of specific trace elements to mental function or behavior. Ideally, the functional tests could be combined with biochemical parameters to provide the needed specificity.

Finally, trace element research methodology is entering a new era. Heretofore, the major emphasis has focused on techniques to provide descriptive data. However, in order to move forward, more effort should be placed on elucidating metabolic roles and mechanisms of action. To provide definitive data, sophisticated state-of-the-art techniques should be coupled with novel approaches. Such experiments will be necessary to reveal more completely the cellular and sub-cellular activities of trace elements and to provide additional clues as to why they are essential nutrients.

REFERENCES

1. Amine, E. K., and Hegsted, D. M. (1971). *J. Nutr.* **101**, 927.
2. Anderson, H. (1987). Unpublished data—personal communication, University of Missouri, Columbia.
3. Anderson, R. A. (1981). *Sci. Total Environ.* **17**, 13.
4. Anderson, R. A., Polansky, M. M., Bryden, N. A., Patterson, K. Y., Veillon, C., and Glinsmann, W. H. (1983). *J. Nutr.* **113**, 276.
5. Anderson, R. A., Bryden, N. A., and Polansky, M. M. (1985). *Am. J. Clin. Nutr.* **41**, 571.
6. Apgar, J. (1977). *J. Nutr.* **107**, 539.
7. Apgar, J., and Fitzgerald, J. A. (1985). *J. Anim. Sci.* **60**, 1530.
8. Aston, B. C. (1924). *N. Z. J. Agric.* **28**, 215, 301; **29**, 14, 84.
9. Baumann, E. (1895–1896). *Z. Physiol. Chem.* **21**, 319; cited by Bing, F. C. (1939). *In* "Mineral Metabolism" (A. T. Shohl, ed.), p. 225. American Chemical Society Monograph Series, Reinhold, New York.
10. Beamish, F. E., and Westland, A. D. (1958). *Anal. Chem.* **30**, 805.
11. Bertrand, G., and Benson, R., (1922). *C.R. Acad. Sci.* **175**, 289.
12. Bertrand, G., and Wakamura, H. (1928). *C.R. Acad. Sci.* **186**, 480.
13. Bing, F. C. (1939). *In* "Mineral Metabolism" (A. T. Shohl, ed.), p. 225. American Chemical Society Monograph Series, Reinhold, New York.
14. Birckner, V. (1919). *J. Biol. Chem.* **38**, 191.
15. Borel, J. S., and Anderson, R. A. (1984). *In* "Biochemistry of the Essential Ultratrace Elements" (E. Frieden, ed.), p. 175. Plenum, New York.
16. Bradfield, R. B., Cordano, A., Baertl, J., and Graham, G. G. (1980). *Lancet* **2**, 343.
17. Buerk, C. A., Chandy, G., Pearson, E., MacAuly, A., and Soroff, H. S. (1973). *Surg. Forum* **24**, 101.
18. Bunge, G. (1902). *In* "Textbook of Physiological and Pathological Chemistry". 2nd English Ed. Blakiston, Philadelphia; cited *in* "Mineral Metabolism" (A. T. Shohl, ed.), p. 6. American Chemical Society Monograph Series, Reinhold, New York, 1939.
19. Burke, R. F. (1976). *In* "Trace Elements in Human Health and Disease. Vol. 2. Essential and Toxic Elements" (A. S. Prasad, ed.), p. 105. Academic Press, New York.
20. Burke, R. F., Lane, J. M., Lawrence, R. A., and Gregory, P. E. (1981). *J. Nutr.* **111**, 690.
21. Butler, J. A., Whanger, P. D., and Tripp, M. J. (1982). *Am. J. Clin. Nutr.* **36**, 15.
22. Chatin, A. (1850–1854). *C.R. Acad. Sci.* 30–39; cited in "Trace Element in Human Nutrition" (E. J. Underwood, ed.), p. 3, 4th Ed. Academic Press, New York, 1977.
23. Chen, X., Yang, G., Chen, J., Chen, X., Wen, Z., and Ge, K. (1980). *Biol. Trace Elem. Res.* **2**, 91.
24. Cotzias, G. C. (1967). *In* "Trace Substances in Environmental Health." *Proc. Univ. Missouri Annu. Conf. 1st* p. 5.
25. Cotzias, G. C., and Foradori, A. C. (1969). *In* "The Biological Basis of Medicine" (E. Bittar and N. Bittar, eds.). Acdemic Press, New York.
26. Dallman, P. R., Yip, R., and Johnson, C. (1984). *Am. J. Clin. Nutr.* **39**, 437.

27. Davies, I. J. T. (1972). "The Clinical Significance of the Essential Biological Metals," p. 1X. Thomas, Springfield, IL.
28. Davis, P. N., Norris, L. C., and Kratzer, F. H. (1962). *J. Nutr.* **78,** 445.
29. Delves, H. T. (1976). *Clin. Chim. Acta* **71,** 495.
30. Delves, H. T., Bunker, V., and Husbands, A. P. (1983). *In* "Trace Element Analytical Chemistry in Medicine and Biology" (P. Bratter and P. Schramel, eds.), Vol. 2, p. 1123. De Gruyter, New York.
31. DeRenzo, E. C., Kaleita, E., Heytler, P. G., Oleson, J. J., Hutchings, B. L., and Williams, J. H. (1953). *Arch. Biochem. Biophys.* **45,** 247.
32. Dutkiewicz, T. (1977). *Environ. Health Perspect.* **19,** 173.
33. Failla, M. L., and Cousins, R. J. (1978). *Biochim. Biophys Acta* **538,** 435.
34. Failla, M. L., and Cousins, R. J. (1978). *Biochim. Biophys Acta* **543,** 293.
35. Fields, M., Ferretti, R. J., Smith, J. C., Jr., and Reiser, S. (1983). *J. Nutr.* **113,** 1335.
36. Fields, M., Ferretti, R. J., Reiser, S., and Smith, J. C., Jr. (1984). *Proc. Exp. Biol. Med.* **175,** 530.
37. Fields, M., Ferretti, R. J., Smith, J. C., and Reiser, S. (1984). *Biol. Trace Elem. Res.* **6,** 379.
38. Fields, M., Ferretti, R. J., Smith, J. C., Jr., and Reiser, S. (1984). *Am. J. Clin. Nutr.* **39,** 289.
39. Fields, M., Ferretti, R. J., Judge, J. M., Smith, J. C., and Reiser, S. (1985). *Proc. Soc. Exp. Biol. Med.* **178,** 362.
39a. Fitzgerald, J. A., Everett, G., and Apgar, J. (1986). *Can. J. Anim. Sci.* **66,** 643.
40. Foley, B., Johnson, S. A., Hackley, B., Smith, J. C., Jr., and Halsted, J. A. (1968). *Proc. Soc. Exp. Biol. Med.* **128,** 265
41. Forster, J. (1873). *Z. Biol.* **9,** 297; cited *in* "Mineral Metabolism" (A. T. Shohl, ed.), p. 5. American Chemical Society Monograph Series, Reinhold, New York, 1939.
42. Fox, M. R. S., and Briggs, G. M. (1960). *J. Nutr.* **72,** 243.
43. Garvey, J. S. (1984). *Environ. Health Perspect.* **54,** 117.
44. Glinsmann, W. H., Feldman, J. F., and Mertz, W. (1966). *Science* **152,** 1243.
45. Golub, M. S., Gershwin, M. E., Hurley, L. S., Baly, D. L., and Hendrickx, A. G. (1984). *Am. J. Clin. Nutr.* **39,** 879.
46. Greenfield, H., and Briggs, G. M. (1971). *Annu. Rev. Biochem.* **40,** 549.
47. Gürson, C. T., and Saner, G. (1978). *Am. J. Clin. Nutr.* **31,** 1162.
48. Halas, E. S., and Kawamoto, J. C. (1984). *In* "The Neurobiology of Zinc, Part B: Deficiency, Toxicity and Pathology" (C. J. Frederickson, G. A. Howell, and E. J. Kasarskis, eds.), p. 91. Liss, New York
49. Halsted, J. A., Hackley, B. M., and Smith, J. C., Jr. (1968). *Lancet* **2,** 278.
50. Halsted, J. A., and Smith, J. C., Jr. (1970). *Lancet* **1,** 322.
51. Halsted, J. A., Ronaghy, H. A., Abadi, P., Haghshenass, M., Amirhakemi, G. H., Barakat, R. A., and Reinhold, J. G. (1972). *Am. J. Med.* **53,** 277.
52. Halsted, J. A., Smith, J. C., Jr., and Irwin, M. I. (1974). *J. Nutr.* **104,** 345.
53. Hambidge, K. M. (1973). *Am. J. Clin. Nutr.* **26,** 1212.
54. Handjani, A. M., Smith, J. C., Jr., Hermann, J. B., and Halsted, J. A. (1974). *Chest* **65,** 185.
55. Hart, E. B., Steenbock, H., Waddell, J., and Elvehjem, C. A. (1928). *J. Biol. Chem.* **77,** 797.
56. Hegsted, D. M. (1971). *In* "Newer Trace Elements in Nutrition" (W. Mertz and W. E. Cornatzer, eds.), p. 19. Dekker, New York.
57. Hegsted, D. M. (1975). *J. Nutr.* **106,** 307.
58. Henkin, R. I. (1971). *In* "Newer Trace Elements in Nutrition" (W. Mertz and W. E. Cornatzer, eds.), p. 255. Dekker, New York.
59. Hubbel, R. B., Mendel, L. B., and Wakeman, A. J. (1937). *J. Nutr.* **14,** 273.

60. Hughes, M. K., Fusillo, M. H., and Smith, J. C., Jr. (1973). *Environ. Biol. Med.* **2**, 23.
61. Hunt, I. F., Murphy, N. J., Cleaver, A. E., Faraji, B., Coulson, A. H., Clark, V. A., Laine, N., Davis, C. A., and Smith, J. C., Jr. (1983). *Am. J. Clin. Nutr.* **37**, 572.
62. Jaskot, R. H., Charlet, E. G., Grose, E. C., Grady, M. A., and Roycroft, J. H. (1983). *J. Anal. Toxicol.* **7**, 86.
63. Jeejeebhoy, K. N., Chu, R. C., Marliss, E. B., Greenberg, G. R., and Bruce-Robertson, A. (1977). *Am. J. Clin. Nutr.* **30**, 531.
64. Kasperek, K., Kiem, J., Iyengar, G. V., and Feinendegen, L. E. (1981). *Sci. Total Environ.* **17**, 133.
65. Kendall, E. C. (1915). *J. Am. Med. Assoc.* **64**, 2042.
66. Keshan Disease Research Group (1979). *Chin. Med. J.* **92**, 477.
67. Kim, W. W., Mertz, W., Judd, J. T., Marshall, M. W., Kelsay, J. L., and Prather, E. S. (1984). *Am. J. Clin. Nutr.* **40**, 1333.
68. Klevay, L. M. (1970). *Am. J. Clin. Nutr.* **23**, 1194.
69. Klevay, L. M. (1980). *In* "Micronutrient Interactions: Vitamins, Minerals, and Hazardous Elements" (O. A. Levander and L. Cheng, eds.). *Ann. N.Y. Acad. Sci.* **355**, 140.
70. Klevay, L. M. (1983). *Biol. Trace Elem. Res.* **5**, 245.
71. Klevay, L. M. (1984). *In* "Metabolism of Trace Metals in Man" (O. M. Rennert and W. Y. Chang, eds.), Vol. 1, p. 129, CRC Press, Boca Raton, FL.
72. Lahey, M. E., Gubler, C. J., Cartwright, G. E., and Wintrobe, M. M. (1953). *J. Clin. Invest.* **34**, 322.
73. Landes, D. R. (1975). *Proc. Soc. Exp. Biol. Med.* **150**, 686.
74. Levander, O. A., Alfthan, G., Arvilommi, H., Gref, C. G., Huttunen, J. K., Kataja, M., Koivistoinen, R., and Pikkarainen, J. (1983). *Am. J. Clin. Nutr.* **37**, 887.
75. Levander, O. A., and Morris, V. C. (1984). *Am. J. Clin. Nutr.* **39**, 809.
76. Life Sciences Research Office (1984). Assessment of the Iron Nutritional Status of the U.S. Population Based on Data Collected in the Second National Health and Nutrition Examination Survey, 1976–1980. Federation of American Societies of Experimental Biology, Rockville, MD.
77. Life Sciences Research Office (1984). Assessment of the Zinc Nutritional Status of the U.S. Population Based on Data Collected in the Second National Health and Nutrition Examination Survey, 1976–1980. Federation of American Societies for Experimental Biology, Rockville, MD.
77a. Pilch, S. M., and Senti, F. R. (1985). *J. Nutr.* **115**, 1393.
78. Linder, M. (1983). *J. Nutr. Growth Cancer* **1**, 27.
79. Liu, V. J. K., and Morris, J. S. (1978). *Am. J. Clin. Nutr.* **31**, 972.
80. Luecke, R. W., Olman, M. E., and Baltzer, B. V. (1968). *J. Nutr.* **94**, 344.
81. Lunin, N. (1881). *Z. Physiol. Chem.;* cited *in* "Mineral Metabolism" (A. T. Shohl, ed.), p. 5. American Chemical Society Monograph Series, Reinhold, New York, 1939.
82. Lutz, R. E. (1926). *J. Indu. Hyg.* **8**, 177.
83. MacCallum, A. B. (1891–1892). *Proc. R. Soc. London* **50**, 277; *J. Physiol. (London)* **16**, 268 (1894); *Q. J. Microsc. Sci.* **38**, 175 (1896); cited *in* "Mineral Metabolism" (A. T. Shohl, ed.), p. 204. American Chemical Society Monograph Series, Reinhold, New York, 1939.
84. Makino, T. (1983). *Clin. Chem.* **29**, 1313 (Letter).
85. Mason, K. E. (1979). *J. Nutr.* **109**, 1979.
86. McHargue, J. S. (1926). *Am. J. Physiol.* **77**, 245.
86a. Mertz, W., and Schwarz, K. (1959). *Am. J. Physiol.* **196**, 614.
87. Mertz, W. (1969). *Physiol. Rev.* **49**, 163.
88. Mertz, W. (1970). *Fed. Proc.; Fed. Am. Soc. Exp. Biol.* **29**, 1482.
89. Mertz, W. (1975). *Clin. Chem.* **21**, 468.

90. Mertz, W., and Kelsay, J. L. (1984). *Am. J. Clin. Nutr.* **40**, (Suppl.), 1323.
91. Mohr, H. E., and Hopkins, L. L., Jr. (1972). *Lab. Anim. Sci.* **22**, 96.
92. Monsen, E. R., Hallberg, L., Layrisse, M., Hegsted, D. M., Cook, J. D., Mertz, W., and Finch, C. A. (1978). *Am. J. Clin. Nutr.* **31**, 134.
93. Moody, J. R. (1982). *Anal. Chem.* **54**, 1358A.
94. Moody, J. R. (1983). *Trends Anal. Chem.* **2**, 116.
95. Newell, J. M., and McCollum, E. V. (1933). *J. Nutr.* **6**, 289.
96. Nielsen, F. H. (1984). *Annu. Rev. Nutr.* **4**, 21.
97. Nielsen, F. H., and Mertz, W. (1984). *In* "Present Knowledge in Nutrition," 5th Ed., p. 607. Nutrition Foundation, Washington, D.C.
98. O'Dell, B. L., and Savage, J. E. (1960). *Proc. Soc. Exp. Biol. Med.* **103**, 304.
99. Osborne, T. B., and Mendel, L. B. (1913). *J. Biol. Chem.* **15**, 311.
100. Osborne, T. B., and Mendel, L. B. (1918). *J. Biol. Chem.* **34**, 131.
101. Patterson, K. Y., Holbrook, J., Bodner, J., Kelsay, J. L., Smith, J. C., Jr., and Veillon, C. (1984). *Am. J. Clin. Nutr.* **40**, 1397.
102. Polansky, M. M., and Anderson, R. A. (1979). *Lab. Anim. Sci.* **29**, 357.
103. Phipps, D. A. (1976). *In* "Metals and Metabolism," p. 17. Claredon, Oxford.
104. Prasad, A. S., Halsted, J. A., and Nadimi, M. (1961). *Am. J. Med.* **31**, 532.
105. Prasad, A. S. (1979). "Zinc in Human Nutrition," p. 16. CRC Press, Boca Raton, FL.
106. Punsar, S., Wolf, W., Mertz, W., and Karvonen, M. J. (1977). *Ann. Clin. Res.* **9**, 79.
107. Recommended Dietary Allowances (1980). 9th Ed., U.S. National Academy of Sciences, Washington, DC.
108. Reiser, S., Smith, J. C., Jr., Mertz, W., Holbrook, J. T., Scholfield, D. J., Powell, A. S., Canfield, W. K., and Canary, J. J. (1985). *Am. J. Clin. Nutr.* **42**, 242.
109. Richards, M. P. (1982). *Poultry Sci.* **61**, 2089.
110. Richards, M. P. (1984). *J. Pediatr. Gastrol. Nutr.* **3**, 128.
111. Richert, D. A., and Westerfeld, W. W. (1953). *J. Biol. Chem.* **203**, 915.
112. Robscheit-Robbins, F. S. (1929). *Physiol. Rev.* **9**, 666.
113. Roginski, E., and Mertz, W. (1974). *J. Nutr.* **104**, 599.
114. Rosenthal, R. W., and Blackburn, A. (1974). *Clin. Chem.* **20**, 1233.
115. Rotruck, J. T., Pope, A. L., Gander, H. E., Swanson, A. B., Hafeman, D. G., and Hoekstra, W. G. (1973). *Science* **179**, 588.
116. Ruberman, W., Weinblatt, E., Goldberg, J. D., Frank, W. N., and Shapiro, S. (1977). *N. Engl. J. Med.* **297**, 750.
117. Sandstead, H. H., Strobel, D. A., Logan, G. M., Jr., Marks, E. O., and Jacob, R. A. (1978). *Am. J. Clin. Nutr.* **31**, 844.
118. Schroeder, H. A., Vinton, W. H., Jr., and Balassa, J. J. (1963). *J. Nutr.* **80**, 39.
119. Schwarz, K., and Foltz, C. M. (1957). *J. Am. Chem. Soc.* **79**, 3292; *J. Biol. Chem.* **233**, 245 (1958).
120. Schwarz, K. (1961). *Fed. Proc., Fed. Am. Soc. Exp. Biol.* **20**, 666.
121. Schwarz, K. (1970). *J. Nutr.* **100**, 1487.
122. Shaw, J. C. L., Bury, A. J., Barber, A., Mann, L., and Taylor, A. (1982). *Clin. Chim. Acta* **118**, 229.
123. Shohl, A. T. (1939). *In* "Mineral Metabolism" (A. T. Shohl, ed.), p. 17. American Chemical Society Monograph Series, Reinhold, New York.
124. Shohl, A. T. (1939). *In* "Mineral Metabolism" (A. T. Shohl, ed.), p. 234. American Chemical Society Monograph Series, Reinhold, New York.
125. Shohl, A. T. (1939). *In* "Mineral Metabolism" (A. T. Shohl, ed.), p. 235. American Chemical Society Monograph Series Reinhold, New York.
126. Shohl, A. T. (1939). *In* "Mineral Metabolism" (A. T. Shohl, ed.), p. 347. American Chemical Society Monograph Series Reinhold, New York.
127. Smith, J. C., and Schwarz, K. (1967). *J. Nutr.* **93**, 182.

128.   Smith, J. C., Jr. (1969). *In* "Trace Substances in Environmental Health II" (D. D. Hemphill, ed.), p. 233. Univ. of Missouri Press, Columbia.

129.   Smith, J. C., Jr., and McDaniel, E. G. (1972). *Invest. Urol.* **9**, 518.

130.   Smith, J. C., Jr., McDaniel, E. G., and Doft, F. S. (1973). *In* "Germfree Research" (J. Heneghan, ed.), p. 285. Academic Press, New York.

131.   Smith, J. C., and Brown, E. D. (1976). *In* "Trace Elements in Human Health and Disease" (A. S. Prasad and D. Oberleas, eds.), p. 315. Academic Press, New York.

132.   Smith, J. C., Jr., Butrimovitz, G. P., and Purdy, W. C. (1979). *Clin. Chem.* **25**, 1487.

133.   Smith, J. C., Jr., Anderson, R. A., Ferretti, R., Levander, O. A., Morris, E. R., Roginski, E. E., Veillon, C., Wolf, W. R., Anderson, J. J. B., and Mertz, W. (1981). *Fed. Proc., Fed. Am. Soc. Exp. Biol.* **40**, 2120.

134.   Smith, J. C., Jr., Morris, E. R., and Ellis, R. (1983). *"In* "Zinc Deficiency in Human Subjects" (A. S. Prasad, A. D., Cavadar, G. J., Brewer, and P. J. Aggett, eds.), p. 147. Liss, New York.

135.   Smith, J. C., Jr., Holbrook, J., and Danford, D. (1985). *J. Am. College Nutr.* **4**, 627.

135a.   Smith, J. C., Jr., Lewis, S., Holbrook, J., Seidel, K., and Rose, A. (1987). *Clin. Chem.* **33**, 814.

136.   Solomons, N. W. (1979). *Am. J. Clin. Nutr.* **32**, 856.

136a.   Solomons, N. W. (1985). *J. Am. College Nutr.* **4**, 83.

137.   Spencer, H. (1973). *In* "Biological Mineralization" (I. Zipkin, ed.), p. 689. Wiley, New York.

137a.   Steele, N., Richards, M., Darcey, S., Fields, M., Smith, J., and Reiser, S. (1986). *Fed. Proc. Fed. Am. Soc. Exp. Biol.* **45**, (Abstr.) 1179.

138.   Stewart, R. D. H., Griffiths, N. M., Thomson, C. D., and Robinson, M. F. (1978). *Br. J. Nutr.* **40**, 45.

139.   Thiers, R. E. (1957). *Methods Biochem. Anal.* **5**, 273.

140.   Thomson, C. D. (1970). *Proc. Univ. Otago Med.* **50**, 31.

141.   Tipton, I. H., and Cook, M. J. (1963). *Health Phys.* **9**, 103.

142.   Todd, W. R., Elvehjem, C. A., and Hart, E. B. (1934). *Am. J. Physiol.* **107**, 146.

143.   Tucker, H. F., and Salmon, W. D. (1955). *Proc. Soc. Exp. Biol. Med.* **88**, 613.

144.   Tucker, D. M., and Sandstead, H. H. (1981). *Physiol. Behav.* **25**, 439.

145.   Tucker, D. M., and Sandstead, H. H. (1984). *In* "The Neurobiology of Zinc, Part B: Deficiency, Toxicity and Pathology" (C. J. Frederickson, G. A., Howell, and E. J. Kasarski, eds.), p. 139. Liss, New York.

146.   Underwood, E. J., and Filmer, J. F. (1935). *Aust. Vet. J.* **11**, 84.

147.   Underwood, E. J., ed. (1962). *In* "Trace Elements in Human and Animal Nutrition," 2nd Ed. Chap. 8, p. 218. Academic Press, New York.

148.   Underwood, E. J. (1971). *In* "Newer Trace Elements in Nutrition" (W. Mertz and W. E. Cornatzer, eds.), p. 9. Dekker, New York.

149.   Van Rij, A. M., Thomson, C. D., McKenzie, J. M., and Robinson, M. F. (1979). *Am. J. Clin. Nutr.* **32**, 2076.

150.   Veillon, C., and Vallee, B. L. (1978). *In* "Methods in Enzymology" (S. Fleisher and L. Packer, eds.), Part E, Vol. 54, p. 446. Academic Press, New York.

151.   Veillon, C., and Alvarez, R. (1983). *In* "Metal Ions in Biological Systems" (H. Sigel, ed.), p. 103. Dekker, New York.

152.   Veillon, C. (1984). *In* "Stable Isotopes in Nutrition" (J. R. Turnlund and P. E. Johnson, eds.), p. 91. ACS Symposium Series No. 258, Washington D.C.

153.   Veillon, C., Lewis, S. A., Patterson, K. Y., Wolf, W. R., Harnly, J. M., Versieck, J., Vanballenberghe, L., Cornelis, R., and O'Haver, T. C. (1985). *Anal. Chem.* **57**, 2106.

154.   Volkman (1874); cited by Voit, C. (1939). *In* "Handbuch der Physiologie" (L. Hermann, ed.); Vol. 6, Part 1 (A. T. Shohl, ed.), p. 17. American Chemical Society Monograph Series, Reinhold, New York.

155. Waddell, J., Elvehjem, C. A., Steenbock, H., and Hart, E. B. (1928). *J. Biol. Chem.* **77**, 777.
156. Waite, J. H., and Nielson, I. L. (1919). *Med. J. Aust.* **1**, 1.
157. Walsh, A. (1955). *Spectrochim. Acta* **7**, 108.
158. Wintrobe, M. M., Cartwright, G. E., and Gubler, C. J. (1953). *J. Nutr.* **50**, 395.
159. Wintrobe, M. M. (1958). *Am. J. Clin. Nutr.* **6**, 75.
160. Wolf, W. R. (1987). *In* "Trace Elements in Human and Animal Nutrition" (W. Mertz, ed.), 5th Ed., Vol. 1, Chap. 3.
161. Wright, N. C., and Papish, J. (1929). *Science* **69**, 78.

3

Quality Assurance for Trace Element Analysis

WAYNE R. WOLF

U.S. Department of Agriculture
Agricultural Research Service
Beltsville Human Nutrition Research Center
Beltsville, Maryland

I. INTRODUCTION

"Most of the trace elements can be accurately measured some of the time. Some of the trace elements can be accurately measured most of the time. Most of the trace elements are not accurately measured most of the time."

Advances in analytical technology have led to development of a variety of very sensitive techniques with measurement capability for the very low levels of trace elements in biological materials, ranging from low parts per million to less than a part per billion. The complexity of required instrumentation and methodology increases and the reliability of the data decreases as these levels become lower. However, the demands on reliability of the analytical data for valid biochemical interpretation become higher as our knowledge of the metabolism of each element becomes more detailed. A knowledge of the practical limitations of the analytical methodology, and application of proper quality assurance practices to obtain the most accurate, reliable data is required to correctly advance and apply our understanding of the biochemical effects of trace elements in nutrition.

Quality assurance is the system of activities whose purpose it is to provide assurance that the overall quality control limits are being met (34). The quality control limits define the required and expected accuracy and precision of the desired analytical results. Without a knowledge of the expected quality control limits and a defined quality assurance program, all analytical results must be suspect. Proper quality assurance practices stem from an understanding of the whole measurement process, including choice and collection of the analytical sample, validation and quality control of the analytical methodology, estimation of uncertainty and accuracy of the analytical results, and appropriate data handling and evaluation. Most aspects of quality assurance are method and analyte independent and can be discussed in generic terms in this chapter.

Widely divergent values have appeared in the literature for trace elements in biological materials (17, 41, 43), and estimation of "normal" values in tissue and body fluids is a difficult and frustrating task. Attempts to explain these variations solely in terms of biological significance have usually failed. A large body of evidence has accumulated that most of these reported variations have been due to nonbiological causes, such as presampling changes, inadequate specimen collection and handling, analytical errors, or deficiencies in handling and evaluation of the analytical data (41). Documentation exists in the analytical literature describing many of these factors which affect reliability of analytical data for trace elements in biological systems. A general awareness of these factors, in addition to awareness of the practical limitations of the analytical methodology, is required to make valid biochemical interpretations of trace element analytical data. Recognition of these factors and generation of many new data have led to a revision of the values cited in previous editions of this book. This updating and revision is a constant process as new techniques and understanding of these factors improve.

A biological system is one in constant change. Different chemical species or forms of the elements are constantly changing and interacting. The levels and interconversions between these species are dependent on a multitude of chemical and physical parameters subject to change in the system. Determination of the utilization or mode of action of these species within the biological system will depend on identification of specific functions and development of analytical methodology and functional tests for these species (24). Reaching this goal will require a concerted, collaborative effort employing techniques and efforts from a variety of disciplines including nutrition, biochemistry, trace element analysis, physiology, analytical separations, coordination chemistry, enzymology, and many others. Reliable measurement of these different chemical species adds another layer of factors on top of the many factors required for reliable analysis of total trace element content. At present, only preliminary studies and documentation of these factors are available (47). The discussion in this chapter will deal with quality assurance for total trace element analysis. Discussion of approaches to analytical determinations of chemical species will have to await future editions of this book.

II. ACCURACY-BASED MEASUREMENT

In discussion of accuracy-based measurement, it is necessary to define and distinguish between the terms accuracy and precision. Precision is an expression of the repeatability of a measurement; it generally reflects the random-error component. Precision is a statistically defined term usually expressed as a standard deviation of the mean or in relative terms to the mean as a coefficient of variation (CV, standard deviation divided by the mean). Accuracy is an expression of the difference in the best estimate of the measured value from the "true" value. Inaccuracy reflects the extent of systematic error in the measurement. If no systematic error exists, the measurement is accurate. A measurement method can have a high degree of precision yet give inaccurate values due to systematic errors. This is seen in Fig. 1, which shows four cases illustrating the concepts of accuracy and precision. Cases A and B show good precision or repeatability; random errors are small. Case A is an accurate analysis; the values are on the "bull's-eye" or "true" value. Case B shows the presence of a systematic error which makes the value inaccurate. The low-precision cases, C and D, show the difficulty of determining the accuracy when the precision is

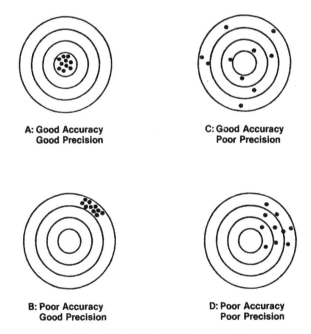

A: Good Accuracy
Good Precision

C: Good Accuracy
Poor Precision

B: Poor Accuracy
Good Precision

D: Poor Accuracy
Poor Precision

Fig. 1. Accuracy versus precision. Adapted from S.S. Brown, *In* "Clinical Chemistry and Toxicity of Metals," p. 381. Elsevier, 1977.

poor. The average value in case C may be close to the "true" value, but the presence of large random errors lowers the certainty with which a statement can be made regarding the accuracy. Likewise the existence of large random errors in case D makes it difficult to assess the degree of systematic error in the method. For case C it is possible to obtain the "true" value by averaging a large number of determinations, but this is usually not feasible in practice. It is necessary to improve the analytical method to an acceptable precision level during development before study and elimination of the systematic errors can be accomplished.

The cases shown in C and D also illustrate the inability to produce quantitative analytical results at the detection limit of the method. Random errors are inherent in any measurement technique which depends on a continuous or analog scale such as the readout of a spectrometer. These have to do with the variations or noise level of electronic or other components of the measurement system among other factors. These errors can be minimized by careful control of the experimental variables, but they always exist. These errors limit method precision and, as an analytical value approaches the inherent variability of the measurement system, quantitation becomes less certain. Below a certain concentration limit defined as the *quantitation limit,* where quantitation within a specified confidence level is not possible, it is possible to make only qualitative claims that the analyte is present or absent at concentrations above the *detection limit.* These parameters have been rigorously defined and are illustrated in Fig. 2 (18).

Measurements are made to communicate information on properties of materials in a purposeful way to accomplish useful goals. Measurement is the process whereby a numerical value is associated with a distinct specific and unique property of a material (2). This process consists of at least two components (37). First, some type of scale is needed to estimate quantitatively the value of the property of the material of interest; second, a method for applying the scale is required. Applying a method plus a scale allows a definitive value to be assigned to the property under consideration. These measurements provide the quan-

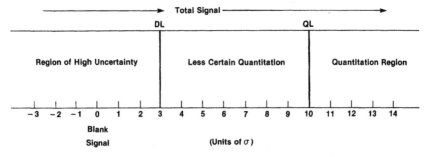

Fig. 2. Relationship of detection limit (DL) and quantitation limit (QL) to signal strength. Adapted from Keith *et al.* (18).

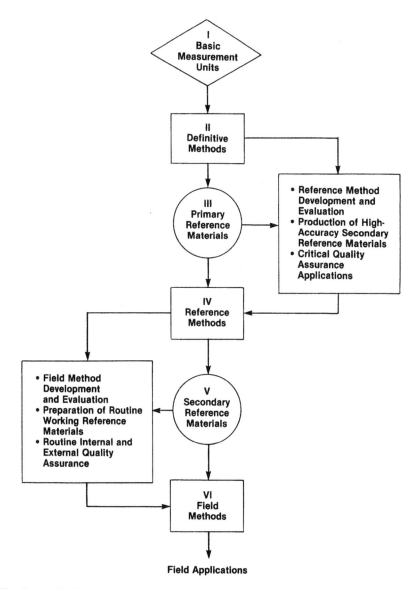

Fig. 3. An idealized accuracy-based measurement system, which illustrates the role of primary reference materials in transferring the accuracy inherent in definitive methods into widespread laboratory use. Reprinted by permission (1).

titative data that allow communication of information among the users of the data. A critical step in the communication of information via data from a measurement system is that the parties involved agree on the results of the measurement and the meaning of the numbers associated with it. When the imprecision and inaccuracies inherent in the measurement process are taken into consideration and the measurements are in agreement, the measurements are said to be compatible. By definition a measurement is accurate when the numbers are both precise and free of systematic error. Under these conditions compatibility is highly probable. Thus, measurement accuracy leads to measurement compatibility (37).

The concepts of accurate measurement and "true" values have been discussed (37) in describing an idealized systems approach to development of an accuracy-based measurement system (1). These discussions define a measurement hierarchy needed to transfer accuracy from the basic measurement units established by the fundamental metrologists to methods used in field or routine applications. This accuracy-based measurement system is shown in Fig. 3. Definitive or absolute methods (methods of highest accuracy) are more closely linked to the basic measurement units and thus subject to less opportunity for error. Although these methods have been sufficiently well tested and evaluated so that they have essentially zero systematic errors and very high precision, they are generally too elaborate and expensive for routine work. Reference methods have known, established accuracy and precision, although inaccuracies are of a greater magnitude than with absolute or definitive methods. These methods are generally faster and less expensive than the absolute methods. The field or routine methods are generally fast and economical, and require analytically nonsophisticated personnel to carry out the method. They are generally used in applications involving large numbers of separate measurements performed rapidly. These methods may be very precise but lacking in accuracy. Accuracy can be transferred throughout this hierarchy of measurements by use of appropriate well-characterized reference materials (37).

III. SAMPLING

For the study of the biochemical and metabolic effects of trace elements in animals and humans, various specific tissue or fluid samples must be removed and analyzed outside the living system. A few samples such as blood, hair, excretory products, and, infrequently, biopsy samples can be obtained from living subjects. Most other specimens are obtained either upon sacrifice of the animal or at autopsy for humans. Upon separation of a biological sample from its "*in situ*" environment, it is subject to possible biological changes due to a variety of extrinsic and intrinsic factors which may influence the trace element concentration. These include a variety of long- and short-term physiological influences and other presampling factors which need to be taken into consideration when selecting and collecting representative samples (16). Often a significant amount of time occurs between sample collection and analysis. Many biological materials are not stable once they have been sampled. Dehydration or bacterial growth may occur. Decomposition of the matrix may lead to formation of volatile compounds of the trace elements. A critical literature survey on sampling and storage for trace metals in environmental materials has been published (23). Specific methods from sampling, subsampling, and homogenization have been reviewed (21).

Due to the very low levels of most trace elements in biological samples, there is high potential for gross contamination during sampling, storage, handling, and analysis. As the level of trace element in the sample decreases to nanogram or picogram levels, at some point the contamination levels will exceed the true levels of the element being measured. The results of these contamination problems are wasted analytical effort and improper conclusions drawn from contaminated samples. Improper sampling has in the past led to general acceptance of ''natural'' levels that are grossly in error (26). For example, Everson and Patterson (4) have shown that all previous literature values for lead in plasma are incorrect due to contamination in sampling. A review by Versieck et al. (42) describes the potential for contamination in clinical sampling procedures for many elements.

While elaborate precautions to obtain and use nonmetal sampling devices, tools, and containers may arouse objection due to the cost and time involved, the potential for totally misleading analytical data exists. Evaluation of trace element contamination during sampling is one of the most difficult tasks in the analytical community. If analytical results do not change when low-contamination techniques are used, then it may be concluded that the sampling scheme is correct (26). But until the low-contamination techniques are tried and evaluated, there is a large degree of uncertainty due to potential of sample contamination. Potential sources of contamination include the environment, containers, collection apparatus, and, probably most important, the chemist (28)! Some aspects of container materials have been studied and recommendations for cleaning these have been made (27). Plastics are highly recommended, with Teflon and polyethylene being the most favorable. There are differences in properties of various plastics. They should not be considered on a generic basis, but rather particular sources of supply should be periodically reviewed to determine continued suitability of the material (26).

One approach being adapted for storage of whole discrete samples is cryostorage. Under very low temperatures, sample deterioration and sample–container interactions become small. This approach is being studied as part of specimen bank programs in the United States (45) and in the Federal Republic of Germany (33).

The area of sample collection and manipulation can best be summarized by the warning given by Thiers (35) that: "Unless the complete history of any sample is known with certainty, the analyst is well advised not to spend his time analyzing it."

IV. GENERATING VALID ANALYTICAL DATA

Three general aspects of generating valid analytical data are (1) developing a precise rugged method of analysis, (2) validating the method for accuracy, and (3) ensuring that the method remains valid throughout its use by establishing an effective quality control (QC) program. The strengths, weaknesses, limitations, and ruggedness of the analytical aspects of the whole procedure, including sample preparation and handling, must be fully understood, critically examined, and validated by appropriate research and development. Potential matrix effects must be identified for each type of biological specimen. For a particular trace element, a method that is valid for one type of biological material may not be valid for another. Certain aspects of the instrumentation and/or sample preparation might cause significant analytical differences for different matrices. Such effects must be identified and corrected in the development of a routine procedure. Once the analytical parameters of the procedure are well known and the method is under control (i.e., reproducible results with high precision can be obtained), the method must be validated to give correct or accurate results of the "true" value of the analyte in the samples. After a method has been developed, verified, and established as routine, it must be monitored by appropriate procedures that are established to ensure that the data obtained during the use of the method remain valid. The key to this QC is the availability and use of appropriate QC materials.

V. DEVELOPMENT OF
ANALYTICAL METHODOLOGY

A. General Laboratory Practices

There are certain requirements in laboratory technique that are generally understood and practiced to some extent by investigators determining trace elements in biological material. The quality of data generated reflects the extent of these practices (18). A most important requirement is the avoidance of contamination (51). Because of the low level of trace elements and the ubiquitous presence of significant levels of these elements in reagents, containers, and airborne particulates, the investigator must maintain an attitude of a ''useful paranoia'' in evaluating and eliminating potential sources of contamination in every step of the analytical methodology. These steps include analytical subsampling, analytical sample preparation, digestion or destruction of the analytical matrix, and obtaining the analytical response in appropriate instrumentation.

Once the sample arrives at the laboratory, it is often necessary to homogenize and take an analytical subsample. This subsampling can be very susceptible to contamination in that extensive homogenization is often carried out in some type of blender open to the atmosphere. Commercial blenders with stainless-steel blades can be a very significant source of contamination of the metals contained in the blades. Stainless steel can contain 5–to 20% chromium, along with percentage amounts of nickel, cobalt, and manganese, in addition to being an obvious source of iron. All of these metals can be abraded or dissolved from blades during a high-speed blending process. The amount of chromium contamination introduced in blending a total diet composite homogenate has been shown to be directly proportional to the blending time (22). If the elements of interest are chromium, nickel, cobalt, manganese, or iron, then alternate homogenization procedures using non-stainless-steel blenders must be developed.

The general laboratory area where any sample handling or transfer is carried out or where the sample is exposed to the atmosphere has to have control of airborne particulates. A minimum requirement is the availability of work stations within laminar-flow clean-air hoods. In these stations the air is passed through a high-efficiency particulate air (HEPA) filter, which removes particulate matter greater than 0.5 μm in diameter to produce an environment classified as class 100 or containing less than 100 particles per cubic meter of air. To put this into perspective, an ordinary laboratory with no air filtration system may have more than a million particles per cubic meter. With modern techniques capable of analysis at the nanogram or picogram level, a single particle of airborne particulate (usually inorganic in nature) falling into a sample can invalidate the observed results. Design of clean laboratories and their use has been described (25). The clean laboratory is considered to be essential in control of analytical blank and reduction of sources of contamination. All apparatus and sample containers with which the sample has contact can be sources of contamination. Purified reagents, especially acids, are required. Most importantly, after obvious sources of contamination have been considered, the chemists must look to their own habits. Smoking leads to trace element contamination of hair, skin, and clothing. Other problem areas include cosmetics, soaps, detergents, and shampoos, which can lead to high levels of specific elements. While some of these may seem trivial, they are examples of areas that are easily overlooked. The best advice is to use common sense, avoid unnecessary contact with potential contamination sources, and maintain a high level of personal hygiene inside and outside the laboratory (26).

A completely satisfactory cleaning of plastic ware for trace element analysis can be accomplished by cleaning with a chelating detergent cleaning agent followed by multiple rinsing with metal-free water. Disposable plastic containers, sample cups, beakers, pipette tips, and other equipment are usually "clean" as received, and require only rinsing with metal-free water to remove surface particles. The use of water is ubiquitous for preparing standards, diluting acids and digest solutions, and cleaning and rinsing containers and apparatus. Veillon and Vallee (39) have summarized the essential aspects of assuring availability of large amounts of high-purity water. Their approach utilizes a deionization pretreatment system feeding a Millipore system that generates 18-MΩ water.

More extensive discussions of general laboratory practices for trace element analysis have been reported elsewhere. Discussion regarding appropriate sample preparation techniques can also be found in these sources (14,49).

B. Analytical Techniques

There is an extensive body of literature available on the determination of trace elements in biological and environmental samples. A number of general reviews give references to detailed reports of analysis of specific elements in specific matrices (14,30,49). A wide range of potential techniques for determining trace elements exists. Most of these techniques inherently have satisfactory accuracy and precision for required studies if senitivity is adequate—that is, when analyses are carried out above the quantitation limit. The available techniques include atomic spectroscopy techniques (atomic absorption, atomic emission, and atomic fluorescence), nuclear-radioactive techniques (neutron activation, X-ray fluorescence, substoichiometric extraction), mass spectrometry (isotope dilution), and spectrophotometry of metal complexes. Choice of a suitable technique to carry out trace element studies probably depends more on the individual preference and expertise of the analyst involved in the project than on inherent superiority of the individual technique. There are advantages and limitations to all techniques. The skill and expertise of the analyst is of utmost importance in carrying out these types of analyses.

Of the many specific analytical techniques reported for use in the analysis of trace elements, the two most utilized are atomic absorption spectrometry (AAS) as a single-element technique and neutron activation analysis (NAA) as a multielement technique. The increasing interest in multielement analysis and the commercial availability of instrumentation using the inductively coupled plasma (ICP) as an atomization source has led to a resurgence in interest in atomic emission spectrometry (AES) as a multielement technique. ICP systems are being set up and utilized in a number of laboratories, and a considerable amount of trace element data will be generated by this method in the coming years. NAA is almost exclusively located in centers of analytical expertise due to the necessity for sophisticated equipment and personnel to handle the radiochemical techniques involved. AAS equipment and techniques, on the other hand, are of modest cost, are widely available to laboratories of all level of analytical expertise, and are often utilized by analytically nonsophisticated personnel to generate trace element data. As a result, more specific detail of general practical limitations of AAS will be given in this chapter. Detailed discussion of other techniques is left to the expert literature in the field.

C. Atomic Absorption Spectrometry

Modern AAS developed from the work of Walsh (44). Beginning in 1960, it achieved rapid commercial success and today is the most popular method for metal determinations in general and the most widely used technique for analysis of trace elements. The popularity of AAS arises from its analytical specificity, good sensitivity and precision, and relatively low cost. The main drawbacks have historically been the limited linear calibration range and the inability to analyze more than one element at a time. Solutions to both of these drawbacks have been reported but have not yet been adopted for commercial use (7,8). The basic principles of AAS are well known and have been detailed in a number of textbooks and literature references. A wide variety of commercial AAS instruments are currently available. In general, each consists of a light source (a hollow-cathode lamp, HCL), an atomization source, and a dispersion–detection device. The most critical component for which the greatest analytical variability exists is the atomization source. Two main types of atomization sources are used: the flame and the graphite furnace.

1. Flame Atomization

Air–acetylene and nitrous oxide–acetylene are the most popular flames used to atomize liquid solutions for AAS. Air–acetylene is the cooler of the two (~2300°C) and is more suitable for elements which ionize easily. Problems with excessive ionization of air–acetylene flames are often overcome by addition of large amounts of easily ionized elements such as lanthanum or cesium to the solutions to be analyzed. The hotter nitrous oxide–acetylene flame (~2700°C) supplies the extra thermal energy needed to atomize those elements which form stable compounds, such as oxides. Detection limits in flame AAS range from 0.001 to 0.1 μg/ml and precisions of 0.1–0.3% can be attained. While these detection limits are very reasonable for many applications, they are often unsuitable for quantitative analysis of many of the trace elements which occur at less than 1 ppm in biological materials. Attempts to analyze these elements at levels below the quantitation limit have often led to false or incorrect values. These detection limits are approximately an order of magnitude worse than those for

ICP-AES and two or three orders of magnitude worse than for furnace AAS. An extensive body of literature exists for flame AAS, and the analyst can almost always find a pertinent reference for each new sample type to be analyzed.

2. Graphite Furnace Atomization

The growth of furnace atomization methods has been quite rapid since the mid-1970s and is just now approaching a period of stability, allowing for more complete characterization of abilities and limitations (20,31). Furnace atomizers are simple in design. A small hollow graphite tube, with a small opening located in the middle for sample introduction, is positioned horizontally in the light beam of the spectrometer. An electrical current passed through the tube produces resistive heating. Sample solution is injected into the tube either by micropipette or by formation of an aerosol sprayed into the tube, which is heated above 100°C. After drying the sample, a higher current is employed to ash the sample and destroy the organic matter. Finally, a high current is applied to produce temperatures as high as 3000°C in order to atomize the sample. Required solution volumes range from 5 to 100 μl for discrete deposition and are comparable to flame atomization for aerosol deposition. The solution may be placed either directly on the inside wall of the furnace or on a "platform." Material deposited directly on the wall heats at the same rate as the wall and is atomized into an atmosphere whose temperature lags behind the wall temperature and is changing rapidly with time. Material deposited on the platform is heated primarily by radiation from the furnace wall and therefore lags behind the wall temperature. When the delayed atomization from the platform occurs, the temperature of the atmosphere in the tube has had a chance to approach the wall temperature and is changing less rapidly. Platform atomization into a more isothermal atmosphere reduces many interferences that have been observed for furnace atomization directly off the wall. Detection limits for furnace atomization range from 0.01 to 1.0 ng/ml, which are the best of any of the atomic spectrometry methods. Due to this sensitivity, furnace AAS is an analytical technique which is difficult to handle reliably. Precision in furnace AAS is not as good as flame, ranging from 1 to 4%, primarily because of inherent variability of atomization from a carbon surface.

Nonspecific background absorption and scattering of light from condensed particles (smoke) or gaseous molecules is a major problem with furnace AAS. This problem is particularly severe with biological samples, due to the large excess of organic material and major electrolytes relative to the very low levels of trace element. Significantly improved background correction methods have become commercially available (3,32). Use of these improvements is required in order to obtain reliable trace element analysis of biological materials by furnace AAS. In addition, severe matrix effects have been reported for furnace AAS. However, the use of platform atomization, faster heating rates, and matrix modification have served to improve the accuracy of furnace determinations. The importance of identifying and correcting for possible interferences cannot be overstressed (9). The literature on furnace atomization AAS is far from complete or conclusive at present. In most cases, the analyst must reevaluate existing methods or develop new methods for each new sample matrix.

D. Criteria for Routine Analytical Procedures

In order to generate the needed level of accurate, quantitative data on content and variation of trace elements in nutritional studies, it is necessary to have adequate methodology for routine analysis of a sizable number of samples. Specific criteria for the establishment of routine analytical procedures include specificity, precision, critical evaluation, and quality control (46).

The method must be specific for the element of interest. False-positive results or interferences should be minimal and easily recognized. For routine analysis, this criterion is usually met by separating the trace element from the organic matrix, followed by atomic spectrometric detection. The separation usually entails ashing or digestion of the organic matter, with subsequent dissolution for transfer of the element to the spectrometer system. AAS is relatively free of spectral interferences, whereas spectral line overlaps and interferences are coming to be recognized as abundant in AES using an ICP source. Although in general NAA is very specific, interferences do occur for specific isotopes. These are generally known and documented. Systems with multielement capability need to be individually characterized for each element of interest.

Precision of the method must be adequate for the projected application of the data. For example, if a serving of food supplies less than 5% of the recommended daily allowance of a nutrient, the analyst does not need to determine the content of that nutrient in that food at a precision of 1%. Because of the natural variability of trace elements in most biological materials, it is usually advantageous to allocate analytical resources to analysis of greater numbers of samples with slightly less precision than to push the methodology to the ultimate precision on every sample.

Sensitivity must be adequate to measure the levels that occur naturally in biological samples. For good routine, quantitative procedures, the analysis must be carried out at a level greater than the quantitation limit of the procedure as defined above. Trace elements occur at parts per million or parts per billion levels, so analytical procedures for these elements must often be sensitive to the subparts per billion level.

E. Sample Digestion

Although several of the "nondestructive" analytical techniques such as NAA and X-ray fluorescence sometimes can be utilized with minimal sample preparation, the more widely used atomic spectrometric techniques almost all require a pretreatment to remove the bulk organic matter. This is usually accomplished by either a dry ashing or wet oxidation–digestion. High-temperature dry ashing consists of placing the sample into a suitable dish, usually platinum or quartz, drying the sample, and placing it into a muffle furnace at no more than 450°C for 24 h or longer. Often ashing aids and/or posttreatment with small amounts of acid are required to destroy the organic material completely. The desirable features of this procedure, ability to ash large numbers of samples with a minimum amount of analyst time and attention, are offset by the potential problems in contamination or loss of analyte. The samples are required to be open in the atmosphere within the furnace during the entire time, and airborne-particulate contamination is impossible to avoid unless strict precautions are taken to provide highly filtered air to the furnace. The potential exists for loss of analyte either by adsorption on the container walls or by volatilization losses. Mercury and selenium can be easily volatilized, and elements such as arsenic, cadmium, chromium, iron, lead, and zinc may be lost if ashing temperatures are not carefully controlled and/or are allowed to exceed 500°C. Halides and other anions in the sample may lead to analyte loss through the formation of volatile compounds (36). Verification of absence of these losses is usually accomplished by use of radiotracers. Low-temperature or plasma ashing utilizes radiofrequency energy to produce an activated-oxygen plasma for destruction of the organic matter (12). Plasma ashing is usually employed when volatility of one of the elements of interest prohibits high-temperature ashing.

One main advantage of acid digestion with mixtures of strong mineral acids such as H_2SO_4 or NHO_3 and strong oxidizing agents such as H_2O_2 or $HClO_4$ is that the resulting sample solution is often ready for spectrometric analysis with no further treatment other than possibly dilution. Wet digestion is usually faster than dry ashing and can be set up to do a number of digestions simultaneously. Acids and reagents used must be of the highest purity, since large amounts are sometimes required. For ultratrace analysis, the control of the blank is often the determinate factor in the level of analysis that can be performed. The analyst must ascertain the purity of each batch of acid or reagents to be used. This is especially true if the acids are shipped or stored for any length of time in glass bottles. Volatile acids and bases like HC and ammonia can be prepared in the laboratory by a process of isothermal distillation (38). Other nonvolatile strong acids like HNO_3, H_2SO_4, and $HClO_4$ can be prepared in high purity by subboiling distillation in Teflon or quartz apparatus (19).

F. Calibration Standards

Atomic spectroscopic analytical methods determine the concentration of samples by comparing their analytical signals to those of a series of calibration standards. The analytical determinations are only as accurate as the standards. A number of schemes are employed for calibration. Matrix matching is the most accurate but is time-consuming and requires detailed knowledge of the sample matrix. Dilution of the samples and standards into a common matrix can eliminate the most obvious interferences and is suitable for large numbers of samples. The method of standard additions uses the sample solution itself as the matrix but assumes equilibrium between the added standard and the endogenous element, an assumption that is not always correct. Pure standard solutions can sometimes be used depending on the nature of the sample and the method of analysis. Multielement standards require compatibility of the elements in solution (no coprecipitation or opposing pH requirements), additional precautions (generally higher acid concentrations) to ensure the long-term stability of the solution, and high purity of each of the components (especially when one component is several orders of magnitude less concentrated than another).

G. Sources of Signal Measurement Error

Accurate and precise measurements of the analytical signal are crucial to every determination. Systematic interference errors arise from spectral sources or from the chemical and physical nature of the sample (29). Spectral interferences, both broad-band and line overlap interferences in atomic spectroscopy, are additive to the analytical signal. Chemical and physical interferences associated with the sample matrix (matrix effects) are multiplicative with the true analytical signal. Types and severity of these interferences are dependent on the method of analysis and the nature of the sample. These interferences must be identified and corrected to obtain accurate determinations (9). Detection or random errors are inherent in the instrument operation and are present in the measurement of every analytical signal, regardless of whether it is a blank, standard, or sample. The uncertainty of a measurement can be evaluated by making repeat measurements of the signal and computing a standard deviation. The signal–noise ratio, or inversely, the noise–signal expressed as a relative standard deviation or coefficient of variation (CV), can be used to establish the reliability of a signal. The random noise in the baseline determines the lowest signal which can be detected. Detection limits and quantitation limits as described above (Fig. 2) are useful for evaluating a method's suitability for a particular analysis, for comparing instruments and methods, and for evaluating instrumental performance between experiments. These limits should be determined and reported in each report of trace element analysis (18).

Single-element methods of analysis can be optimized for all analytical parameters such as sample treatment, standards preparation, the signal measurement process, and handling and evaluation of the data. Multielement methods, however, require compromises in almost all analytical parameters. The sample size, the dilution factor, and the instrumental parameters cannot be optimized for all elements simultaneously. The analyst can choose parameters that compromise all elements equally, favor the element with the worst signal–noise ratio, or are weighted in any manner desired. Handling and evaluation of the data are compromised with respect to time. Since the data are produced at a faster rate in a multielement method, the analyst often no longer critically examines each result for accuracy in computation. This activity is turned over to a computer. The extent of compromise is dependent on the method of analysis and elements to be analyzed. The analyst must understand the optimum conditions for each individual element and have a full knowledge of the compromises involved in applying multielement techniques.

VI. VALIDATION OF ANALYTICAL METHODOLOGY

Every analytical method must be critically validated with respect to accuracy. The first test usually conducted in proof of accuracy is recovery of added analyte. Recovery studies are done by adding known amounts of the element to homogenized aliquots of the sample. These spiked aliquots are then analyzed along with unspiked aliquots in order to ascertain that the analytical value accounts for close to 100% of the added element. Recovery is a necessary but not a sufficient proof of accuracy. There is no guarantee that the added elements are in the same chemical form or interact in the same way with the sample preparation, analysis procedure as the endogenous element in the sample.

Validation of the accuracy of the analytical method is best done by verification of the results using an independent method or by obtaining accurate results for a certified reference material (CRM) of the same composition as the samples to be analyzed. Less conclusive methods are the comparison of results to values reported in the literature or the comparison of results with other laboratories for the analysis of an exchange sample.

The establishment of two independent methods for determining the same elements is the ideal procedure for validating analytical results. Independent methods, each based on different physical principles, seldom suffer from the same systematic biases, or interferences. If different sample preparation methods are used, then analytical agreement of the two methods allows the analyst to place a great deal of confidence in the result. This approach is not usually considered except in the case of development of primary reference materials, since it is beyond the capability and resources of most laboratories. The development of a single state-of-the-art method can be quite a drain on time, money, and resources, especially for trace elements. Development of additional methods is not usually considered except in analytical centers whose purpose is often to produce highly accurate values for appropriate reference materials.

Accurate determination of a CRM is the most common means of establishing accuracy of a routine analytical method. CRM are issued by a recognized official standards agency and are accompanied by a certificate which gives the reference value of the component plus confidence limits. These materials are primary standards carefully prepared for homogeneity and stability. They have been characterized by at least two independent analytical methods. Their purpose is to transfer accuracy, validate routine methodology, and establish compatibility within the measurement system for which that specific CRM is an appropriate matrix for the constituents certified. These materials also provide a common sample for generation of data or constituents not certified (6). Information on the development and distribution of biological CRM and other homogeneous and well-characterized materials for improvement of analytical methods is consolidated in a book on this topic (48). Over the past decade, the primary sources of biological CRM include the National Bureau of Standards (United States) and the International Atomic Energy Agency (Vienna, Austria). Several other agencies have issued or are in the process of preparing additional materials (48). The major problem associated with biological CRM is finding one with the same matrix as the sample to be analyzed. At time of writing of this chapter, readily available biological CRM are limited mainly to a number of botanical materials, with only a few additional materials with trace element levels near natural occurrence levels. Interest and activity in this area are increasing rapidly and a number of new materials are becoming available, including much needed materials for human tissue and fluids (48). Cooperative efforts between agencies are helping to meet the needs for these materials. For example, a bovine serum pool has been developed and well characterized for major, minor, and trace element content by the Human Nutrition Research Center, (Beltsville, Maryland, USDA, Ref. 40). The composition and physical properties of bovine serum (viscosity, moisture, specific gravity, total protein, fat, glucose, electrolytes, trace elements) are virtually identical to human serum. Potential problems of hepatitis and other health risks in handling human serum and potential problems in trace element contamination in collection of a large pool of human serum are alleviated with bovine serum. This material should be a good substitute for a reference material for trace element analysis of human serum material and is available to the public through the Office of Standard Reference Materials, National Bureau of Standards, as Reference Material 8419. Use of this bovine serum and future availability of other reference materials will be very helpful in improving the quality of trace element analytical data in biological materials (50).

Agreement of results with published values lends support to the accuracy of the analytical method, but disagreements can be very dissatisfying. In most cases, there is no way to resolve differences between the data. It is often the case that these differences arise from differences in the samples and are not due to the method.

Analysis of a common exchange sample by a group of laboratories is another means of evaluating analytical results. Homogeneity and stability of the exchange sample are critical to the success of this round-robin approach. If the group exchanging samples is not sufficiently large, or if each laboratory employs the exact same analytical method, as is done in testing "official" methods of analysis, the consensus result may have good interlaboratory reproducibility but may not be accurate. Common systematic errors may be present within the laboratories or in the common method. Agreement between laboratories is much more significant if more than one independent method, including sample preparation, are employed.

VII. QUALITY CONTROL OF ANALYTICAL METHODOLOGY

After a method has been developed, validated, and implemented, it is necessary to ensure that the analytical results continue to be valid throughout its use. The best way to establish this quality control is to analyze periodically a sample of known value. Primary or certified reference materials can be used for this control sample, but they can be expensive and are not usually available in large quantities. These materials are generally used only for the initial validation of the method and for infrequent revalidation. For routine QC usage on a batch or daily basis, it is necessary that a secondary reference material or pool sample be prepared for QC purposes. This pool or QC sample consists of a particularly large amount of an individual sample or several samples pooled together. The QC pool is carefully characterized using CRM (if possible) and the best quality assurance methods to determine the "true" value of the component of interest and the normal range of variation of the "true" value. The QC samples are used in a number of ways (10). Routine analytical methods require recalibration at frequent intervals. A QC sample should be analyzed with each calibration. At least two QC samples should also be run with each batch of unknown samples. The results of the QC samples are compared with the predetermined "true" values and ranges of variation. Poor accuracy and precision of a QC sample throws doubt on the validity of all the samples analyzed in that batch, and usually lead to rejection of the results for those samples. Rejection criteria must be established by the analyst ahead of the time of actual analysis (10).

VIII. DATA HANDLING AND EVALUATION

Advances in computer technology have had a dramatic effect on all aspects of data handling and evaluation. After an analysis, the raw data from instrument readings are converted into concentrations in the analytical sample. A calibration function is fit to the calibration standards and the sample concentrations computed. The results should be evaluated with respect to their reliability and to the experimental design. Calibration functions historically have been restricted to the linear range, either by hand plotting of data or use of desktop calculators. With today's computer technology, complex equations can be used for calibration (10), and statistical methods of analysis can be applied to the raw data and computed concentrations. Inspection of the calibration data should show good agreement between repeat determinations. Systematic changes in standard values indicate drift in one or more of the analytical parameters. The standards must have an acceptable signal–noise ratio, and progress logically in the order and proportion of their concentration. The most accurate means of calibration is to use a pair of standards whose concentrations bracket that of the sample as tightly as possible. This eliminates any concern about linearity and reduces the time interval between the determination of the standards and the sample, minimizing fluctuation or drift errors. This approach is not practical with samples that cover a wide concentration range, or when doing multielement analysis on a large number of samples.

Computation of sample concentrations is never the final step in the analytical process. Even though the analyst may have no further involvement in the project, the results will always be used (or misused) by someone. All too often the end users of analytical data lack an analytical background and tend to interpret the results as absolute, forgetting the inherent uncertainty in each value. It is the duty of the analyst to assign a value of uncertainty to the data in order to aid their proper usage.

Systematic errors can be evaluated and corrected as discussed above. Random errors remain and can only be evaluated statistically. Repeated determinations of the same sample can be used to compute a standard deviation that reflects the random measurement error. The significance of the standard deviation can be expended by including more variables, such as multiple determinations made on different days by different analysts of different sample preparations. For this reason, the standard deviations of the control or QC samples are usually the most accurate measurement of analytical uncertainty.

The use of the signal–noise ratio to define "reliable" data has found wide acceptance (18). Specific ratios such as detection limit (CV >33%) and quantitation limit (CV <10%) are now defined (15,18) as discussed above. These limits divide all analytical data into three categories: undetectable (less than the detection limit), detected or semiquantitative (between the detection and quantitation limits), and quantitative (above the quantitation limit). Such a classification can create problems when repeat determinations of a sample fall into different categories. If the true value of sample falls close to either the detection or quantitation limit, then the normal distribution for repeat determinations will yield results falling on both sides of the limit. If this limit is applied as a filter to determine whether data are to be reported, then a fraction of the determinations will be discarded. Averaging the remaining fraction will produce a result which is biased high (11,13).

The weakest link in the analytical process is fast becoming the interpretation of the data, especially with increased emphasis on rapid multielement determinations. Development of the field of study of "chemometrics" has offered great potential for strengthening this data–evaluation link. This discipline uses mathematical and statistical methods to design optimal measurement procedures and to provide maximum chemical information by analyzing chemical data (5). A number of chemometric methods are available for evaluation of multicomponent data such as those generated in multielement trace element studies. These techniques can be very useful in determining and sorting out multifunctional relationships within group of samples.

Data which previously kept the analyst busy for weeks can now be processed in a matter of minutes using computerized mathematical approaches. This computerization does create additional quality assurance problems for the analyst. Once a computer program is established it treats each datum in a very rigid identical fashion. Deviations from expected data patterns can lead to erroneous results. For example, whereas an analyst might notice if a calibration standard were reading slightly high or low, the computer, unless specifically programmed, will accept the data as absolute. These types of problems are compounded in multielement procedures where most of the collected raw data are never examined by an analyst. Unless computer programs are designed to detect a wide variety of possible errors, significant analytical errors can result. To automate the data reduction process completely, each step, which was previously performed by the analyst, must be translated into a computer algorithm, or a series of algorithms. It is imperative that the analyst be familiar with how the data are handled by the computer program. This allows the analyst to inspect the data and results, detect errors, and anticipate conditions for which the algorithms are not appropriate. The analyst must try to evaluate the computer program's performance under as many circumstances as possible. In most cases analysis of reference materials and the QC samples is the best method for evaluating the accuracy of both the analytical methodology and the data reduction by the computer programs.

IX. CONCLUSION

As implied in the quotation at the beginning of this chapter, potential capability exists for accurate analysis of most trace elements of nutritional interest. However, this potential is often not attained. Many analysts have not taken full advantage of the new tools and advanced methods in instrumentation, computerization, and automation which are available in the analytical community. In order effectively to apply these advances for routine determinations of trace elements, total analytical procedures must be properly defined, critically tested, and validated, and appropriate QC procedures established. The end purpose of generating accurate analytical data is not the data themselves. The end purpose is a more correct detailed understanding of the metabolism and biochemistry of trace elements. Success in accomplishment of this purpose will come with close collaborative effort by nutritionists, biochemists, and analysts in thoughtful application of analytical experience and expertise to answer challenging biochemical questions.

REFERENCES

1. Alvarez, R., Rasberry, S. D., and Uriano, G. A. (1982). *Anal. Chem.* **54,** 1226A.
2. Cali, P. *et al.* (1975). The Role of Standard Reference Materials in Measurement Systems, *Nat. Bur. Stand. Monogr.* (148).
3. DeGalan, L. (1982). *Trends Anal. Chem.* **1,** 203–205.
4. Everson, J., and Patterson, C. C. (1980). *Clin. Chem.* **26,** 1603–1607.
5. Frank, I. E., and Kowalski, B. R. (1982). *Anal. Chem.* **54,** 232R.
6. Gladney, E. S. *et al.* (1984). 1982 Compilation of Elemental Concentration Data for NBS Biological, Geological and Environmental Reference Materials. *Nat. Bur. Stand.* 260–88.
7. Harnly, J. M., and O'Haver, T. C. (1981). *Anal. Chem.* **53,** 1291.
8. Harnly, J. M., O'Haver, T. C., Golden, B. M., and Wolf, W. R. (1979). *Anal. Chem.* **51,** 2007.
9. Harnly, J. M., and Wolf, W. R. (1984). *In* "Analysis of Foods and Beverages: Modern Techniques" (G. Charalambous, ed.), pp. 451–481. Academic Press, New York.
10. Harnly, J. M., and Wolf, W. R. (1984). "Analysis of Foods and Beverages: Modern Techniques" (G. Charalambous, ed.), pp. 483–504. Academic Press, New York.
11. Harnly, J. M., Wolf, W. R., and Miller-Ihli, N. J. (1984). *In* Modern Methods of Food Analysis. *IFT Basic Symp. Ser.*
12. Hollahan, J. R. (1974), *In* "Techniques and Applications of Plasma Chemistry" (J. R. Hollahan and A. T. Bell, eds.), pp. 229–253. Wiley, New York.
13. Horwitz, W. (1983). *J. Assoc. Off. Anal. Chem.* **66,** 1295–1301.
14. IAEA (1980). Elemental Analysis of Biological Materials: Current Problems and Techniques with Special Reference to Trace Elements. Technical Report No. 197, IAEA Vienna.
15. IUPAC Commission on Spectrochemical and Other Optical Procedures (1976). *Pure Appl. Chem.* **44,** 2050–2056.
16. Iyengar, G. V. (1982). *Anal. Chem.* **54,** 554A.
17. Iyengar, G. V., Kollmer, W. E., and Bowen, H. J. M. (1978). "The Elemental Composition of Human Tissues and Body Fluids." Verlag Chemie, Weinheim.
18. Keith, L. H. *et al.* (1983). *Anal. Chem.* **55,** 2210–2218.
19. Keuhner, E. C., Alvarez, R., Paulsen, P. J., and Murphy, T. J. (1972). *Anal. Chem.* **44,** 2050–2056.
20. Koirtyohann, S. R., and Kaiser, M. (1982). *Anal. Chem.* **54,** 1515A.
21. Kratochvil, G., and Taylor, J. K. (1981). *Anal. Chem.* **53,** 924A–938A.
22. Kumpulainen, J. T., Wolf, W. R., Veillon, C., and Mertz, W. (1979). *J. Agric. Food Chem.* **27,** 490–494.

23. Maienthal, E. J., and Becker, D. A. (1976). *Nat. Bur. of Stand. Tech. Note* (929).
24. Mertz, W. (1984). *Proc. Int. Symp. Health Effects Interact. Essential Toxic Elem., Lund, Sweden (Nutr. Res.,* Suppl. 1), pp. 5169–5174.
25. Moody, J. R. (1982). *Anal. Chem.* **54,** 1358A–1376A.
26. Moody, J. R. (1983). *Trends Anal. Chem.* **2,** 116.
27. Moody, J. K., and Lindstrom, R. M. (1977). *Anal. Chem.* **49,** 2264–2267.
28. Murphy, T. J. (1976). *Proc. IMR Symp., 7th (Nat. Bur. Stand.* **422,** 509–539).
29. O'Haver, T. C. (1976). *In* "Trace Analysis, Spectroscopic Methods for Elements" (J. D. Winefordner, ed.), pp. 15–62. Wiley, New York.
30. Parsons, M. L., Major, S., and Forster, A. R. (1983). *Appl. Spectrosc.* **37,** 411–418.
31. Slavin, W. (1982). *Anal. Chem.* **54,** 685A.
32. Smith, S. B., and Hieftje, G. M. (1983). *Appl. Spectrosc.* **37,** 419–424.
33. Stoeppler, M., *et al.* (1983). *In* "Specimen Banking and Monitoring as Related to Banking" (R. Lewis, N. Stein, and C. W. Lewis, eds.), pp. 95–107. Martinus Nijhoff, The Hague.
34. Taylor, J. K. (1981). *Anal. Chem.* **53,** 1588–1596.
35. Thiers, R. E. (1957). *In* "Methods of Biochemical Analysis" (D. Glick, ed.), p. 274. Wiley (Interscience), New York.
36. Thiers, R. E. (1957). "Trace Analysis, p. 636. Wiley, New York.
37. Uriano, G., and Cali, P. (1977). *In* "Validation of the Measurement Process" (J. R. DeVoe, ed.). *Am. Chem. Soc. Symp. Ser.* **63,** 140–161.
38. Veillon, C., and Reamer, D. C. (1981). *Anal Chem.* **53,** 549–550.
39. Veillon, C., and Vallee, B. L. (1978). *In* "Methods in Enzymology," (S. Fleischer and L. Packer, eds.), Vol. 54, pp. 446–484. Academic Press, New York.
40. Veillon, C., Lewis, S. A. *et al.* (1985). *Anal. Chem.* **57,** 2106–2109.
41. Versieck, J. (1984). *Trace Elem. Med.* **1,** 2–212.
42. Versieck, J., Barbier, F., Cornelis, R., and Hoste, J. (1982). *Talanta* **29,** 973.
43. Versieck, J., and Cornelis, R. (1980). *Anal. Chim. Acta* **116,** 217.
44. Walsh, A. (1955). *Spectrochim. Acta* **7,** 108.
45. Wise, S. A. and Zeisler, R. (1984). *Environ. Sci. Technol.* **18,** 302A–307A.
46. Wolf, W. R. (1980). *In* "Nutrient Analysis of Foods: The State of the Art for Routine Analysis," pp. 64–85. Association of Official Analytical Chemists, Washington, D.C.
47. Wolf, W. R. (1986). *In* "The Importance of Chemical Speciation in Environmental Processes." *Dahlem Conf., Berlin* (M. Bernhand, F. Brinckman, and P. Sadler, eds.), pp. 39–58. Springer-Verlag, Berlin.
48. Wolf, W. R. (1985). "Biological Reference Materials; Availability Uses and Need for Validation of Nutrient Measurement." Wiley (Interscience), New York.
49. Wolf, W. R., and Harnly, J. M. (1984), *In* "Developments in Food Analysis Techniques - 3" (R. D. King, ed.), pp. 69–97. Elsevier, London.
50. Wolf, W. R., and Stoeppler, M., eds. (1987). *Proc. 2nd. Int. Symp. Biological Reference Materials* (Fresenius Z. Anal. Chem. **326,** 597–745).
51. Zeif, M., and Mitchell, J. W. (1976). "Contamination Control in Trace Element Analysis." Wiley (Interscience), New York.

4

Iron

EUGENE R. MORRIS

U.S. Department of Agriculture
Agricultural Research Service
Beltsville Human Nutrition Research Center
Beltsville, Maryland

I. IRON IN ANIMAL TISSUES AND FLUIDS

A. Total Content and Distribution

The total iron (Fe) content of the animal body varies with age, sex, nutrition, state of health, and species. Normal adult humans are estimated to contain 4–5 g of iron (65) or 60–70 $\mu g/g$ of the whole body of a 70-kg individual. The adult rat contains approximately 50 $\mu g/g$ iron in the whole body (542), while levels of ≤ 40 $\mu g/g$ are normal for suckling rats (334). Most of the body iron exists in complex forms bound to protein, either as porphyrin or heme compounds, particularly hemoglobin and myoglobin, or as nonheme protein-bound complexes such as transferrin, ferritin, and hemosiderin. The hemoprotein and flavoprotein enzymes together constitute <1% of the total-body iron. Free ionic iron is present in negligible quantities.

Hemoglobin iron occupies a dominant position in all healthy animals, although in myoglobin-rich species, such as the horse and dog, the proportion is lower than in humans. Thus Hahn (227) estimates blood hemoglobin iron to be 57% and myoglobin iron to be 7% of total-body iron in the adult dog, compared with 60–70% and 3%, respectively, in the adult human. Approximately 50% of total-body iron consists of heme iron in normal suckling rats (334).

Species differences in total-body iron concentrations occur in the newborn but become much less pronounced in the adult, as shown in Table I. The differences at birth reflect differences in liver iron stores and blood hemoglobin levels. For example, the pig has relatively little iron in its body at birth because it is normally born with low liver iron stores and has no polycythemia of the newborn as does the human infant. The newborn rabbit, by contrast, has exceptionally high total-body iron concentration because of its large liver iron stores (603). Manipulation of maternal iron metabolism, creating either iron-loaded or iron-depleted mothers during pregnancy and lactation, does not appear to affect greatly the iron status of the suckling young (312,412,466).

Female rats have a higher total-body iron than males and accumulate more iron in their livers on the same diet (449). Female mice and birds also carry greater concentrations of liver iron than males, but no such sex differences are apparent in rabbits or guinea pigs (605). Lower values for total-body iron would be expected in women than in men because of their normally lower blood hemoglobin, muscle myoglobin levels, and body iron stores (65).

Among the body organs, the liver and spleen usually carry the highest iron concentrations, followed by the kidney, heart, skeletal muscles, and brain, which contain only one-half to one-tenth the levels in the liver and spleen. Individual variation in the iron levels in liver, kidney, and spleen can be very high. In most species the liver has a remarkably high storage capacity for iron. Large increases in the total iron content of the liver, up to a total of 10 g, may occur in cases of human malignancy and chronic infection (277). In the final stages of hemochromatosis, as much as 50 g of iron may accumulate in the human body (152). Iron overload in storage organs such as the liver and spleen is also characteristic of copper deficiency in swine (330) and in chicks (273), but not always in rats (170,598). The iron content of these organs, and of the bone marrow, is reduced below normal in dietary iron deficiency and in hemorrhagic anemia.

Table I
Iron Content of Bodies of Different Species[a]

Age	Iron content[b]						
	Human	Pig	Cat	Rabbit	Guinea pig	Rat	Mouse
Adult	74	90	60	60	—	60	—
Newborn	94	29	55	135	67	59	66

[a] From Widdowson (603).
[b] Given in micrograms per gram of fat-free tissue.

B. Iron in Blood

Iron occurs in blood as hemoglobin in the erythrocytes and as transferrin in the plasma in a ratio of nearly 1000 : 1. Small quantities of iron as ferritin are present in the erythrocytes of human blood (47,173,553), in serum (289), and in the leukocytes, especially the monocytes (553). The levels of ferritin in serum vary with the iron status of the individual and with certain disease states, as will be seen (5,288,615). The iron present in this form represents only 0.2–0.4% of the serum iron normally present in the adult.

1. Hemoglobin

Hemoglobin is a complex of globin and four ferroprotoporphyrin or "heme" moieties. A three-dimensional picture of the molecule and the nature of the bond between iron and globin were established in 1956 (281). This union stabilizes the iron in the ferrous state and allows it to be reversibly bonded to oxygen, thus permitting hemoglobin to function as an oxygen carrier. The molecule is similar in size (molecular weight close to 65,000) in all animal species and has an iron content close to 0.35%.

The synthesis of heme and its attachment to globin take place in the later stages of red cell development in the bone marrow. The biosynthesis of heme includes the condensation of two molecules of α-aminolevulnic acid (ALA) to form porphobilinogen (487), a process that involves the enzyme ALA dehydrogenase (282). Iron is carried to the bone marrow in the ferric form as transferrin, reduced to the ferrous form, and detached from transferrin with the aid of reducing substances, thus facilitating its transfer to protoporphyrin (217,487). The process of transport between iron uptake by the membrane surface of the reticulocyte and its incorporation into heme is considered in Section II,D.

The normal ranges of blood hemoglobin levels in the adults of different species are as follows: human and rat, 13–17 g/dl; dog, 13–14 g/dl; cow and rabbit, 11–12 g/dl; and pig, sheep, goat, and horse, 10–11 g/dl. The total hemoglobin in mammals is proportional to body weight. In five species examined by Drabkin (152), the mean proportion of hemoglobin was found to be 12.7 g/kg body weight.

The levels of hemoglobin in the blood vary with age, sex, nutrition, pregnancy, lactation, altitude, and disease. In human infants, the level falls rapidly from about 18 to 19 g/dl at birth to about 12 g at 3–4 months, at which level it usually remains until the child is about a year old, when a slow rise to adult values normally begins. A rise occurs at puberty in males and the higher levels of the male continue through the life span. This is evidently a real sex difference, since a significant rise in the hemoglobin level of females does not take place after the menopause or a hysterectomy, when menstrual blood losses no longer occur

(577). Sex differences similar to those in humans do not occur in rats (302), or in cattle (86).

Blood hemoglobin levels decline in late pregnancy in normal rats and in healthy women but not in ewes (376). In both species the decline arises primarily from a rise in plasma volume greater than the rise in red cell volume, but iron deficiency can be a contributing factor. This point is discussed later when considering iron deficiency in human adults. The problem of the "anemia" of pregnancy has been assessed by de Leeu *et al.* (332) and by Hall (230).

2. Serum Transferrin

The iron of the serum is bound to a specific protein, designated transferrin or siderophilin (269,509). Transferrin occurs in the blood of all vertebrate species (293). Transferrin is a glycoprotein with two identical or almost identical iron-binding sites, each capable of binding one atom of ferric iron (159). The exact molecular weight is uncertain, but 76,000 is considered an acceptable value for human transferrin (174) although values from 73,000 to 90,000 have been reported (52). The total carbohydrate content is 5.3% and is arranged in the form of two identical branched side chains, each branch ending in a molecule of *N*-acetylneuraminic acid (sialic acid), giving four sialic acid residues per molecule (292). Transferrins from other species have similar but not identical physical and chemical properties to those of human transferrin (52,175). Many genetically controlled transferrin variants occur in human blood (42).

Transferrin serves as the principal carrier of iron in the blood and therefore plays a central role in iron metabolism, although serum ferritin is also believed to serve a transport function (528). Transferrin also has a second important function, that of participating in the defense mechanisms of the body against infection (82). In normal individuals only 30–40% of the transferrin carries iron, the remainder being known as the latent iron-binding capacity. Serum iron and total and latent iron-binding capacity vary greatly among species, among individuals of the same species, and in various disease states. In humans a well-marked diurnal rhythm occurs. Vahlquist (576) found 15 normal men to have a mean serum iron of 135 ± 10.6 μg/dl at 8 AM and 99.9 ± 9.2 μg/dl at 6 PM. High morning and low evening values have since been observed by several investigators, except in night workers in which the diurnal rhythm is reversed and diminished (252,452). Pigs display a unique diurnal variation in serum iron, with the highest values at midnight and 3 AM (192). Diurnal fluctuations in plasma iron in dogs appear to be the result of variable partitioning of iron between the rapid early and slower late phases in the release of iron from the reticuloendothelial (RE) cells (180).

Individual variability in serum iron and total iron-binding capacity (TIBC) is high in all species studied (460). Species differences are small, with some evidence of higher average values in pigs, sheep, and cattle (192,195,460, 559,574) than in humans. In zebu cattle both serum iron and TIBC values are at a maximum during early adulthood and decline to lower levels between 8 and 17 years of age (559). In male birds and chicks both serum iron and TIBC concentrations are similar to those of humans. In hens and ducks during the laying season serum iron is markedly elevated by a factor of almost five (473). TIBC levels are also raised, though not to the same extent, and the iron-binding capacity is fully saturated (459,461). The transport of iron in the laying fowl presents special characteristics involving the existence of mechanisms auxiliary to that of transferrin (459,461). The elevation of plasma iron in hens during egg-laying is due to the appearance of a specific phosphoprotein, phosvitin, which binds about two-thirds of the iron present in plasma and transports iron to the ovocytes and egg yolk (9,451).

Representative values for serum iron, TIBC, and percentage saturation in several species are presented in Table II. The differences in human serum iron that appear in the table can be understood if the iron of the serum is conceived as a pool into which iron enters, leaves, and is returned at varying rates for the

Table II

Serum Iron, Total Iron-Binding Capacity (TIBC), and Percentage Saturation in Various Species

Species and condition	No. of cases	Serum iron[a] (μg/dl)	TIBC[a] μg/dl	Mean Saturation (%)	References
Human adults					
Normal male	35	127 (67–191)	333 (253–416)	33	93
Normal female	35	113 (63–202)	329 (250–416)	37	93
Iron deficiency anemia	35	32 (0–78)	428 (204–705)	7	93
Late pregnancy	106	94 (22–185)	532 (373–712)	18	93
Hemochromatosis	14	250 (191–290)	263 (205–330)	96	93
Infections	11	47 (30–72)	260 (182–270)	20	93
Bovine adults					
Normal cows	10	146 (89–253)	553 (388–724)	26	574
Normal bulls	10	145 (92–270)	432 (332–521)	33	574
Ovine adults					
Normal ewes	12	182 (102–304)	353 (278–456)	51	574
Normal rams	12	152 (114–191)	353 (248–455)	43	574
Castrate males	65	180 (108–268)	331 (264–406)	56	196
Adult pigs	91	123 (49–197)	540 (374–635)	22	192
Birds					
Chicks	90	102 ± 6	1239 ± 4	43	459
Nonlaying hens	40	158	258	61	459
Laying hens	38	516	333	100	459

[a] Range given in parentheses.

synthesis and resynthesis of hemoglobin, ferritin, and other iron compounds. For instance, in iron deficiency low serum iron levels result from low intake, depletion of body stores, and reduced hemoglobin destruction accompanying the anemia. In fact, there is evidence that the serum iron level begins to fall before the iron stores are completely mobilized (371), but decreased serum iron and transferrin saturation is not specific for depleted stores. The high serum and percentage saturation values characteristic of pernicious anemia and aplastic anemia (94) can be associated with increased iron absorption and a bone marrow block to hemoglobin synthesis in the presence of adequate iron stores. The high serum iron and percentage saturation of the iron-binding protein in hemochromatosis can be related to excessive absorption and deposition of iron in the body. The low serum iron and TIBC levels in the nephrotic syndrome (93) may be explained by losses of bound iron in the urine in the presence of considerable proteinuria (95).

Not all the changes in serum iron and TIBC levels can be explained so easily. For instance, the low serum iron and TIBC levels of malignancy and infections, other than hepatitis, where serum iron levels are elevated (64), present a problem, although it is known that the decrease in serum iron is not due to the reduction in iron-binding capacity (44,46,47). Beisel et al. (44) have produced evidence indicating that infections induce a redistribution of iron (and zinc) within the body, initiated by a hormonelike protein factor which is released from phagocytizing cells. This factor, leukocytic endogenous mediator (LEM), stimulates the liver to take up iron (and zinc) from serum.

The reason for the variations in TIBC levels in various disease states and for the rise in the third trimester of human pregnancy are far from clear. Such changes are not necessarily correlated with the absorption of iron, the magnitude of the iron stores, or the need of the body for greater or lesser transport of iron (65). However, Morgan (415) has demonstrated a good inverse relationship between TIBC and hemoglobin in the rat. He suggests that "the main factor regulating plasma transferrin concentration is the balance between tissue supply and requirements for oxygen, i.e. relative oxygen supply."

3. Serum Ferritin

The development of an immunoradiometric method for quantitating serum ferritin (5) led to studies of serum ferritin concentrations in normal men and women and in patients with anemia and disorders of iron metabolism. Initial reports indicated the mean concentration in men is two to three times that in women, suggesting that the level reflects total-body stores (288,591). A high degree of correlation has since been demonstrated between serum ferritin concentration and storage iron in normal subjects (115,121,340,591) (see Fig. 1), and serum ferritin assay has been shown to be a useful tool in the evaluation of

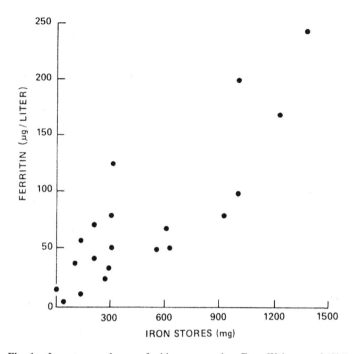

Fig. 1. Iron stores and serum ferritin concentration. From Walters *et al.*(591).

iron status in humans, including children. Thus Siimes and co-workers (528) found the serum ferritin concentration in 573 infants and children to parallel known changes in iron stores during development. The median ferritin concentration was 101 ng/ml at birth, rose to 356 ng/ml at 1 month, and then fell rapidly to a median value near 30 ng/ml (range 7–142) between 6 months and 15 years of age. Median values for normal adults reported in this investigation were 30 ng/ml in the female and 140 ng/ml in the male. In 13 children with iron deficiency anemia the serum ferritin was ≤9 ng/ml. Lipschitz *et al.* (340) obtained a geometric mean value of 4 ng/ml for 32 patients with uncomplicated iron deficiency anemia and 2930 ng/ml for patients with iron overload. Among subjects with anemia from causes other than iron deficiency the mean serum ferritin level was 180 ng/ml.

Liver disease and increased red cell turnover elevate serum ferritin concentration to a degree disproportionate to iron stores, as shown in Table III. The data of Walters *et al.* (591), obtained with normal subjects in which iron was measured by quantitative phlebotomy, suggest that 1 ng/ml of serum represents about 8 mg of storage iron. In a series of 75 healthy males and 44 healthy females in which the mean serum ferritin concentrations were 69 and 35 ng/ml, respectively (288),

Table III

Relation between Serum Ferritin and Bone Marrow Hemosiderin

Marrow iron	Control		Patients with inflammation		Patients with liver disease	
	Serum (n)	Ferritin[a] (ng/ml)	Serum (n)	Ferritin[a] (ng/ml)	Serum (n)	Ferritin[a] (ng/ml)
Absent	12	6 (1–37)	2	21 (16–28)	6	61 (25–91)
Diminished	8	51 (21–163)	6	146 (47–296)	2	182 (91–368)
Moderate	5	159 (60–253)	9	581 (129–1338)	8	622 (304–1257)
Increased	2	589 (442–669)	8	922 (290–1358)	7	1631 (1201–2077)

[a] Geometric mean, Lipschitz et al. (340).

the mean iron stores would thus be 552 mg in the men and 280 mg in the women. However, changes in serum ferritin and storage iron may not correlate well at low concentrations (285). Somewhat higher mean normal serum ferritin values were obtained by Cook et al. (115) in a group of subjects in whom iron-deficient erythropoiesis was excluded on the basis of transferrin saturation and red cell protoporphyrin. The serum ferritin was log-normally distributed, with a geometric mean of 94 ng/ml in 174 males and 34 ng/ml in 152 females. Serum ferritin levels are markedly elevated in patients with leukemia (615). This increase in ferritin probably derives from the leukemic cells themselves, since these cells, especially the monocytes, are high in ferritin compared with the red cells (553). However, ferritin is modified during its entry into the plasma, and even in cases of iron overload the iron content of serum ferritin may be low (614).

4. Hemopexin and Haptoglobin

Intravascular hemolysis releases free hemoglobin into the plasma. The proteins hemopexin and haptoglobin function to return free heme or hemoglobin to the liver for catalysis and conservation of the iron, hemopexin binding heme and haptoglobin binding hemoglobin. Hemopexin is a glycoprotein (20% carbohydrate) secreted by the liver (304). Hepatic uptake of heme is a receptor-mediated process in which the heme–hemopexin complex is bound, but the hemopexin protein is conserved and returned to the circulation (537,538). Haptoglobin is also a glycoprotein secreted by the liver (253,254). It is tetrameric in structure, with amino acid sequences of the peptide chains similar to serine proteases (220,318,350). The hemoglobin–haptoglobin complex is incorporated into rat liver parenchymal cells through specific receptor sites on the cell surface (266,308).

C. Storage Iron Compounds

The reserve or storage iron of the body occurs predominantly as the two nonheme compounds ferritin and hemosiderin. These occur widely in the tissues, with the highest concentrations normally present in the liver, spleen, and bone marrow. The two compounds are chemically dissimilar though intimately related in function. Chemical methods for their estimation are available based on the fact that ferritin is soluble in water while hemosiderin is insoluble (195,302). Results obtained by such means are supported by immunochemical and radiotracer techniques (195).

In the crystalline state ferritin is a brown compound containing up to 20% iron. It consists of a central nucleus of iron stored in six micelles arranged at the corners of a regular octahedron and surrounded by a shell of protein approximately spherical in shape (423). The colorless, iron-free protein, apoferritin, is a physiochemically homogeneous globulin with a molecular weight of 460,000 (212,497). Hemosiderin is a relatively amorphous compound which may contain up to 35% iron, consisting mainly of ferric hydroxide condensed into an essentially protein-free aggregate (525). It exists in the tissues as a brown, granular, readily stainable pigment.

Histochemical examination of aspirated samples of bone marrow for hemosiderin provides a useful index of body iron stores (549) and is a valuable aid in diagnosing iron deficiency anemia (51). Shoden and Sturgeon (526) have emphasized the value of using both staining and chemical methods which estimate the unstainable soluble (ferritin) as well as the insoluble (hemosiderin) forms of storage iron. In a study of 130 human necropsy specimens involving histological and chemical estimates of hepatic and splenic storage iron, Morgan and Walters (419) found a general agreement between the two estimates, but there was considerable variation. Histological examination of these tissues gave only a very approximate idea of storage iron levels. The situation is similar in respect to histochemical examinations of bone marrow (305).

The factors affecting the amounts and proportions of ferritin and hemosiderin in the liver and spleen of rats, rabbits, and humans have been extensively studied (414,419,523,524,526). Up to certain levels and rates of iron storage, iron is deposited in the liver and spleen readily and in roughly the same amounts as ferritin and hemosiderin. Iron is utilized equally readily from these two compounds for the demands of erythropoiesis and placental iron transfer to the fetus. A 3-fold increase in liver and spleen total-iron storage was achieved in rats without a change in the relative proportions of ferritin and hemosiderin, compared with those of normal rats not fed additional iron (414). Moreover, the ferritin iron–hemosiderin iron ratio changed little in these organs in induced chronic or acute hemolytic anemia. Davis *et al.* (140) similarly observed little change after the first week in the ferritin–hemosiderin ratio in the livers of chicks during a depletion period on an iron-deficient diet.

The main factor affecting the relative distribution of iron between ferritin and hemosiderin in mammals is the total storage iron concentration. When the iron level in the liver and spleen of rats and rabbits increases beyond about 2000 $\mu g/g$, hemosiderin begins to predominate. At iron levels beyond 3000–4000 $\mu g/g$, additional storage iron is deposited quantitatively as hemosiderin (414,523). From their study of iron storage in human necropsy samples, Morgan and Walters (419) concluded that with total storage iron in liver and spleen below 500 μ/g of tissue, more iron was stored as ferritin than as hemosiderin. With levels above 1000 $\mu g/g$, more was stored as hemosiderin. The situation may not be comparable in avian species. Larger amounts of hemosiderin than ferritin exist in the livers of chicks at much lower levels of total iron than those quoted for the rat and human (140). In human diseases such as hemochromatosis and transfusional siderosis, which are characterized by extremely high levels of iron in the tissues, most of the iron is present as hemosiderin (152,419). A similar situation also exists with respect to the heavy iron deposits in the liver and spleen in nutritional siderosis in cattle (256) and copper and cobalt deficiencies in sheep (407,572).

The ratio of ferritin iron to hemosiderin iron is affected further by the rate of storage. When iron is injected at very high rates or administered in a form such as saccharated iron which is rapidly cleared from the serum, hemosiderin is deposited rather than ferritin. With equivalent injections of iron dextran, which remains in the serum for a relatively long time, ferritin production is greater and hemosiderin smaller (525). This suggests that there is a limit to the amount of apoferritin which is present in, or can be produced by, the liver of the rabbit in a given interval of time. These findings apply only to very high rates of iron storage. Over a wide range of iron depletion and storage rates and levels, the distribution of iron between ferritin and hemosiderin remains relatively constant and the iron moves readily from one storage form to the other. Apoferritin synthesis in liver is stimulated by iron independently of a general increase in total protein synthesis (622).

The bone marrow and the muscles contain considerable amounts of nonheme iron. The storage iron concentration of the bone marrow in normal humans is given by Hallgren (248) as about 100 $\mu g/g$. If the total active bone marrow weight is taken as 3000 g, it can be calculated that the bone marrow of the human body would contain approximately 300 mg of storage iron, or about one-third to one-fifth of the total estimated storage iron. The concentration of nonheme iron in the muscles is low (248,566), but because of their large mass, the total amount is high. Torrance et al. (566) have shown that in human subjects with normal body stores, the total amount of nonheme iron in the muscles is at least equal to that in the liver. In subjects with iron overload the iron concentrations in muscles are raised but to a much smaller extent than in the liver. In rats this muscle iron represents a relatively nonmiscible pool which responds little to acute changes in the iron environment. In rats, two electrophoretically and structurally different ferritins have been demonstrated in cardiac and skeletal muscle, the ratio of which is sex dependent (590).

D. Iron in Milk

The iron content of milk varies with the species and stage of lactation and is resistant to changes in the level of dietary iron. Individual variation within the species is high, but some of the reported variation probably reflects analytical inadequacies and insufficient care to avoid contamination. Contamination from metal receptacles can more than double the iron content of cow's milk (177,283).

The average iron concentration is very similar in human, cow's, and goat's milk. A high proportion of the most acceptable values fall between 0.2 and 0.6 µg/ml (96,177,529). The level of iron in colostrum is three to five times higher than that of true milk. Studies have shown a decline in milk iron concentration as lactation progressed in women, from 0.6 to 0.3 mg/liter in one study (529) and from about 1 to 0.76 mg/liter in another (176). After the initial fall there is no evidence of a significant decline throughout lactation in milk cows (172) or sows (588). The milk of sows and of several other species is appreciably richer in iron than that of the species just cited. Mean iron levels of 1.4 and 1.2 µg/ml were obtained for the milk of sows iron-supplemented and unsupplemented, respectively, during the first 3 weeks of lactation (467). Rabbit's milk is still higher in iron, with levels mostly lying between 2 and 4 µg/ml (560). After 4 months lactation, mare's milk contained 0.5 µg/g of iron (570). Rat's milk is exceptionally rich in this element, and so is the milk of the Australian marsupial *Setonyx brachyurus* (344). In all species studied, the iron content of milk is highest at the beginning of lactation and is reduced toward the end (347).

A marked effect of stage of lactation on milk iron is apparent in rats. In one study the iron concentration declined from the high level of 13.5 to 8.1 µg/ml in the first 4 days and then more slowly to a mean of 3.0 µg/ml at 24 days, with no change thereafter (172,303). Similar findings were reported by Loh (343), but the rate and extent of decline is dependent on dietary iron concentration (14).

Administration of supplementary iron to lactating cows, sows, and women is ineffective in raising the iron content of their milk to levels above normal (313,457,467,588). In Nigerian mothers, no relationship could be demonstrated between milk iron concentration and iron status (431). This is of interest because of the low iron content of the milk of those species relative to the needs of the suckling. In the rat the levels of iron in the milk can be raised by iron loading of the lactating animal (172,14). Whether the iron content of milk is reduced below normal in the iron-deficient animal is not clear, although Ezekiel and Morgan (172) observed a decrease in the iron level of the milk of rats whose iron stores and hemoglobin levels were depleted by repeated bleeding.

Iron occurs in milk in combination with several proteins. Lactoferrin, an iron protein compound first named ferrilactin or "red protein" by Groves (222), occurs in milk and in many other body fluids including saliva and sweat (363). The milk from several species contains transferrin as well as lactoferrin (364). The nature of the iron-binding sites and their reactions with iron are very similar in transferrin and lactoferrin (7), but the two proteins differ in immunological properties (144), amino acid and peptide composition (209,546), and electrophoretic mobility (364). Rabbit milk whey has a high iron-binding capacity, 5–10 times that of serum, due to the presence of transferrin (30,298). The physiological role of the higher concentration of transferrin in rabbit milk is obscure. It does not appear to aid the transfer of iron from serum to milk in the lactating rabbit or to aid iron absorption by the suckling young (298). Perhaps, as Baker and co-workers (28) have suggested, it possesses some antibacterial function, but Cole *et al.* (105) found no effect on the numbers of coliform streptococci and lactobacilli recovered from the intestine of neonatal rabbits given milk with saturated iron-binding capacity. Specific binding of lactoferrin to brush border receptors in intestine of monkeys has been demonstrated (138), and the possible role of lactoferrin in iron absorption, particularly in infants, cannot be dismissed summarily at present.

A variable portion of the total iron in milk is not carried by the specific iron-binding proteins lactoferrin and transferrin. In cow's milk a fat globule membrane carries a significant fraction of the "nonwhey" iron (486), and a high proportion of the iron of whole rat milk occurs in association with the casein (345). Loh and Kaldor (344) have shown that in the rat and marsupial *(Setonix brachyurus)*, two species with particularly high milk iron content, the amount of iron in the cream and whey fractions is small. The remainder of the iron, which gives the milk its iron-rich character, is associated with casein. Human milk fat contains about 30% of the whole-milk iron, the greater fraction of which is in the outer fat globule membrane (189).

E. Iron in the Avian Egg

An average hen's egg contains close to 1 mg of iron (147,565), or approximately 20 μg/g of the edible portion. A high proportion of this iron is present in the yolk, bound to the protein phosvitin (219). Phosvitin is a glycophosphoprotein, although some avian species may produce carbohydrate-free phosphoprotein constituents, and the carbohydrate of ostrich and crocodile differs from hen or duck carbohydrate component (104,454). Phosvitin promotes the oxidation of ferrous ion *in vitro* and binds ferric ion strongly (556).

The presence in egg white of the iron-containing glycoprotein conalbumin was demonstrated some years ago (8). Conalbumin has since been shown to differ from chicken transferrin only in its carbohydrate content (606) and to have similar reactions with iron (175). It probably acts as a bacteriostatic agent, but other physiological functions, if any, remain to be determined.

F. Iron in Hair and Wool

Iyengar *et al.* (284) summarized several studies reporting iron in human hair with means ranging from 5 to 68 μg/g. A mean iron level of 29±10 μg/g is given for normal women by Eatough *et al.* (156). Levels generally less than 20 μg/g were obtained for women in late pregnancy with hematocrits below 38%. Baumslag *et al.* (40) report mean iron levels of 22.1 and 30.9 μg/g for maternal and neonatal scalp hair, respectively. The washing procedure may contribute to considerable variation in hair iron values (483), and there may be marked ethnic influences (39).

The average iron concentration of cleaned wool from 50 fleeces was found to be 50 μg/g (85). Apparently a great deal of this iron is in the grease and the suint, because separate analyses of 8 of these fleeces gave the following mean iron concentrations: wool fiber, 10 μg/g; grease, 700 μg/g; and suint, 600 μg/g.

II. IRON METABOLISM

A. Factors Affecting Iron Absorption

Because of the limited capacity of the body to excrete iron (372), iron home-ostasis is maintained primarily by adjusting iron absorption to bodily needs. The absorption of iron is affected by (1) the age, iron status, and state of health of the animal or individual; (2) conditions within the gastrointestinal tract; (3) the amount and chemical form of the iron ingested; and (4) the amounts and propor-tions of various other components of the diet, both organic and inorganic.

In monogastric species iron absorption takes place mainly in the duodenum (77) and, it was thought, in the ferrous form (440). Generally the iron of ferrous salts is absorbed to a greater extent than from ferric salts, but absorption may not depend on valency, since some ferric compounds are more bioavailable than some ferrous compounds (190). Iron of nonheme, iron–protein compounds ap-parently needs to be released from conjugation for effective absorption, whereas the iron in heme compounds is absorbed as the heme moiety into the mucosal cells of the intestine, without the necessity of release from its bound form (247). The absorption of food iron thus involves two independent systems. Normal gastric secretion is necessary for optimal absorption of iron by rats (430), and the administration of HCl can increase iron absorption in achlorhydrics (290). In-organic iron is able to form complexes with normal gastric juice at a low pH (287). These complexes remain soluble when the pH is raised to neutrality and may enable the iron to be available in a suitable state for absorption. Absorption of nonheme iron from bread correlated with the ability of the patient's gastric juice to form soluble iron *in vitro* (53).

The absorption of inorganic and nonheme iron is much more sensitive to changes in the intestinal milieu than is the absorption of heme iron. Most of the studies with humans since the mid-1960s have attempted to define bioavailability of nonheme iron using the radiolabel extrinsic-tag technique (60–62,114,325). The technique is a valid measure of iron absorption from a meal if the extrinsic tag equilibrates with the nonheme iron pool of a meal or food. Direct comparison of extrinsic- and intrinsic-labeled salts or foods, revealed a tendency toward greater absorption of the extrinsic label, but the mean ratio extrinsic–intrinsic for several different studies was generally no greater than 1.1. Specific instances in which the extrinsic label may exhibit incomplete equilibration with the total nonheme content are the presence of insoluble or contaminant iron (238), the iron of purified ferritin (327), some rice preparations (62,562), and some isolated protein preparations (540). Except for these specific examples, the extrinsic-tag technique appears to provide a valid measure of nonheme iron absorption in humans, and most of the studies discussed below have used it. The ratio extrinsic–intrinsic was 1.14 in one rat study (398) and 1.2 in a second (592), both using intrinsically labeled wheat, corn, or soybeans.

Ascorbic acid, which can both reduce ferric iron and chelate iron, increases iron absorption under a wide range of conditions (75,79,150,329,580,582). Ascorbic acid has no such effect on the absorption of hemoglobin iron (247), one confirmation of the separate mechanisms for absorption of heme and nonheme iron. Growing rats fed egg yolks as the dietary iron source in a prophylactic bioassay maintained higher hemoglobin concentration with than without ascorbic acid (422). In a depletion–repletion, therapeutic bioassay of egg yolk iron, Miller and McNeal (388) observed only a slight enhancing effect of ascorbic acid, but using a prophylactic bioassay, ascorbic acid in the yolk preparation did show an enhancing effect on iron utilization by rats (389). In contrast, ascorbic acid did not improve utilization of the iron from wheat by rats (420). Wheat iron is highly bioavailable to rats and, perhaps only in iron-deficiency or for poorly bioavailable iron sources will ascorbic acid stimulate iron utilization by rats. Lynch and Cook (351) reviewed in 1980 the interaction of Vitamin C and iron absorption in human nutrition. Salient points are that ascorbic acid is effective in promoting iron absorption by humans when taken as crystalline supplement or present in fruits and vegetables of the meal, and it must be consumed with the meal to be effective. Absorption of nonheme iron from 20 different Western-type lunch and dinner meals was greatest from a vegetarian meal high in ascorbic acid (241) and, when similar-type meals were warmed at 75°C for 4 hr, the ascorbic acid content was decreased and nonheme iron absorption was depressed compared to freshly prepared meals [see Table IV (246)]. Studying both Western-

Table IV

Effect of Prolonged Warming of a Meal on Ascorbic Acid Content
and Nonheme Iron Absorption[a]

Measured parameter	Before warming	After warming
Ascorbic acid (mg)	80	16
Iron absorption (%)	34	23

[a]Mean of data from two different meals, adapted from Hallberg *et al.* (246).

type meals and a Latin-American meal of maize, rice, and black beans, Hallberg and Rossander demonstrated with single-meal absorption tests that the ascorbic acid content was a major factor correlating with iron absorption (243,244). The ratio of ascorbic acid to iron that most effectively promotes iron absorption has not been extensively studied. A marked break in the dose–response curve occurred at molar ratio of 7.5 mol ascorbic acid to iron in the study by Cook and Monsen (118), representing 100 mg of ascorbic acid in a meal with 3 mg of iron. Absorption of iron from infant cereal by multiparous women was stimulated 6-fold by about 10 mg ascorbic acid per milligram of iron (145). The amount of ascorbic acid in a meal is a component of one scheme for estimating absorbable iron in a meal (399,401). Studies of the physicochemistry of ascorbate–iron relationships in foods may aid in predicting bioavailability of the iron in these foods (210,433). In the studies with egg yolk fed to rats (422), other reducing compounds stimulated hemoglobin concentration only if they possessed complexing potential. Hungerford and Linder (272) studied interactions of pH and ascorbate in the rat intestine and stress the importance of iron solubility in iron absorption measurements. A principal action of ascorbic acid in iron absorption may be maintaining a soluble iron complex when pH of the lumen contents rise in the duodenum (108).

Other organic acids, including citric, lactic, malic, pyruvic, succinic, and tartaric, also enhance iron absorption (75,202,241,463). Absorbability of iron from certain vegetables and meals containing these vegetables correlates with the content of some of these organic acids (202,241). Concomitant administration of iron dose and acetylsalicylic acid by gavage enhanced iron uptake by rats (205).

A second major dietary component having an enhancing effect on iron absorption is the ''meat factor'' or content of meat, fish, and poultry, or MFP (241,399,401). Studies with intrinsically radioiron-labeled single foods indicated higher absorbability of iron from animal than from vegetable foods (322). When veal muscle and corn or black beans were tested together, absorption of the veal muscle iron was slightly depressed, but the vegetable iron absorption was almost doubled (359). Cook and Monsen (118) compared the effects of different animal proteins on nonheme iron absorption when substituted for beef in a high-bioavailability meal or for ovalbumin in a low-bioavailability semisynthetic meal. Lamb, pork, liver, fish, and chicken in the high-bioavailability meal did not change iron absorption significantly, but milk, cheese, or egg substitution each produced a severalfold depression in relative absorption of iron. Opposite effects were found when ovalbumin in the low-bioavailability semisynthetic meal was replaced by the respective meat sources, and neither milk, cheese, nor egg substitution resulted in any significant change. The mechanism whereby the MFP effect is expressed is uncertain. Björn-Rasmussen and Hallberg (58) concluded that meat acts by counteracting luminal factors that inhibit absorption, probably by formation of a luminal iron transport carrier. Amino acids could conceivably be such a carrier, and studies by Layrisse and co-workers indicate that cysteine (either free or peptide form), but not histidine, may be a partial explanation of the MFP factor effect (326,361). Hallberg *et al.* (236) found that the absorption of iron from tablets taken with a meal was not influenced by meal content of meat or fish.

Other compounds and dietary components influence iron absorption. Amine and Hegsted (11) found the retention of iron by rats to be affected by carbohydrate in the order lactose $>$ sucrose $>$ glucose $>$ starch. However, Garretson and Conrad (197) found no effect of starch or sucrose. Simple sugars such as fructose (286) and sorbitol (264) can, by contrast, increase iron absorption, and ferric fructose facilitated iron retention by guinea pigs in comparison with ferric gluconate (38). Absorption by humans from a therapeutic dose (50 mg) of iron as polymaltose complex was not different from ferrous sulfate, with or without food (291). The effect, if any, of type of carbohydrate in iron absorption by humans is not clear.

Several amino acids are effective in enhancing ferrous iron uptake from isolated intestinal segments of fasting rats (315) or when given by gavage (103). Histidine, lysine, and cysteine increase absorption of ferric iron when administered in solution by gavage (286,463,580,581,583). The ability of these amino acids to form tridentate chelates is essential to their effectiveness in enhancing iron uptake (580). Miller interpreted a hematological response to methionine and lysine supplementation of diets marginal in proteins as improved iron utilization (387). In fasting children El-Hawary *et al.* (160) observed greater serum iron increase to orally administered solutions of ferrous sulfate with valine or histidine than the iron salt alone or with cysteine or glutamic acid. Martínez-Torres and Layrisse (358) gave a meal of black beans mixed with each of 16 different amino acids, and only cysteine or cysteine plus methionine enhanced iron absorption, mimicking the meat factor effect discussed above. Both animal and human studies, therefore, agree that certain amino acids are effective in promoting iron absorption from a solution, but dietary effectiveness is questionable, with the possible exception of cysteine.

Desferrioxamine, a bacterial siderophore, is a potent inhibitor of nonheme iron absorption (112,317). The compound is useful in treatment of iron overload (65); however, it and other bacterial siderophores are not known to be present in human diets in sufficiently high amounts to be of concern nutritionally. Although the ferric salt of ethylenediaminetetracetic acid (EDTA) has been studied as an iron fortification compound (324,589), Cook and Monsen observed 50% depression in nonheme iron absorption from either a semisynthetic or regular meal at 2 : 1 EDTA–iron molar ratio (117). Ascorbic acid could overcome the effects of EDTA in absorption of iron from an isolated rat intestinal loop (403), but EDTA may be a potential problem in some situations. The material studied by Layrisse and Martínez-Torres (324) and Viteri et al. (589) was prepared from the sodium salt of EDTA and is a water-soluble preparation that either will act as or is miscible with an extrinsic tag. Iron absorption was greater from vegetables fortified with the Fe(III)–EDTA complex than when fortified with ferrous sulfate (362). MacPhail et al. (355) found absorption from Fe(III)–EDTA was marginally greater than the intrinsic iron of some cereal meals, but concluded that it forms a common pool with the intrinsic nonheme iron and that the compound may be useful for fortification of cereals. Approximately 1% of the administered dose of Fe(III)–EDTA is excreted in the urine by humans and swine (91).

In studies by Hallberg and Sölvell (247) and Hussain and Patwardhan (276), the addition of sodium phytate reduced iron absorption in humans. Thus, the decreased chemical iron balance observed by Widdowson and McCance (604) when adults consumed bread made from high-extraction wheat flour, or reduced whole-blood retention of radioiron from white bread with added wheat bran (56), was attributed to the effect of phytate. Added sodium phytate does not reduce iron utilization by rats (123,161,274), and rats utilize the iron from wheat bran very well (208, 420). The iron extracted from wheat bran by aqueous salt or ammonium sulfate solutions is monoferric phytate, and the iron in this compound has a high biological availability to rats (420). Mössbauer parameters of the iron in ^{57}Fe-enriched bran are the same as for monoferric phytate prepared in vitro (367). Monoferric phytate is soluble in water and in dilute salt solutions over the pH range encountered in the small intestine; however, in the absence of excess phytate ion, diferric phytate will precipitate from a solution of monoferric phytate at pH < 3, or a solution high in Cl^- ion (161,421,477). In contrast to monoferric phytate, the iron of diferric phytate is utilized very poorly by rats and from tetraferric phytate to an intermediate degree (see Table V). Radioiron-labeled monoferric phytate equilibrates with the miscible nonheme iron pool of a meal and is absorbed to the same extent as the miscible pool by dogs (342) and by humans (530). In the latter study, enzymatically dephytinized bran inhibited iron absorption to the same extent as whole bran, indicating that the inhibitory action of bran previously cited (56) may not reside in the phytate content per se. Hallberg and Rossander found that bran dephytinized by washing with HCl did not inhibit iron absorption, but inhibition was restored when sodium phytate was

Table V
Comparison of Iron Phytates as Dietary Iron
Source for Rats

Iron source[a]	Hemoglobin (g/dl)
$Fe(NH_4)_2(SO_4)_2 \cdot 6H_2O$	10.3 ± 1.2
Monoferric phytate	10.6 ± 1.1
Diferric phytate	5.4 ± 1.1
Tetraferric phytate	7.4 ± 0.5

[a] Diets contained 18 mg/kg of iron. Adapted from Ellis and Morris (161).

added (242). Poor iron bioavailability was noted by Gillooly *et al.* from wheat germ, butter beans, and lentils, foods with high phytate contents, and added sodium phytate decreased iron absorption from a broccoli meal (202). Thus, the effect of phytate on absorption of nonheme iron is not clearly defined. The available data indicate that (1) rats utilize the iron from wheat and other cereals very well, but (2) wheat bran tends to reduce nonheme iron absorption by humans. The insoluble iron phytate complex is a poor source of bioavailable iron, and this fact likely explains the poor efficacy of iron phytate administered to humans (406) and veal calves (74). Balance studies with humans do not differentiate between dietary heme and nonheme iron. Short-term studies of 2–4 weeks duration have indicated no adverse effect on iron balance when humans consumed 16–22 g/day of wheat bran, but in some instances the fractional fecal excretion of iron increased when bran was consumed (17,505,585). In another study, an initially excessive fecal loss of iron when whole bran was consumed did not persist, and iron balance was most positive when whole bran compared to enzymatically dephytinized bran was consumed for 15 days (421).

Studies over a decade ago with intrinsically radioiron-labeled soybeans indicated that, among plant foods, the iron from soybeans was more bioavailable for humans than iron from corn or wheat (20,322). In 1981, Cook *et al.* (119) reported that substitution of egg albumen by several soy products in a semi-synthetic low iron bioavailability meal or inclusion of soy products in a high-bioavailability meal substantially reduced nonheme iron absorption when tested with adult males. The report caused concern because of increasing use of soy products in human diets worldwide. Hallberg and Rossander (245) also observed decreased percentage absorption when soy protein was added to a hamburger meal, but because the soy–hamburger meal contained more iron, there was no change in the amount of nonheme iron absorbed. Later studies have indicated that the inhibitory effect of soy products may be modified slightly by the cooking procedure, and the addition of ascorbic acid to the meal greatly improves relative nonheme iron absorption (409). Evidence has since been obtained that substitution of soy products in a meal containing meat promotes the absorption of heme iron (352). Rats utilized iron from mature soybeans to a greater extent than from immature seeds with one-third the phytate content (599); however, dephytinizing did not modify the inhibitory action of soy flour in human studies (245). Rats did not use iron from isolated soy proteins in a hemoglobin repletion assay as well as from ferrous sulfate added to a casein diet (548). Iron-deficient rats absorbed less iron from an extrinsically labeled test meal containing soy isolate and casein than from an isonitrogenous meal of casein (564). In another study, iron-deficient rats retained less iron from labeled meals containing soy products than from a gavage of ferrous sulfate, but there was no difference in iron-repleted rats (592). The iron-deficient rat in the above studies seemed more nearly to mimic the observed effects of soy on iron absorption by humans, but Schricker et al. observed no effect of iron status (511).

Several varieties of clay reduced iron absorption from $^{59}FeSO_4$ and ^{59}Fe-labeled hemoglobin in normal and iron-deficient subjects (392); the greater the cation exchange capacity of the clays, the greater their effect on iron absorption. This effect could be a contributing factor in the production of anemia in geophagia. In a cohort of mentally retarded individuals, mean hemoglobin values did not differ for those practicing pica, but other iron status indicators, plasma iron, TIBC, and ferritin, indicated those practicing pica were in generally poorer iron status (137).

High levels of phosphate reduce iron absorption (76,260) and may block it almost completely in humans (455). Calcium may be an interacting factor with phosphate, as shown by Monsen and Cook using a semisynthetic meal (400), but the extent to which this may be a factor with usual human diets is not known. When 1-g supplements of phosphorus were given as either orthophosphate or hexametaphosphate, they increased fecal excretion of iron in a balance study with either low (356 mg) or high (1166 mg) intakes of calcium (71). High intakes of cobalt, zinc, cadmium, copper, and manganese also interfere with iron absorption, presumably through competition for absorption binding sites. These interactions with iron are considered further in the appropriate chapters that follow. The mutual antagonism between cobalt and iron at the absorptive level reported by Forth et al. (187) is of particular interest. These workers found that a 10-fold cobalt excess reduced ^{59}Fe absorption from jejunal loops of the rat by nearly two-thirds, and a 100-fold excess depressed iron absorption almost completely. It appears that cobalt shares with iron at least part of the same intestinal mucosal transport pathway—a pathway in which acceleration of transport of both elements is apparently governed by the same mechanism. For example, cobalt absorption as well as iron absorption is significantly enhanced in iron-deficient rats (462), in iron deficiency in humans (579), in patients wih portal cirrhosis with iron overload, and in those with idiopathic hemochromatosis (578,579). Studying the time course of uptake and release of cobalt and iron from intestinal segments and the subcellular distribution, Becker *et al.* concluded that cobalt interferes with iron absorption by an interaction with release of iron at the serosal side, not by uptake from the intestinal lumen (43).

The findings of Settlemire and Matrone (515) with rats fed diets very high in zinc are also pertinent. These workers obtained evidence that the zinc reduced iron absorption by interfering with the incorporation of iron into or release from ferritin. The increase in iron requirement brought about by the high zinc intakes was enhanced by a shortening of life of the red blood cells, resulting in a faster turnover of iron. Eltohamy *et al.* (165) found depressed liver iron concentration in chicks fed diets containing 1000 mg/kg zinc, but they give no growth or hematological information. Addition of zinc to perfusate inhibited iron uptake and transfer by intestinal loops of iron-deficient mice (250).

Two commonly consumed beverages consumed worldwide can inhibit iron absorption: tea and coffee. Tea is the more potent inhibitor and is inhibitory for iron in solution or in meals consumed by humans (151). Iron binds to phenolic groups in extracts of tea (593), and 500 mg of tannic acid greatly reduced nonheme iron absorption from a high-bioavailability broccoli meal (202). Other vegetables and foods high in tannin (polyphenol) content depress iron absorption and may be an important factor in diets that contain sorghum (202,471,474). Anemic subjects absorbed more iron from low-tannin than from high-tannin sorghum varieties (471). A prophylactic role of tea in preventing excessive iron absorption in the thalassemia syndromes has been suggested (143). Coffee also inhibits iron absorption in a concentration-dependent fashion when consumed with or 1 hr after a meal, but taken 1 hr before the meal did not produce inhibition (410).

Iron is poorly absorbed from most diets, with better absorption from foods of animal than of plant origin. Only some 5–15% of food iron was absorbed by adult humans from ordinary mixed diets (405,458). Absorption of food iron may be increased to twice this level or more in children and in iron deficiency. In the rat the high absorption of iron during suckling drops abruptly at weaning to the low level characteristic of the adult (186). Josephs (299) reported that 2–20% of an oral dose of radioiron was absorbed in normal human subjects, compared with 20–60% in patients with iron deficiency anemia. Iron absorption values obtained in another study were $27 \pm 11.8\%$ for those with iron deficiency anemia (541). Subsequent studies employing the double-radioiron isotope method of Hallberg and Björn-Rasmussen (234) have disclosed lower iron absorption levels than those just described. In a study of young men consuming Swedish diets, 2–4% of the nonheme iron was reported to be absorbed (59). Iron absorption from a diet of rice, vegetables, and spices consumed by Thai subjects and containing almost 10 mg iron approximated only 0.4% in normal individuals. When fruit or meat or both was added, the absorption increased but to values still below 2% in normal subjects. In a few iron-deficient individuals, iron absorption up to nearly 10% was observed (239). A satisfactory explanation of such seemingly low values has not appeared. Since the subjects had no malabsorption of an iron salt, the authors suggested that the diets contained some unknown compound or compounds markedly inhibiting the absorption of nonheme iron from the diets. A partial explanation was derived from studies with whole foods compared to homogenized meals consumed by the Thai subjects. Absorption was greater from the whole-food meals than from the homogenized meals (237). The role of the above-discussed bioavailability factors in maintaining iron balance in humans is referred to in the section discussing iron in human dietaries.

The efficiency of iron absorption changes in various disease and deficiency states. Increased absorption occurs in iron deficiency (512), hemolytic anemia, aplastic anemia, pernicious anemia, pyridoxine deficiency, and hemochromatosis (94), and is related to increased erythropoiesis (596), depletion of body iron stores (383), and hypoxia (382). Decreased iron absorption occurs in transfusional polycythemia (68) and has been related to tissue iron overload (109). Markedly reduced iron absorption occurs in severely copper-deficient rats, presumably a response to the raised iron stores in the animals (512), and also occurs in nickel deficiency in rats, as discussed in Chapter 8 (see Nielsen, Chapter 8 in this volume).

B. Mechanism and Regulation of Iron Absorption

The mechanisms by which the body increases the efficiency of iron absorption during periods of iron need and decreases this absorption during times of iron overload, or in other words regulates iron absorption in accordance with body iron needs, are not completely understood. According to the mucosal block theory advanced in 1943 by Hahn and co-workers (228), and elaborated by Granick (211), the intestinal mucosa absorbs iron during periods of need and rejects it when stores are adequate. This was explained as follows: iron taken into mucosal cells is converted into ferritin, and when the cells become physiologically saturated with ferritin, further absorption is impeded until the iron is released from ferritin and transferred to plasma. Conrad and Crosby (107,109) later provided evidence that in the rat, the ultimate regulator of iron absorption may be the iron concentration in the epithelial cells of the upper intestine. In normal rats with moderate intestinal iron concentrations only a small part of the ingested iron taken up by mucosal cells is transferred to the bloodstream and retained by the animal; the remainder stays in the mucosal cells and is lost into the gut lumen when the cells are sloughed from the tips of the intestinal villi. In iron-deficient rats with decreased intestinal iron concentration, most of the ingested iron is absorbed directly into the bloodstream with very little remaining in the mucosal cells. In rats given excessive iron stores parenterally the epithelial cells are "loaded from the rear" and are therefore unable to accept the ingested iron. This mechanism apparently does not apply to humans (10). The nonheme iron concentrations of duodenal mucosa obtained from normal subjects did not differ from those of iron-deficient or iron-loaded subjects. Furthermore, these concentrations showed no significant correlations with simultaneous measurements of serum iron levels or with radioiron absorption.

Charlton et al. (100) observed decreased absorption of iron 18 hr after repeated intravenous injections of ferric ammonium citrate in rats. Release of radioiron from isolated intestinal mucosal cells was greater in the presence of serum from iron-deficient compared to serum from nondeficient rats; however, an acute intravenous infusion of unsaturated transferrin did not produce an increase in whole-body absorption of an oral dose of ^{59}Fe (336). Rosenmund et al. (496) concluded that regulation was not mediated by a humoral factor in plasma. They exchange-transfused plasma from iron-deficient or iron-loaded donor rats to normal or iron-deficient rats and observed no change in iron absorption by the recipient animals. Bhargava and Gabbe (54) observed increased intestinal iron absorption by rats after 8 days exposure to lowered dietary iron levels and within 4 hr of an acute bleeding equivalent to approximately 15% of total blood volume. Thus, they suggest that in rats, intestinal iron absorption is stimulated by iron uptake by erythropoietic activity of the erythron. But, how regulated iron transfer from the mucosal cells occurs remains an enigma.

Examination of homogenates of intestine and intestinal mucosa cells led to the postulate of an intracellular iron transport system within the mucosal cell (78,169,187), and early studies indicated several separable iron-binding components. One high-molecular-weight component is identified as ferritin (249,296, 620). Other separated high-molecular-weight fractions may be intracellular or brush border membrane fractions (249,425), and some low-molecular-weight fractions that may be amino acid or other organic chelates have also been partially characterized (78,516). Not all investigations have clearly shown the low-molecular-weight fraction, and it may not be involved in the transmucosal transport of iron. A fraction that was labeled as transferrin-like in early investigations (78,187,464) is now thought to be mucosal transferrin (78,162,296) but its source is unknown. Decreased incorporation of radioiron into mucosal transferrin was observed in (1) anemia of protein deficiency, correlating with the absence of increased iron absorption (163), and (2) when excess copper was placed in duodenal segments, depressing uptake of iron (164). Savin and Cook found the mucosal transferrin concentration to vary directly with the total mucosal uptake of radioiron by closed intestinal loops (508), and Johnson et al. detected doubling of the mucosal transferrin in response to iron deficiency (296). The mucosal transferrin exhibited some immunochemical differences from serum transferrin and is resistant to proteolytic action of trypsin (296,508).

Processes that occur at the luminal surface of the intestine to effect iron absorption are uncertain. Savin and Cook (507) concluded that initial uptake of iron by isolated intestinal mucosal cells was by passive adsorption. The *in vitro* binding of iron was pH and temperature dependent, and was blocked by iron-chelating agents in a concentration-dependent fashion, but was unaffected by inhibitors of cellular respiration. Slightly more than one-half the uptake by the isolated cells was bound by the brush borders. Muir *et al.* (425) found that iron uptake *in vitro* by brush border membrane vesicles exhibited saturation kinetics, evidence for a carrier-mediated process under biological control. Further experiments with purified brush border membrane vesicles showed greater uptake of Fe(II) ascorbate by preparation from proximal compared to distal small intestine segments and increased uptake by only the proximal vesicle preparations from iron-deficient mice (424). Simpson *et al.* (531) reported uptake of ferric iron by brush border membrane vesicles. Huebers *et al.* (271) proposed that apomucosal transferrin is secreted from mucosal cells, loaded with iron in the intestinal lumen, and the iron–transferrin is taken into the cell where the iron is released and the apotransferrin returns to the brush border to be recycled. They injected ^{125}I-labeled diferric transferrin into tied-off gut segments and demonstrated intracellular appearance of unchanged transferrin and a small amount in the blood. There was no absorption of diferric transferrin from the ileum, but high absorption from the duodenal and jejunal segments. However, Idzerda *et al.* (279a) could not detect transferrin mRNA in mucosal cells and they postulate the liver as source of mucosal transferrin by way of bile secretion. Johnson *et al.* (296) found the adsorption of iron in *in situ* ligated intestinal loops (mucosal uptake) was uniform throughout the length of the small intestine, but absorption (carcass uptake) showed a steep decreasing gradient from the duodenum to the ileum. Morgan (417) examined the role of plasma transferrin in iron absorption and found some transcapillary exchange of the plasma transferrin. The rate of transcapillary exchange was insufficient, however, to account for the movement of iron into the blood from the intestine.

Difference in the mechanism of iron absorption occur within as well as among species depending on the chemical form in which the iron is ingested. As mentioned earlier, the absorption of the iron of hemoglobin, unlike inorganic iron, is not affected by ascorbic acid or nonabsorbable chelating agents (231). This suggests that iron is not released from heme in the gut lumen. Instead, the complex is taken up directly by the intestinal epithelial cell, subsequently appearing in the plasma in nonheme form (110,247). Weintraub *et al.* (597) demonstrated in duodenal mucosal homogenates from the dog the presence of an enzymelike substance which is capable of releasing iron from hemoglobin *in vitro*. Raffin *et al.* (472) found evidence that mucosal heme oxygenase may be the enzyme. The rate at which the heme-splitting substance worked *in vivo* appeared to be increased by the removal of the nonheme iron end product from the epithelial cell to the plasma. It seemed, therefore, that the labile nonheme iron content of the intestinal cell determines its ability to accept heme from the lumen in dogs, as well as ionized iron from the lumen in rats. Studies by Wheby *et al.* suggested that the release of iron from heme in the mucosal cell and transfer to the plasma, not the uptake from intestinal lumen, is the rate-limiting step in absorption of hemoglobin iron by dogs (601,602). Porcine upper small intestine was found to bind radioactive heme *in vitro,* the site being localized to the brush border (213). Binding of heme was not inhibited by ferrous sulfate. The heme receptor or binding factor could also be demonstrated in human duodenal tissue.

Two components of the gastric secretions have been investigated for roles in iron absorption. Patients with achylia gastrica have decreased absorption of iron, and may exhibit iron deficiency if the condition exists for a prolonged time (97,112). Normal gastric secretion provides an acidic milieu thought to be important in promoting and maintenance of soluble low-molecular-weight iron complexes (108,287). A study by Bezwoda *et al.* (53) showed that the ability of a patient's gastric juice to solubilize iron from bread and the presence of a low-molecular-weight gel chromatography fraction from *in vitro* incubations, corre-lated well with *in vivo* absorption of the nonheme iron from bread. Studies with cimetidine, a selective inhibitor of gastric acid secretion, suggest that nonheme iron absorption is severely depressed only after 80% or more decrease in acid secretion (536). A high-molecular-weight iron-binding glycoprotein (gastrofer-rin) is found in normal gastric juice (141,142,287,500) and has been proposed as a gastric factor in the mechanism and regulation of iron absorption. Davies and associates (141) observed that it was absent in hemochromatosis. In iron defi-ciency anemia caused by blood loss, the concentration of gastroferrin in gastric juice was reduced and returned to normal levels when hemoglobin values were restored (349). The hypothesis was advanced that gastroferrin production is involved in the regulation of iron absorption—normal levels acting to inhibit the absorption of excessive iron intakes and reduced levels permitting enhanced absorption in iron deficiency. It was further proposed that failure to produce gastroferrin through an inborn error of metabolism is a causal factor in hemo-chromatosis. Other workers have proposed that increased iron absorption in iron deficiency or hemochromatosis is not due to gastroferrin deficiency but to a factor in the gastric juice that increases absorption (428,569). Physiocochemical characterization of gastroferrin has been described (427,499), but its role in iron absorption is uncertain.

The excess absorption and deposition of iron characteristic of hemochromato-sis has also been related to a primary pancreatic defect, but uncertainties exist for this hypothesis. Davis and Biggs (139) showed that iron absorption can be significantly reduced in hemochromatosis by the addition of a pancreatic extract with the oral dose of iron. A similar inhibitory effect of a pancreatic extract on iron absorption was demonstrated in the intact rat and with the isolated loop of the rat jejunum (55,139). Feeding of pancreas powder or pancreatectomy did not, however, produce changes in iron absorption in rats (429,532). Bile did not appear to exert an influence on iron absorption in dogs (600). Heinrich *et al.* (261,262) concluded that absorption of iron, either inorganic or food iron, was controlled by iron stores, not by a pancreatic factor, in pancreatic exocrine insufficiency of cystic fibrosis. The full importance of the gastric and pancreatic secretions to the mechanism and regulation of iron absorption remains to be determined.

C. Excretion of Iron

The limited ability of the body to excrete iron has been abundantly confirmed since the original observations of McCance and Widdowson (372). Even in hemolytic anemia and in the treatment of polycythemia with phenylhydrazine, when large amounts of iron are liberated in the body from the destruction of red cells, less than 0.5% of this iron appears in the urine and feces (373). Although absorbed iron is retained with great tenacity and, in the absence of bleeding, excretion is very small, the amounts lost cannot be neglected and are of nutritional importance, as Moore (405) has stressed.

The total iron in the feces of normal human adults usually lies between 6 and 16 mg/day, depending on the amounts ingested. Most of this consists of unabsorbed food iron. True excretory iron is estimated at about 0.2 mg/day by chemical balance studies (280) and 0.3–0.5 mg/day by a radioiron technique (153,216). This iron is derived from desquamated cells and bile. Iron occurs in the bile, mostly from hemoglobin breakdown, to the extent of about 1 mg/day. Most of this may be reabsorbed and does not reach the feces. Hall and Seymonds (229) observed a 40% decrease in iron concentration in bile of steers being infused with large amounts of manganese. This situation is unlikely to occur except in the unusual circumstance of high manganese intakes. The hemoglobin concentration in feces of long-distance runners was found to increase 4-fold following a race (550). The amount was sufficient in 7 of 21 individuals to account for 2 mg or more of iron loss and indicated gastrointestinal bleeding. For ostensibly normal individuals, Green *et al.* (216) estimated 0.5 mg fecal iron loss per day, 0.1 mg in desquamated cells and 0.4 from blood. Postabsorptive (i.e., >2 weeks following dose of radioactive or stable isotope) fecal loss of iron has been observed in humans (57,295). The loss was closely correlated with iron stores (57) or inversely correlated with absorption (295). Refsum and Schreiner (482) described iron-excreting goblet cells in duodenal and jejunal biopsy material. These latter observations may represent a regulatory mechanism for control of body iron balance.

The mean urinary excretion of iron by normal adult men and women was reported many years ago to be 0.2–0.3 mg/day (35). Only about half this amount has been observed in young New Zealand women (490). In two adults studied by Man and Wadsworth (357), the amount of urinary iron was directly influenced by the level of dietary iron, but did not exceed the amounts noted above. Urinary iron losses can be increased significantly to as much as 10 mg/day, by injection or infusion of chelating agents such as desferrioxamine alone or concomitantly with ascorbic acid in the treatment of iron overload (111,179). Increased urinary iron losses may also occur in nephrotic syndrome (609), and streptozotocin-diabetic rats exhibit increased urinary iron loss (320). A large fraction of chelated iron injected parenterally was excreted in the urine of rats if the binding constant of the chelate equaled or exceeded that of iron transferrin (23).

In addition to the iron excreted in the urine and feces, there is a continual dermal loss in the sweat, hair, and nails. Most of this occurs in desquamated cells, but cell-free sweat contains some iron. In one investigation an average iron content of 0.3 μg/ml was found in sweat low in cells and 7.1 μg/ml for sweat high in cells (4). Hussain and Patwardhan (277) collected the sweat from the forearm of healthy Indian men and women and from iron-deficient anemic women. The "cell-rich" sweat of the healthy individuals averaged 1.15 and 1.61 μg iron per milliliter compared with 0.34 and 0.44 μg/ml for the "cell-free" sweat of men and women, respectively. In the anemic women the average iron content of the cell-rich sweat was 0.44 μg/ml and the cell-free fraction had no detectable iron. The total amount of iron lost daily in the sweat will depend on the individual, the ambient temperature, and the cell content. Losses as high as 6.5 mg/day have been estimated in some circumstances (4,188). The average loss of iron through the skin of a healthy adult has been assessed as about 0.5 mg/day (395). Sweat collected from exercising athletes by Paulev et al. (453) contained less than 0.3 μg/ml, but daily losses were estimated at up to 1 mg of iron. Iron lost by this route could be much greater in the tropics where the volume of sweat can be as much as 5 liters per day. Such losses have been proposed as a factor contributing to the high incidence of iron deficiency anemia in tropical areas (188).

The total quantity of body iron lost in the urine, feces, and sweat amounts to 0.6–1.0 mg/day in most individuals. Bothwell and Finch (67) estimate these basal losses of iron as 14 μg/kg/day, which is close to 0.8–1.0 mg/day for women and men of average size. A loss of this magnitude is appreciable when it is realized that the average amount of iron absorbed from ordinary mixed diets is only 1.0–1.5 mg/day and may be much lower for some diets (239). In women the problem of iron balance is more precarious because they are subject to regular additional losses in the menstrual blood, from the menarche to the menopause, apart from the loss from time to time with the newborn infant and its adnexa and in the milk during lactation. Despite the low iron content of milk, lactation is of some significance with respect to iron loss. If the average iron concentration in human milk is taken as 0.5 μg/ml and 800 ml are secreted daily, the loss of iron would amount to 0.4 mg/day. In some women the loss of iron in the milk would, of course, be much greater.

Menstrual blood losses are extremely variable between women, with a much smaller variation from period to period in the same woman (106,415,501). Average blood losses of 35–70 ml containing 16–32 mg of iron have been reported (297). In a study of 12 young women over 12 months, the mean individual iron losses ranged from 4 to 26 mg per period (240) and in adolescent girls ranged from 2 to 59 mg/day (221). Higher losses have been reported in other investigations (207), with 60–80 ml of blood (27–37 mg iron) regarded as the upper limit of normal loss (501). It is evident that most women between puberty and the menopause lose 0.5–0.8mg/day of iron as a consequence of the menstrual flow, and some lose considerably more.

D. Physiological Chemistry of Iron

A high proportion of absorbed iron is continuously redistributed throughout the body in several metabolic circuits of which the cycle of plasma → erythroid marrow → red cell → senescent red cell → plasma, is quantitatively the most important. Subsidiary metabolic circuits exist, including the cycles, plasma–ferritin and hemosiderin–plasma, and plasma–myoglobin and iron-containing enzymes–plasma. The iron of the plasma (transferrin) provides the link between the cycles and regulates the distribution of iron in the body.

The hemoglobin cycle dominates the physiological chemistry of iron. Normally more than 70% of plasma iron turnover goes to the bone marrow (65). In the pregnant animal near term a relatively large proportion of plasma iron turnover may also go to the placenta (67). Some 21–24 mg of endogenous iron is liberated daily from the destruction of hemoglobin, if the survival period of hemoglobin is taken as the same as that of the erythrocyte in humans, namely, 120–125 days. This far transcends the amount of iron absorbed daily from ordinary diets and indicates the magnitude of the hemoglobin cycle. The removal of hemoglobin or nonviable erythrocytes, the breakdown of the heme moiety and release of iron, and the return of this iron to the plasma is performed by RE cells of the liver, spleen, and bone marrow. The recircuiting of iron from senescent red cells is an important link in internal iron exchange. After an initial processing period within the RE cell, the iron is either rapidly returned to the circulation ($t_{1/2}$ 34 min) or transferred to a slowly exchanging pool of storage iron within the RE cell ($t_{1/2}$ for release to plasma, 7 days) (180,341).

In adult humans 25–40 mg of total iron are transported in the plasma every 24 hr, even though only 3–4 mg are present in the whole plasma volume at any one time. This iron has a rapid turnover rate with a normal half-time of only 90–100 min (465). The large and rapid flow of plasma iron to the bone marrow is reflected in the completeness and promptness with which tracer doses of iron are used for hemoglobin synthesis. Tagged hemoglobin can be identified in the peripheral blood within 4–8 hr of administration of tracer iron. Within 7–14 days, 70–100% of the isotope is found in circulating hemoglobin (404). The above two paragraphs constitute a very brief summary of ferrokinetics (65).

During the process of incorporation of iron into hemoglobin in the reticulocyte, the first step is the binding of the transferrin in the cells (294,418). Reticulocytes, but not mature erythrocytes, bind iron transferrin to the surface and return apotransferrin to the medium (294,416). Most of this iron is incorporated into heme in the mitochondria (506). Transferrin molecules move into reticulocytes by a two-stage process involving first binding at membrane receptor sites and then movement into the reticulocyte by a slower temperature-dependent process, the mechanism of uptake being endocytosis (19). Fielding and Speyer (178,545) have obtained several iron-binding components from human reticulocytes by chromatographic fractionation and investigated their role in intracellular iron transport. Their results suggest that iron moves initially from iron transferrin receptor complex B2 (MW 230,000) to membrane components B1 (MW$\sim 10^6$), from which it may diverge into membrane components of high particle size, or follow the main pathway through cytosol component C to hemoglobin.

The rapid process of incorporation of plasma iron into ferritin in the cells of the liver, spleen, and bone marrow is dependent on energy-yielding reactions for the continued synthesis of ATP which, together with ascorbic acid, reduced the ferric iron of transferrin to the ferrous state, thus releasing it from its bond to protein and rendering it available for incorporation into ferritin (368–370). It has been claimed that the reverse process, the release of iron from hepatic ferritin to the plasma, is mediated by the molybdenum and iron-containing enzyme xanthine oxidase acting as a dehydrogenase (369,370,446). However Osaki and Sirivech (448) found that in liver homogenates xanthine oxidase substrates did not produce a significant release of iron from ferritin, nor did milk xanthine oxidase or chicken liver xanthine dehydrogenase. Investigations with horse spleen ferritin have shown that only the reduced riboflavin and riboflavin derivatives can reduce ferritin Fe(III) at a rate and to an extent that is likely to be significant physiologically (533). In the reduced state thus produced, the iron of ferritin dissociates from its bond to protein and is accepted by transferrin.

Mobilization of iron from iron stores also requires the presence of the copper-containing enzyme of the plasma ceruloplasmin (ferroxidase I), as discussed in Chapter 10. The demonstration of the iron oxidase activity of ceruloplasmin (446) led to the suggestion that this enzyme may be important in normal iron metabolism. This hypothesis is supported by *in vivo* experimental evidence (494) and by studies carried out by Osaki and co-workers (447) with perfused porcine and canine livers. These workers concluded that ''ferroxidase activity results in the substantial elimination of Fe(II), generating a maximum concentration gradient from the iron stores to the capillary system, thus promoting a rapid iron efflux from the iron storage cells of the perfused liver.'' The redistribution of body iron involving increased movement of iron to the liver from the serum that occurs in acute and chronic infections was discussed earlier in Section I,B,2.

The iron of plasma is deposited in the liver and spleen as freely as ferritin or hemosiderin and is released just as readily from both compounds for utilization by the tissues. This occurs over a wide range of rates of iron deposition and depletion. Examination of subcellular distribution of radioactivity showed that ^{59}Fe from labeled transferrin moved directly into hepatic ferritin regardless of iron status in rats and at no time was associated with lysosomes or mitochondria (607). Shoden and Sturgeon (527) showed that the iron of transferrin is not incorporated directly into hemosiderin in rabbits. In this species ferritin has first to be formed, so that liver ferritin may be regarded as the immediate precursor of liver parenchymal cell hemosiderin.

The central role which transferrin holds in iron transport has stimulated research on the transferrin receptor and the mechanism of cellular iron exchange (6,124,439,441,568,621). The transcellular processes mentioned above appear to be mediated by the transferrin receptor, but mechanistic details are not entirely clear. Receptors are apparently present in all cell membranes except mature erythrocytes and function to provide iron to the cell, an obligatory need for every cell.

Placental iron transport is unidirectional (69) and increases rapidly as pregnancy progresses, with the plasma iron turnover of the fetus in late pregnancy being much greater than that of the newborn (378,379). In animals with the hemochorial type of placenta, including the rat, rabbit, guinea pig, and human, the rate of iron transfer across the placenta from maternal plasma transferrin is sufficient to account for all iron accumulated by the fetus (65). Wong and Morgan (611) have shown that this is an active process dependent on cellular metabolism in which maternal plasma transferrin is taken up by the placenta, followed by removal of iron from the transferrin and transfer to the fetal blood. Subtracting the iron that had dissociated and bound to transferrin revealed that almost none of the iron of injected [^{59}Fe]iron dextran transferred to the fetus of near-term rats (557). Intestinal iron absorption increased during the course of pregnancy in guinea pigs to the extent that the absorption rate equaled placental transfer at day 55 (584). Fetal uptake of ^{59}Fe from radiolabeled transferrin was depressed by injected nicotine at doses relevant to the amount consumed by heavy smokers (198). A progesterone-inducible purple protein possessing phosphatase activity, uteroferrin, has been isolated from porcine placental tissue and is proposed to have a role in transfer of iron to the fetal piglet (81,154). In animals with endotheliochorial types of placentas, such as the cat and the dog, the rate of iron transfer from maternal plasma to fetus is much less than that required for fetal needs (29,513). In the cat, for example, transfer from maternal plasma is insignificant, and the major source of fetal iron is the maternal erythro-cytes (597). This process probably occurs, as explained by Wong and Morgan (612), "by extravasation of maternal erythrocytes into the uterine lumen, fol-lowed by their phagocytosis and digestion by paraplacental chorionic epithelial cells, and transfer of the iron released from hemoglobin to fetal capillaries in the chorionic membrane and thence to the fetuses." Since each milliliter of blood contains 300–500 times as much hemoglobin iron as transferrin iron, this is clearly an effective and economical transfer process.

III. IRON DEFICIENCY

A. Manifestations of Iron Deficiency

Iron deficiency in human adults is manifested clinically by listlessness and fatigue, palpitations on exertion, and sometimes by a sore tongue, angular stom-atitis, dysphagia, and koilonychia. In children anorexia, depressed growth, and decreased resistance to infection are commonly observed as in young, growing, iron-deficient animals, but the oral lesions and nail changes are rare. A significant reduction in physical activity and performance by rats and humans (157,158,206,442), resistance to infection with *Salmonella typhimurium* (22) in iron-deficient anemic rats, and prolonged cardiorespiratory recovery periods after exercise in anemic women (16) have been observed.

Prolonged iron deficiency results in the development of an anemia of the hypochromic microcytic type accompanied by a normoblastic, hyperplastic bone marrow containing little or no hemosiderin. Serum iron and serum ferritin levels are subnormal, TIBC is above normal, and there is a decreased saturation of transferrin. Studies of patients with iron deficiency anemia have shown that 16% saturation of plasma transferrin or less implies an inadequate supply of iron to the erythroid tissue and is associated in time with hypochromic, microcytic anemia (27). The percentage saturation of transferrin is thus a good criterion of iron-deficient erythropoiesis. The value of serum ferritin levels as an indicator of body iron stores has been discussed (Section I,B,3). However, the rat develops decreased hematocrit before iron stores in the liver and spleen are exhausted (131), and in conditions such as inflammation, serum ferritin values may be abnormally elevated. In addition, considerable overlap of values for iron status indices may occur in humans (130), and anemia may be caused by other nutrient deficiencies. Therefore, a series of measurements must be applied to determine definitively iron deficiency in an individual (113,132,181,619).

Lipid abnormalities with elevated serum triglyceride levels associated with iron deficiency of nutritional origin have been observed in rats (224,337) and chicks (12). The accumulated lipid may be a reflection of increased rate of synthesis or a decrease in the rate of clearing. The latter seemed probable in light of the observation by Lewis and Iammarino (337) that serum and tissue lipoprotein lipase activity is decreased in iron-deficient rats. However, Sherman *et al.* (517,521) detected greater synthesis of lipids from deuterated water and glucose in the adipose tissue of both iron-deficient pups and adults. Tissue carnitine level is decreased in iron-deficient rat pups, and Bartholmey and Sherman (36) propose that the consequent alteration in lipid metabolism may cause the hyperlipidemia. Decreased levels of certain unsaturated fatty acids occur in tissues of iron-deficient pups (476,520), which did not correlate with milk lipids but may be related to decreased activity of stearoyl CoA-desaturase (475). The hyperlipidemia of iron-deficient rats appears to be reversible with iron supplementation (518). Blood lipid alterations may occur in several types of human anemias not limited to iron deficiency, but no consistent trend is apparent as with rats. Ohira *et al.* could determine no relationship between serum triglyceride and hemoglobin levels in 97 Sri Lankan subjects, the majority being women (443).

Low serum folic acid levels may be observed in iron deficiency anemia (489,567) which can be restored to normal by iron therapy. The decrease in serum folic acid is not due to reduced absorption and is apparently secondary to increased requirements for this vitamin, as a consequence of a significant decrease in red cell survival and a nearly 3-fold increase in heme catabolism in the iron-deficient animals (567). Decreased red cell survival (275,323) and ineffective erythropoiesis (80,491) have both been reported in iron deficiency. The stress of lactation superimposed on iron deficiency may cause lowered folate levels in milk and result in decreased liver folate activity in 18-day-old deficient rat pups (312).

Abnormalities of the gastrointestinal tract, including gastric achlorhydria and varying degrees of gastritis and atrophy, can occur in iron deficiency anemia (257,329). A high incidence of blood loss and loss of plasma proteins into the intestinal lumen (608) and a diffuse and reversible enteropathy (432) have been reported in infants and children with iron deficiency. Gastric achlorhydria and impaired absorption of xylose and vitamin A, together with duodenitis and mucosal atrophy, were observed. Most of the abnormalities returned to normal following treatment with iron and were generally absent from children with anemias not due to iron deficiency.

Muscle myoglobin, long believed to remain inviolate, can be reduced in iron deficiency. A profound depression in myoglobin concentration has been observed in young iron-deficient piglets and puppies (223), chicks (140), and rats (127,129). Young animals are more susceptible to loss of myoglobin than mature ones (223), and skeletal muscles appear to be more susceptible to this loss than cardiac or diaphragmatic muscle (127,226). Hargreaves *et al.* detected significantly lower myoglobin concentration in the serum of iron-deficient compared to nondeficient children (255).

A reduction in the levels and activities of the heme enzymes in the blood and tissues of iron-deficient animals is well documented. Beutler (49,50) was the first to show clearly that the body does not necessarily accord priority to the heme enzymes over hemoglobin under conditions of restricted iron supply. In iron-deficient rats a marked decrease in cytochrome c was found in the liver and kidneys, in succinic dehydrogenase activity in the cardiac muscle and kidneys but not in the liver, and in cytochrome oxidase activity in the kidneys but not in the heart. Dallman (129) showed that cytochrome c can be reduced to as low as half the normal concentrations in the skeletal muscles, heart, liver, kidney, and intestinal mucosa of young iron-deficient rats, with a smaller reduction in the brain. Grassmann and Kirchgessner (214,215) later reported a marked reduction in the catalase activity of the blood of iron-deficient rats, piglets, and calves. The reduction in enzyme activity was greater than the fall in hemoglobin, indicating that, during the development of the anemia, priority for iron use was given to hemoglobin synthesis over catalase synthesis. The reduced aerobic metabolic capacity of homogenates and subcellular organelles from muscle of iron-deficient rats and mice (136,354,377,444) is likely related to reduced oxidoreductase and dehydrogenase enzyme activity of skeletal muscle mitochondria (2,445). Iron-deficient rats exposed to cold became hypothermic and expressed altered triiodothyronine metabolism (101,149). The hypothermia responds to correction of the anemia by transfusion and is apparently not related to the changes in thyroid hormone metabolism (41). Thyroidectomized rats have increased liver ferritin levels (610). The full consequence of the interrelationship of anemia, iron deficiency, and neuroendocrine function, particularly in humans, is not clear.

The large body of literature relating iron and microbial growth, particularly infectious organisms, is reviewed by Weinberg (594), Neilands (438), and Barclay (34). The hypoferremia of inflammation and infection, thought to be initiated by LEM (468), may be a primary defense mechanism against infection. The concept that low iron stores and the resultant low transferrin saturation are desirable to prevent infection is contraindicated, however, as other immune functions are depressed in iron deficiency (45,519). Srikantia et al. (547) noted impaired cell-mediated immune response and bactericidal activity of leukocytes in children when hemoglobin fell to 10 g/dl or less. Iron-deficient rat pups exhibit impairment of antibody formation (311), and this may be related to inadequate protein synthesis (495,598). Iron availability is postulated to affect the successful establishment of neoplastic disease (335,595). The role of iron and iron status in infection and immunity continues an active research field. The multinutrient aspects of clinically evident nutritional deficiency syndromes renders the singular role of iron in immunity difficult to ascertain (98).

B. Iron Deficiency in Infants, Children, and Adolescents

Iron deficiency anemia can arise in the rapidly growing suckling animal as a consequence of inadequate amounts of storage iron at birth and low iron concentrations in maternal milk. Even in the rat, whose milk is richer in iron than that of women and most other species, such an anemia develops (171). The anemia of human infants, which occurs most frequently between 4 and 24 months of age, is an expression of iron deficiency, since it usually responds to iron and is characterized by the typical abnormalities of erythrocyte morphology (551). A depletion of iron reserves occurs during the period of rapid growth of the infant, despite the considerable "store" of iron contained in the plethora of hemoglobin in the blood at birth. The blood of the newborn child contains 18–19 g/dl of hemoglobin, which, if the blood volume is taken as 3 dl, represents 180–190 mg iron or about six times the amount usually stored in the liver (374). At 6 months of age the average baby weighs 7 kg and has a blood hemoglobin level of about 12 g/dl. It therefore contains about 280 mg iron in the form of hemoglobin, of which about 2/3 was present at birth. If this iron plus the amount stored in the liver and absorbed from the milk were all retained in the body, it would be difficult to visualize iron deficiency arising at all, except in premature, low birth weight infants with low body stores. However, Cavell and Widdowson (6), in a study of babies at 1 week of age, found significant negative iron balances due to the excretion of approximately 10 times as much iron in the feces as was ingested daily in the breast milk. If such excess iron excretion were to continue beyond the early neonatal period, it could be an important cause of iron depletion.

In several studies of infants in the United States an incidence of iron deficiency anemia has been reported from less than 5% (191) to as high as 64% (551). Some of this variation undoubtedly stems from differences in the socioeconomic groups under investigation and the criteria of anemia employed. Including several criteria of iron status improves the accuracy of predicting hematological response of infants to iron therapy over the measurement of hemoglobin level only (481). Adjustments in laboratory diagnostic values for age, but not race, are important in accurately establishing iron deficiency anemia (480,618). Estimates in a 1984 study give an overall 5.7% incidence of anemia among infants in the United States, the majority of which is iron-deficiency anemia (135). Serum ferritin values correlate well with iron nutrition within groups of infants and with known developmental changes in iron stores, but should not be relied on as the sole criterion of iron deficiency (146,225,503).

Some of the information on iron absorption from infant formulas was obtained using infants (488), but much is derived from studies with adults (145,203,411). Different results using infant, compared to juvenile monkeys (346), suggest that adult humans may not be an appropriate model to study iron absorption from infant formulas.

Several workers have demonstrated significant increases in hemoglobin levels in infants receiving oral iron or intramuscular injections of iron dextran (534). Sturgeon (551) stressed regular need for supplementation with iron or with iron-fortified foods, even with infants born of nonanemic mothers who have received ample iron during pregnancy. However, in the study of Fuerth (191), none of the objective criteria revealed any difference between the iron-supplemented group and the children receiving the placebo. Woodruff *et al.* found no difference in the incidence of mild anemia and biochemical iron deficiency in breast-fed infants compared to formula-fed infants consuming more iron-containing cereals and foods (613). In a study by Burman (84), an iron supplement of 10 mg/day iron in the form of colloidal ferric hydroxide raised the hemoglobin levels of infants of lower birth weight, presumably due to their lower body iron stores, but made no difference to the incidence of infection. This worker contends that efforts to raise hemoglobin levels should not be made unless it can be shown that maximal hemoglobin is beneficial to infants. He maintains further that "levels above 9.5 or 10 g Hb/dl should be considered nonpathological not only at 6–8 weeks of age but also for the remainder of the first 2 years." This concept is not universally accepted, however. One report indicated deficits in cognitive indices for infants no older than 24 months with hemoglobin values less than 10.5 g/dl and two of three biochemical measurements indicating iron deficiency (348). Despite criticisms of methodology used in this latter study (21), the question of cognitive deficits in iron-deficient infants and children merits further investigation.

The nonheme iron in brains of rats reared to 28 or 48 days of age on an iron-deficient regimen was 27% and 22% below the value for control rats. A deficit persisted into the adult animal, even though iron intake was adequate following the period of iron deficiency (133).

The incidence of iron deficiency in preadolescents also varies considerably with geographic and socioeconomic factors. Of 270 children in the Faroe Islands, 4–13 years old, about 1% were found to have latent iron deficiency and none exhibited iron deficiency anemia (391). The incidence of iron deficiency anemia was nearly 10% in a group of Indonesian children 2–12 years old (575). The differential role of diet and parasitism in causing iron deficiency is not known, but parasitism may be a major factor in many geographic locations (148,279). Improvement in iron status indicators may occur as result of changes in general health care and overall nutritional status (390,456). Hematological indices of vegetarian children suggested mild iron deficiency (155), but definitive ascertainment could not be accomplished in absence of serum ferritin measurements. The use of several biochemical measurements now available rather than anemia per se, must be used to assess iron deficiency correctly (265).

Menarche marks the beginning of menstrual blood losses in the adolescent female, and the incidence of iron deficiency becomes greater than for adolescent males. In one study, the mean serum ferritin for pre- and postmenarchal girls was 34.5 versus 25.8 ng/ml, while the boys had a mean value of 38.9 ng/ml (24). The same investigators reported slightly greater than 10% incidence of anemia and biochemical indices indicative of iron deficiency in adolescent girls from an urban low-income population, about three times the incidence in boys from the same population group (26). The estimated incidence of anemia for females ages

15–17 years in the United States is 5.9% and less than 3% for males (135), most of which is considered iron deficiency anemia. Only one case of iron-deficient erythropoiesis and no iron deficiency anemia was detected in 990 male adolescents in Singapore (563).

C. Iron Deficiency in Human Adults

In adult men and postmenopausal women, a principal contributing factor in iron deficiency is chronic blood loss due to infections, malignancy, bleeding ulcers, and parasitism. However, anemia of dietary origin is found among institutionalized elderly (338,380). Because of their iron losses in menstruation, pregnancy, parturition, and lactation, as discussed earlier, iron deficiency is much more common in women during their fertile years than in men (31,87,130,306).

Surveys have revealed a widely varying incidence of low body iron stores (31) and anemia in women, the latter attributed to iron deficiency (87,306,380,501). In developing countries, where the population relies heavily on vegetable foods and where infections and excessive sweating are common, the incidence of iron deficiency anemia is generally higher than in the more industrialized and temperate climate areas of the world. About two decades ago an incidence as high as 50% had been reported in India (426,587), and 20–25% in Sweden (501) or 10–20% in England (87,306). Later data include 44–50% of Indian women living in Durban, South Africa under 45 years of age with serum ferritin <12 ng/ml (356), 35–40% of 172 mature Senegalese women with indicators of poor iron status (498), and 50% of female Sri Lankan tea workers found to be anemic (514). In comparison, <6% of women in Göteborg, Sweden, were estimated to be anemic (333), and an overall incidence of <6% is estimated in women of the United States between 15 and 44 years of age (135). Not all anemia is due to iron deficiency (384), but iron deficiency anemia is of much lower incidence in the latter two populations. On the other hand, there may be low iron stores without anemia (113,183,543), and 18% of women in a British study (561) and 30% in a U.S. study (385) had serum ferritin levels <10 ng/ml. Although serum ferritin was not determined, a group of vegetarian women in Canada consuming adequate protein and energy seemed at no greater risk of poor iron status than their cohort consuming omnivore diets (15).

Premenopausal women escape menstruation only by pregnancy. From the standpoint of evading iron loss, this can be an unprofitable exercise because some 350–450 mg iron are lost in the fetus and its adnexa, compared with 200–300 mg yearly in the menstrual flow. Some compensation for the iron demands of pregnancy is achieved by increased absorptive efficiency, but this is insufficient to prevent signs of iron deficiency in many women in late pregnancy. In the third trimester of pregnancy, hemoglobin levels normally fall due mainly to a rise in plasma volume greater than the rise in red cell volume (see Section I,B,1).

Inadequate intakes of iron or intakes of low bioavailability can be a contributory cause, and supplementary iron often increases blood hemoglobin levels (332,413,552). The intramuscular injection of a single dose of 1000 mg or multiple injections of 100 mg of iron dextran is effective for this purpose (332,470). The therapeutic value of iron preparations containing ascorbic acid, which potentiates the absorption of ferrous sulfate, has been stressed (375), while ferric fructose may be particularly effective in the treatment of anemia in pregnant women (156). The greatest percentage of response is obtained with multinutrient treatment (544) or an overall nutrition intervention program (25).

While there is no doubt that iron deficiency anemia and low body iron stores are common in women in many populations and that an increase in hemoglobin is usually obtained in response to appropriate iron therapy, it is less certain that such increases, in cases of mild anemia, confer significant benefits in terms of improved well-being or activity, or lower morbidity or mortality (87,167,558). However, when the iron status is very poor, particularly in instances of parasitism and/or overall suboptimal nutrition, correction by appropriate intervention is beneficial (25,319,470,514).

D. Iron Deficiency in Pigs

Iron deficiency anemia occurs in baby pigs and in older animals fed rations very high in copper to promote growth. Piglets denied access to sources of iron other than sow's milk develop anemia within 2–4 weeks of birth. Blood hemoglobin levels fall from a normal of about 10 g/dl to as low as 4 g/100 dl. Breathing becomes labored and spasmodic (hence the name "thumps"), appetite declines and growth is poor, or the piglets lose weight and some die. Surviving piglets begin a slow spontaneous recovery at 6–7 weeks, when they begin to eat the sow's food and undertake such foraging as is permitted by their conditions of housing. The condition is usually an uncomplicated iron deficiency responsive to iron (588) and with no significant additional response from copper (263,309) or vitamin B12 (309).

The baby pig is particularly susceptible to iron deficiency because of its high growth rate and poor endowment with iron at birth. Piglets normally reach four to five times their birth weight at the end of 3 weeks and eight times their birth weight by 8 weeks. A growth rate of this magnitude imposes iron demands much greater than can be supplied by the sow's milk. At birth the pig has relatively low concentrations of total-body iron and liver iron stores (Table I) and has no polycythemia of the newborn. These sources of endogenous iron are therefore denied to the baby pig, and anemia is inevitable unless the piglet has access to additional iron.

Feeding supplementary iron to the sow before or after farrowing is ineffective (309), because such treatment does not significantly increase the iron stores of the piglet at birth (263,588), or the iron content of the sow's milk (467). Successful treatment involves direct increase in the iron intake of the piglets. Oral or parenteral administration of iron to the piglets within the first few days of life is a routine practice, where pigs are maintained free from soil and grass. Injection of such compounds as iron dextran, or dextrin ferric oxide (Table VI) is effective in promoting maximum hemoglobin levels (339,450,466).

When an iron complex containing 100 mg iron is injected at 2–4 days, a second injection within 2 weeks is necessary to increase hemoglobin levels at 3–4 weeks of age (339,450). Similarly, oral administration of iron tablets or solutions containing 300–400 mg iron within 4 days of birth promotes growth and prevents mortality, but a second such dose is required for maximum hemoglobin levels. A single oral dose of 150 mg iron given as iron tartrate on the third day (18), or 315 mg iron as iron dextran tablets on the fourth day (309), maintained the hemoglobin level of the piglets at >10 g/dl for a period of 15–17 days following administration. Ku et al. found no significant changes in serum electrolyte concentrations from intramuscular injections of up to 200 mg iron as iron dextran in 4-day-old nursing pigs (316).

Anemia due to iron deficiency can occur in older pigs fed rations very high in copper to promote growth and increase the efficiency of feed use, unless these rations are supplemented with iron at levels well above those that are otherwise adequate (83,204,259). Gipp and co-workers (204) showed that the hypochromic, microcytic anemia induced by high dietary copper is due to an impairment of iron absorption and that this impairment is ameliorated by ascorbic acid. Dietary copper levels up to 60 μg/g had no effect on liver iron concentration, but 120 μg/g copper resulted in 50% decrease in liver iron (72).

Piglet anemia is often associated with *Escherichia coli* infection, inducing piglet edema disease. Decreased resistance to the endotoxin of *E. coli* has been

Table VI
Treatment of Piglet Anemia by Injection of Iron Complexes[a]

Treatment	No. of pigs	Hemoglobin (g)		Body weight (1b)	
		0	3 weeks	0	3 weeks
Untreated	24	7.5	4.8	3.8	9.6
Iron dextran at 2–4 days (100 mg iron)	21	7.5	9.0	3.6	11.3
Dextrin ferric oxide at 2–4 days (100 mg iron)	22	8.0	7.8	3.6	11.9
Dextrin ferric oxide at 2 and 14 days (100 mg iron each dose)	16	8.3	10.5	3.8	11.5

[a] From Linkenheimer et al. (339).

demonstrated in anemic piglets (270). On the other hand, enhanced susceptibility to experimentally induced bacterial infection has been demonstrated in pigs given an injection of iron (184,301,310). Significance of these contrasting observations to practical swine production practice remains to be fully ascertained.

E. Iron Deficiency in Lambs and Calves

There is no convincing evidence that iron deficiency ever occurs in sheep or cattle grazing under natural conditions, except in circumstances involving blood loss or disturbance in iron metabolism as a consequence of parasitic infestation or disease. Heavy infestation with helminth intestinal parasites results in an iron deficiency type of anemia in lambs and calves (90,485). Iron deficiency anemia also occurs in young calves reared for veal on milk-based rations and in lambs similarly raised (1,48). Intramuscular injections of 150 mg iron as iron dextran into newborn lambs, or iron at 12 mg/lb body weight into newborn calves, can produce significant improvements in hemoglobin levels and some improvement in body weight over several weeks (92,484). Subcutaneous injections in calves of 1000 mg of iron the first week postpartum produced increases in hemoglobin concentration that persisted until 12 weeks postpartum, but noninjected controls did not become anemic and no difference in weight gain was observed (479). Oral iron supplements can be equally effective. Thus Bremner and Dalgarno (74) found that adding iron at the rate of 30 μg/kg to a fat-supplemented skim milk ration improved the hematological status of calves provided that the iron was supplied as ferrous sulfate, ferric citrate, or ferric EDTA. The iron in soy flour or fish concentrate fed to veal calves was much less bioavailable than ferrous sulfate (586). One study with calves found high serum iron and low-binding capacity to be positively correlated with adequate defense against *Salmonella* infection (586).

IV. IRON REQUIREMENTS

A. Humans

The demands of the body for iron are greatest during three periods: the first 2 years of life, the period of rapid growth and hemoglobin increase of adolescence, and throughout the childbearing period in women. The physiological requirement of the infant for iron averages approximately 0.6 mg/day, but may be as high as 1 mg/day during the first year of life according to Josephs (300). At puberty and the following few years the requirement increases temporarily in males due to the rapid growth and hemoglobin accretion that occurs during this time. An estimate of a 13 μg/kg iron requirement by adults is based primarily on long-term measurements of whole-body radioiron loss (65,216) and amounts to about 1 mg daily for a 70-kg man and 0.8 mg for a 55-kg woman (233). To provide the iron needs of 95% of normal menstruating women, enough iron must be consumed to permit absorption of 2 mg/day (402). The absorbability of food iron must be then taken into account to estimate the amount of dietary iron to provide the physiological requirement. Recommended daily dietary allowances for iron given by the U.S. National Academy of Sciences in 1980 are shown in Table VII (434).

Conversion of the physiological requirements for iron into dietary requirements is made difficult by variations among individuals in absorptive capacity and among foods and various combinations of foods in the bioavailability of the iron for absorption. The position is further complicated by the ability of the body to increase iron absorption in iron deficiency. Normal subjects may absorb 5–10% of the iron of mixed diets, and iron-deficient individuals 15–20% or more of this iron, but considerable divergence from these values can occur. Numerous workers have shown that iron absorption is significantly lower, and dietary iron requirement therefore higher, in vegetal diets than in foods from animal sources, mixed diets, and iron salts (239,278,325). This is particularly apparent from a study of Venezuelan diets (325) in which absorption of 3–4 mg of vegetal iron was found to be increased about twice by the addition of 50 g of meat, about three times by 100 g fish, and about five times by 150 g papaya containing 66 mg ascorbic acid. The ingredients of meals and diets clearly affect dietary iron

Table VII
Recommended Daily Allowance for Iron[a]

	Age (years)	Weight (kg)	Amount (mg)
Infants	0.0–0.5	6	10
	0.5–1.0	9	15
Children	1–3	13	15
	4–6	20	10
	7–10	28	10
Males	11–14	45	18
	15–18	66	18
	19–51+	67–70	10
Females	11–50	44–55	18
	51+	55	10
Pregnant			18+[b]
Lactating			18

[a] From National Academy of Sciences, U.S.A. (434).
[b] This increased requirement cannot be met by habitual diets. The use of supplemental iron is therefore recommended.

requirements profoundly through their effects on absorption. This aspect was discussed in Section II and is considered further in Section V.

B. Pigs

The minimum iron requirements of pigs for satisfactory growth and hemoglobin formation cannot be given with precision. Braude and co-workers (73) estimated that the piglet must retain 21 mg iron per kilogram of body weight increase to maintain a satisfactory level of iron in the body. This is equivalent to less than the 7–11 mg iron daily for this purpose obtained by Venn *et al.* in an earlier study (588). The dietary concentrations of iron required to supply these quantities vary with the percentage absorption of the iron in the diet and with the levels of other elements in the diet, such as copper, zinc, and manganese, which affect this absorption. Ullrey and co-workers (571) found 125 μg iron per gram of dietary solids to be necessary for full growth and hemoglobin production in baby pigs, while the results of Matrone *et al.* (366) indicate that 60 ppm is adequate. Later work by Miller *et al.* (386) led them to conclude that both germ-free and conventional pigs fed a condensed-milk diet require 50–100 μg iron per gram of solids to maintain desirable hematological traits for 4 weeks. Nursing pigs dosed with 8 mg iron orally as ferric citrate incorporated into the red blood cells a mean of 59% of the doses as early as 3 days of age (193). Serum ferritin concentration may be used to monitor iron stores in growing pigs (194,539). The marked increase in the iron requirements of older pigs brought about by feeding high-copper rations is discussed in Chapter 10.

C. Poultry

The iron requirements of chicks during the first 4 weeks of life have been estimated as 75–80 μg/g of the diet (140). These estimates were based on growth, blood data, myoglobin levels, liver iron stores, and succinic dehydrogenase levels, and are higher than the 50 μg/g iron suggested earlier as adequate for chicks by Hill and Matrone (267). The U.S. National Research Council recommends 80 μg/g in the diet of chicks through 8 weeks of age, then 40 μg/g to 18 weeks of age (435). Adequacy of dietary protein influences utilization of dietary iron by chicks, a low-protein diet leading to decreased absorption and excessive accumulation of iron in liver and muscle, but decreased erythropoiesis (394). A dystrophic strain of New Hampshire chickens is thought to have a higher iron requirement than the normal birds of the New Hampshire breed (102). Ascorbic acid at 0.1% of the diet increased total-body iron retention by chicks fed diets considered marginal in iron (40 μg/g) but had no effect if the diet contained 100 μg/g of iron (393).

The demands of egg production for iron are large compared with those of the nonlaying mature bird. Since an average hen's egg contains about 1 mg iron, a heavy layer will lose close to 6 mg/week in the eggs alone, so that more than this needs to be absorbed from the diet. With the onset of laying the efficiency of iron absorption is increased and serum iron levels are raised. This is usually accompanied by a small fall in blood hemoglobin levels but no significant reduction in storage iron (473). Using a purified-type diet, Morck and Austic (408) determined the iron requirement for maintenance of hematocrit in white leghorn hens was 35–45 μg/g, but 55 μg/g were required for maximum hatchability of fertile eggs. The increased requirements for iron imposed by egg-laying are apparently met by the usual laying rations without the need for iron supplements.

D. Cattle and Sheep

Little is known of the iron requirements of adult sheep and cattle based on definitive experiments. In male calves raised on a milk diet containing 1 μg/g iron (dry basis), normal growth and hemoglobin levels were maintained for 40 weeks from birth with supplemental iron at either 30 mg or 60 mg daily (365). This suggests that the minimum iron requirement for calves is not more than 30 mg/day. A later study with male calves reared on fat-supplemented skim milk indicates that a dietary iron intake of 40 μg/g of dry diet is sufficient to prevent all but a very mild anemia, provided the supplemental iron is present in a soluble form (74). Suttle (554) estimated growing calves require 63 mg of iron per day to support growth rate of 1 kg carcass gain per day and that the requirement for maintenance is small relative to the requirement for growth. Experiments with growing–finishing lambs indicate that 10 μg/g iron is inadequate and that their minimum dietary iron requirements lie between 25 and 40 μg/g (321).

V. SOURCES OF IRON

A. Iron in Human Foods and Dietaries

The iron content of most foods may vary greatly from sample to sample, as a reflection of varietal differences and differences in the soil and climatic conditions under which the foods are grown or produced. The richest sources of total iron are the organ meats (liver and kidney), egg yolk, dried legumes, cocoa, cane molasses, and parsley. Poor sources include milk and milk products, unless contaminated in processing, white sugar, white flour and bread (unenriched), polished rice, sago, potatoes, and most fresh fruit. Foods of intermediate iron content are the muscle meats, fish and poultry, nuts, green vegetables, and whole-meal flour and bread. Boiling in water can reduce the levels of iron in vegetables by as much as 20% if the cooking water is discarded (535), while milling of wheat lowers the iron content in the resulting white flour. For example, in a study of North American wheats and the flours milled from those wheats, the mean iron content of the whole grain was 43 $\mu g/g$ and that of the flour 10.5 $\mu g/g$ (128). The effect of refining processes on the iron content of foods is further evident from a study of the concentration of various elements in different types of cane sugars (251). The iron concentrations reported were, for Barbados brown sugar, 49 $\mu g/g$ (dry basis); Demerara sugar, 8; refined sugar, 11; and granulated (white) sugar, 0.1.

Overall dietary iron intakes vary greatly with the total amounts of foods and beverages consumed and with the proportion of iron-rich and iron-poor foods that they contain. The degree of contamination with iron to which the foods and beverages are exposed in processing, storing, and cooking may also be important. Prabhavathi and Rao (469) surveyed market samples of cereals and legumes in India and concluded that 13–47% of the total iron might be surface contamination. Extremely high iron intakes are reported for Ethiopian women, part of which is contamination iron (200). Average U.S. dietaries were estimated 50 years ago to supply 15–22 mg iron per man daily (522), and Australian diets 20–22 mg iron per adult male daily (437). A 1984 estimate of mean iron intakes in the United States is about 14.6 and 10.2 mg/day for men and women, respectively, 15–64 years of age, from a U.S. Department of Agriculture food consumption survey (478). Daily intakes by Japanese adults as a whole have been estimated to be close to 19 mg (616). In a study of total daily intakes from composite English diets, a mean iron level of 23.2 $\pm$ 1.1 mg/day was obtained (251). This is more than twice the mean of 8.9 mg reported in 1985 as being consumed by young women in Britain (33). Lower total iron intakes are common from the diets high in milled rice or corn and low in meat, of underprivileged groups in many of the developing countries (239,258,325).

Typical Western diets may provide as much as 7 mg per 1000 calories, but the more acceptable value is 6 mg (380,478). The importance of total caloric intake as a determinant of iron intake is evident from several North American studies. Monsen and co-workers (402) found that the iron intakes of 13 young women averaged only 9.2 mg/day when consuming their normal diets. These diets were adequate in protein but provided only 1600 calories per day. In a 1984 report, the mean calculated daily intakes by adult men and women, respectively, over 1 year of self-chosen diets were 2760 and 1850 calories, providing 16.6 and 12 mg of iron, about 6 mg per 1000 calories (307). In the British study of young women previously cited (33), those "on a diet" were consuming 7.6 mg daily compared to 8.9 mg for those "not on a diet."

B. Efficacy of Iron in Human Dietaries

Bioavailability factors affecting utilization of dietary iron were discussed in Section II,A. The concept of inhibitors and enhancers of nonheme iron absorption was well established by the mid-1970s. In the fourth edition of this book, Underwood wrote:

> Great caution needs to be exercised in relating total dietary intakes of iron to deficiency or sufficiency of this element because of the marked differences in iron availability from different foods. The amounts of iron ingested, therefore, do not necessarily reflect the net amounts absorbed by individuals. The lower absorption of iron from most vegetable foods than from most foods of animal origin was mentioned earlier when considering iron absorption. The two types of food interact in the ingesta so that the inclusion of foods of animal origin raises the absorption of iron from foods of plant origin and vice versa. The better absorption of the iron of cereals and comparable foods when combined with foods of animal origin, other than egg (168), than when given alone is of great practical importance (239,325,328,360). The iron of egg yolk is very poorly utilized except when combined with sufficient ascorbic acid (422). Liver is a particularly valuable dietary component because of its high iron content, high absorbability, and enhancing effect on the absorption of vegetal iron (360). The inclusion of fruit juices rich in ascorbic acid to improve iron absorption from such diets, or from bread, can also be important (88,168,239). In fact the regular provision of some meat, fish, or fruit in the diet is of paramount importance to ensure a reasonable utilization of nonheme iron.

In a consensus effort to consolidate the then-known bioavailability factors into a useful tool, Monsen et al. (401) proposed in 1978, a model for estimating "absorbable iron" in a meal. Refinements have been made in some aspects of the original proposal (399), but the basic information needed for the estimation is the amount of heme and nonheme iron and the amounts of enhancing factors, ascorbic acid and MFP (see Section II,A) in a meal. No methodology is established for direct measurement of heme iron in foods. Therefore, heme iron is assumed to be 40% of the total iron present in MFP, but accuracy and applicability of this value is uncertain. Schricker et al. (510) measured total and nonheme iron in raw pork, lamb, and beef and, assuming the difference was heme iron, reported 49, 57, and 62%, respectively, of the total iron as heme iron. All food iron except the heme iron estimated in MFP is assumed to be nonheme iron. Dependent on the amount of the enhancing factors, milligrams of ascorbic acid, and/or grams of cooked MFP, an absorbability percentage is assigned to nonheme iron in the meal. The percentage absorption will be dependent on iron status (see Section II,A). Calculation for a reference individual with 500 mg of storage iron, for example, would assume 23% absorption of heme iron and 3–8% (logarithmic relationship) absorption of nonheme iron with 75 units of enhancing factors in the meal.

Using this model, Anderson et al. (15) estimated a daily mean of 0.68 mg bioavailable iron (range 0.4–1.1 mg) in diets of Canadian vegetarian women. Serum ferritin was not determined, but the majority of these individuals had normal indices of iron status (serum iron, TIBC, and transferrin), even though most did not meet the Canada Dietary Standards of 1.6 and 0.9 mg/day of absorbed iron for pre- and postmenopausal women. In a subsequent survey of 196 women, also Canadian, but not vegetarians (201), 73% of the pre- and 30% of postmenopausal women had less than the recommended requirement of bioavailable iron using the 500-mg stores, but only 41% of the premenopausal women failed the recommended standard if 250-mg stores was assumed. No biochemical data were given for the latter study, and correlation with iron status cannot be made. Calculation of bioavailable iron has been applied to data from two national surveys in the United States. The iron intake adjusted for bioavailability did not provide a better predictor of low hemoglobin in the "National Health and Nutrition Examination Survey 1971–74" than did total iron intake (436). Bioavailable iron values of intake data from the U.S. Department of Agriculture Nationwide Food Consumption Survey ranged from 5.9 to 7.7% (478) using the refined model of Monsen and Balintfy (399). At first thought, correlation of the absorbable-iron concept with iron status does not appear too promising. However, iron stores may correlate well with the estimated absorbable iron value. No valid measure of iron stores is available for the above studies. Also, although stores of 500 mg can be assumed when simply comparing the predictable efficacy of different dietaries, lower stores, 250 mg or possibly none for premenopausal women, should be assumed when attempting to correlate hematological data and diet. Hallberg studied iron absorption from a number of different complete meals and calculated bioavailable iron density: the amount of iron (in milligram) absorbed from a meal per unit energy (1000 kcal) by subjects who are borderline iron deficient (232). The values, based on actual absorption measurements, correlated well with our understanding of the individual bioavailability factors and meal composition. A value of 0.9 mg/day was obtained as an average bioavailable iron density of the Swedish diet.

The extrinsic-tag model for measuring absorption of iron in the miscible nonheme pool of a meal was discussed in Section II,A. In direct comparison, the extrinsic tag tends to be absorbed at slightly higher percentage than a concomitantly administered intrinsic tag, and thus, would overestimate absorption of nonheme iron in a meal. The differences observed have not been great, and even an extrinsic–intrinsic ratio of 1.1 is not of great practical significance. If the nonheme iron pool of a meal is not totally miscible, the extrinsic tag will greatly overestimate absorption, for example, when the meal contains insoluble contaminant iron. Nonmiscible iron can be estimated by an *in vitro* technique and appropriate correction made to provide a more precise measure of iron absorption from meals that may contain insoluble or contamination iron (235,469).

Collective intelligence indicates, therefore, that consideration of bioavailability factors should enable fabrication of diets that will prove efficacious in preventing iron deficiency. In single-meal studies by Hallberg and Rossander (244), nonheme iron absorption from a maize, rice, black bean meal was increased from 0.17 to 0.58 or 0.41 mg by addition of cauliflower or ascorbic acid. Few long-term studies have been conducted, however. Bates *et al.* (37) observed no changes in plasma iron parameters with ascorbic acid supplements of up to 70 mg above the basal intakes of about 35 mg/day in lactating women in a West

African rural community. The study period was only 5 weeks in duration, and little information was provided on diet. In a 6-month study, men, women, and children consumed soy-extended ground beef, and no change in iron status was observed (63). Consumption of the soy-extended ground beef attempted to evaluate practical conditions in the United States. A subsequent report by Lynch *et al.* (352) that heme iron absorption is stimulated by soy protein suggests that soy protein in an omnivore diet will not adversely affect iron status. Cook *et al.* (122) provided 17 adults with 2 g of ascorbic acid daily for 4 months during which they ate their usual self-chosen omnivore diets and observed no change in serum ferritin values. Over an additional 20 months of supplementation of four iron-depleted individuals, only one individual showed a positive response in serum ferritin. The individuals were not refractory to the single-meal effect of ascorbic acid on iron absorption after the initial 4 months of supplementation. These investigators concluded that altering the bioavailability of nonheme dietary iron has little effect on iron status when the diet contains substantial amounts of meat. Acosta *et al.* (3) determined daily absorption of iron from diets typical of lower socioeconomic class Latin Americans. "Differences in iron absorption from meals up to 7-fold, could be attributed to the varying contents of absorption enhancers, e.g., in meat, and of inhibitors in tea, vegetables, and wheat or maize bread." Iron-deficient individuals in this study tended to fulfill the physiological requirements, but the normal individual did not.

C. Iron Fortification and Supplementation

With normal men and postmenopausal women there is little reason for concern about iron intakes. Infants and women in their fertile period are in a more precarious position because of their larger iron needs and often lower calorie consumption by women. For some women an otherwise adequate diet may be, and often is, inadequate in iron. Some form of iron supplementation or food fortification may be necessary.

In the infant this can readily be accomplished through fortification of special food items, the use of which is confined to infancy. An iron intake of 10 mg/day from proprietary milk and infant cereals fortified with iron is effective in preventing anemia in infants (396). Bioavailability of iron in human breast milk is high (199,381,504), but the quantity is insufficient to maintain iron stores of the infant (134,199,502). Common pediatric practice is to recommend iron-fortified formula if not breast-fed, and iron-fortified cereal in order to provide the recommended intake of iron (185,617).

For iron intakes of adults and children to be adequately improved through a fortification program, the foods chosen must be widely used and in sufficient quantity to make a worthwhile contribution to total iron intake. The food most extensively fortified is white flour from wheat, but other food vehicles have been proposed and investigated at least in preliminary nature, including sugar, salt, rice, and processed cereals (120). The technology and food chemistry of iron fortification is complex, and problems of choosing a bioavailable fortificant that will not adversely affect organoleptic properties of foods to which it is added may be myriad (116,331) and will not be discussed herein.

The original intent of white flour enrichment in the United States was to restore the level of iron in white bread to that in whole-wheat bread. Enrichment of flour with iron at 15 mg/lb (~10 mg/lb in bread) was started in the United States in 1945, but the program had little effect in reducing the incidence of anemia, or in Britain where a similar enrichment program was initiated. This was due at least in part to low bioavailability of the forms of iron salts used for enrichment (116,166). A proposal was made in 1974 to increase the iron fortification level in the United States to 40 mg/lb of flour or 25 mg/lb in bread. The proposal was opposed on the grounds of efficacy and safety and was not finalized. The safety aspects involved the increase in dietary iron intake and consequently more rapid development of iron overload by individuals with idiopathic hemochromatosis, Laennec's cirrhosis, and iron-loading anemias (87,125). Swiss and Beaton (555), using Canadian consumption and food composition data, computed the predictable benefit of an iron fortification program. They concluded that an appreciable reduction in the risk of iron deficiency in the target population (menstruating women) could be achieved by iron fortification of bread, without an exorbitant increase in iron intakes by other groups; but "reduction of risk of deficiency to zero in the target population by iron fortification alone would require high levels of fortification, much higher than would seem either prudent or technologically feasible." Hallberg, however, credits the Swedish iron fortification program with contributing to the decrease in incidence of anemia in young women in that nation (232).

D. Iron in Animal Feeds

The iron content of animal feeds is highly variable. The level of iron in herbage plants is basically determined by the species and type of soil on which the plants grow and can be greatly affected by contamination with soil and dust. In a study of New Zealand pastures, iron levels ranged from 111 to 3850 μg/g (dry basis, mean ± 163), with the highest levels probably due to soil contamination (89). Leguminous pasture plants usually contain 200–400 μg/g iron (dry basis). Values as high as 700–800 μg/g have been recorded for uncontaminated lucerne (alfalfa) and as low as 40 μg/g or less for some grasses grown on sandy soils (see Underwood, Ref. 573). Most cereal grains contain 30–60 μg/g, and species differences appear to be small. The leguminous and oilseeds are richer in iron than the cereal grains and may contain 100–200 μg/g iron.

Feeds of animal origin, other than milk and milk products, are rich sources of iron. Meat meals and fish meals commonly contain 400–600 μg/g and blood meals more than 3000 μg/g iron. Ground limestone, oyster shell, and many forms of calcium phosphate used as mineral supplements frequently contain 2000–5000 μg/g iron.

All the above figures refer to total iron. Little is known of the availability of this iron to herbivorous animals, particularly ruminants with a digestive and absorptive apparatus very different from that of humans or other monogastric species. However, orally administered ferrous sulfate, ferrous carbonate, and ferric chloride are equally available to calves and sheep on the basis of tissue ^{59}Fe deposition and the iron in ferric oxide significantly less available on this basis (13). It seems, therefore, that the solubility of the chemical form of iron affects its availability to ruminants as it does in other species.

VI. IRON TOXICITY

Data which permit the delineation of toxic levels of dietary iron in a manner analogous to iron requirement levels are unavailable. Iron intakes of 50 mg/day (397) or 25–75 mg/day (182) by humans have been cited as safe, but many individuals have taken iron medication at these or higher levels for extended periods without reported harm. More information on maximum safe human tolerances to iron in different forms is obviously needed.

Accidental ingestion of iron supplement tablets is not an infrequent occurrence in young children (218,314). The consequences may be fatal, and treatment should be as prompt as possible (32,492).

Long-term iron overload has been observed in malnourished Bantus in South Africa (Bantu siderosis) (353). The native diet may supply 200 mg iron daily due to contamination from iron cooking vessels and a high consumption of Kaffir beer. The beer is prepared in iron pots and had been reported to contain from 15 to 120 mg iron per liter (70,353). This iron is in a soluble form and may supply as much as 2–3 mg of absorbed iron daily (99). The highest dietary intakes of iron appear to be in Ethiopia, where the staple cereal teff has a high iron content and becomes heavily contaminated with iron in grinding, storing, and cooking. An average iron intake of some 470 mg/day has been estimated, of which about three-quarters comes from contamination (493). This iron must be largely unavailable because tissue siderosis is not common, and examination of the storage iron of individuals dying from accidents has shown no difference from that found in a comparable Swedish population (268).

Idiopathic hemochromatosis is a genetic disorder in which excessive amounts of iron are absorbed from a normal diet. The usual control of iron absorption that occurs in the iron-replete condition fails to function, leading to tissue iron overload (65,126). The degree to which increased iron intake accelerates the process in a susceptible individual is uncertain (66).

The importance of the availability of the iron in a particular diet in determining maximum safe levels, or the minimum levels at which toxic effects can be expected, is further apparent from numerous studies with animals involving varying intakes of other elements with which iron interacts at the absorptive level. High intakes of cobalt, copper, zinc, manganese, and deficient intakes of nickel, depress iron absorption, as discussed elsewhere. Levels of iron intake which would produce signs of toxicity in the animal would clearly need to be much higher under conditions of abnormally high intakes of one or more of these interacting elements than when such intakes are low or normal. This is merely an expression for iron of the general principle for trace elements enunciated in Chapter 1 (Mertz, this volume), namely, that "a series of 'safe' dietary levels of potentially toxic trace elements has been established, depending on the extent to which other elements which affect their absorption and retention are present."

REFERENCES

1. Abdelrahim, A. I., Wensing, T., Franken, P., and Schotman, A. J. H. (1983). *Zentralbl. Vet. Med. A* **30**, 325.
2. Ackrell, B. A. C., Maguire, J. J., Dallman, P. R., and Kearney, E. B. (1984). *J. Biol. Chem.* **259**, 10053.
3. Acosta, A., Amar, A., Cornbluth-Szarfarc, S. C., Dillman, E., Fosil, M., Biachi, R. G., Grebe, G., Hertrampf, P. E., Kremenchuzky, S., Layrisse, M., Martínez-Torres, C., Morón, C., Pizarro, F., Reynafarje, C., Stekel, A., Villavicencio, D., and Zuniga, yH. (1984). *Am. J. Clin. Nutr.* **39**, 953.
4. Adams, W. S., Leslie, A., and Levin, M. H. (1950). *Proc. Soc. Exp. Biol. Med.* **74**, 46.
5. Addison, G. M., Beamish, M. R., Hales, C. N., Hodkins, M., Jacobs, A., and Llewelin, P. (1972). *J. Clin. Pathol.* **25**, 326.
6. Aisen, P. (1983). *In* "Biological Aspects of Metals and Metal-Related Diseases" (B. Sarkar, ed.), p. 67. Raven, New York.
7. Aisen, P., and Leibman, A. (1972). *Biochim. Biophys. Acta* **257**, 314.
8. Alderton, G., Wald, W. H., and Fevold, H. L. (1946). *Arch. Biochem.* **11**, 9.
9. Ali, K. E., and Ramsay, W. N. M. (1968). *Biochem. J.* **110**, 36P.
10. Allgood, J. W., and Brown, E. B. (1967). *Scand. J. Haematol.* **4**, 217.
11. Amine, E. K., and Hegsted, D. M. (1971). *J. Nutr.* **101**, 927.
12. Amine, E. K., and Hegsted, D. M. (1971). *J. Nutr.* **101**, 1575.
13. Ammerman, C. B., Wing, J. M., Dunavant, B. G., Robertson, W. K., Feaster, J. P., and Arrington, L. R. (1947). *J. Anim. Sci.* **26**, 404.
14. Anaokar, S. G., and Garry, P. J. (1981). *Am. J. Clin. Nutr.* **34**, 1505.
15. Anderson, B. M., Gibson, R. S., and Sabry, J. H. (1981). *Am. J. Clin. Nutr.* **34**, 1042.
16. Anderson, H. T., and Barkue, H. (1970). *Scand. J. Clin. Lab. Invest.* **25** (Suppl.), 114.
17. Andersson, H., Navert, B., Bingham, S. A., Englyst, H. N., and Cummings, J. H. (1983). *Br. J. Nutr.* **50**, 503.
18. Anke, M., Hennig, A., Hoffman, G. Dittrich, G., Grün, M., Ludke, H., Groppel, B., Gartner, P., Schuler, D., and Schwarz, S. (1972). *Arch. Tierernaehr.* **22**, 357.
19. Appleton, T. C., Morgan, E. H., and Baker, E. (1971). Proc. Int. Symp. Erythropoieticum, 1970 (T. Travnicek and J. Neuwert, eds.), p. 310. Univ. Karlova, Prague.
20. Ashworth, A., Milner, P. F., and Waterlow, J. C. (1973). *Br. J. Nutr.* **29**, 269.

21. Avramidis, L. (1983). *J. Pediatr.* **103**, 339.
22. Baggs, R. B., and Miller, S. A. (1973). *J. Nutr.* **103**, 1554.
23. Bagley, D. H., Zapolski, E. J., Rubin, M., and Princiotto, J. V. (1971). *Clin. Chim. Acta* **35**, 311.
24. Bailey, L., Ginsburg, J., Wagner, P., Noyes, W., Christakis, G., and Dinning, J. (1982). *J. Pediatr.* **101**, 774.
25. Bailey, L. B., O'Farrell-Ray, B., Mahan, C. S., and Dimperio, D. (1983). *Nutr. Res.* **3**, 783.
26. Bailey, L. B., Wagner, P. A., Christakis, G. J., Davis, C. G., Appeldorf, H., Araujo, P. E., Dorsey, E., and Dinning, J. S. (1982). *Am. J. Clin. Nutr.* **35**, 1023.
27. Bainton, D. F., and Finch, C. A. (1964). *Am. J. Med.* **37**, 62.
28. Baker, E., Jordan, S. M., Tuffery, A. A., and Morgan, E. H. (1969). *Life Sci.* **8**, 89.
29. Baker, E., and Morgan, E. H. (1973). *J. Physiol. (London)* **232**, 485.
30. Baker, E., Shaw, D. C., and Morgan, E. H. (1968). *Biochemistry* **7**, 1371.
31. Banerji, L., Hood, S. K., and Ramalingaswami, V. (1968). *Am. J. Clin. Nutr.* **21**, 1139.
32. Banner, W., and Czajka, P. A. (1981). *Am. J. Dis. Child.* **135**, 485.
33. Barber, S. A., Bull, N. L., and Buss, D. H. (1985). *Br. Med. J.* **290**, 743.
34. Barclay, R. (1985). *Med. Lab. Sci.* **42**, 166.
35. Barer, A. P., and Fowler, W. M. (1937). *J. Lab. Clin. Med.* **23**, 148.
36. Bartholmey, S. J., and Sherman, A. R. (1985). *J. Nutr.* **115**, 138.
37. Bates, C. J., Prentice, A. M., Prentice, A., Lamb, W. H., and Whitehead, R. G. (1983). *Int. J. Vitam. Nutr. Res.* **53**, 68.
38. Bates, G. W., Boyer, J., Hegenauer, J. C., and Saltman, P. (1972). *Am. J. Clin. Nutr.* **25**, 983.
39. Baumslag, N., and Petering, H. G. (1976). *Arch. Environ. Health* **31**, 254.
40. Baumslag, N., Yeager, D., Levin, L., and Petering, H. G. (1974). *Arch. Environ. Health* **29**, 186.
41. Beard, J., Green, W., Miller, L., and Finch, C. A. (1984). *Am. J. Physiol.* **247**, R114.
42. Bearn, A. G., and Parker, W. C. (1966). *In* "Glycoproteins, Their Structure and Function" (A. Gottschalk, ed.), p. 413, Elsevier, Amsterdam.
43. Becker, G., Huebers, H., and Rummel, W. (1979). *Blut* **38**, 397.
44. Beisel, W. R., Pekarek, R. S., and Wannemacher, R. W. (1974). *In* "Trace Element Metabolism in Animals" (W. G. Hoekstra, J. W. Suttie, H. E. Ganther, and W. Mertz, eds.), Vol. 2, p. 217. Univ. Park Press, Baltimore, MD.
45. Beisel, W. R. (1982). *Am. J. Clin. Nutr.* **35**, 417.
46. Benstrup. P. (1953). *Acta Med. Scand.* **145**, 315.
47. Bernard, J., Boiron, M., and Paoletti, C. (1958). *Rev. Fr. Etud. Clin. Biol.* **3**, 367.
48. Bernier, J. F., Fillion, F. J., and Brisson, G. J. (1984). *J. Dairy Sci.* **67**, 2369.
49. Beutler, E. (1957). *Am. J. Med. Sci.* **234**, 517; (1959). *Acta Haematol.* **21**, 317; (1959). *J. Clin. Invest.* **38**, 1605.
50. Beutler, E., and Blaisdell, R. K. (1958). *J. Lab. Clin. Med.* **52**, 694; (1960). *Blood* **15**, 30.
51. Beutler, E., Robson, M. J., and Buttenweiser, E. (1958). *Ann. Intern. Med.* **48**, 60.
52. Bezkorovainy, A. (1980). "Biochemistry of Nonheme Iron." Plenum, New York.
53. Bezwoda, W., Charloton, R., Bothwell, T. H., Torrance, J., and Mayet, F. (1978). *J. Lab. Clin. Med.* **92**, 108.
54. Bhargava, M., and Gabbe, E. E. (1984). *J. Nutr.* **114**, 1060.
55. Biggs, J. C., and Davis, A. E (1963). *Lancet* **1**, 814; (1966). *Australas. Ann. Med.* **15**, 36.
56. Björn-Rasmussen, E. (1974). *Nutr. Metab.* **16**, 101.
57. Björn-Rasmussen, E., Carneskog, J., and Cederblad, Å. (1980). *Scand. J. Haematol.* **25**, 124.
58. Björn-Rasmussen, E., and Hallberg, L. (1979). *Nutr. Metab.* **23**, 192.

59. Björn-Rasmussen, E., Hallberg, L., Isaksson, B., and Arvidsson, B. (1974). *J. Clin. Invest.* **53,** 247.
60. Björn-Rasmussen, E., Hallberg, L., Magnusson, B., Rossander, L., Svanberg, B., and Arvidsson, B. (1976). *Am. J. Clin. Nutr.* **29,** 772.
61. Björn-Rasmussen, E., Hallberg, L., and Walker, R. B. (1972). *Am. J. Clin. Nutr.* **25,** 317.
62. Björn-Rasmussen, E., Hallberg, L., and Walker, R. B. (1973). *Am. J. Clin. Nutr.* **26,** 1311.
63. Bodwell, C. E., Miles, C. W., Morris, E. R., Mertz, W., Canary, J. J., and Prather, E. S. (1983). *Fed. Proc., Fed. Am. Soc. Exp. Biol.* **42,** 529.
64. Bolin, T., and Davis, A. E. (1968). *Am. J. Dig. Dis.* **13,** 16.
65. Bothwell, T. H., Charlton, R. W., Cook, J. D., and Finch, C. A. (1979). "Iron Metabolism in Man." Blackwell, London.
66. Bothwell, T. H., Derman, D., Bezwoda, W. R., Torrance, J. D., and Charlton, R. W. (1978). *Hum. Genet. Suppl.* **1,** 131.
67. Bothwell, T. H., and Finch, C. A. (1968). *Symp. Swed. Nutr. Found.* **6,** cited by Hallberg *et al.* (122).
68. Bothwell, T. H., Pirzio-Brioli, G., and Finch, C. A. (1958). *J. Lab. Clin. Med.* **51,** 24.
69. Bothwell, T. H., Pribilla, W. F., Mebust, W., and Finch, C. A. (1958). *Am. J. Physiol.* **193,** 615.
70. Bothwell, T. H., Seftel, H., Jacobs, P., Torrance, J. D., and Brauneslog, N. (1964). *Am. J. Clin. Nutr.* **14,** 47.
71. Bour, N. J. S., Soullier, B. A., and Zemel, M. B. (1984). *Nutr. Res.* **4,** 371.
72. Bradley, B. D. Graber, G., Condon, R. J., and Frobish, L. T. (1983). *J. Anim. Sci.* **56,** 625.
73. Braude, R., Chamberlain, A., Kotarbinski, M., and Mitchell, K. G. (1962). *Br. J. Nutr.* **16,** 427.
74. Bremner, I., and Dalgarno, A. C. (1973). *Br. J. Nutr.* **29,** 229.
75. Brise, H., and Hallberg, L. (1962). *Acta Med. Scand.* **171,** *Suppl.* 376, 51, and 59.
76. Brock, A. B., and Diamond, L. M. (1934). *J. Pediatr.* **4,** 445.
77. Brown, E. B., and Justus, B. W. (1958). *Am. J. Physiol.* **192,** 319.
78. Brown, E. B., and Rother, M. L. (1963). *J. Lab. Clin. Med.* **62,** 357.
79. Brown, E. B., and Rother, M. L. (1963). *J. Lab. Clin. Med.* **62,** 804.
80. Brumstrom, G. M., Karabus, C., and Fielding, J. (1968). *Br. J. Haematol.* **14,** 525.
81. Buhi, W. C., Ducsay, C. A., Bazer, F. W., and Roberts, R. M. (1982). *J. Biol. Chem.* **257,** 1712.
82. Bullen, J. J., Rogers, H. J., and Griffiths, E. (1972). *Br. J. Haematol.* **23,** 389.
83. Bunch, R. J., Speers, V. C., Hays, V. M., and McCall, J. T. (1963). *J. Anim. Sci.* **22,** 56.
84. Burman, D. (1972). *Arch. Dis. Child.* **47,** 261.
85. Burns, R. H., Johnston, A., Hamilton, J. W., McColloch, R. J., Duncan, W. E., and Fisk, H. G. (1964). *J. Anim. Sci.* **23,** 5.
86. Byers, J. H., Jones, I. R., and Haag, J. R. (1953). *J. Dairy Sci.* **35,** 661.
87. Callender, S. T. (1973). *In* "Nutritional Problems in a Changing World" (D. Hollingsworth and M. Russell, eds.), p. 205. Appl. Sci. Publ., London.
88. Callender, S. T., and Warner, G. T. (1968). *Am. J. Calin. Nutr.* **21,** 1170.
89. Campbell, A. G., Coup, M. R., Bishop, W. H., and Wright, D. E. (1974). *N.Z. J. Agric. Res.* **17,** 393.
90. Campbell, J. A., and Gardiner, A. C. (1960). *Vet. Rec.* **72,** 1006.
91. Candela, E., Camacho, M. V., Martínez-Torres, C., Perdomo, J., Mazzarri, G., Acurero, G., and Layrisse, M. (1984). *J. Nutr.* **114,** 2204.
92. Carlson, R. H., Swenson, M. J., Ward, G. M., and Booth, N. H. (1961). *J. Am. Vet. Med. Assoc.* **139,** 457.

93. Cartwright, G. E., and Wintrobe, M. M. (1949). *J. Clin. Invest.* **28,** 86.
94. Cartwright, G. E., and Wintrobe, M. M. (1954). In "Modern Trends in Blood Diseases" (J. F. Wilkinson, ed.), p. 183. Butterworth, London.
95. Cartwright, G. E., Gubler, C. J., and Wintrobe, M. M. (1954). *J. Clin. Invest.* **33,** 685.
96. Cavell, P. A., and Widdowson, E. M. (1964). *Arch. Dis. Child* **39,** 496.
97. Celada, A., Rudolf, H., Herreros, V., and Donath, A. (1978). *Acta Haematol.* **60,** 182.
98. Chandra, R. K. (1985). *J. Am. Coll. Nutr.* **4,** 5.
99. Charlton, R. W., Bothwell, T. H., and Seftel, H. C. (1973). *Clin. Haematol.* **2,** 383.
100. Charlton, R. W., Jacobs, P., Torrance, J. D., and Bothwell, T. H. (1965). *J. Clin. Invest.* **44,** 543.
101. Chen, S. C. H., Shirazi, M. R. S., and Orr, R. A. (1983). *Nutr. Res.* **3,** 91.
102. Chio, L. F., Bunden, K., Vohra, P., and Kratzer, F. H. (1976). *Poult. Sci.* **55,** 808.
103. Christensen, J. M., Ghannam, M., and Ayres, J. W. (1984). *J. Pharm. Sci.* **73,** 1245.
104. Clark, R. C. (1976). *Int. J. Biochem.* **7,** 569.
105. Cole, C. B., Scott, K. J., Henschel, M. J., Coates, M. E., Ford, J. E., and Fuller, R. (1983). *Br. J. Nutr.* **49,** 231.
106. Cole, S. K., Thomson, A. M., Billewicz, W. Z., and Black, A. E. (1971). *J. Obstet. Gynaecol. Br. Commonw.* **78,** 933; Cole, S. K., Billewicz, W. Z., and Thomson, A. M. (1972). *J. Obstet. Gynaecol. Br. Commonw.* **79,** 994.
107. Conrad, M. E., and Crosby, W. H. (1963). *Blood* **22,** 406.
108. Conrad, M. E., and Schade, S. G. (1968). *Gastroenterology* **55,** 35.
109. Conrad, M. E., Weintraub, L. R., and Crosby, W. H. (1964). *J. Clin. Invest.* **43,** 963.
110. Conrad, M. E., Weintraub, L. R., Sears, D. A., and Crosby, W. H. (1966). *Am. J. Physiol.* **211,** 1123.
111. Conte, D., Brunelli, L., Ferrario, L., Mandelli, C., Quatrini, M., Velio, P., and Bianchi, P. A. (1984). *Acta Haematol.* **72,** 117.
112. Cook, J. D., Brown, G. M., and Valberg, L. S. (1964). *J. Clin. Invest.* **43,** 1185.
113. Cook, J. D., and Finch, C. A. (1979). *Am. J. Clin. Nutr.* **32,** 2115.
114. Cook, J. D., Layrisse, M., Martínez-Torres, C., Walker, R., Monsen, E. R., and Finch, C. A. (1972). *J. Clin. Invest* **51,** 805.
115. Cook, J. D., Lipschitz, D. A., Miles, L. E., and Finch, C. A. (1974). *Am. J. Clin. Nutr.* **27,** 681.
116. Cook, J. D., Minnich, V., Moore, C. V., Rasmussen, A., Bradley, W. B., and Finch, C. A. (1973). *Am J. Clin. Nutr.* **26,** 861.
117. Cook, J. D., and Monsen, E. R. (1976). *Am. J. Clin. Nutr.* **29,** 614.
118. Cook, J. D., and Monsen, E. R. (1976). *Am. J. Clin. Nutr.* **29,** 859.
119. Cook, J. D., Morck, T. A., and Lynch, S. R. (1981). *Am. J. Clin. Nutr.* **34,** 2622.
120. Cook, J. D., and Reusser, M. (1983). *Am J. Clin. Nutr.* **38,** 648.
121. Cook, J. D., and Skikne, B. S. (1982). *Am J. Clin. Nutr.* **35,** 1180.
122. Cook, J. D., Watson, S. S., Simpson, K. M., Lipschitz, D. A., and Skikne, B. S. (1984). *Blood* **64,** 721.
123. Cowan, J. W., Esfahani, M., Salji, J. P., and Azzam, S. A. (1966). *J. Nutr.* **90,** 423.
124. Crichton, R. R. (1984). *TIBS* **9,** 283.
125. Crosby, W. H. (1970). *Arch. Intern. Med.* **126,** 911.
126. Crosby, W. H. (1977). *Semin. Hematol.* **14,** 135.
127. Cusack, R. P., and Brown, W. D. (1965). *J. Nutr.* **86,** 383.
128. Czerniejewski, C. P., Shank, C. W., Bechtel, W. G., and Bradley, W. B. (1964). *Cereal Chem.* **41,** 65.
129. Dallman, P. R. (1969). *J. Nutr.* **97,** 475.

130. Dallman, P. R. (1984). *Am. J. Clin. Nutr.* **39**, 937.
131. Dallman, P. R., Reeves, J. D., Driggers, D. A., and Lo, E. Y. T. (1981). *J. Pediatr.* **98**, 376.
132. Dallman, P. R., Refino, C., and Yland, M. J. (1982). *Am. J. Clin. Nutr.* **35**, 671.
133. Dallman, P. R., Siimes, M. A., and Manies, E. C. (1975). *Br. J. Haematol.* **31**, 209.
134. Dallman, P. R., Siimes, M. A., and Stekel, A. (1980). *Am. J. Clin. Nutr.* **33**, 86.
135. Dallman, P. R., Yip, R., and Johnson, C. (1984). *Am. J. Clin. Nutr.* **39**, 437.
136. Davies, K. J. A., Donovan, C. M., Refino, C. J., Brooks, G. A., Parker, L., and Dallman, P. R. (1984). *Am. J. Physiol.* **246**, E535.
137. Danford, D. E., Smith, J. C., Jr., and Huber, A. M. (1982). *Am. J. Clin. Nutr.* **35**, 1958.
138. Davidson, L. A., and Lönnerdal, B. (1985). *Fed. Proc., Fed. Am. Soc. Exp. Biol.* **44**, 1673.
139. Davis, A. E., and Biggs, J. C. (1964). *Australas. Ann. Med.* **13**, 201; *Gut* **6**, 140 (1965).
140. Davis, P. N., Norris, L. C., and Kratzer, J. H. (1968). *J. Nutr.* **94**, 407.
141. Davis, P. S., Luke, C. G., and Deller, D. J. (1966). *Lancet* **2**, 1431; *Nature (London)* **214**, 1126 (1967).
142. Davis, P. S., Multani, J. S., Cepurncek, C. P., and Saltman, P. (1969). *Biochem. Biophys. Res. Commun.* **37**, 532.
143. De Alcorn, P. A., Donovan, M. E., Forbes, G. B., Landaw, S. A., and Stockman, J. A., III. (1979). *N. Engl. J. Med.* **300**, 5.
144. Derechin, S. S., and Johnson, P. (1962). *Nature (London)* **194**, 473.
145. Derman, D. P., Bothwell, T. H., MacPhail, A. P., Torrance, J. D., Bezwoda, W. R., Charlton, R. W., and Mayet, F. (1980). *Scand. J. Haematol.* **25**, 193.
146. Derman, D. P., Lynch, S. R., Bothwell, T. H., Charlton, R. W., Torrance, J. D., Brink, B. A., Margo, G. M., and Metz, J. (1978). *Br. J. Nutr.* **39**, 383.
147. Dewar, W. A., Teague, P. W., and Downie, J. N. (1974). *Br. Poult. Sci.* **15**, 119.
148. Dewey, K. G. (1983). *Am. J. Clin. Nutr.* **37**, 1010.
149. Dillman, E., Gale, C., Green, W., Johnson, D. G., Mackler, B., and Finch, C. A. (1980). *Am. J. Physiol.* **239**, R377.
150. Disler, P. B., Lynch, S. R., Charlton, R. W., Bothwell, T. H., Walker, R. B., and Mayet, F. (1975). *Br. J. Nutr.* **34**, 141.
151. Disler, P. B., Lynch, S. R., Charlton, R. W., Torrance, J. D., Bothwell, T. H., Walker, R. B., and Mayet, F. (1975). *Gut* **16**, 193.
152. Drabkin, D. L. (1951). *Physiol. Rev.* **31**, 345.
153. Dubach, R., Moore, C. V., and Callender, S. T. (1955). *J. Lab. Clin. Med.* **45**, 599.
154. Ducsay, C. A., Buhi, W. C., Bazer, F. W., Roberts, R. M., and Combs, G. E. (1984). *J. Anim. Sci.* **59**, 1303.
155. Dwyer, J. T., Dietz, W. H., Jr., Andrews, E. M., and Suskind, R. M. (1982). *Am. J. Clin. Nutr.* **35**, 204.
156. Eatough, D. J., Mineer, W. A., Christensen, J. J., Izatt, R. M., and Mangelson, N. F. (1974). *in* "Trace Element Metabolism in Animals" (W. G. Hoekstra, J. W. Suttie, H. E. Ganther, and W. Mertz, eds.), Vol. 2, p. 659. Univ. Park Press, Baltimore, MD.
157. Edgerton, V. R., Bryant, S. L., Gillespie, C. A., and Gardner, G. W. (1972). *J. Nutr.* **102**, 381.
158. Edgerton, V. R., Ohira, Y., Hettiarachchi, J., Senewiratne, B., Gardner, G. W., and Barnard, R. J. (1981). *J. Nutr. Sci. Vitaminol.* **27**, 77.
159. Ehrenberg, A., and Laurell, C. B. (1955). *Acta Chem. Scand.* **9**, 68.
160. El-Hawary, M. F. S., El-Shobaki, F. A., Kholeif, T., Sakr, R., and El-Bassoussy, M. (1975). *Br. J. Nutr.* **33**, 351.
161. Ellis, R., and Morris, E. R. (1979). *Nutr. Rep. Int.* **20**, 739.
162. El-Shobaki, F. A., and Rummel, W. (1977). *Res. Exp. Med.* **171**, 243.
163. El-Shobaki, F. A., and Rummel, W. (1978). *Res. Exp. Med.* **173**, 119.

164. El-Shobaki, F. A., and Rummel, W. (1979). *Res. Exp. Med.* **174**, 187.
165. Eltohamy, M. M., Takahara, H., and Okamoto, M. (1979). *J. Fac. Agric. Kyosho Univ.* **24**, 65.
166. Elwood, P. C. (1963). *Br. Med. J.* **1**, 224.
167. Elwood, P. C. (1973). *Am. J. Clin. Nutr.* **26**, 958.
168. Elwood, P. C., Newton, D., Eakins, J. D., and Brown, D. A. (1968). *Am. J. Clin. Nutr.* **21**, 1162.
169. Evans, G. W., and Grace, C. I. (1974). *Proc. Soc. Exp. Biol. Med.* **147**, 687.
170. Evans, J. L., and Abraham, P. A. (1973). *J. Nutr.* **103**, 196.
171. Ezekiel, E. (1967). *J. Lab. Clin. Med.* **70**, 138.
172. Ezekiel, E. and Morgan, E. H. (1963). *J. Physiol. (London)* **165**, 336.
173. Faber, M., and Falbe-Hansen, I. (1963). *Nature (London)* **184**, 1043 (1959).
174. Feeney, R. E., and Allison, R. G. (1969). "Evolutionary Biochemistry of Proteins." Wiley (Interscience), New York.
175. Feeney, R. E., and Komatsu, S. K. (1966). *Struct. Bond. (Berlin)* **1**, 149.
176. Feeley, R. M., Eitenmiller, R. R., Jones, J. B., Jr., and Barnhart, H. (1983). *Am. J. Clin. Nutr.* **37**, 443.
177. Fenillon, Y. M., and Plumier, M. (1952). *Acta Pediatt. (Stockholm)* **41**, 138.
178. Fielding, J., and Speyer, B. E., (1974). *Biochim. Biophys. Acta* **363**, 387.
179. Figueroa, W. G. (1960). *In* "Metal-Binding in Medicine" (M. J. Seven and L. A. Johnson, eds.), p. 146. Lippincott, Philadelphia.
180. Fillet, G., Cook, J. D., and Finch, C. A. (1974). *J. Clin. Invest.* **53**, 1527.
181. Finch, C. A., and Cook, J. D. (1984). *Am. J. Clin. Nutr.* **39**, 471.
182. Finch, C. A., and Monsen, E. R. (1972). *J. Am. Med. Assoc.* **219**, 1462.
183. Finch, S., Haskins, D., and Finch, C. A. (1950). *J. Clin. Invest.* **29**, 1078.
184. Flossman, K.-D., Muller, G., and Heilmann, P. (1983). *Arch. Exp. Vet. Med.* **37**, 217.
185. Fomon, S. J., Filer, L. J., Anderson, T. A., and Ziegler, E. E. (1979). *Pediatrics* **63**, 52.
186. Forbes, G. B., and Reina, J. C. (1972). *J. Nutr.* **102**, 647.
187. Forth, W., Huebers, H., Huebers, E., and Rummel, W. (1973). *In* "Trace Elements in Environmental Health—VI" (D. D. Hemphill, ed.), p. 121. Univ. of Missouri Press, Columbia.
188. Foy, H., and Kondi, A. (1957). *J. Trop. Med. Hyg.* **60**, 105.
189. Fransson, G. B., and Lönnerdal, B. (1984). *Am. J. Clin. Nutr.* **39**, 185.
190. Fritz, J. C., Pla, G. W., Roberts, T., Boehne, J. W., and Hove, E. L. (1970). *J. Agric. Food Chem.* **18**, 647.
191. Fuerth, J. (1972). *J. Pediatr.* **80**, 974.
192. Furugouri, K. (1971). *J. Anim. Sci.* **37**, 667.
193. Furugouri, K., and Kawabata, A. (1975). *J. Anim. Sci.* **41**, 1348.
194. Furugouri, K., Miyata, Y., Shijimaya, K., and Narasaki, N. (1983). *J. Anim. Sci.* **59**, 960.
195. Gabrio, B. W., Shoden, A. W., and Finch, C. A. (1953). *J. Biol. Chem.* **204**, 815.
196. Gardiner, M. R. (1965). *J. Comp. Pathol.* **75**, 397.
197. Garretson, F. D., and Conrad, M. E. (1967). *Proc. Soc. Exp. Biol. Med.* **126**, 304.
198. Garrett, R. J. B. (1975). *Experientia* **31**, 486.
199. Garry, P. J., Owen, G. M., Hooper, E. M., and Gilbert, B. A. (1981). *Pediatr. Res.* **15**, 822.
200. Gebre-Medhin, M., and Gobezie, A. (1975). *Am. J. Clin. Nutr.* **28**, 1322.
201. Gibson, R. S., Martinez, O., and MacDonald, A. C. (1984). *Nutr. Res.* **4**, 315.
202. Gillooly, M., Bothwell, T. H., Torrance, J. D., MacPhail, A. P., Derman, D. P., Bezwoda, W. R., Mills, W., Charlton, R. W., and Mayet, F. (1983). *Br. J. Nutr.* **49**, 331.
203. Gillooly, M., Torrance, J. D., Bothwell, T. H., MacPhail, A. R., Dorman, D. P., Mills, W., and Mayet, F. (1984). *Am J. Clin. Nutr.* **40**, 522.

204. Gipp, W. F., Pond, W. G., Kallfelz, F. A., Tasker, J. B., Van Campen, D. R., Krook, L., and Visek, W. J. (1974). *J. Nutr.* **104**, 532.
205. Glikman, P., Gutnisky, A., Gimeno, M. F., and Gimeno, A. L. (1981). *Experientia* **37**, 589.
206. Glover, J., and Jacobs, A. (1972). *Br. Med. J.* **2**, 627.
207. Goltner, E., and Gailer, H. J. (1964). *Zentralbl. Gynaekol* **34**, 1177.
208. Gordon, D. T., and Chao, L. S. (1984). *J. Nutr.* **114**, 526.
209. Gordon, W. G., Groves, M. L., and Basch, J. J. (1963). *Biochemistry* **2**, 817.
210. Gorman, J. E., and Clydesdale, F. M. (1983). *J. Food Sci.* **48**, 1217.
211. Granick, S. (1951). *Physiol. Rev.* **31**, 489.
212. Granick, S., and Michaelis, L. (1942). *Science* **95**, 439.
213. Gräsbeck, R., Kouvonen, I., Lundberg, M., and Tenhunen, R. (1979). *Scand. J. Haematol.* **23**, 5.
214. Grassman, E., and Kirchgessner, M. (1973). *Zentralbl. Veterinaermed. Reih A* **20**, 481.
215. Grassman, E., and Kirchgessner, M. (1973). *Z. Tierphysiol. Tierenaehr. Futtermittelkd.* **31**, 38.
216. Green, R., Charlton, R., Seftel, H., Bothwell, T., Mayet, F., Adams, B., Finch, C. A., and Layrisse, M. (1968). *Am. J. Med.* **45**, 336.
217. Green, S., Saha, A. K., Carleton, A. W., and Mazur, A. (1958). *Fed. Proc., Fed. Am. Soc. Exp. Biol.* **17**, 233.
218. Greenblatt, D. J., Allen, M. D., and Koch-Weser, J. (1976). *Clin. Pediatr.* **15**, 835.
219. Greengard, O., Sentenac, A., and Mendelsohn, N. (1964). *Biochim. Biophys. Acta* **90**, 406.
220. Greer, J. (1980). *Proc. Natl. Acad. Sci. U.S.A.* **77**, 3393.
221. Greger, J. L., and Buckley, S. (1977). *Nutr. Rep. Int.* **16**, 639.
222. Groves, M. L. (1960). *J. Am. Chem. Soc.* **82**, 3345.
223. Gubler, C. J., Cartwright, G. E., and Wintrobe, M. M. (1958). *J. Biol. Chem.* **224**, 533.
224. Guthrie, H. A., Froozani, M., Sherman, A. R., and Barron, G. P. (1974). *J. Nutr.* **104**, 1273.
225. Haga, P. (1980). *Acta Paediatr. Scand.* **69**, 637.
226. Hagler, L., Askew, E. W., Neville, J. R., Mellick, P. W., Coppes, R. I., Jr., and Lowder, J. R., Jr. (1981). *Am. J. Clin. Nutr.* **34**, 2169.
227. Hahn, P. F. (1937). *Medicine (Baltimore)* **16**, 249.
228. Hahn, P. F., Bale, W. F., Ross, J. F., Balfour, W. M., and Whipple, G. H. (1943). *J. Exp. Med.* **78**, 169.
229. Hall, E. D., and Symonds, H. W. (1980). *Br. J. Nutr.* **45**, 605.
230. Hall, M. H. (1974). *Br. Med. J.* **2**, 661.
231. Hallberg, L. (1981). *Annu. Rev. Nutr.* **1**, 123.
232. Hallberg, L. (1981). *Am J. Clin. Nutr.* **34**, 2242.
233. Hallberg, L. (1984). *In* "Present Knowledge in Nutrition," 5th Ed., p. 459. Nutrition Foundation, Washington, D.C.
234. Hallberg, L., and Björn-Rasmussen, E. (1972). *Scand. J. Haematol.* **9**, 193.
235. Hallberg, L., and Björn-Rasmussen, E. (1981). *Am. J. Clin. Nutr.* **34**, 2808.
236. Hallberg, L., Björn-Rasmussen, E., Ekenved, G., Garby, L., Rossander, L., Pleehachinda, R., Suwanik, R., and Arvidsson, B. (1978). *Scand. J. Haematol.* **21**, 215.
237. Hallberg, L., Björn-Rasmussen, E., Rossander, L., and Suwanik, R. (1977). *Am. J. Clin. Nutr.* **30**, 539.
238. Hallberg, L., Björn-Rasmussen, E., Suwanik, R., Pleehachinda, R., and Tuntawiroon, M. (1983). *Am. J. Clin. Nutr.* **37**, 272.
239. Hallberg, L., Garby, L., Suwanik, R., and Björn-Rasmussen, E. (1974). *Am. J. Clin. Nutr.* **27**, 826.
240. Hallberg, L., and Nilsson, L. (1964). *Acta Obstet. Gynecol. Scand.* **43**, 352.
241. Hallberg, L., and Rossander, L. (1982). *Am. J. Clin. Nutr.* **35**, 502.

242. Hallberg, L., and Rossander, L. (unpublished). *In* "The Effects of Cereals and Legumes on Iron Availability." International Nutritional Anemia Consultative Group, Nutrition Foundation, Washington, D.C.

243. Hallberg, L., and Rossander, L. (1982). *Scand. J. Gastroenterol.* **17,** 151.

244. Hallberg, L., and Rossander, L. (1984). *Am J. Clin. Nutr.* **39,** 577.

245. Hallberg, L., and Rossander, L. (1982). *Am. J. Clin. Nutr.* **36,** 514.

246. Hallberg, L., Rossander, L., Persson, H., and Svahn, E. (1982). *Am. J. Clin. Nutr.* **36,** 846.

247. Hallberg, L., and Slövell, L. (1967). *Acta Med. Scand.* **181,** 335.

248. Hallgren, B. (1954). *Acta Soc. Med. Ups.* **59,** 79.

249. Halliday, J. W., Powell, L. W., and Mack, V. (1976). *Br. J. Haematol.* **34,** 237.

250. Hamilton, D. L., Bellamy, J. E. C., Valberg, J. D., and Valberg, L. S. (1978). *Can. J. Physiol. Biochem.* **56,** 384.

251. Hamilton, E. I., and Minski, M. J. (1972–1973). *Sci. Total Environ.* **1,** 375.

252. Hamilton, L. D., Gubler, C. J., Cartwright, G. E., and Wintrobe, M. M. (1950). *Proc. Soc. Exp. Biol. Med.* **73,** 65.

253. Hangen, T. H., Hanley, J. M., and Heath, E. C. (1981). *J. Biol. Chem.* **256,** 1055.

254. Hanley, J. M., Hangen, T. H., and Heath, E. C. (1983). *J. Biol. Chem.* **258,** 7858.

255. Hargreaves, R. M., Street, M., Hoy, T., Jacobs, A., and McLaren, C. (1981). *Br. J Haematol.* **47,** 399.

256. Hartley, W. J., Mullins, J., and Lawson, E. M. (1959). *N.Z. Vet. J.* **7,** 99.

257. Hawksley, J. C., Lightwood, R., and Bailey, U. M. (1934). *Arch. Dis. Child.* **9,** 359.

258. Health Bull. No. 23. (1951). Govt. India Press, Delhi.

259. Hedges, J. O., and Kornegay, E. T. (1973). *J. Anim. Sci.* **37,** 1147.

260. Hegsted, D. M., Finch, C. A., and Kinney, T. D. (1949). *J. Exp. Med.* **90,** 147.

261. Heinrich, H. C., Bender-Götze, C., Gabbe, E. E., Bartels, H., and Oppitz, K. H. (1977). *Klin. Wochenschr.* **55,** 587.

262. Heinrich, H. C., Gabbe, E. E., Bartels, H., Oppitz, K. H., Bender-Götze, C., and Pfau, A. A. (1977). *Klin. Wochenschr.* **55,** 595.

263. Hemingway, R. G., Brown, N. A., and Luscombe, J. (1974). *In* "Trace Element Metabolism in Animals" (W. G. Hoekstra, J. W. Suttie, H. E. Ganther, and W. Mertz, eds.), Vol. 2, p. 601. Univ. Park Press, Baltimore.

264. Herndon, J. G., Rice, T. G., Tucker, R. G., Van Loon, E. J., and Greenberg, S. M. (1958). *J. Nutr.* **64,** 615.

265. Hershko, C., Bar-Or, D., Naparstek, E., Konijn, A. M., Grassowicz, N., Kaufmann, N., and Izak, G. (1981). *Am. J. Clin. Nutr.* **34,** 1600.

266. Higa, Y., Oshiro, S., Kino, K., Tsunoo, H., and Nakajima, H. (1981). *J. Biol. Chem.* **256,** 12322.

267. Hill, C. H., and Matrone, G. (1961). *J. Nutr.* **73,** 425.

268. Hofvander, G. (1968). *Acta Med. Scand. Suppl.* **494.**

269. Holmberg, C. G., and Laurell, C. B. (1947). *Acta Chem. Scand.* **1,** 944.

270. Horvath, Z. (1970). *In* "Trace Element Metabolism in Animals" (C. F. Mills, ed.), Vol. 1, p. 328. Livingstone, Edinburgh.

271. Huebers, H., Huebers, E., Csiba, E., Rummel, W., and Finch, C. A. (1983). *Blood* **61,** 283.

272. Hungerford, D. M., and Linder, M. C. (1983). *J. Nutr.* **113,** 2615.

273. Hunt, C. E., Landesman, J., and Newberne, P. M. (1970). *Br. J. Nutr.* **24,** 607.

274. Hunter, J. E. (1981). *J. Nutr.* **111,** 841.

275. Huser, J. H., Rieber, E. E., and Berman, A. R. (1967). *J. Lab. Clin. Med.* **69,** 405.

276. Hussain, R., and Patwardhan, V. N. (1959). *Indian J. Med. Res.* **47,** 676.

277. Hussain, R., and Patwardhan, V. N. (1959). *Lancet* **1,** 1073. Hussain, R., Patwardhan, V. N., and Sriranachari, S. (1960). *Indian J. Med. Res.* **48,** 235.

278. Hussain, R., Walker, R. B., Layrisse, M., Clark, P., and Finch, C. A. (1965). *Am. J. Clin. Nutr.* **16**, 464.
279. Hussein, L., Elnaggar, B., Gaafar, S., and Allaam, H. (1981). *Nutr. Rep. Int.* **23**, 901.
279a. Idzerda, R. L., Huebers, H., Finch, C. A., and McKnight, G. S. (1986). *Proc. Natl. Acad. Sci. U.S.A.* **83**, 3723.
280. Ingalls, R. L., and Johnston, F. A. (1954). *J. Nutr.* **53**, 351.
281. Ingram, D. J. E., Gibson, J. F., and Peratz, M. F. (1956). *Nature (London)* **178**, 906.
282. Iodice, A. A., Richert, D. A., and Schulman, M. P. (1958). *Fed. Proc., Fed. Am. Soc. Exp. Biol.* **17**, 248.
283. Itzerott, A. G. (1942). *J. Aust. Inst. Agric. Sci.* **8**, 119.
284. Iyengar, G. V., Kollmar, W. E., and Bowen, H. J. M. (1978). "The Elemental Composition of Human Tissues and Body Fluids." Verlag Chemie, Weinheim.
285. Jacob, R. A., Sandstead, H. H., Klevay, L. M., and Johnson, L. K. (1980). *Blood* **56**, 786.
286. Jacobs, A., and Miles, P. M. (1969). *Br. Med. J.* **4**, 778.
287. Jacobs, A., and Miles, P. M. (1969). *Gut* **10**, 226.
288. Jacobs, A., Miller, F., Worwood, M., Beamish, M. R., and Wardrop, C. A. (1972). *Br. Med. J.* **4**, 206.
289. Jacobs, A., and Worwood, M. (1975). *N. Engl. J. Med.* **292**, 951.
290. Jacobs, P., Bothwell, T., and Charlton, R. W. (1964). *J. Appl. Physiol.* **19**, 187.
291. Jacobs, P., Wormald, L. A., and Gregory, M. C. (1979). *S. Afr. Med. J.* **55**, 1065.
292. Jamieson, G. A., Jett, M., and DeBernardos, S. L. (1971). *J. Biol. Chem.* **246**, 3686.
293. Jandl, J. H., Inman, J. K., Simmons, R. L., and Allen, D. W. (1959). *J. Clin. Invest.* **38**, 161.
294. Jandl, J. H., and Katz, J. H. (1963). *J. Clin. Invest.* **42**, 314.
295. Johnson, P. E. (1984). *Fed. Proc., Fed. Am. Soc. Exp. Biol.* **43**, 847.
296. Johnson, G., Jacobs, P., and Purves, L. R. (1983). *J. Clin. Invest.* **71**, 1467.
297. Johnston, F. A., and McMillan, T. J. (1952). *J. Am. Diet. Assoc.* **28**, 633.
298. Jordon, S. M., Kaldor, I., and Morgan, E. H. (1967). *Nature (London)* **215**, 76.
299. Josephs, H. W. (1958). *Blood* **13**, 1.
300. Josephs, H. W. (1959). *Acta Paediatr. (Stockholm)* **48**, 403.
301. Kadis, S., Udeze, F. A., Polanco, J., and Dressen, D. W. (1984). *Am. J. Vet. Res.* **45**, 255.
302. Kaldor, I. (1954). *Aust. J. Exp. Biol. Med. Sci.* **32**, 437.
303. Kaldor, I., and Ezekiel, E. (1961). *Nature (London)* **196**, 175.
304. Katz, N. R., Goldfarb, V., Liem, H., and Muller-Eberhard, U. (1985). *Eur. J. Biochem.* **146**, 155.
305. Kerr, L. M. H. (1957). *Biochem. J.* **67**, 627.
306. Kilpatrick, G. S. (1970). *In* "Iron Deficiency: Pathogenesis, Clinical Aspects, Therapy" (L. Hallberg, H. G. Harwerth, and A. Vannotti, eds.), p. 441. Academic Press, New York.
307. Kim, W. W., Kelsay, J. L., Judd, J. T., Marshall, M. W., Mertz, W. M., and Prather, E. S. (1984). *Am. J. Clin. Nutr.* **40**, 1327.
308. Kino, K., Tsunoo, H., Higa, Y., Takami, M., and Nakajima, H. (1982). *J. Biol. Chem.* **257**, 4828.
309. Kirchgessner, M., and Pallauf, J. (1973). *Sonderdruck Zuechtungskd.* **45**, 245, 249; Kirchgessner, M., and Weigand, E. (1975). *Z. Tierphysiol. Tierernaehr. Futtermittelkd.* **34**, 205.
310. Knight, C. D., Klasing, K. C., and Forsyth, D. M. (1983–1984). *J. Anim. Sci.* **57**, 387; **59**, 1519.
311. Kochanowski, B. A., and Sherman, A. R. (1985). *Am. J. Clin. Nutr.* **41**, 278.
312. Kochanowski, B. A., Smith, A. M., Picciano, M. F., and Sherman, A. R. (1983). *J. Nutr.* **113**, 2471.

313. Kraus, W. E., and Washburn, R. G. (1936). *J. Biol. Chem.* **114**, 247.
314. Krenzelok, E. P., and Hoff, J. V. (1979). *Pediatrics* **63**, 591.
315. Kroe, D., Kinney, T. D., Kaufman, N., and Klavins, J. V. (1963). *Blood* **21**, 546.
316. Ku, P. K., Miller, E. R., and Ullrey, D. E. (1983). *J. Anim. Sci.* **57**, 638.
317. Kuhn, I. N., Layrisse, M., Roche, M., Martínez-Torres, C., and Walker, R. B. (1968). *Am. J. Clin. Nutr.* **21**, 1184.
318. Kurosky, A., Barnett, D. R., Lee, T.-H., Touchstone, B., Hay, R. E., Arnott, M. S., Bowman, B. H., and Fitch, W. M. (1980). *Proc. Natl. Acad. Sci. U.S.A.* **77**, 3388.
319. Latham, M. C., Stephenson, L. S., Hall, A., Wolgemuth, J. C., Elliot, T. C., and Crompton, D. W. T. (1983). *Trans. R. Soc. Trop. Med. Hyg.* **77**, 41.
320. Lau, A. L., and Failla, M. L. (1984). *J. Nutr.* **114**, 224.
321. Lawlor, M. J., Smith, W. H., and Beeson, W. M. (1965). *J. Anim. Sci.* **24**, 742.
322. Layrisse, M., Cook, J. D., Martinez, C., Roche, M., Kuhn, I. N., Walker, R. B., and Finch, C. A. (1969). *Blood* **33**, 430.
323. Layrisse, M., Linares, J., and Roche, M. (1965). *Blood* **25**, 73.
324. Layrisse, M., and Martínez-Torres, C. (1977). *Am. J. Clin. Nutr.* **30**, 1166.
325. Layrisse, M., Martínez-Torres, C., and Gonzalez, M. (1974). *Am. J. Clin. Nutr.* **27**, 152.
326. Layrisse, M., Martínez-Torres, C., Leets, I., Taylor, P., and Ramirez, J. (1984). *J. Nutr.* **114**, 217.
327. Layrisse, M., Martínez-Torres, C., Renzy, M., and Leets, I. (1975). *Blood* **45**, 689.
328. Layrisse, M., Martínez-Torres, C., and Roche, M. (1968). *Am. J. Clin. Nutr.* **21**, 1175.
329. Lee, F., and Rosenthal, F. D. (1958). *Q. J. Med.* **27**, 19.
330. Lee, G. R., Nacht, S., Lukens, J. N., and Cartwright, G. E. (1968). *J. Clin. Invest.* **47**, 2058.
331. Lee, K., and Clydesdale, F. M. (1979). *CRC Crit. Rev. Food Sci. Nutr.* **11**, 117.
332. Leeuw, N. K., de, Lowenstein, L., and Hsieh, Y. S. (1966). *Medicine (Baltimore)* **45**, 219.
333. Lennartsson, J., Bengtsson, C., Hallberg, L., and Tibblin, E. (1979). *Scand. J. Haematol* **22**, 17.
334. Leslie, A. J., and Kaldor, L. (1971). *Br. J. Nutr.* **26**, 469.
335. Letendre, E. D. (1985). *TIBS* **10**, 166.
336. Levine, P. H., Levine, A. J., and Weintraub, L. R. (1972). *J. Lab. Clin. Med.* **80**, 333.
337. Lewis, M., and Iammarino, R. M. (1971). *J. Lab. Clin. Med.* **78**, 547.
338. Lind, D. E. (1973). *J. Geriatr.* **4**, 19.
339. Linkenheimer, W. H., Patterson, E. L., Milstrey, B. A., Brochman, J. A., and Johnston, D. D. (1960). *J. Anim. Sci.* **19**, 763.
340. Lipschitz, D. A., Cook, J. D., and Finch, C. A. (1974). *N. Engl. J. Med.* **290**, 1213.
341. Lipschitz, D. A., Simon, M. O., Lynch, S. R., Dugard, J., Bothwell, T. H., and Charlton, R. W. (1971). *Br. J. Haematol.* **21**, 289.
342. Lipschitz, D. A., Simpson, K. M., Morris, E. R., and Cook, J. D. (1979). *J. Nutr.* **109**, 1154.
343. Loh, T. T. (1970). *Proc. Soc. Exp. Biol. Med.* **134**, 1070.
344. Loh, T. T. and Kaldor, I. (1973). *Comp. Biochem. Physiol.* **B44**, 337.
345. Loh, T. T., and Kaldor, I. (1974). *J. Dairy Sci.* **50**, 339.
346. Lönnerdal, B., Davidson, L., and Keen, C. L. (1985). *Fed. Proc., Fed. Am. Soc. Exp. Biol.* **44**, 1850.
347. Lönnerdal, B., Keen, C. L., and Hurley, L. S. (1981). *Annu. Rev. Nutr.* **1**, 149.
348. Lozoff, B., Brittenham, G. M., Viteri, F. E., Wolf, A. W., and Urrutia, J. J. (1982). *J. Pediatr.* **101**, 948.
349. Luke, C. G., Davis, P. S., and Deller, D. J. (1967). *Lancet* **1**, 926.
350. Lustbader, J. W., Arcoleo, J. P., Birken, S., and Greer, J. (1983). *J. Biol. Chem.* **258**, 1227.
351. Lynch, S. R., and Cook, J. D. (1980). *Ann. N.Y. Acad. Sci.* **355**, 32.

352. Lynch, S. R., Dassenko, S. A., Morck, T. A., Beard, J. L., and Cook, J. D. (1985). *Am. J. Clin. Nutr.* **41**, 13.
353. MacDonald, R. A., Becker, B. J. P., and Picket, G. S. (1963). *Arch. Intern. Med.* **3**, 315.
354. Mackler, B., Grace, R., and Finch, C. A. (1984). *Pediatr. Res.* **18**, 499.
355. MacPhail, A. P., Bothwell, T. H., Torrance, J. D., Derman, D. P., Bezwoda, W. R., Charlton, R. W., and Mayet, F. (1981). *Br. J. Nutr.* **45**, 215.
356. MacPhail, A. P., Bothwell, T. H., Torrance, J. D., Derman, D. P., Bezwoda, W. R., Charlton, R. W., and Mayet, F. G. H. (1981). *S.A. Med. J.* **59**, 939.
357. Man, Y. K., and Wadsworth, G. R. (1969). *Clin. Sci.* **36**, 479.
358. Martínez-Torres, C., and Layrisse, M. (1970). *Blood* **35**, 669.
359. Martínez-Torres, C., and Layrisse, M. (1971). *Am. J. Clin. Nutr.* **24**, 531.
360. Martínez-Torres, C., Leets, I., Renzi, B., and Layrisse, M. (1974). *J. Nutr.* **104**, 983.
361. Martínez-Torres, C., Romano, E., and Layrisse, M. (1981). *Am. J. Clin. Nutr.* **34**, 322.
362. Martínez-Torres, C., Romano, E., Renzi, M., and Layrisse, M. (1979). *Am J. Clin. Nutr.* **32**, 809.
363. Masson, P. L., and Heremans, J. F. (1966). *Protides Biol. Fluids Proc. Colloq.* **14**, 115.
364. Masson, P. L., and Heremans, J. F. (1971). *Comp. Biochem. Physiol. B* **39**, 119.
365. Matrone, G., Conley, C., Wise, G. H., and Waugh, R. K. (1957). *J. Dairy Sci.* **40**, 1437.
366. Matrone, G., Thomason, E. L., and Bunn, C. R. (1960). *J. Nutr.* **72**, 459.
367. May, L., Morris, E. R., and Ellis, R. (1980). *J. Agric. Food Chem.* **28**, 1004.
368. Mazur, A., and Carleton, A. (1965). *Blood* **26**, 317.
369. Mazur, A., Green, S., and Carelton, A. (1960). *J. Biol. Chem.* **235**, 595.
370. Mazur, A., Green, S., Saha, A., and Carelton, A. (1958). *J. Clin. Invest.* **37**, 1809.
371. McCall, M. G., Newman, G. E., O'Brien, J. R., and Witts, L. J. (1962). *Br. J. Nutr.* **16**, 305.
372. McCance, R. A., and Widdowson, E. M. (1937). *Lancet* **2**, 680. (1938). *J. Physiol. (London)* **94**, 138.
373. McCance, R. A., and Widdowson, E. M. (1943). *Nature (London)* **152**, 326.
374. McCance, R. A., and Widdowson, E. M. (1951). *Br. Med. Bull.* **7**, 297.
375. McCurdy, P. R., and Dern, R. J. (1968). *Am. J. Clin. Nutr.* **21**, 284.
376. McDougal, E. I. (1947). *J. Agric. Sci.* **37**, 337.
377. McLane, J. A., Fell, R. D., McKay, R. H., Winder, W. W., Brown, E. B., and Holloszy, J. O. (1981). *Am. J. Physiol.* **241**, C47.
378. McLaurin, L. P., and Cotter, J. R. (1967). *Am. J. Obstet. Gynecol.* **98**, 931.
379. McLean, F. W., Cotter, J. R., Blechner, J. N., and Noyes, W. D. (1970). *Am. J. Obstet. Gynecol.* **106**, 699.
380. McLennan, W. J., Andrews, G. R., Macleod, C., and Caird, F. I. (1973). *Q. J. Med.* **42**, 1.
381. McMillan, J. A., Landaw, S. A., and Oski, F. A. (1976). *Pediatrics* **58**, 682.
382. Mendel, G. A. (1961). *Blood* **18**, 727.
383. Mendel, G. A., Wiler, R. J., and Mangalik, A. (1963). *Blood* **22**, 450.
384. Meyers, L. D., Habicht, J. P., Johnson, C. L., and Brownie, C. (1983). *Am. J. Public Health* **73**, 1042.
385. Miles, C. W., Collins, J. S., Holbrook, J. T., Patterson, K. Y., and Bodwell, C. E. (1984). *Am J. Clin. Nutr.* **40**, 1393.
386. Miller, E. R., Waxler, G. L., Ku, P. K., Ullrey, D. E., and Whitehair, C. K. (1982). *J. Anim. Sci.* **54**, 106.
387. Miller, J. (1974). *Nutr. Rep. Int.* **10**, 1.
388. Miller, J., and McNeal, L. S. (1983). *J. Nutr.* **113**, 115.
389. Miller, J., and Nnanna, I. (1983). *J. Nutr.* **113**, 1169.
390. Miller, V., Swaney, S., and Deinard, A. (1985). *Pediatrics* **75**, 100.

391. Milman, N., Cohn, J., and Pedersen, N. S. (1984). *Eur. J. Pediatr.* **142**, 89.
392. Minnich, V., Okçuoğlu, A., Tarcon, Y., Arcasoy, A., Cin, S., Yörükoğlu, O., Renda, F., and Demirağ, B. (1968). *Am J. Clin. Nutr.* **21**, 78.
393. Miski, A. M.-A., and Kratzer, F. H. (1976). *Poult. Sci.* **55**, 454.
394. Miski, A. M.-A., and Kratzer, F. H. (1977). *J. Nutr.* **107**, 24.
305. Mitchell, H. H., and Edman, M. (1962). *Am. J. Clin. Nutr.* **10**, 163.
396. Moe, P. J. (1963). *Acta Paediatr. (Stockholm), Suppl.* **150,**
397. Monsen, E. R. (1971). *J. Nutr. Educ.* **2**, 152.
398. Monsen, E. R. (1974). *J. Nutr.* **104**, 1490.
399. Monsen, E. R., and Balintfy, J. L. (1982). *J. Am. Diet. Assoc.* **80**, 307.
400. Monsen, E. R., and Cook, J. D. (1976). *Am. J. Clin. Nutr.* **29**, 1142.
401. Monsen, E. R., Hallberg, L., Layrisse, M., Hegsted, D. M., Cook, J. D., Mertz, W. M., and Finch, C. A. (1978). *Am J. Clin. Nutr.* **31**, 134.
402. Monsen, E. R., Kuhn, I. N., and Finch, C. A. (1967). *Am. J. Clin. Nutr.* **20**, 842.
403. Monsen, E. R., and Page, J. F. (1978). *J. Agric. Food Chem.* **26**, 223.
404. Moore, C. V. (1951). *Harvey Lect.* **55**, 67.
405. Moore, C. V. (1955). *Am J. Clin. Nutr.* **3**, 3.
406. Moore, C. V., Minnich, V., and Dubach, R. (1943). *J. Am. Diet. Assoc.* **19**, 841.
407. Moore, H. O. (1938). *Aust. C.S.I.R.O. Bull.* **133.**
408. Morck, T. A., and Austic, R. E. (1981). *Poult. Sci.* **60**, 1497.
409. Morck, T. A., Lynch, S. R., and Cook, J. D. (1982). *Am. J. Clin. Nutr.* **36**, 219.
410. Morck, T. A., Lynch, S. R., and Cook, J. D. (1983). *Am. J. Clin. Nutr.* **37**, 416.
411. Morck, T. A., Lynch, S. R., Skikne, B. S., and Cook, J. D. (1981). *Am. J. Clin. Nutr.* **34**, 2630.
412. Morgan, E. H. (1961). *J. Physiol. (London)* **158**, 573.
413. Morgan, E. H. (1961). *Lancet* **1**, 9.
414. Morgan, E. H. (1961). *Aust. J. Exp. Biol. Med.* **39**, 361, 371; (1962). *J. Pathol. Bacteriol.* **84**, 65.
415. Morgan, E. H. (1962). *Q. J. Exp. Physiol. Cogn. Med. Sci.* **47**, 57; (1963). **48**, 170.
416. Morgan, E. H. (1964). *Br. J. Haematol.* **10**, 442.
417. Morgan, E. H. (1980). *Q. J. Exp. Physiol.* **65**, 239.
418. Morgan, E. H., and Laurell, C. B. (1963). *Br. J. Haematol.* **9**, 471.
419. Morgan, E. H., and Walters, M. N. I. (1963). *J. Clin. Pathol.* **16**, 101.
420. Morris, E. R., and Ellis, R. (1976). *J. Nutr.* **106**, 753.
421. Morris, E. R., and Ellis, R. (1982). *In* "ACS Symposium Series, No. 203 "Nutritional Bioavailability of Iron" (C. Kies, ed.), p. 121. American Chemical Society, Washington, D.C.
422. Morris, E. R., and Greene, F. E. (1972). *J. Nutr.* **102**, 901.
423. Muir, A. R. (1960). *Q. J. Exp. Physiol. Cogn. Med. Sci.* **45**, 192.
424. Muir, A. R., and Hopfer, U. (1985). *Am. J. Physiol.* **248**, G376.
425. Muir, W. A., Hopfer, U., and King, M. (1984). *J. Biol. Chem.* **259**, 4896.
426. Mukherjee, C., and Mukherjee, S. K. (1953). *J. Indian Med. Assoc.* **22**, 345.
427. Multani, J. S., Cepurneek, C. P., Davis, P. S., and Saltman, P. (1970). *Biochemistry* **9**, 3970.
428. Murray, J., and Stein, N. (1968). *Lancet* **1**, 614.
429. Murray, J., and Stein, N. (1968). *Am. J. Dig. Dis.* **3**, 527.
430. Murray, J., and Stein, N. (1970). *In* "Trace Element Metabolism in Animals" (C. F. Mills, ed.), Vol. 1, p. 321. Livingstone, Edinburgh.
431. Murray, M. J., Murray, A. B., Murray, N. J., and Murray, M. B. (1978). *Br. J. Nutr.* **39**, 627.

432. Naiman, J. L., Oski, F. A., Diamond, L., Vawter, G. F., and Schwachman, H. (1964). *Pediatrics* **33,** 83.
433. Nojeim, S. J., and Clydesdale, F. M. (1981). J. Food Sci. **46,** 606
434. National Academy of Sciences, U.S.A. (1980). "Recommended Dietary Allowances," 9th Ed.
435. National Academy of Sciences, U.S.A. (1977). "Nutrient Requirements of Poultry," 7th Ed.
436. National Center for Health Statistics: Singer, J. D., Granahan, R., Goodrich, N. N. *et al.* (1982). Diet and Iron Status, A Study of Relationships: United States, 1971–1974, DHHS Publ. No. (PHS) 83-1679. Public Health Service, Washington, D.C., U.S. Government Printing Office.
437. National Health, Natl. Health Med. Res. Counc. (Aust.) (1945). Spec. Rep. No. 1.
438. Neilands, J. B. (1981). *Annu. Rev. Nutr.* **1,** 27.
439. Newman, R., Schneider, C., Sutherland, R., Vodinelich, L., and Graves, M. (1983). *TIBS* **7,** 397.
440. Niccum, W. L., Jackson, R. L., and Stearns, G. (1953). *Am. J. Dis. Child.* **86,** 553.
441. Octave, J.-N., Schneider, Y.-J., Trouet, A., and Crichton, R. R. (1983). *TIBS* **8,** 217.
442. Ohira, Y., Edgerton, V. R., Gardner, G. W., Gunawardena, K. A., Senewiratne, B., and Ikawa, S. (1981). *J. Nutr. Sci. Vitaminol.* **27,** 87.
443. Ohira, Y., Edgerton, V. R., Gardner, G. W., and Senewiratne, B. (1980). *J. Nutr. Sci. Vitaminol.* **26,** 375.
444. Ohira, Y., and Gill, S. L. (1983). *J. Nutr.* **113,** 1811.
445. Ohira, Y., Hegenauer, J., Strause, L., Chen, C.-S., Saltman, P., and Beinert, H. (1982). *Br. J. Haematol.* **52,** 623.
446. Osaki, S., Johnson, D. A., and Frieden, E. (1966). *J. Biol. Chem.* **241,** 2746.
447. Osaki, S., Johnson, D. A., and Frieden, E. (1971). *J. Biol. Chem.* **246,** 3018.
448. Osaki, S., and Sirivech, S. (1971). *Fed. Proc., Fed. Am. Soc. Exp. Biol.* **30,** 1292.
449. Otis, L., and Smith, M. C. (1940). *Science* **91,** 146.
450. Pallauf, J., and Kirchgessner, M. (1973). *Sonderdruck Zuechtungskd.* **45,** 119.
451. Panic, B. (1970). *In* "Trace Element Metabolism in Animals" (C. F. Mills, ed.), Vol. 1, p. 324. Livingstone, Edinburgh.
452. Patterson, J. C. S., Marrack, D., and Wiggins, H. S. (1953). *J. Clin. Pathol.* **6,** 105.
453. Paulev, P.-E., Jordal, R., and Pedersen, N. S. (1983). *Clin. Chim. Acta* **127,** 19.
454. Perlmann, G. E. (1973). *Isr. J. Chem.* **11,** 393.
455. Peters, T., Apt, L., and Ross, J. F. (1971). *Gastroenterology* **61,** 315.
456. Petersen, K. M., and Brant, L. J. (1984). *Am. J. Clin. Nutr.* **39,** 460.
457. Picciano, M. F., and Guthrie, H. A. (1976). *Am. J. Clin. Nutr.* **29,** 242.
458. Pirzio-Biroli, G., Bothwell, T. H., and Finch, C. A. (1958). *J. Lab. Clin. Med.* **51,** 37.
459. Planas, J. (1967). *Nature (London)* **215,** 289.
460. Planas, J., and de Castro, S. (1960). *Nature (London)* **187,** 1126.
461. Planas, J., de Castro, S., and Recio, J. M. (1961). *Nature (London)* **189,** 668.
462. Pollack, S., George, J. N., Raba, R. C., Kaufman, R. M., and Crosby, W. H. (1965). *J. Clin. Invest.* **44,** 1470.
463. Pollack, S., Kaufman, R. M., and Crosby, W. H. (1964). *Blood* **24,** 557.
464. Pollack, S., and Lasky, F. D. (1976). *J. Lab. Clin. Med.* **87,** 670.
465. Polycove, M. (1958). *In* "Iron in Clinical Medicine" (R. O. Wallerstein and S. R. Mettier, eds.), p. 43. Univ. of California Press, Berkeley.
466. Pond, W. G., Lowrey, R. L., Maner, J. H., and Loosli, J. K. (1960). *J. Anim. Sci.* **19,** 1286.
467. Pond, W. G., Veum, T. L., and Lazar, V. A. (1965). *J. Anim. Sci.* **24,** 668.
468. Powanda, M. C., and Beisel, W. R. (1982). *Am J. Clin. Nutr.* **35,** 762.
469. Prabhavathi, T., and Rao, B. S. N. (1981). *Indian J. Med. Res.* **74,** 37.

470. Prema, K., Ramalakshmi, B. A., Madhavapeddi, R., and Babu, S. (1982). *Indian J. Med. Res.* **75**, 534.
471. Radhakrishnan, M. K., and Sivaprasada, J. (1980). *J. Agric. Food Chem.* **28**, 55.
472. Raffin, S. B., Woo, C. H., Roost, K. T., Price, D. C., and Schmid, R. (1974). *J. Clin. Invest.* **54**, 1344.
473. Ramsay, W. N. M., and Campbell, E. A. (1958). *Biochem. J.* **58**, 313.
474. Rao, B. S. N., and Prabhavathi, T. (1982). *J. Sci. Food Agric.* **33**, 89.
475. Rao, G. A., Crane, R. T., and Larkin, E. C. (1983). *Lipids* **18**, 573.
476. Rao, G. A., Manix, M., and Larkin, E. C. (1980). *Lipids* **15**, 55.
477. Rao, K. S., and Rao, B. S. N. (1983). *Nutr. Rep. Int.* **28**, 771.
478. Raper, N. R., Rosenthal, J. C., and Woteki, C. E. (1984). *J. Am. Diet. Assoc.* **84**, 783.
479. Reece, W. O., Self, H. L., and Hotchkiss, D. K. (1984). *Am. J. Vet. Res.* **45**, 2119.
480. Reeves, J. D., Driggers, D. A., Lo, E. Y. T., and Dallman, P. R. (1981). *Am. J. Clin. Nutr.* **34**, 2154.
481. Reeves, J. D., Driggers, D. A., Lo, E. Y. T., and Dallman, P. R. (1981). *J. Pediatr.* **98**, 894.
482. Refsum, S. B., and Schreiner, B. (1984). *Scand. J. Gastroenterol.* **19**, 867.
483. Reilly, A., and Harrison, R. J. (1979). *Hum. Nutr.* **33**, 248.
484. Rice, R. W., Nelms, G. E., and Schoonover, C. O. (1967). *J. Anim. Sci.* **26**, 613.
485. Richard, R. M., Shumard, R. F., Pope, A. L., Phillips, P. H., Herrick, C. A., and Bohstedt, G. (1954). *J. Anim. Sci.* **13**, 274 and 674.
486. Richardson, T., and Guss, P. L. (1965). *J. Dairy Sci.* **48**, 523.
487. Rimington, C. (1959). *Br. Med. Bull.* **15**, 19.
488. Rios, E., Hunter, R. E., Cook, J. D., Smith, N. J., and Finch, C. A. (1975). *Pediatrics* **55**, 686.
489. Roberts, P. D., St. John, D. J. B., Sinha, R., Stewart, J. S., Baird, I. M., Coghill, N. F., and Morgan, J. O. (1971). *Br. J. Haematol.* **20**, 165.
490. Robinson, M. F., McKenzie, J. F., Thomson, C. D., and van Rij, A. L. (1973). *Br. J. Nutr.* **30**, 195.
491. Robinson, S. H. (1969). *Blood* **33**, 909.
492. Robotham, J. L., and Lietman, P. S. (1980). *Am. J. Dis. Child* **134**, 875.
493. Roe, D. A. (1966). *N.Y. State J. Med.* **66**, 1233.
494. Roeser, H. P., Lee, G. R., Nacht, S., and Cartwright, G. E. (1970). *J. Clin. Invest.* **49**, 2408.
495. Rosch, L. M., and Sherman, A. R. (1985). *Fed. Proc., Fed. Am. Soc. Exp. Biol.* **44**, 1511.
496. Rosenmund, A., Gerber, S., Huebers, H., and Finch, C. A. (1980). *Blood* **56**, 30.
497. Rothen, A. (1944). *J. Biol. Chem.* **152**, 679.
498. Rougereau, A., Goré, J., N'diaye, M., and Person, O. (1982). *Am J. Clin. Nutr.* **36**, 314.
499. Rudzki, Z., Baker, R. J., and Deller, D. J. (1973). *Digestion* **8**, 53.
500. Rudzki, Z., and Deller, D. J. (1973). *Digestion* **8**, 35.
501. Rybo, G. (1966). *Acta Obstet. Gynecol. Scand.* **45** (Suppl.), 7.
502. Saarinen, U. M. (1978). *J. Pediatr.* **93**, 177.
503. Saarinen, U. M., and Siimes, M. A. (1978). *Acta Paediatr. Scand.* **67**, 745.
504. Saarinen, U. M., Siimes, M. A., and Dallman, P. R. (1977). *J. Pediatr.* **91**, 36.
505. Sandberg, A.-S., Hasselblad, C., Hasselblad, K., and Hultén, L. (1982). *Br. J. Nutr.* **48**, 185.
506. Sano, S., Inoue, S., Tanabe, Y., Sumiya, C., and Koike, S. (1959). *Science* **129**, 275.
507. Savin, M. A., and Cook, J. D. (1978). *Gastroenterology* **75**, 688.
508. Savin, M. A., and Cook, J. D. (1980). *Blood* **56**, 1029.
509. Schade, A. L., Reinhart, R. W., and Levy, H. (1949). *Arch. Biochem.* **20**, 170.
510. Schricker, B. R., Miller, D. D., and Stouffer, J. R. (1982). *J. Food Sci* **47**, 740.

511. Schricker, B. R., Miller, D. D., and Van Campen, D. (1983). *J. Nutr.* **113**, 996.
512. Schwarz, F. J., and Kirchgessner, M. (1974). *Int. J. Vitam. Nutr. Res.* **44**, 116.
513. Seal, U. S., Sinha, A. A., and Doe, R. P. (1972). *Am. J. Anat.* **134**, 263.
514. Senewiratne, B., Hettiarachchi, J., and Senewiratne, K. (1974). *J. Trop. Med. Hyg.* **77**, 177.
515. Settlemire, C. T., and Matrone, G. (1967). *J. Nutr.* **92**, 153, 159.
516. Sheehan, R. G., and Frenkel, E. P. (1972). *J. Clin. Invest.* **51**, 224.
517. Sherman, A. R. (1978). *Lipids* **13**, 473.
518. Sherman, A. R. (1979). *Lipids* **14**, 888.
519. Sherman, A. R. (1984). *In* "Nutrition, Disease Resistance and Immune Function" (R. R. Watson, ed.), p. 251. Dekker, New York.
520. Sherman, A. R., Bartholmey, S. J., and Perkins, E. G. (1982). *Lipids* **17**, 639.
521. Sherman, A. R., Guthrie, H. A., Wolinsky, I., and Zulak, I. M. (1978). *J. Nutr.* **108**, 152.
522. Sherman, H. C. (1937). "Chemistry of Food and Nutrition," 5th Ed., p. 323. Macmillan, New York.
523. Shoden, A., Gabrio, B. W., and Finch, C. A. (1953). *J. Biol. Chem.* **204**, 823.
524. Shoden, A., and Sturgeon, P. (1958). *Am. J. Pathol.* **34**, 113.
525. Shoden, A., and Sturgeon, P. (1960). *Acta Haematol.* **23**, 376; (1962). **27**, 33; (1961). *Nature (London)* **189**, 846.
526. Shoden, A., and Sturgeon, P. (1960). *Proc. Int. Congr. Histochem. Cytochem. 1st.*
527. Shoden, A., and Sturgeon, P. (1963). *Br. J. Haematol.* **9**, 471.
528. Siimes, M. A., Addiego, J. E., Jr., and Dallman, P. R. (1974). *Blood* **43**, 581; Siimes, M. A., and Dallman, P. R. (1974). *Br. J. Haematol.* **28**, 7.
529. Siimes, M. A., Vuori, E., and Kuitunen, P. (1979). *Acta Paediatr. Scand.* **68**, 29.
530. Simpson, K. M., Morris, E. R., and Cook, J. D. (1981). *Am. J. Clin. Nutr.* **34**, 1469.
531. Simpson, R. J., Raja, K. B., and Peters, T. J. (1985). *Biochim. Biophys. Acta* **814**, 8.
532. Sinniah, R., Bell, T. K., and Neill, D. W. (1973). *J. Clin. Pathol* **26**, 130.
533. Sirivech, S., Frieden, E., and Osaki, S. (1974). *Biochem. J.* **143**, 311.
534. Sisson, T. R. C. (1964). *Fed. Proc., Fed. Am. Soc. Exp. Biol.* **23**, 879.
535. Skeets, O., Frazier, E., and Dickins, D. (1931). *Miss. Agric. Exp. Stn. Bull.* **291**.
536. Skikne, B. S., Lynch, S. R., and Cook, J. D. (1981). *Gastroenterology* **81**, 1068.
537. Smith, A., and Morgan, W. T. (1979). *Biochem. J.* **182**, 47.
538. Smith, A., and Morgan, W. T. (1981). *J. Biol. Chem.* **256**, 10902.
539. Smith, J. E., Moore, K., Boyington, D., Pollmann, D. S., and Schoneweis, D. (1984). *Vet. Pathol.* **21**, 597.
540. Smith, K. T. (1983). *Food Technol.* **37**, 115.
541. Smith, M. D., and Mallet, B. J. (1957). *Clin. Sci.* **16**, 23.
542. Smythe, C. V., and Miller, R. C. (1929). *J. Nutr.* **1**, 209.
543. Sood, S. K., Banerji, L., and Ramalingaswami, V. (1968). *Am. J. Clin. Nutr.* **21**, 1149.
544. Sood, S. K., Ramachandran, K., Mathur, M., Gupta, K., Ramalingaswamy, V., Swarnabai, C., Ponniah, J., Mathan, V. I., and Baker, S. J. (1975). *Q. J. Med.* **44**, 241.
545. Speyer, B. E., and Fielding J. (1974). *Biochim. Biophys. Acta* **332**, 192.
546. Spik, G., and Montreuil, J. (1966). *C. R. Seances Soc. Biol. (Ses Fil)* **160**, 94.
547. Srikantia, S. G., Prasad, J. S., Bhaskaram, C., and Krishnamachari, K. A. V. R. (1976). *Lancet* **1**, 7973.
548. Steinke, F. H., and Hopkins, D. T. (1978). *J. Nutr.* **108**, 481.
549. Stevens, A. R., Coleman, D. H., and Finch, C. A. (1953). *Ann. Intern. Med.* **38**, 199.
550. Stewart, J. G., Ahlquist, D. A., McGill, D. B., Ilstrup, D. M., Schwartz, S., and Owen, R. M. (1984). *Ann. Intern. Med.* **100**, 843.
551. Sturgeon, P. (1956). *Pediatrics* **17**, 341; (1956). **18**, 267.
552. Sturgeon, P. (1959). *Br. J. Haematol.* **5**, 31.

553. Summers, M., Worwood, M., and Jacobs, A. (1974). *Br. J. Haematol.* **28**, 19.
554. Suttle, N. F. (1979). *Br. J. Nutr.* **42**, 89.
555. Swiss, L. D., and Beaton, G. H. (1974). *Am. J. Clin. Nutr.* **27**, 373.
556. Taborsky, G. (1963). *Biochemistry* **2**, 266.
557. Takahashi, S., Kubota, Y., and Matsuoka, O. (1983). *J. Radiat. Res.* **24**, 137.
558. Takkunen, H., and Aromaa, A. (1974). *Am. J. Clin. Nutr.* **27**, 323.
559. Tartour, G. (1973). *Res. Vet. Sci.,* **15**, 389.
560. Tarvydas, H., Jordan, S. M., and Morgan, E. H. (1968). *Br. J. Nutr.* **22**, 565.
561. Taylor, D. J., Mallen, C., McDougall, N., and Lind, T. (1982). *Br. J. Obstet. Gynaecol.* **89**, 1000.
562. Thein-Than, Thane-Toe, Mg-Mg-Thwin, and Aung-Than-Batu (1977). *J. Lab. Clin. Med.* **90**, 231.
563. Teo, C. G., Seet, L. C., Ting, W. C., and Ong, Y. W. (1984). *Pathology* **16**, 141.
564. Thompson, D. B., and Erdman, J. W., Jr. (1984). *J. Nutr.* **114**, 307.
565. Tolan, A., Robertson, J., Orton, C. A., Head, M. J., Christie, A. A., and Millburn, B. A. (1974). *Br. J. Nutr.* **31**, 185.
566. Torrance, J. D., Charlton, R. W., Schmaman, A., Lynch, S. R., and Bothwell, T. H. (1968). *J. Clin. Pathol.* **21**, 495.
567. Toskes, P. P., Smith, G. W., Bensinger, T. A., Giannela, R. A., and Conrad, M. E. (1974). *Am. J. Clin. Nutr.* **27**, 335.
568. Trowbridge, I. S., Newman, R. A., Domingo, D. L., and Sauvage, C. (1984). *Biochem. Pharmacol.* **33**, 925.
569. Turnberg, L. A. (1968). *Lancet* **1**, 921.
570. Ullrey, D. E., Ely, W. T., and Covert, R. L. (1974). *J. Anim. Sci.* **38**, 1276.
571. Ullrey, D. E., Miller, E. R., Thompson, O. A., Ackerman, I. M., Schmidt, D. A., Hoefer, J. A., and Luecke, R. W. (1960). *J. Nutr.* **70**, 187.
572. Underwood, E. J. (1934). *Aust. Vet. J.* **10**, 87.
573. Underwood, E. J. (1966). *"The Mineral Nutrition of Livestock."* FAO/CAB, Publ. Central Press, Aberdeen.
574. Underwood, E. J., and Morgan, E. H. (1963). *Aust. J. Exp. Biol. Med. Sci.* **41**, 247.
575. Untario, S., Juwana, M. T., Netty, R. H. T., Permono, B., and Harsono, N. (1978). *Trop. Geogr. Med.* **30**, 337.
576. Vahlquist, B. (1941). *Acta Paediatr. (Stockholm)* **28** (Suppl.), 5.
577. Vahlquist, B. (1950). *Blood* **5**, 874.
578. Valberg, L. S. (1971). In "Intestinal Absorption of Metal Ions, Trace Elements and Radionuclides" (S. C. Skoryna and D. Waldron-Edward, eds.), p. 257. Pergamon, Oxford.
579. Valberg, L. S., Ludwig, J., and Olatunbosun, D. (1969). *Gastroenterology* **56**, 241.
580. Van Campen, D. (1972). *J. Nutr.* **102**, 165.
581. Van Campen, D. (1973). *J. Nutr.* **103**, 139.
582. Van Campen, D., (1974). *Fed. Proc., Fed. Am. Soc. Exp. Biol.* **33**, 100.
583. Van Campen, D., and Gross, E. (1969). *J. Nutr.* **99**, 68.
584. Van Dijk, J. P., Van Kreel, B. K., and Heeren, J. W. A. (1983). *J. Dev. Physiol.* **5**, 195.
585. Van Dokkum, W. Wesstra, A., and Schippers, F. (1982). *Br. J. Nutr.* **47**, 451.
586. Van Weerden, E. J., Huisman, J., and Sprietsma, J. E. (1978). *Z. Tierphysiol. Tierernaehr. Futtermittelkd.* **40**, 209.
587. Venkatachalam, P. S. (1968). *Am. J. Clin. Nutr.* **21**, 1156.
588. Venn, J. A. J., McCance, R. A., and Widdowson, E. M. (1947). *J. Comp. Pathol. Ther.* **57**, 314.
589. Viteri, F. E., Garcia-Ibañez, R., and Torún, B. (1978). *Am. J. Clin. Nutr.* **31**, 961.
590. Vulimiri, L., Linder, M. C., and Munro, H. N. (1977). *Biochim. Biophys. Acta* **497**, 280.

142

Eugene R. Morris

591. Walters, G. O., Miller, F. M., and Worwood, M. (1973). *J. Clin. Pathol.* **26**, 770.
592. Weaver, C. M., Nelson, N., and Elliott, J. G. (1984). *J. Nutr.* **114**, 1042.
593. Weber, G., and Schwedt, G. (1984). *Z. Lebensm. Unters. Forsch.* **178**, 110.
594. Weinberg, E. D. (1978). *Microbiol. Rev.* **42**, 45.
595. Weinberg, E. D. (1983). *Nutr. Cancer* **4**, 223.
596. Weintraub, L. R., Conrad, M. E., and Crosby, W. H. (1965). *Br. J. Haematol.* **11**, 432.
597. Weintraub, L. R., Conrad, M. E., and Crosby, W. H. (1968). *J. Clin. Invest.* **47**, 531.
598. Weisenberg, E., Halbreich, A., and Mager, J. (1980). *Biochem. J.* **188**, 633.
599. Welch, R. M., and Van Campen, D. R. (1975). *J. Nutr.* **105**, 253.
600. Wheby, M. S., Conrad, M. E., Hedberg, S. E., and Crosby, W. H. (1962). *Gastroenterology* **42**, 319.
601. Wheby, M. S., and Spyker, D. A. (1981). *Am. J. Clin. Nutr.* **34**, 1686.
602. Wheby, M. S., Suttle, G. E., and Ford, K. T. (1970). *Gastroenterology* **58**, 647.
603. Widdowson, E. M. (1950). *Nature (London)* **166**, 626; Spray, C. M., and Widdowson, E. M. (1951). *Br. J. Nutr.* **4**, 332; Widdowson, E. M., and Spray, C. M. (1951). *Arch. Dis. Child.* **26**, 205.
604. Widdowson, E. M., and McCance, R. A. (1942). *Lancet* **1**, 588.
605. Widdowson, E. M., and McCance, R. A. (1948). *Biochem. J.* **42**, 488.
606. Williams, J. (1962). *Biochem. J.* **83**, 355.
607. Wilms, J. W., and Batey, R. G. (1983). *Am. J. Physiol.* **244**, G138.
608. Wilson, J. F., Heiner, D. C., and Lahey, M. E. (1962). *J. Pediatr.* **60**, 787.
609. Wiltink, W. F., van Eijk, H. G., Bobeck-Rutsaert, M. M., Gerbrandy, J., and Leijnse, B. (1972). *Acta Haematol.* **47**, 269.
610. Winkelmann, J. C., Mariash, C. N., Towle, H. C., and Oppenheimer, J. H. (1981). *Science* **213**, 569.
611. Wong, C. T., and Morgan, E. H. (1973). *Q. J. Exp. Physiol. Cogn. Med. Sci.* **58**, 47.
612. Wong, C. T., and Morgan, E. H. (1974). *Aust. J. Exp. Biol. Med. Sci.* **52**, 413.
613. Woodruff, C. W., Latham, C., and McDavid, S. (1976). *J. Pediatr.* **90**, 36.
614. Worwood, M., Aherne, W., Dawkins, S., and Jacobs, A. (1975). *Clin. Sci. Mol. Med.* **48**, 441.
615. Worwood, M., Summers, M., Miller, F., Jacobs, A., and Whittaker, J. A. (1974). *Br. J. Haematol.* **28**, 27.
616. Yamagata, N., and Yamagata, T. (1964). *Bull. Inst. Public Health (Tokyo)* **13**, 11.
617. Yeung, D. L., Pennell, M. D., Leung, M., Hall, J., and Anderson, G. H. (1981). *Can. Med. Assoc. J.* **125**, 999.
618. Yip, R., Johnson, C., and Dallman, P. R. (1984). *Am. J. Clin. Nutr.* **39**, 427.
619. Yip, R., Schwartz, S., and Deinard, A. S. (1983). *Pediatrics* **72**, 214.
620. Yoshino, Y., and Hiramatsu, Y. (1974). *J. Biochem.* **75**, 221.
621. Youngs, S., and Bomford, A. (1984). *Clin. Sci.* **67**, 273.
622. Zähringer, J., Konijn, A. M., Baliga, B. S., and Munro, H. N. (1975). *Biochem. Biophys. Res. Commun.* **65**, 583.

5

Cobalt

RICHARD M. SMITH

CSIRO Division of Human Nutrition
Kintore Avenue
Adelaide, South Australia, Australia

I. COBALT AS AN ESSENTIAL TRACE ELEMENT

Among the trace elements essential for humans and animals, cobalt (Co) is unique in that the requirement by the animal is not for an ionic form of the metal to be converted in the body into one or more catalytically active species, but for a preformed cobalt compound, vitamin B_{12}, that is itself the catalytically active entity. As indicated in Fig. 1, a direct requirement for cobalt is found only in certain bacteria and algae, and the only purpose of this cobalt is for the production of the cobamides needed for the metabolic processes of the organism. Cobamides are not produced by higher plants or by animals, and the fulfillment of the absolute requirements of animals for vitamin B_{12} reflects a strong but sometimes indirect symbiotic relationship between microorganisms and animals. The relationship is most direct in ruminants in which vitamin B_{12} produced by certain rumen microorganisms is absorbed further down the gastrointestinal tract of the animal. In those carnivores and mixed feeders that consume ruminant flesh, the ultimate source of vitamin B_{12} is still the microorganisms of the rumen, but the symbiotic relationship is once removed. Other herbivores and those mixed feeders that do not consume ruminant flesh receive their vitamin B_{12} through other food chains, but in all cases the first members of the chain are found to be microorganisms, and it is only the microbial synthesis of vitamin B_{12} that is vulnerable to cobalt deficiency.

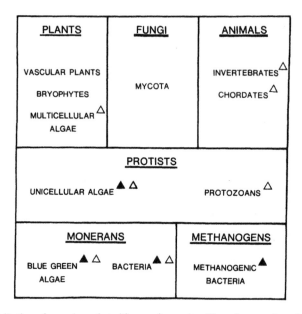

Fig. 1. Distribution of organisms that either produce cobamide and so require cobalt (▲), or that require preformed cobamides (△). With the exception of the animal kingdom, where the requirement for vitamin B_{12} is thought to be universal, only a few organisms in each classification may show the indicated properties. Among the chordates, the ruminant presents a special case. The vitamin B_{12} required by the animal is normally synthesized by bacteria in the rumen and absorbed further down the alimentary tract. Since these bacteria are supplied with cobalt solely through the diet of the animal, cobalt becomes a dietary essential for the ruminant itself.

The production of cobamides from inorganic cobalt by microorganisms in nature is widespread but erratic. Of the six separate kingdoms of living organisms that are now recognized (65,79), vitamin B_{12} or related cobamides are produced by representatives of at least three (Fig. 1). These are the methanogens (118,245), the monerans, including bacteria (3,44,76,84,93,182) and blue-green algae (35,78,97,181), and the protists (35,181). Some multicellular red and brown algae contain vitamin B_{12}, but this may be derived from the environment (63,181). Other members of these groups that do not produce cobamides may be found to require them, and these requirements are met by symbiotic associations with those that do. Thus, seawater itself contains from 0.1 to 50 ng cobamides per liter, sufficient to meet the needs of most organisms that live there.

Vitamin B_{12} is needed by all animal species that have been critically studied, but except for certain algae, members of the plant kingdom have been thought neither to produce nor to require cobamides. The discovery by Poston of two vitamin B_{12}-dependent enzymes, leucine 2,3-amino mutase and methylmalonyl coenzyme A mutase, in potato tubers (179) has rendered this generalization

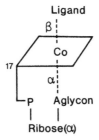

Structure I.   General form of cobamides.

insecure. Leucine 2,3-amino mutase was also detected in extracts of bean seedlings, where its activity was found to be suppressed by the specific vitamin B_{12} binder intrinsic factor, an observation suggesting the presence in the plant enzyme of a vitamin B_{12}-like coenzyme or prosthetic group (178). Notwithstanding these findings, the amounts of any cobamides in plant tissues are so small as to be nutritionally insignificant (201).

Vitamin B_{12}, or cobalamin, is the only identifiable cobamide normally found in animal tissues, but it is only one of many cobalt-containing corrinoids produced by bacterial and algal cells. The nomenclature of the corrinoids is described in Ref. 102 and, using the conventions adopted there, Structure I represents the general form of the natural cobalt-containing corrinoids or cobamides. Structure I shows the cobalt atom at the center of the corrin ring, the α ligand which characterizes the specific cobamide, and the variable β ligand which is directly involved in the catalytic activities of the molecule. In commercial vitamin B_{12} (**II**), the β ligand is a cyano group and the α ligand (aglycone) is 5,6-dimethylbenzimidazole. This substance, also known as cyanocobalamin, is defined as vitamin B_{12}, although the cyano group is not generally present in the living cell and is introduced during isolation of cobamides to confer stability. Table I shows the more important bacterial and algal cobamides with an indication of their biological activity for animals (201).

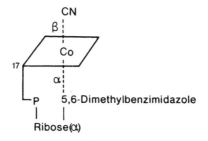

Structure II.   Vitamin B_{12}.

Table I

Structure and Activity of Natural Corrinoids Prepared as Cyano Compounds

Common names	α-Ligandyl (β-ligandyl) cobamide	Activity in animal assays relative to vitamin $B_{12}{}^a$
Vitamin B_{12}, cyanocobalamin	5,6-Dimethylbenzimidazolyl (Coβ cyano) cobamide	Fully active
	Benzimidazolyl (Coβ cyano) cobamide	Partly active
B_{12}-Factor (III)	5-Hydroxybenzimidazolyl (Coβ cyano) cobamide	Partly active
ψ-Vitamin B_{12}	Adenyl (Coβ cyano) cobamide	Inactive
Factor A	2-Methyladenyl (Coβ cyano) cobamide	Very slightly active
Factor B	Cobinamide (no ligands or side chains)	Inactive
Factor C	Guanyl (Coβ cyano) cobamide	Unknown, but inactive for *O. malhamensis*

^{a} Data from Smith (201). In general activity in animal systems is paralleled by activity for *Ochromonas malhamensis*.

The catalytically active forms of vitamin B_{12} present in animal cells are adenosylcobalamin (**III**) and methylcobalamin (**IV**). The unique chemical properties of these substances, each of which contains a covalent carbon–cobalt bond, is described briefly in Section VII, which shows the mechanism of action of cobalt in the vitamin B_{12} coenzymes.

The production of cobamides by representatives of so many different classes of microorganisms suggests that the essentiality of cobalt may have emerged early in the evolution of life. The widespread network of symbiotic relationships involving the cobamides that still persists, suggests that the molecular evolutionary niche that was then filled by cobalt was a crucial one. However, apart from the animal kingdom where a requirement for vitamin B_{12} seems to be universal,

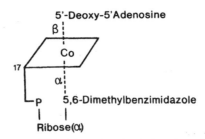

Structure III.   Adenosylcobalamin (coenzyme B_{12}).

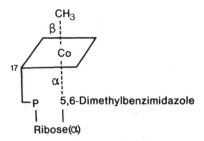

Structure IV. Methylcobalamin.

most other broad groups of organisms contain some members that do and some that do not require either cobalt or a cobamide.

The dichotomy between vitamin B_{12}-dependent and vitamin B_{12}-independent pathways in microorganisms has been examined by Dickman (49), who has proposed that a requirement for cobamides, and hence for cobalt, may have evolved very early indeed in the phylogenetic sequence. Thus, of the two microbial pathways that bring about reduction of ribonucleoside phosphates to deoxyribonucleoside phosphates, one involves vitamin B_{12} and employs cobalt while the other does not. Dickman points out that, in general, those organisms that use the vitamin B_{12}-dependent pathway, many of them anaerobes and blue-green algae, are phylogenetically more primitive than those that do not. The fact that some strains of the strictly anaerobic *Clostridia* produce corrinoids but do not produce porphyrins has led to the further suggestion that the requirement for cobalt may have emerged before the requirement for iron (49).

Another line of evidence has appeared that is also consistent with the very early evolution of a biochemical role for cobalt-containing corrinoids. The classification of the methanogenic bacteria as a new and separate kingdom as shown in Fig. 1 emerged as a result of oligonucleotide sequence analyses of RNA from 10 methanogenic bacteria (65). Close sequence homologies were found between certain RNA segments from all of the methanogens, but these patterns differed so sharply from those common to bacteria and blue-green algae that an entirely separate classification was indicated (65,79). The methanogens are a morphologically diverse group of organisms that can grow anaerobically on the energy derived from the conversion of hydrogen and carbon dioxide into methane. An involvement of cobamides in the reaction sequence was suggested initially by the presence of cobamides in many methanogens and by the observation that methyl cobalamin could serve as a precursor for methane production by extracts of *Methanosarcina barkeri* (31), but no general role for vitamin B_{12} in the production of methane emerged for some time. Then, following the recognition that the relatively large quantities of cobamides found in *M. barkeri* were not present as vitamin B_{12} itself, but predominantly as the closely related analog

B_{12}-factor III (175) (see Table I), a resolution of the biochemical role of co-bamides in methane synthesis by *M. barkeri* was achieved (221,240,245). It remains to be seen whether other methane producers share this mechanism, but the presence of cobamides in many methanogens (118,133,216), the existence of the cobamide pathway in *M. barkeri*, and the inhibition by corrinoid antagonists of methane synthesis by methanogenic bacteria other than *M. barkeri* (107) make this not unlikely.

The methanogens may represent the vestige of a class of archaic anaerobes that, in parallel with the monerans, evolved from a common but much earlier precursor (65). If so, their origins are very ancient indeed, yet in both its form and function their use of cobalt as an essential trace element differs only slightly from that of the modern mammalian cell employing methyl cobalamin to synthe-size the methyl group of the amino acid methionine. The biochemical roles of cobalt seem to be very specific and very highly conserved.

II. COBALT IN ANIMAL TISSUES AND FLUIDS

The report of the Task Group on Reference Man (101) puts the total cobalt content of the body of adult humans at <1.5 mg, with the liver containing 0.11 mg, skeletal muscle 0.2 mg, bone 0.28 mg, hair 0.31 mg, and adipose tissue 0.36 mg. The values used were based on multiple measurements in several laboratories, using a variety of methods, and they were generally carried out prior to the advent of atomic absorption spectrophotometry. In fact, the latter technique has found little favor in the field of tissue cobalt analysis, which tends now to be performed by neutron activation techniques (103,105,241). Tissue levels of cobalt are very low in comparison with other essential metals, and measurements are consequently more uncertain. Values listed for the cobalt concentration of human tissues in the compilation of Iyengar *et al.* (104) show ranges for many tissues that cover two, and even three, orders of magnitude. Such large differences probably reflect methodological errors rather than true variations in cobalt content and have rendered necessary selection of those values used for reference man. Based on those values for the normal adult, accumula-tion of cobalt does not occur in any particular organ or tissue, but liver, heart, and bone contain the highest concentrations of this element and kidney may also show relatively high levels (228).

A similar distribution of retained cobalt has been demonstrated in the tissues of mice, rats, rabbits, pigs, dogs, chicks, sheep, and cattle, with little distinction between species (34,40,122,123,190,218). The concentrations found in normal human tissues are similar to those in other species (101,228), and there is little evidence to suggest any accumulation of cobalt in human tissues with age (194,228). Thus the concentration of cobalt in six newborn human livers was found to be 0.16 ± SD 0.08 µg/g (dry weight) (5) compared with 0.22 µg/g (dry weight) assessed for reference man (101). Table II shows representative values for ovine tissues compared both with those of reference human (101) and with the data of Tipton and Cook for human tissues measured by emission spectrography (228).

As shown in Table II, concentrations of cobalt in the tissues are below normal in cobalt deficiency (20). Prior to the recognition that the most suitable index of the cobalt status of grazing animals was the concentration in the liver of vitamin B_{12}, much careful work was carried out in determining the ranges of liver cobalt levels in normal and clinically cobalt-deficient sheep (154,236) and cattle (77,144). McNaught (144) suggested that 0.04–0.06 ppm cobalt (dry basis) or less in the livers of sheep and cattle indicated cobalt deficiency and that 0.08–0.12 ppm cobalt indicated a satisfactory cobalt status. Unlike iron or copper, cobalt does not normally accumulate in the fetal liver of ruminants or humans (5,144), but the cobalt (and vitamin B_{12}) content of the liver of the newborn lamb and calf is reduced below normal if the mother has received a cobalt-deficient diet. Prepartum treatment of the mother with oral cobalt can restore the cobalt level of the fetal liver to normal (14,166).

Liver and kidney concentrations can be increased well above normal levels by cobalt injections without this cobalt being available for vitamin B_{12} synthesis (6,28,100,153,174), and, conversely, ruminants on cobalt-deficient diets can be maintained in health by injections of vitamin B_{12} without raising their liver cobalt levels to normal. For these reasons liver cobalt concentration, though of some practical value, is not a reliable criterion of the nutritional cobalt status of the animal.

The proportion of liver cobalt that occurs as vitamin B_{12} varies with the cobalt status of the animal. Under grazing conditions where there is adequate cobalt in the pasture, most of the liver cobalt can be accounted for as vitamin B_{12}, but in

Table II
Cobalt Concentration in Tissues

Species	Cobalt concentration (ppm, dry basis)				
	Liver	Spleen	Kidney	Heart	Pancreas
Normal human[a]	0.22	<0.16	0.06	0.11	<0.08[d]
Normal human[b]	0.18	0.09	0.23	0.10	0.06[d]
Healthy sheep[c]	0.15	0.09	0.25	0.06	0.11
Cobalt-deficient sheep[c]	0.02	0.03	0.05	0.01	0.02

[a] From "Report of the Task Group on Reference Man" (101).
[b] From Tipton and Cook (228).
[c] From Askew and Watson (20).
[d] Derived from the same data.

cobalt deficiency only about one-third of the liver cobalt exists in this form (12,13). This indicates that in cobalt deficiency, liver vitamin B_{12} becomes depleted faster than other forms of cobalt. The converse situation occurs on oral supplementation of cobalt-deficient lambs with cobalt, when the increase of liver cobalt was found to be up to 5-fold greater than could be accounted for by the increase in vitamin B_{12} (12,13). There is little information on the forms in which cobalt exists in the tissues other than as vitamin B_{12}, although such forms have been detected in several species (122,123,161).

As with other tissues, early reports of the concentrations of cobalt in normal human blood showed extreme variation (104,114,224,244). The cobalt concentrations proposed for reference man (101) are as follows: for whole blood 0.31 ng/g, for blood plasma 0.45 ng/g, and for erythrocytes 0.14 ng/g. Later reported values for human plasma cobalt tend to range from 0.1 to 0.2 ng/ml (105,241), although 10-fold higher values were reported in newborn serum (163). Even the lowest values reported (105,241), however, are at least 5-fold greater than would be required to account for normal adult levels of plasma vitamin B_{12} (0.6 ng/ml of vitamin B_{12} would contribute 0.024 ng/ml of cobalt). Few measurements of blood cobalt levels have been reported for ruminants, but confirmation is needed of values of 60–80 ng/ml plasma reported in 1982 for unsupplemented lactating dairy cows (137).

Normal cow's milk analyzed by microcolorimetric methods showed concentrations of cobalt ranging from 0.4 to 1.1 μg/liter, with a mean near 0.5 μg/liter (18,138,211,225). This value has been borne out in work using neutron activation analysis (103) which showed a pooled sample of cow's milk to contain 5.1 ± 0.55 ng cobalt per gram of milk solids. This is equivalent to about 0.4 μg cobalt per liter of milk. Analyzed similarly, pooled human milk contained only 1.5 ± 0.5 ng/g milk solids, equivalent to about 0.1 μg/liter of human milk (103). Cow's colostrum is higher in cobalt than mature milk (138). Supplementing the normal diet of the cow with cobalt can increase the level in milk (18), but this cobalt cannot be converted into vitamin B_{12} by suckling animals until a suitable ruminal population of microbes has developed. It has been reported that supplementary cobalt is ineffective in raising the vitamin B_{12} concentration in cow's milk if the ration already contains sufficient cobalt (90). However, cobalt supplementation of cobalt-deficient or marginal diets significantly increases the vitamin B_{12} content of both milk and colostrum (88,166).

III. COBALT METABOLISM

A. Absorption

Orally administered cobalt is well absorbed by small laboratory animals and by humans. Thus Toskes *et al.* (231) found that normal mice absorbed 26.2% of an oral dose of labeled cobalt, and balance studies in humans suggest that a substantial but variable amount ranging from 20 to 97% of dietary cobalt is absorbed from the intestine (56,87). The absorption of labeled cobaltous chloride diminishes when it is administered after a meal or complexed with protein (171).

Cobalt absorption, as well as iron absorption, is significantly enhanced in iron deficiency in the rat (176) and human (239). Increased cobalt absorption also parallels increased iron absorption in patients with portal cirrhosis or with idiopathic hemochromatosis (239) (see Fig. 2). These findings led Valberg and coworkers to postulate that cobalt shares a common intestinal mucosal transport pathway with iron, in which acceleration of transport of both elements is governed by the same mechanism (226,238). The idea of a common absorptive pathway for both elements involving competition for absorption sites is compatible with the demonstration of a mutual antagonism between the two elements at the absorptive level. Forth and Rummel (64) showed that the absorption of ^{59}Fe from jejunal loops of iron-deficient rats is reduced by almost two-thirds in the presence of a 10-fold higher cobalt than iron concentration. A further increase to a 100-fold excess of cobalt suppressed absorption of ^{59}Fe almost completely. In the reverse experiment a 10-fold excess of iron decreased cobalt-58 (^{58}Co) absorption to about one-half the control value and a 100-fold excess to less than one-fifth the control value. The mechanism by which iron deficiency enhances

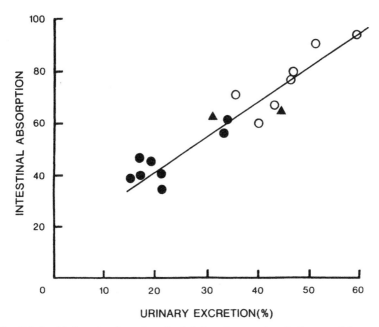

Fig. 2.   Relationship between absorption of cobalt from the gastrointestinal tract and the excretion of cobalt in human urine. Shown are the effects of normal iron repletion (●), iron depletion (○), and idiopathic hemochromatosis (▲). From Valbert (238).

cobalt absorption has still not been resolved. The increased transport observed in closed duodenal loops of iron-deficient rats *in vivo,* however, did not appear to involve binding of cobalt to cytoplasmic proteins in the mucosal cells (195). Consistent with this finding is the observation that the mode of inhibition of duodenal iron absorption by cobalt ions relates to the release of iron from the mucosal cells into the bloodstream and not to uptake of iron by the cells themselves (29).

Absorption of soluble cobalt by the ruminant is much less efficient than that in simple-stomached animals. Following oral or intraruminal administration of labeled cobalt to sheep or cattle, 84–98% appeared in the feces within 5–14 days (40,136,161,202). The poor absorption may be related to the rapid binding of cobalt by rumen microorganisms that takes place (40,190,202,230).

In most experimental animals the tissue distribution of injected radiocobalt is similar, with initial rapid uptake by liver and kidney and, to a lesser extent, by the spleen, pancreas, and parts of the gastrointestinal tract (40,122,161,167). With elapsed time this label is fairly rapidly lost, and the distinction between tissues becomes less marked (167). Loss of injected labeled cobalt by mice was found to be especially rapid, with only 4% of the initial dose remaining after 6 days (64). In chicks following intravenous injection of cobalt-60 (^{60}Co), the wall of the large intestine was found to be especially radioactive, and it was concluded that ^{60}Co was actively secreted there (123).

B. Excretion

The major route of excretion of cobalt in humans is the urine (87,99,109,194,239), and a direct relationship has been observed between the proportion of an oral dose of cobalt that is absorbed from the intestine and the proportion that is excreted in the urine (238) (see Fig. 2). Small amounts of cobalt are also lost by way of the feces (238), sweat (42), and hair (194). Studies with experimental animals show that a proportion of fecal cobalt represents cobalt that has been reexcreted via the bile and to a small extent through the intestinal wall (38,40,167), but the pancreatic juice is not a significant route of excretion of cobalt (196).

In animals as well as in humans, the majority of systemic cobalt is eliminated via the urine (40,167).

IV. COBALT IN RUMINANT NUTRITION

A. Discovery of the Need for Cobalt

Restricted areas in several parts of the world were long known to be unsatisfactory for the raising of sheep and cattle, in spite of apparently satisfactory pastures. Horses and other nonruminants thrived in these areas, whereas sheep and cattle became weak and emaciated, progressively anemic, and usually died. The condition was essentially similar in each locality and could be controlled only by periodic removal of the animals to healthy areas for varying periods.

The first serious scientific work on any of these diseases began in New Zealand at the end of the last century. A series of investigations led by Aston (21–24) culminated in the claim that "bush sickness" in cattle was due to a deficiency of iron. This was based on the occurrence of anemia in affected animals, the low iron content of "bush sick" pastures and soils compared with those of healthy areas, and the effectiveness of crude iron salts and ores in preventing or curing the disease. The iron deficiency theory received support from other parts of the world where iron compounds were also found effective in preventing or curing similar maladies (169). Subsequent work in New Zealand, showing insignificant differences in the iron content of healthy and "bush sick" pastures and little correlation between the iron content of different compounds and their curative effects (81), cast some doubt on the iron deficiency hypothesis. Indisputable evidence against this hypothesis was obtained by Filmer and Underwood (58,60,61,233,235) in their investigation of a wasting disease of sheep and cattle (enzootic marasmus) occurring in the south coastal areas of Western Australia. These workers could find little relation between the size of an effective dose and the amount of iron it supplied. Furthermore, they discovered that (1) the liver and spleen of affected animals contained large stores of iron, (2) whole liver was curative in doses that supplied insignificant amounts of iron, and (3) an *iron-free* extract of one of the curative compounds (limonite, $Fe_2O_3 \cdot H_2O$) was just as potent as whole limonite.

The hypothesis was advanced that enzootic marasmus was due to a deficiency in the soils and herbage of some trace element which occurred as a contaminant of the iron compounds successfully used. Underwood and Filmer (235) then chemically fractionated limonite and, after some misleading tests with nickel suggested by the large amounts of that element present, found that the potency of the limonite resided in the cobalt that it contained. Normal growth and health of sheep and cattle on the deficient pastures were then secured by the administration of small oral doses of a cobalt salt, and the soils, pastures, and livers of affected animals were shown to contain subnormal cobalt concentrations (236).

While the above investigations were proceeding, studies of "coast disease" of sheep occurring in the calcareous sandy dunes of South Australia were under way. The possibility that this disease was due to a deficiency of a mineral element was recognized early. Supplements of phosphorus and copper proved ineffective, and the relatively small doses of the iron compounds used produced only a transitory improvement in the condition of "coasty" sheep (155). A mineral mixture supplying small amounts of iron, copper, boron, manganese, cobalt, nickel, zinc, arsenic, bromine, fluorine, and aluminium, on the other hand, was highly effective (155). The fact that coast disease is accompanied by anemia, coupled with Waltner and Waltner's earlier finding (242) that cobalt stimulates hematopoiesis in rats, led R.G. Thomas to suggest that cobalt might be the element responsible for the beneficial effects of the mineral mixture. Experiments by Marston (148) and by Lines (134) revealed a dramatic improvement in the condition of coasty sheep from the oral administration of 1 mg cobalt per day. These findings preceded by a few months the outcome of the methodical approach by Underwood and Filmer (235). Subsequent experiments disclosed that supplementation with copper, as well as with cobalt, was necessary for completely successful treatment of coast disease.

Within a few years of these discoveries, cobalt supplements were found to be equally effective in the cure and prevention of all the diseases previously shown to respond to massive doses of iron compounds, and the soils and herbage of the affected areas were shown to contain subnormal levels of cobalt (see Russell and Duncan, Ref. 192). Subsequently it became apparent that larger areas existed in many countries where the deficiency was less severe. We now know that all of the consequences of cobalt deficiency in ruminants are due to a deficiency of vitamin B_{12} in the tissues and can be averted by injection of vitamin B_{12} (96,153,210).

B. Manifestations of Cobalt Deficiency

The appearance of a severely cobalt-deficient animal is one of emaciation and rheumy-eyed listlessness, indistinguishable from that of a starved animal except that the visible mucous membranes are blanched and the skin is pale and fragile. The emaciation or wasting of the musculature (marasmus) results from the failure of appetite, which is an early and conspicuous feature of the disease, along with paleness of the skin and mucous membranes from the anemia which, in the field, usually develops progressively with the severity of the deficiency.

The cobalt deficiency syndrome varies from this acute and fatal condition through a series of less acute stages to a mild, ill-defined, and often transient state of unthriftiness that is difficult to diagnose. When ruminants are confined to cobalt-deficient pastures or are fed cobalt-deficient rations, there is a characteristic response. At first they thrive and grow normally for a period of several weeks or months, depending on their age, previous history, and the degree of deficiency of the diet. During this time they are drawing on the vitamin B_{12} reserves in the liver and other tissues. This period is followed by a gradual loss of appetite and failure of growth or loss of body weight, succeeded by extreme inappetence, rapid wasting, and anemia, culminating in death (58,155). Loss of appetite and low plasma glucose levels have been reported to be the best indicators of developing cobalt deficiency in steers (146), but fasting plasma glucose levels were not depressed in cobalt-deficient sheep (204). In some areas the cobalt deficiency is less severe, so that the acute disease conditions do not arise. Milder manifestations characterized by unthriftiness in young stock, and in mature stock also by diminution of lactation and birth of weak lambs and calves that do not survive long, are the only evidence of cobalt deficiency in such areas. In these mild or marginal areas the unthriftiness can be apparent in some years and absent in others (124).

At autopsy the body of severly affected animals presents a picture of extreme emaciation, often with a total absence of body fat. The liver is fatty, the spleen hemosiderized, and in some animals there is hypoplasia of the erythrogenic tissue in the bone marrow (58). The red cell numbers and blood hemoglobin levels are always subnormal and sometimes very low. The anemia was reported by Filmer (58) to be normocytic and hypochromic in lambs and microcytic and hypochromic in calves. The nature of the anemia in cobalt-deficient calves does not appear to have been reinvestigated. Later studies characterized the anemia in lambs as normocytic and normochromic (73,209). However, it is not the often mild anemia that is responsible for the main signs of cobalt deficiency. Inappetence and marasmus invariably precede any considerable degree of anemia. The first discernible response to cobalt feeding, or vitamin B_{12} injections, is a rapid improvement in appetite and body weight. Improvement in the blood picture may be equally dramatic but is sometimes delayed.

These clinical and pathological manifestations of cobalt deficiency in ruminants are accompanied by characteristic biochemical changes in the tissues. The decline in cobalt concentrations in the liver and kidney has already been mentioned (Table II). The cobalt level in the rumen fluid also falls, as would be expected from the subnormal dietary intake of the element. When this has fallen below a critical level, tentatively set at <0.5 ng/ml (202), vitamin B_{12} synthesis by the rumen microorganisms is inhibited and the levels of the vitamin decline in the rumen, blood, liver, and other tissues (12,13,47,95,111,112,151) (see Table III). The tissue vitamin B_{12} depletion is accompanied by a marked depression of appetite, impaired propionate utilization (as discussed in Section IV,F), and a metabolic inefficiency (203). The latter was reflected in a 30% faster loss of

Table III

Vitamin B_{12} Activity of Blood, Liver, and Rumen Ingesta of Sheep[a]

Sample	Cobalt-deficient		Cobalt-sufficient (full fed)		Cobalt-sufficient (limited fed)	
	n	Mean ± SD	n	Mean ± SD	n	Mean ± SD
Whole blood (ng/ml)	16	0.47±0.11	6	2.3±0.6	3	4.3±1.5
Liver (μg/g wet wt)	9	0.05±0.01	6	0.93±0.26	3	1.24±0.20
Rumen ingesta (μg/g dry wt)	4	0.09±0.06	5	1.3±0.4	3	1.3±0.9

[a] From Hoekstra et al. (95).

body weight, a higher excretion of fecal nitrogen, and a higher fasting energy expenditure than was observed in pair-fed sheep injected with vitamin B_{12}. On the other hand, retention of combustible energy from the diet by vitamin B_{12}-deficient animals was not significantly different from that of pair-fed animals treated with cobalt or vitamin B_{12} (203). The low plasma glucose and elevated plasma pyruvate levels reported in one study of cobalt-deficient sheep (145,147) was not encountered in an earlier study (204), although the rate at which plasma glucose rose following an infusion of propionate was significantly depressed in deficient sheep in comparison with their pair-fed controls (204).

C. Cobalt and Vitamin B_{12} Requirements

Under grazing conditions, lambs are the most sensitive to cobalt deficiency, followed by mature sheep, calves, and mature cattle, in that order (7). Field experience suggests that species differences among ruminants in cobalt requirements are small (165). Early evidence from Australia (60,61,149) and New Zealand (143) indicated that 0.07 or 0.08 ppm cobalt in the dry diet was just adequate for sheep and cattle. This level of dietary cobalt therefore became accepted as the minimum requirement for these species. Later studies placed the minimum level of "pasture-associated" cobalt required by growing lambs appreciably higher, namely, 0.11 ppm on the dry basis (15). In a later study of a marginally cobalt-deficient area in New Zealand, Andrews (8) assessed the position as follows: "Mean (pasture) values of 0.11 ppm Co or more would probably exclude the likelihood of cobalt deficiency. Mean values approaching 0.08 ppm would suggest but not prove actual or potential existence of the disease." More precise estimates of minimum cobalt requirements applicable under all grazing conditions are difficult because of the influence of many variables such as seasonal changes in herbage cobalt concentrations, selective grazing habits, and soil contamination. Lee and Marston (130) provide evidence that for sheep grazing grossly cobalt-deficient pastures, the total intake of cobalt to ensure optimum growth and hemoglobin production is 0.08 mg/day when supplementary cobalt is given three times each week. In growing lambs the requirement was stated to be higher. These findings apply to sheep consuming pasture, that is, on high-roughage diets. On low-roughage diets there is a significant reduction in vitamin B_{12} production (220).

The results of pen-feeding experiments with sheep consuming purified diets carried out by Somers and Gawthorne (213) point to a minimum cobalt requirement closer to the earlier level quoted above—namely, 0.07–0.08 ppm—than to a level of 0.11 ppm. These figures were obtained with sheep that were 18 months of age at the beginning of an experiment designed for other purposes and lasting only 39 weeks. In a study designed specifically to determine the cobalt and vitamin B_{12} requirements of sheep, Marston (150) found that sheep confined to pens and given a cobalt-deficient ration which supplied about 30 μg cobalt per day, required for maintenance of normal growth rate a cobalt supplement approaching 40 μg administered by mouth daily; for maintenance of what appeared to be the maximum vitamin B_{12} status of sheep, namely 3 ng vitamin B_{12} per milliliter of serum and 1.4 μg vitamin B_{12} per gram of liver tissue, a cobalt supplement of between 0.5 and 1.0 mg/day was necessary. Natural rations high in cereal grains do not necessarily supply adequate cobalt. Growth responses to supplementary cobalt have been demonstrated in steers fed a fattening ration based on barley grain (183), sorghum grain, and silage (162).

The requirements of ruminants for parenterally administered or absorbed vitamin B_{12} are of the same order on a body weight basis as those found for the rat, pig, and chick but are higher than those for humans (see Smith and Loosli, Ref. 211). The minimum total requirement of sheep for parenteral or absorbed cobalamin has been estimated to be 11 ± 2 μg/day (150,202)—a level consistent with a later estimate (92). Injections of 150 μg once every 2 weeks were found to be adequate for lambs fed a cobalt-deficient diet (209), and lambs grazing a cobalt-deficient pasture grew as well with injections of 100 μg of vitamin B_{12} at weekly intervals as they did with ample oral cobalt. By contrast, the oral requirements of ruminants for vitamin B_{12} are substantially higher than those of other species. Young dairy calves fed a synthetic milk ration have been reported to require between 20 and 40 μg vitamin B_{12} per kilogram of dry matter consumed (120). Andrews and Anderson (10) fed crystalline vitamin B_{12} to lambs grazing a cobalt-deficient pasture at the rate of 1000 μg/week for 16 weeks. The growth response was much smaller than that obtained from either oral cobalt or injected vitamin B_{12}. Marston (150) also found oral doses of 100 μg vitamin B_{12} per day to be quite inadequate for young sheep. Extrapolation from these experiments suggests an oral requirement for growing lambs of some 200 μg/day, which is about 10 times the reported oral requirement of other species per unit of food intake. The reasons for this high requirement are discussed in Section IV,F.

D.   Diagnosis of Cobalt–Vitamin B_{12} Deficiency

The milder forms of cobalt deficiency in ruminants are impossible to diagnose with certainty on the basis of clinical and pathological observations alone. The only evidence of the deficiency is a state of unthriftiness, and there is usually no sign of anemia. A secure diagnosis of cobalt–vitamin B_{12} deficiency can be achieved in these circumstances by measuring the response in temperament, appetite, and live weight that follows cobalt feeding or vitamin B_{12} injections. However, if the ration of grazing consistently contains <0.08 ppm cobalt, cobalt–vitamin B_{12} deficiency can be predicted with confidence.

The level of cobalt in the livers of sheep and cattle is sufficiently responsive to changes in cobalt intake to have some value in the diagnosis of cobalt deficiency, but must be used with caution for the reasons given earlier (Section II). Liver vitamin B_{12} concentration is a more sensitive and reliable criterion than liver cobalt concentration. The criteria for sheep in Table IV were suggested by Andrews *et al.* (12,13) and can tentatively be applied to cattle.

The diagnostic value of liver vitamin B_{12} levels, and still more of kidney vitamin B_{12} levels, may be realized if the cobalt deficiency is coexistent with other diseases or conditions resulting in loss of appetite (11). From an examination of numerous controlled trials with sheep, Andrews (8) emphasized the limitations of liver vitamin B_{12} assays as an aid to diagnosis of mild or intermittent forms of cobalt deficiency, but stated that "values of 0.10 μg/g or less for livers from individual sheep can be accepted as clearly diagnostic of cobalt deficiency disease." Marston (150) also found that loss of appetite occurred when the concentration of vitamin B_{12} in the liver was reduced to about 0.1 μg/g wet weight.

Serum vitamin B_{12} assays have obvious advantages over liver or kidney determinations. With the onset of cobalt deficiency, serum vitamin B_{12} levels fall markedly and the levels can be related to the amount of cobalt ingested (12,13,47,95,111,112,151). Dawbarn *et al.* (47) concluded that "signs of Co deficiency may be expected to supervene when the mean vitamin B_{12} activity in the plasma falls to about 0.2 ng/ml." Andrews and Stephenson (14) reached similar conclusions from experiments with grazing ewes and lambs, although considerable individual variability was observed. Incipient stages of cobalt deficiency were associated with mean serum B_{12} values of 0.26 ng/ml for ewes and 0.30 ng/ml for lambs. When the cobalt deficiency had become marked in ewes and acute in lambs, mean serum B_{12} concentrations fell in both cases to levels <0.20 ng/ml. Similar conclusions were reached by Somers and Gawthorne (213), although they observed plasma vitamin B_{12} concentrations of ≤0.2 ng/ml for a few weeks in some sheep without the appearance of any clinical signs of deficiency.

Table IV
Criteria of Vitamin B_{12} Deficiency in Sheep[a]

Condition of animal	Vitamin B_{12} concentration in liver (μg/g wet wt)
Severe cobalt deficiency	<0.07
Moderate cobalt deficiency	0.07–0.10
Mild cobalt deficiency	0.11–0.19
Cobalt sufficiency	>0.19

[a] From Andrews *et al.* (12,13).

E. Prevention and Treatment of Cobalt Deficiency

Cobalt deficiency in ruminants can be controlled either by treatment of the pastures with cobalt-containing fertilizers, by direct oral administration of cobalt to the animal, or by injections of vitamin B_{12}. Subcutaneous depot injections of vitamin B_{12} are emerging as a practical means of prophylaxis, especially in lambs. The effective period of a single injection is 8-14 weeks for young growing sheep (86). Vitamin B_{12b} (hydroxocobalamin) is as effective as vitamin B_{12} itself (cyanocobalamin) in correcting cobalt deficiency in lambs when injected at the rate of 100–125 μg/week (113), but folinic acid at levels of 71 μg, 5 mg, or 15 mg daily produces no such response (111). Large oral doses of dried or fresh whole liver were also shown to be highly curative (58), a finding which led Filmer and Underwood (61) as early as 1937 to propose that "the potency of liver may be due to the presence of a stored factor and that cobalt may function through the production of this factor within the body." Subsequent attempts to cure cobalt deficiency with liver were unsuccessful (153), presumably because of differences in the vitamin B_{12} or cobalt contents of the liver preparations employed.

The most economical means of ensuring continuous and adequate supplies of cobalt to sheep and cattle grazing cobalt-deficient areas is by treatment or "top dressing" of the pastures with fertilizers to which a small proportion of cobalt salts or ores has been added. In this way the concentration of cobalt in the herbage can be raised for extended periods to levels adequate for the requirements of the animal. In some areas improvement in the growth of the legume component of the pasture can also be achieved by such treatment (170,180). The latter effect can be traced to the requirement for cobalt by symbiotic microorganisms (*Rhizobia*) in the root nodules of legumes. The cobalt is needed for synthesis of the cobamides required for rhizobial growth and so for nitrogen fixation (3).

In order to maintain adequate cobalt levels in pasture by top dressing, as little as 100–150 g of cobalt sulfate per acre applied annually or biennially is sufficient for most deficient areas (19,189). A single aerial dressing of 500–600 g of cobalt sulfate per acre has maintained satisfactory pasture cobalt levels for 3 years in New Zealand (9). The most efficient quantity and frequency of top dressing with cobalt depends on the degree of deficiency and the soil type and has to be determined for each area. Cobalt uptake by plants on highly calcareous soils, or on soils high in manganese which fixes the cobalt in unavailable forms (1), can be so low as to render treatment with "cobaltized" fertilizers inefficient or ineffective.

Direct administration of cobalt to stall-fed animals is usually achieved by incorporating cobalt oxide or salts into the mineral mixtures normally fed as supplements. The inclusion of cobalt in such mixtures is common, even where there is no clear evidence that the unsupplemented rations are cobalt deficient. Cobalt may also be supplied successfully to either stall-fed or grazing sheep and cattle in the form of cobalt-containing salt "licks," but variable consumption of the lick, and therefore of cobalt, is a disadvantage inherent in this form of treatment. Oral dosing or "drenching" of animals with solutions of a cobalt salt is still practiced in many areas and can be completely successful if the dosing is frequent enough.

Dosing sheep with 2 mg cobalt twice weekly or 7 mg cobalt weekly (124) is completely adequate, even under acutely cobalt-deficient conditions. Since individual drenching is a tedious and labor-demanding operation, the effect of larger doses at longer intervals has been investigated. Lee (124) found that 35 mg cobalt administered once in 5 weeks "merely delays the onset of symptoms." Filmer (59) observed that monthly doses of 140 mg cobalt kept ewes and lambs alive on cobalt-deficient land, but the sheep did not thrive as well as those treated twice weekly or weekly with the usual small doses. Andrews and co-workers (16) gave monthly doses of 300 mg cobalt, either as soluble cobalt sulfate or insoluble cobalt oxide, to lambs grazing cobalt-deficient pastures for 5 months. Their growth rates and vitamin B_{12} levels in blood and liver were compared with those of untreated lambs and lambs dosed with 7 mg cobalt weekly. The two forms of cobalt were equally effective, but neither of the monthly treatments prevented cobalt deficiency entirely. Both permitted growth at suboptimal rates and greatly reduced mortality but were less effective than the weekly dosing. Lee and Marston (130) showed further that the equivalent of 1 mg cobalt per day is as effective for 2–3 years when given weekly as when given more frequently. Over longer periods the body weight response is slightly but significantly in favor of the more frequent administration.

The necessity for regular and frequent dosing arises partly from the fact that cobalt, unlike iron and copper, is not readily stored in the liver or elsewhere in the body, but mainly because the need for cobalt exists in the rumen and the supply must be continuous. The cobalt which is present in the tissues does not easily pass into the ruminoreticular region of the digestive tract where it is needed for vitamin B_{12} synthesis. Injections of cobalt are therefore largely ineffective. Where large amounts are injected some improvement in the condition of cobalt-deficient sheep has been observed (112), presumably as a consequence of small amounts of cobalt reaching the rumen in the saliva or via the rumen wall. Studies with radioactive cobalt indicate that the element only reaches the rumen, reticulum, and omasum when large amounts are injected, and even then only in very small quantities (40). Cobalt injections are capable of increasing the vitamin B_{12} content of the cecum and large intestine of cobalt-deficient lambs, probably as a result of bacterial synthesis following the excretion of cobalt into the duodenum via the bile. The vitamin so produced is apparently not absorbed, since blood and body tissue vitamin B_{12} levels are not improved (112). Phillipson and

Mitchell (174) placed cobalt salts directly into the abomasum or duodenum by means of appropriate fistulas, and found them to be less effective than cobalt given by mouth, although there was evidence of some movement of cobalt into the rumen. It is apparent that any treatment with cobalt, if it is to be fully successful, must be capable of maintaining adequate cobalt concentrations in the rumen fluid.

Many of the disadvantages of other forms of treatment can be avoided by the use of cobalt pellets or "bullets," first devised by Dewey and co-workers (48) in 1958. The small, dense pellets composed of cobalt oxide and finely divided iron (specific gravity 5.0), when delivered into the esophagus with a balling gun, lodge in the reticulorumen where they usually remain to yield a steady supply of cobalt to the rumen liquor. The usefulness of cobalt pellet therapy in the prevention of cobalt deficiency has been established in sheep and cattle (11–13,131,200,243). Two problems have arisen with this form of treatment: rejection of the pellets by regurgitation in a proportion of animals, and surface coating of the pellets with calcium phosphate so that the rate of cobalt release into the rumen is reduced. Millar and Andrews (158) prepared radioactive cobaltic oxide pellets so that they could be detected in live sheep and followed their retention in grazing ewes and lambs over a 16-month period. By the end of the period about one-third of the ewes and lambs had rejected their pellets, and a proportion of those retained were coated with calcium phosphate. Furthermore, it was reported that the administration of two pellets or one pellet plus a 0.5-in. grub screw grinder was ineffective in reducing the calcium phosphate coating. This is contrary to Australian experience, in which abrasion between the two objects has been quite successful in keeping the pellet surface clean and maintaining a steady supply of cobalt to the rumen of sheep for >5 years (131).

Newer methods of achieving the slow release of cobalt in the rumen include the use of soluble cobalt-containing glasses. The composition of these can be adjusted so as to provide appropriate amounts of several trace elements simultaneously (112a,222).

F. Mode of Action of Cobalt

Production of vitamin B_{12} in the rumen depends on a number of factors, including the cobalt and the roughage content of the diet and the total feed intake (92,220). When the concentration of cobalt in the rumen fluid falls below a critical level, placed at <5 ng/ml (202), the rate of vitamin B_{12} synthesis by the rumen organisms is reduced below the animal's needs. When the stores of vitamin B_{12} laid down mainly in the liver become depleted below a critical level close to 0.1 ng/g wet weight (8,150), and when the plasma vitamin B_{12} levels fall below a critical level of about 0.2 ng/ml as a result of inadequate production and absorption combined with near-exhaustion of the body stores, then characteristic signs of cobalt deficiency begin to appear in most sheep and cattle (14,47,150,213). The length of time taken for clinical signs of deficiency to become apparent depends primarily on the degree of dietary cobalt deficiency, the magnitude of the vitamin B_{12} stores previously built up, and the age and rate of growth of the animal. There is no evidence that vitamin B_{12} synthesis is possible within the body tissues or that the vitamin B_{12} synthesized by bacteria in the cecum and large intestine can be absorbed (110,112,202). The ruminant is therefore ultimately dependent on the synthetic capacity of its rumen organisms.

Although changes in the bacterial population of the rumen have been reported to occur in cobalt-deficient lambs (64), there is no evidence that the normal fermentative processes in the rumen are changed as a result (152,203). No direct evidence has been obtained implicating particular organisms in vitamin B_{12} synthesis in the rumen or demonstrating a preferential diminution of the numbers of such organisms as a consequence of a lack of dietary cobalt. Thus, although some methanogenic bacteria and some propionic acid bacteria produce large amounts of cobamides, neither propionic acid production (152) nor methane production (203) seems to be impaired in the cobalt-deficient rumen. Dryden *et al.* (52) established that several pure strains of bovine rumen bacteria each produced a mixture of vitamin B_{12} and its analogs when cultured in a cobalt-adequate medium.

In addition to vitamin B_{12} itself, the mixed rumen organisms normally produce many cobamides that have no physiological activity in the body tissues (30,70,71,94,177). These include factors A, B, and C as well as pseudovitamin B_{12} (see Table I). The relative proportions of the cobamides that are produced in the rumen depend on both the nature of the diet and the concentration of cobalt in the rumen (70,92,202). The proportions of vitamin B_{12} to total cobamides can be assessed by comparative assays with *Escherichia coli* or *Lactobacillus leichmanii* and with *Ochromonas malhamensis*. The latter organism responds only to vitamin B_{12} itself. Alternatively the component cobamides can be determined individually after electrophoretic separation (30,69). Under cobalt deficiency conditions, both total cobamides and *Ochromonas* vitamin B_{12} activity fall sharply. Most observers have found the fall in total cobamide activity to be greater than that in vitamin B_{12} itself (47,70,94), but in the experiments of Hedrich *et al.* (92), activity in rumen fluid showed the opposite trend, with additional cobalt tending to produce a higher proportion of vitamin B_{12} in total cobamides. The latter effect was even more pronounced in duodenal fluid from the same animals (30,92). There is evidence from comparative microbiological assays that rumen contents from pasture-fed animals contain a higher proportion of vitamin B_{12} than rumen contents from animals fed a high-roughage, cobalt-deficient diet, either with or without additional cobalt (202). The factors that control the relative rates of synthesis of the various cobamides in the rumen are evidently complex, and not all of them have been identified.

Despite the variable mix of cobamides in the rumen, vitamin B_{12} activity as measured by microbiological activity in blood plasma of sheep is almost entirely accounted for as cobalamin (47,70,177), with no more than 1–2% present as other identified cobamides or as cobinamide. The emergence of radioisotope dilution assays (121) for vitamin B_{12}, however, has led to the discovery of substances in blood plasma and tissues of several species (including ruminants) that are neither cobalamin itself nor any of the known cobamides (115,227). The nature and function (if any) of these substances, referred to as cobalamin analogs, remains unknown. They may be degradation products of cobalamin that arise systemically or in the gastrointestinal tract.

In the normal human the predominant circulating forms of vitamin B_{12} are the two coenzymes methylcobalamin (IV) and adenosylcobalamin (III), with a trace of hydroxocobalamin, but there is virtually no circulating cyanocobalamin (135). These proportions are not the same in all animals, however, and the relative amounts circulating in ruminant plasma are unknown.

Two factors that contribute to the relatively high cobalt requirement of the ruminant are the inefficiency of production of vitamin B_{12} from cobalt in the rumen and the relative inefficiency of absorption of vitamin B_{12} in the small intestine. Estimates of the amounts of vitamin B_{12} produced in the rumen at various cobalt intakes differ somewhat, and this may reflect methodological differences as well as intrinsic variation in rates of production under different dietary regimes (172). Large week-to-week variations in the vitamin B_{12} content of rumen fluid were found by Gawthorne (70), notwithstanding constant cobalt intake. In experiments using a high-roughage, cobalt-deficient diet with and without supplementary cobalt, Smith and Marston (202) found the vitamin B_{12}–lignin ratio to reach a minimum value 4h after feeding, and they used these values to estimate minimum production rates from the known daily flow of lignin. On full rations, minimum rates of vitamin B_{12} production by this method were 50–119 μg/day under cobalt deficiency conditions, and 400–700 μg/day when 1 mg cobalt was given daily. Using the flow of abomasal contents as a basis for the estimate, Elliott and co-workers (55) found daily production rates on two cobalt-sufficient diets to range from 450 to 2240 μg/day with an average of around 1000 μg/day. Later estimates of this group, based on the flow of duodenal contents, yielded mean values of 37 μg/day, 1006 μg/day, and 1553 μg/day, at daily mean cobalt intakes of 0.047 mg, 0.41 mg, and 0.83 mg of cobalt, respectively. Based on these values, the efficiency of vitamin B_{12} production from the cobalt passing through the rumen ranged from 3% in the cobalt-deficient rumen to just under 10% under conditions of cobalt sufficiency. From simultaneous measurements of cobalt and vitamin B_{12} in whole-rumen contents, Smith and Marston (202) found the opposite trend, with about 13% of the cobalt present as cobalamin under deficiency conditions and about 3% in vitamin B_{12} when a daily cobalt supplement was given.

There is evidence that in the sheep, as in other animals, absorption of vitamin B_{12} occurs primarily from the small intestine (202), but estimates of absorptive efficiency vary widely. The experiments of Rickard and Elliot (186) show that when labeled vitamin B_{12} (as [^{57}Co]cyanocobalamin) was introduced directly into the duodenum and was accompanied by repeated large flushing doses of vitamin B_{12} given intramuscularly, then between 10 and 32% of the label, depending on the diet, could be recovered in the urine, the higher recovery occurring with the diet containing least cobalt. Estimates based on the difference between duodenal and terminal ileal flow of vitamin B_{12} also suggest an absorptive efficiency in the small intestine of around 20% (92). However, estimates of either production or absorption of vitamin B_{12} that are based in part on the rate of duodenal flow may include a contribution from the vitamin B_{12} enterohepatic cycle (46) and, to the extent that this occurs, may be overestimates. A transient rise in the duodenum of the concentration of vitamin B_{12} measured along the gastrointestinal tract of sheep has been observed in several studies (94,112,202).

The minimum requirement of adult sheep for absorbed vitamin B_{12} is around 11 μg/day (92,150,202) and this must be capable of being supplied by the minimum cobalt intake of 70 μg/day. The latter could, if it were all converted, give rise to some 1630 μg of vitamin B_{12}, so that by any estimate the overall efficiency with which the animal obtains vitamin B_{12} from dietary cobalt is very low. The two experimental solutions to this problem so far obtained suggest, on the one hand, an efficiency of ruminal production of around 13% combined with an absorptive efficiency of about 5% (202), and on the other hand, an efficiency of ruminal production of around 3% combined with an absorptive efficiency of about 20% (92). Both sets of values contain elements of uncertainty.

The marked failure of appetite that is such a conspicuous feature of cobalt–vitamin B_{12} deficiency in ruminants is not nearly so noticeable in vitamin B_{12} deficiency in humans or other species. This difference can probably be related to the means by which the two types of animals derive their energy. The main source of energy to ruminants is not glucose but acetic and propionic acids, together with smaller quantities of butyric and other short-chain fatty acids produced by fermentation in the rumen. The biochemical pathways by which acetate and butyrate are metabolized are simple and of equal importance in ruminants and nonruminants. Thus the tricarboxylic acid (Krebs) cycle itself consists in the oxidation of acetate. The pathway by which propionate is metabolized is more complex (62) and, though present in the nonruminant, is quantitatively of much less significance. Propionate as propionyl coenzyme A is first carboxylated to form D-methylmalonyl coenzyme A. The latter, after conversion to the L form by the action of a racemase, undergoes an unusual rearrangement that involves the crossover migration of two groups on adjacent carbon atoms (see Section VII). The enzyme methylmalonyl coenzyme A mutase catalyses the reaction:

L-methylmalonyl coenzyme A⇌succinyl coenzyme A

Succinyl coenzyme A is a mainline intermediate in the tricarboxylic acid cycle and a direct precursor of glucose (26). Because of ruminal fermentation, the ruminant derives little glucose directly from the diet and has to synthesize its needs in the liver, where propionate is the chief precursor. Metabolism of propionate is therefore of unique significance to the ruminant.

The discovery that the enzyme methylmalonyl coenzyme A mutase required a coenzyme form of vitamin B_{12}, adenosylcobalamin, for its activity followed directly from observations of impaired propionate metabolism in the liver of vitamin B_{12}-deficient sheep (151,205). Discovery of the involvement of vitamin B_{12} in the reaction represented the first demonstration of a direct enzymic function for vitamin B_{12} in an animal tissue and was quickly confirmed (83). Following this discovery, Marston and associates (151) proceeded to investigate the possibility that (a) a breakdown in propionate utilization at this point in the metabolic pathway might be the primary defect in cobalt–vitamin B_{12} deficiency in ruminants, and (b) the depression of appetite might be due to an increased level in the blood of propionic acid or some other metabolic product stemmed back by the reduced capacity of the B_{12}-deficient animal to metabolize propionate. The production and absorption of propionic and other fatty acids were found to proceed more or less normally in the cobalt-deficient sheep (152). As the deficiency state progressed, the rate of disappearance from the blood of injected propionate fell (152). Examination of liver homogenates from vitamin B_{12}-deficient sheep revealed (a) a failure to convert propionate efficiently to succinate, (b) an accumulation of the intermediate methylmalonyl coenzyme A, and (c) prevention of this accumulation by the addition of 5′-deoxyadenosylcobalamin (151). Methylmalonic acid (MMA) was also shown to accumulate *in vivo* in the livers of vitamin B_{12}-deficient sheep (208).

Subsequently Somers (212) found that propionate and acetate clearance rates from the blood of sheep were increasingly adversely affected as vitamin B_{12} deficiency intensified. These effects were greater than the effects of depressed feed intake alone, and both variables had a more pronounced effect on the rate of clearance of propionate than of acetate. Marston and co-workers (152) attributed the impaired acetate clearance to the elevated levels of circulating propionate, since propionate was shown to have a marked inhibitory effect on acetate metabolism in deficient sheep. Failure to metabolize acetate is therefore probably not a primary consequence of vitamin B_{12} deficiency. Further evidence of depressed propionate metabolism in cobalt–vitamin B_{12}-deficient sheep was obtained by Gawthorne (72), who found that the excretion of MMA in the urine of severely vitamin B_{12}-deficient sheep was 5–12 times greater than that of pair-fed, B_{12}-injected controls and was restored to normal within 3 weeks by injections of the vitamin. A significant increase in urinary MMA excretion occurred only when the sheep were severely affected by the deficiency. Urinary excretion of MMA may therefore have only limited value as an index of incipient cobalt deficiency (159), although it does have the advantage of reflecting directly the crucial metabolic block. In the early stages of the deficiency it seems that the potential accumulation and augmented excretion of MMA are offset by the reduced feed intake, which reduces the amount of propionate presented for metabolism. The most significant evidence linking inappetence with impaired propionate metabolism is an inverse relationship between voluntary feed intake of deficient sheep and the half-time for propionate clearance (152) (Fig. 3). Propionate clearance was shown to be already fairly severely impaired at the mildly deficient feed

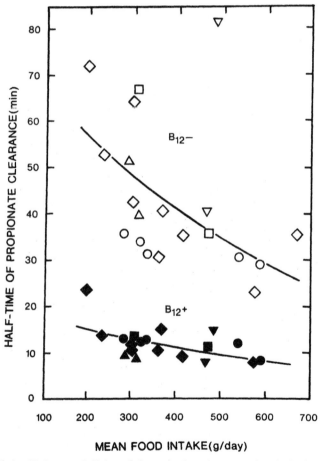

Fig. 3. Relationship between half-time of clearance of propionate and food intake for vitamin B_{12}-deficient ($B_{12}-$) and vitamin B_{12}-treated ($B_{12}+$) sheep. From Marston *et al.* (152).

intake of 700 g/day. Impairment of propionate clearance is clearly an early consequence of vitamin B_{12} deficiency.

Further effects attributable to the depleted activity of methylmalonyl coenzyme A mutase in the tissues of cobalt-deficient sheep are a reduced rate of incorporation of labeled propionate into glucose in liver slices (173) and an increased proportion of odd-numbered fatty acids in liver and depot lipids of neonatal lambs born to cobalt-deficient ewes (53).

Some of the effects of vitamin B_{12} deficiency in sheep can be attributed to depressed activity of a second vitamin B_{12}-containing enzyme, in addition to the effects of a critical depression in activity of methylmalonyl coenzyme A mutase, as just discussed. This second enzyme is 5-methyltetrahydrofolate : homocysteine methyltransferase, which catalyzes the reformation of methionine from homocysteine, thus permitting the recycling of methionine following loss of its labile methyl group, in accordance with the reaction:

$$5\text{-methyltetrahydrofolate} + \text{homocysteine} \rightleftharpoons \text{methionine} + \text{tetrahydrofolate}$$

The activity of the methyltransferase is depressed in the liver of vitamin B_{12}-deficient sheep and is accompanied by subnormal levels of S-adenosylmethionine (74) and a fatty liver with a depleted choline content (207). Both defects are corrected by injection of extra methionine. Since the rate of turnover of the methyl group of methionine in the liver is rapid (2), restriction of methyltransferase activity in vitamin B_{12} deficiency could lead to a deficiency of available methionine. This provides a possible basis for the impaired nitrogen retention found in vitamin B_{12}-deficient sheep (203) and could be a critical limiting factor in both wool and body growth.

A further consequence of cobalt-vitamin B_{12} deficiency is a marked depletion of liver folate stores. This occurs in rats (119) as well as in sheep (206), and in both species the levels of liver folate can be restored by treatment with L-methionine (75,207). In fact methionine, like vitamin B_{12}, increases the concentrations of all the major classes of folate in the liver of vitamin B_{12}-deficient sheep (207). It appears, therefore, that the effect of vitamin B_{12} on folate metabolism, as well as on liver lipids, is exerted via methionine, but the way in which methionine affects folate stores remains unclear. Thus although the increase in the proportion of the tissue folates present as methyltetrahydrofolate may well have resulted from a lowered activity of the methyltransferase (207), it is by no means clear why this effect in itself should have led to a loss of tissue stores or why methionine should reverse the trend (206,207).

The existence of a systemic folate deficiency in vitamin B_{12}-deficient sheep is readily demonstrated by the presence in urine of formiminoglutamic acid which accumulates as a result of the lack of available folate. However, the relationship of this event to the primary cobalt deficiency is very indirect, and it is not surprising that urinary excretion of excess formiminoglutamate as an index of cobalt–vitamin B_{12} status has not found general acceptance (191,217).

G. White Liver Disease

Ovine white liver disease (WLD) was first recognized as an enzootic entity in New Zealand in 1974 (45,156). Subsequent reports have come from that country (39,219), from Australia (157,160,185), from Ireland (142), and from Europe (232). The disease occurs mainly in lambs from 2 to 6 months of age, although occasionally older sheep are affected (157,160,185,219,232). The condition is one of ill thrift followed by sudden inappetence with rapid loss of weight that may proceed to a fatal conclusion. The percentage of lambs affected in an outbreak varies widely, but both morbidity and mortality can be very high. In a report from the Albany district of Western Australia, 350 lambs died of a flock of 420 (185). Acute clinical signs generally appear 7–14 days before death. The disease can be arrested and reversed at almost any stage by oral administration of cobalt or, more rapidly, by injection of vitamin B_{12} (142,160,185,219,232).

White liver disease generally occurs in regions of endemic cobalt deficiency, but because the levels of cobalt in pasture are not always found to be low after the outbreak is established (39,157,185), and because some of the symptoms and pathology of the disease have been thought not to be characteristic of simple cobalt deficiency, doubts have been expressed as to whether cobalt deficiency is a primary cause of WLD and, if it is, whether pasture toxins may not also be obligatorily involved (160,185,219). Whenever examined, however, both liver cobalt (160,185) and liver or serum vitamin B_{12} levels (39,142,160,185,219) have been characteristic of acute cobalt deficiency, and there is now no doubt that severe dietary cobalt deficiency is an essential precondition of WLD. The question of whether or not pasture toxins must also be involved is less well resolved.

The symptoms of WLD in lambs include ill thrift, listlessness, inappetence, loss of weight, a bilateral serous ocular discharge, a mild anemia, and the pale liver that gives rise to the name. All of these symptoms occur also in simple cobalt–vitamin B_{12} deficiency, and all are rapidly responsive to injected vitamin B_{12}. The additional symptoms in WLD that are not commonly found in cobalt-deficient sheep include the "photosensitivity," which consists of a serous crusty lesion on ears, nose, lips, or back (39,142,157,185,219,232), and certain neurological symptoms that may include ataxia, tonic spasms, tremors of the head and neck, or blindness (185,219,232). The neurological symptoms are rather rare (185,219,232) and are often not observed or not recorded (39,157,160). The incidence of photosensitivity in outbreaks of WLD is much higher. Sutherland et al. (219) reported that in 31 outbreaks, whereas overall morbidity ranged from 6 to 100%, the incidence of photosensitivity ranged from 6 to 60%. This reflects the general experience, although Mitchell and co-workers did not report photosensitivity in seven outbreaks in Victoria, Australia (160), and McLaughlin et al. (142) found photosensitivity in only 10% of 374 affected lambs in Ireland. Photosensitivity has not been recognized as a consequence of simple cobalt deficiency, and it occurs with sufficient frequency to warrant caution in attributing WLD solely to cobalt deficiency. However, the agents responsible for this complication are not the common ones and have not been identified, and it is possible that susceptibility to them is itself a consequence of cobalt–vitamin B_{12} deficiency. Certainly all symptoms of WLD, including photosensitivity, respond to vitamin B_{12} therapy.

The occasional appearance of neurological symptoms is also unexplained, although in this case the incidence is too low for the symptoms to be regarded as an essential feature of WLD. Cerebrocortical necrosis, associated with thiamine deficiency and sometimes accompanied by blindness, was reported to result from experimental cobalt deficiency in sheep by McPherson et al. (147), and although this interpretation was subsequently criticized (54), the observation that simple dietary cobalt deficiency of long standing could lead to pathological changes in the brain appears to stand and might represent a condition related to the occasional neurological changes in WLD.

The specific liver pathology reported for WLD, and often regarded as pathognomonic for the condition, consists of three changes: a fatty infiltration of hepatocytes confined initially to the centrilobular (periacinar) regions, a proliferation of bile duct epithelium and mesenchymal cells in portal areas, and the presence of ceroid pigment in periacinar macrophages and elsewhere (160,219). The first of these liver effects is frequently encountered in uncomplicated ovine vitamin B_{12} deficiency accompanied by higher concentrations of lipid and lower concentrations of choline and S-adenosylmethionine than in control animals receiving vitamin B_{12} (74,207). A centrilobular location of fatty changes in the liver accompanied by bile duct proliferation is characteristic of the fatty liver associated with experimental choline deficiency in rats (91). In humans, the condition known as nutritional liver disease results from a diet deficient in lipotrophic substances such as choline and methionine and exhibits similar liver changes with centrilobular fatty infiltration and bile duct proliferation, sometimes with the appearance of ceroid pigment (197). The liver abnormality of WLD might therefore be due to a depletion of choline and methionine in the liver, effects directly attributable to lowered activity of the vitamin B_{12} enzyme 5-methyltetrahydrofolate : homocysteine methyltransferase (74,207). The damage to liver cells that results from acute fatty liver is presumably responsible for the appearance in blood plasma of the range of hepatic enzymes recorded in WLD (142,160,185,219).

H. Cobalt and *Phalaris* Staggers

In certain areas sheep and, to a lesser extent, cattle grazing pastures containing the perennial grass *Phalaris tuberosa* or certain other *Phalaris* species can develop either a peracute disease from which they quickly die, or a chronic condition with nervous disorders characterized by a marked incoordination of gait and changes in the central nervous system (CNS). An acute form of *Phalaris* poisoning with nervous symptoms but without specific pathological changes is also recognized (57). The cause of death in the peracute (sudden death) disease is acute heart failure. One of the distinctions between acute *Phalaris* poisoning and chronic *Phalaris* staggers is that whereas recovery from the former can be brought about quickly by transfer to sound pasture, the chronic form generally represents a permanent disability with specific neuronal changes and damage to spinal cord myelin. Chronic *Phalaris* staggers can be prevented by an adequate and continuous supply of cobalt in the rumen (127).

Phalaris staggers, which designates the chronic condition, was first described by McDonald (139,140) in South Australia and has been observed elsewhere in Australia. An identical staggers syndrome occurs in sheep and cattle grazing *Phalaris arundinacea* in New Zealand (198), Ronpha grass (*P. tuberosa* × *P. arundinacea*) in Florida (190a), and various *Phalaris* species in California and elsewhere (89,184).

Evidence has been obtained that the tryptamine alkaloids, closely related to serotonin and shown to be present in *P. tuberosa* (43,67), are responsible for the "sudden death" syndrome and for some features of acute *Phalaris* poisoning. However, the acute toxicity is not always related to the tryptamine alkaloid concentration in the plant (168), and it is improbable that these alkaloids produce the persistent neurological disorders associated with chronic *Phalaris* staggers (125). The CNS pathology associated with chronic *Phalaris* staggers includes the appearance of pigmented material in the cytoplasm of nerve cell bodies in the brain stem (89) and evidence of degeneration of myelin in the spinal cord (66,139) (see Fig. 4).

Lee and Kuchel (126) first demonstrated the association of this condition with the cobalt status of the animal and secured its complete prevention in sheep grazing affected pastures by regular oral dosing with cobalt salts. This finding has been confirmed (129,214) and extended to include the protective action of cobalt pellets. As cobalt does not appear to be effective against acute and peracute *Phalaris* poisoning, it seems that cobalt must in some way be concerned with preventing the development of degenerative structural changes in the CNS. This might occur by an action of cobalt to promote the destruction of some unknown phytotoxin in the rumen itself. There is no preventive effect of cobalt given intravenously (128).

The action of cobalt to protect against *Phalaris* staggers is different from and additional to its action in preventing cobalt deficiency in ruminants. Thus administration of vitamin B_{12} was ineffective in preventing the disease under conditions where small daily oral doses of cobalt were fully effective (127,128).

Where chronic staggers does not develop in animals grazing *Phalaris* pastures, it is contended that the soils maintain sufficient concentrations of cobalt in the

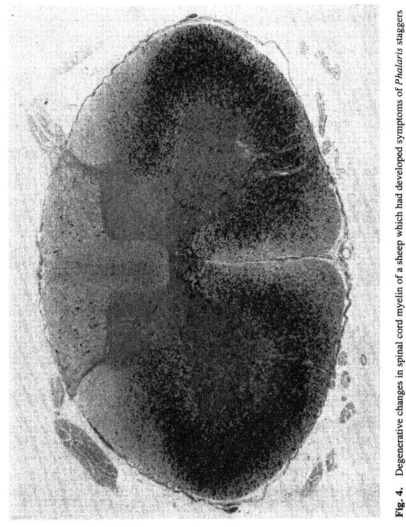

Fig. 4. Degenerative changes in spinal cord myelin of a sheep which had developed symptoms of *Phalaris* staggers 15 days before slaughter. Marchi degeneration (66) is seen in ventral and ventrolateral columns of white matter (×16). Courtesy of H.J. Lee and F.J. Fraser.

pastures to meet the normal requirements of ruminants for this element, plus the extra requirements needed to prevent the degenerative CNS changes.

V. COBALT IN THE NUTRITION OF HUMANS AND OTHER NONRUMINANTS

A. Humans

Cobalt must be supplied in the human diet and that of other monogastric species entirely in its physiologically active form, vitamin B_{12}. The tissues of these animals are unable to synthesize the vitamin from dietary cobalt, and their intestinal microflora have an extremely limited capacity to effect this vital transformation at a point in the digestive tract where the vitamin can be absorbed. In these unique circumstances the cobalt status of human foods and dietaries is relatively unimportant—it is their vitamin B_{12} content that is crucial.

Reported data for the cobalt content of human foods and total diets are both meager and highly variable. Some of the variation undoubtedly stems from analytical errors or inadequacies, but some also reflects soil and climatic differences directly affecting the cobalt content of foods of plant origin and indirectly those of animal origin. Thus Murthy and co-workers (164), in a study of the diets of children from 28 widely separated institutions in the United States, found the cobalt concentration in the total diets to vary from 0.25 to 0.69 mg/kg and the total cobalt intakes to range from 0.30 to 1.77 mg/day, with a mean close to 1 mg/day. These levels are much higher than the 0.16–0.17 mg/day for two adults consuming North American diets estimated by Tipton et al. (229), the 0.16–0.58 mg/day estimated by Schroeder et al. (194), the mean of 20 μg/day given for Japanese adults (246), or the daily intakes of college women on a mixed diet of 6.0–7.6 μg/day cobalt reported by Harp and Scoular (87). It is difficult to believe that differences of this magnitude are not mainly of analytical origin.

Among individual types of foods the green leafy vegetables are the richest and most variable in cobalt content, while dairy products, refined cereals, and sugar are the poorest. Typical values for the former group are 0.2–0.6 ppm (dry basis), and for the latter 0.01–0.03 ppm cobalt (dry basis). Plant products have been estimated to contribute up to 88% of the total cobalt of Japanese diets (246). Normal cow's milk is very low in cobalt, with most values lying close to 0.5 μg/liter, as discussed in Section I. The organ meats, liver and kidney, commonly contain 0.15–0.25 ppm (dry basis), and the muscle meats about half those levels. These foods contain much more cobalt that can be accounted for as vitamin B_{12}. Fruits, vegetables, and cereals contain virtually none of their cobalt as vitamin B_{12}.

The well-known effect of the cobalt ion to induce erythropoiesis in humans and experimental animals is mediated by increased renal production of erythropoietin (106). Stimulation of hematopoiesis occurs despite an inhibitory effect of cobalt (probably as cobalt protoporphyrin) on the activity of 5-aminolevulinate synthase, one of the enzymes involved in heme biosynthesis (199).

Cobalt has been used as a nonspecific erythropoietic stimulant in the treatment of the anemias of nephritis and infection (68), and in that arising from long-term hemodialysis (33), and several reports of responses to cobalt, in addition to iron, in the treatment of iron deficiency anemia in children (223) and pregnant women (85) have appeared. The amounts of cobalt required to elicit the hematopoietic response are large (20–30 mg/day), so that serious toxic manifestations can occur, including thyroid hyperplasia, myxedema, and congestive heart failure in infants (117,187). Cobalt therefore occupies a very restricted place in the treatment of human anemias.

Some evidence for an effect of varying intakes of cobalt on the thyroid gland has been obtained. Cobalt, and also manganese, have been reported to be necessary for the synthesis of the thyroid hormone in rats (32). The addition of physiological doses of cobalt to a diet which was not naturally high in iodine or cobalt did not affect the weight of the gland but caused definite histological changes, including a decrease in the size of the follicles and an increase in the height of the epithelial cells. It was concluded that "the appearance of endemic disturbances of thyroid gland function in people inhabiting biogeochemical provinces with a low-iodine and a low-cobalt content depends not only on the level of iodine and cobalt, but also on the ratio of these elements in the environment" (32). Other workers in the Soviet Union have observed an inverse correlation between the cobalt levels in the foods, waters, and soils in certain areas and the incidence of goiter in humans and farm animals (116). A possible relationship between the cobalt status of the environment and the incidence of goiter warrants investigation in other areas, and further examination of the suggestive cobalt–iodine interaction in animals is desirable (32,188).

B. Nonruminant Animals

Cobalt deficiency has never been demonstrated in laboratory animals. Rabbits have been successfully maintained on diets reported to supply only 0.1 μg cobalt per day (225), and rats were found to grow as well on diets supplying 0.5 μg cobalt per day (234), or 0.3 μg/day (98), as when those diets were supplemented with cobalt. Horses thrive on pastures so low in cobalt that sheep and cattle dependent on such pastures soon waste and die.

Monogastric animals consuming all-plant rations, which directly supply little or no vitamin B_{12}, obviously need some dietary cobalt so that their intestinal flora can synthesize sufficient vitamin B_{12} to meet the metabolic needs of the tissues. The diets consumed generally contain sufficient cobalt for this purpose. However, Dinusson et al. (51) observed a small but significant increase from additions of cobalt in the rate of gain and efficiency of feed use of pigs fed a corn–soybean meal diet. In some experiments a similar improvement was observed from supplements of vitamin B_{12} or meat scraps, with a response from cobalt even when those rations contained 5% meat scraps. Intestinal synthesis of vitamin B_{12} is not always adequate for the needs of growing pigs and poultry consuming all-plant rations. Additional supplies of this vitamin can be obtained by coprophagy, and by the consumption of litter and refuse in which fermentation by bacteria voided in the feces or contaminated from the soil has occurred. Where these adventitious sources of vitamin B_{12} are denied to the animals, growth responses can be obtained from supplements rich in the vitamin.

The "animal protein factor" effect with pigs and poultry receiving all-plant diets is due partly and in some cases largely to the vitamin B_{12} supplied. Cobalt, of itself, is rarely important in this respect.

VI. COBALT TOXICITY

Cobalt has a low order of toxicity in all species studied, including humans. Daily cobalt doses of 3 mg/kg body weight, which approximate 150 ppm cobalt in the dry diet or some 1000 times normal levels, can be tolerated by sheep for many weeks without visible toxic effects (27). With cobalt doses of 4 or 10 mg/kg body weight, appetite and body weight are severely depressed, the animals become anemic, and some deaths occur at the higher level. The anemia may arise from a depression in iron absorption by the very high intakes of cobalt, as discussed in Section III,A. Andrews (6) estimated that a single cobalt dose of 300 mg/kg body weight as a soluble salt would usually be lethal to sheep, and that single doses as small as 4–60 mg/kg body weight would occasionally be fatal. There are some indications that cattle are less tolerant of high cobalt intakes than sheep (50,141).

Rats and various species other than the adult ruminant fed large amounts of cobalt as cobalt salts develop a true polycythemia accompanied by hyperplasia of the bone marrow, reticulocytosis, and increased blood volume (80,106,242). The oral intakes of cobalt necessary to induce significant polycythemia approximate 200–250 ppm of the total diet and are therefore clearly many times greater than those that could conceivably be obtained from normal foods and beverages. For rats LD_{50} values for cobalt range from 200 to 500 mg/kg body weight, depending on the cobalt compound employed (215).

Under certain circumstances, as yet unexplained, cobalt intakes substantially lower than the 20–30 mg/day mentioned earlier as producing toxic manifestations when used as an erythropoietic stimulant, can be toxic to humans. Cobalt has been incriminated as the precipitating factor in several outbreaks of severe cardiac failure in heavy beer drinkers. Cobalt was suspected because of the high incidence of polycythemia, thyroid epithelial hyperplasia, and colloid depletion noted in the fatalities, in addition to congestive heart failure (17). Cobalt salts had been added to the beer to improve its foaming qualities at concentrations of 1.2–1.5 ppm cobalt, a practice no longer in use. At such concentrations the reported formidable consumption of 24 pints (approximately 5 liters per day) of beer would supply about 8 mg of cobalt sulfate, an amount well below that which can be taken with impunity by normal individuals. It seems that high cobalt and high alcohol intakes are both necessary to induce the distinctive cardiomyopathy, plus a third factor which may be low dietary protein (4) or perhaps thiamine deficiency (82), although a combination of cobalt, beer, and thiamine deficiency did not produce a comparable cardiomyopathy in pigs (36).

Toxicological effects of cobalt include vasodilation and flushing in humans, (132), a centrally mediated decrease in heat production in the mouse (37), and a striking reduction in myocardial hypoxic contracture in the rat (41). The most suitable model for study of cobalt-induced cardiomyopathy has proved to be the dog, however, where aberrations in β- adrenergic function occur on administration of cobalt to animals on a low-protein, low-thiamine diet (237). Ultrastructural changes accompanying cobalt-induced cardiomyopathy in dogs have been described (193). Cardiomyopathy following industrial exposure to cobalt in humans has also been recorded (108).

VII.   MODE OF ACTION OF COBALT IN VITAMIN B_{12}

Two basic reaction types are catalyzed by the cobalt atom in vitamin B_{12}, those involving methylcobalamin (IV) and those involving adenosylcobalamin (III). The adenosylcobalamin-dependent reactions comprise a group of 10 intramolecular rearrangements that include the important mammalian (and bacterial) enzyme, methylmalonyl coenzyme A mutase (Section IV,F). The basic reaction sequence in all 10 is similar and involves the transfer of a covalently bonded moiety from one carbon atom to an adjacent carbon atom on a carbon chain accompanied by reverse transfer of the appropriate hydrogen atom between the same two carbon atoms. The latter step takes place without exchange with hydrogen atoms from solvent water to indicate a tightly bonded intermediate. Extensive studies on the molecular mechanisms involved make it likely that the reaction involves homolytic splitting of the covalent bonds concerned to produce free-radical intermediates that are stabilized by the cobalt atom as shown in Fig. 5 (see Ref. 25). The unusual character of this mechanism is made possible by the unique properties of the carbon–cobalt bond in adenosylcobalamin, which readily undergoes reversible homolytic splitting.

A similar bond is present in methylcobalamin, the intermediate involved both in the synthesis of methane and in the enzyme methyltetrahydrofolate : homocysteine methyltransferase responsible for synthesis of methionine. It is not unlikely

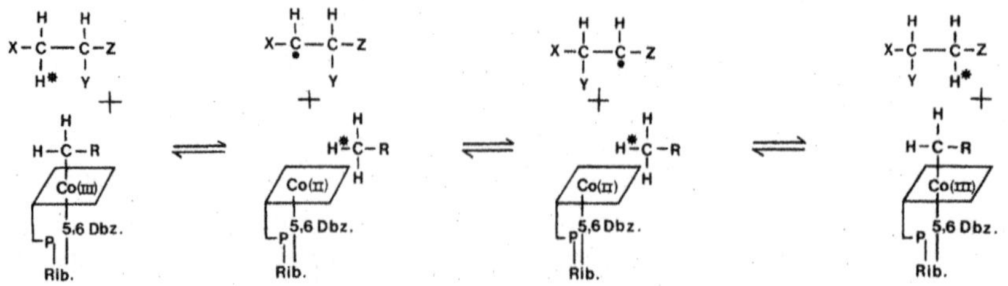

Fig. 5. Mechanism of the several mutase-type reactions catalyzed by adenosylcobalamin-requiring enzymes, including methylmalonyl coenzyme A mutase, the B_{12} enzyme involved in ruminant and other mammalian tissues. As the substrate molecule (CH_2X—$CHYZ$) approaches the enzyme-bound 5′-deoxy-5′-adenosylcobalamin (adenosylcobalamin) molecule, the carbon–cobalt bond of the latter breaks homolytically to form a Co(II) radical and a deoxyadenosyl radical. The latter extracts and holds a hydrogen radical from the substrate while the Co(II) atom stabilizes the residual substrate radical. The latter is then able to undergo the rearrangement shown wherein the Y group migrates reversibly between two adjacent carbon atoms. Reintroduction of the bound hydrogen atom from the adenosyl group produces the product $CHXY$—CH_2Z or else regenerates the substrate CH_2X—$CHYZ$. Simultaneously the carbon–cobalt bond of adenosylcobalamin is re-formed to regenerate adenosylcobalamin. 5,6-Dbz, 5,6-dimethylbenzimidazole.

that these reactions too involve free-radical intermediates, but this has not been so well established.

One unique property of the cobalt atom that is not shared by other transition metals is its capacity to exist in a biological milieu in three stable oxidation states, Co(III), Co(II), and Co(I). The Co(II) free-radical state of vitamin B_{12} is referred to as Cob(II)alamin and is represented in Fig. 5. The single unpaired electron is localized in the $3d_x^2$ orbital, which lies in the direction of the β ligand (25) and facilitates the sequence shown. Cob(I)alamin, a very powerful nucleophile, contains a pair of unshared electrons in the $3d_x^2$ orbital, and is without a ligand at either α or β position. Cob(I)alamin appears to be involved in reactions that employ methylcobalamin, but the involvement is not stoichiometric and is not fully understood. It may be concerned with the regeneration of Cob(II)alamin following its occasional autoxidation.

REFERENCES

1. Adams, S. N., Honeysett, J. L., Tiller, K. G., and Norrish, K. (1969). *Aust. J. Soil Res.* **7,** 29–42.
2. Aguilar, T. S., Benevenga, N. J., and Harper, A. E. (1974). *J. Nutr.* **104,** 761–771.
3. Ahmed, S., and Evans, H. J. (1961). *Proc. Natl. Acad. Sci. U.S.A.* **47,** 24–36.
4. Alexander, C. S. (1969). *Ann. Intern. Med.* **70,** 411–413.
5. Alexiou, D., Grimanis, A. P., Grimani, M., Papaevangelou, G., Koumantakis, E., and Papadatos, G. (1977). *Pediatr. Res.* **11,** 646–648.

6. Andrews, E. D. (1965). *N.Z. Vet. J.* **13**, 101–103.
7. Andrews, E. D. (1956). *N.Z. Agric.* **92**, 239–244.
8. Andrews, E. D. (1965). *N.Z. Agric. Res.* **8**, 788–817.
9. Andrews, E. D. (1953). *N.Z. J. Sci. Technol. Sect. A* **35**, 301–310.
10. Andrews, E. D., and Anderson, J. P. (1954). *N.Z. J. Sci. Technol. Sect. A* **35**, 483–488.
11. Andrews, E. D., and Hart, L. I. (1962). *N.Z. J. Agric. Res.* **5**, 403–408.
12. Andrews, E. D., Hart, L. I., and Stephenson, B. J. (1959). *N.Z. J. Agric. Res.* **2**, 274–282.
13. Andrews, E. D., Hart, L. I., and Stephenson, B. J. (1960). *N.Z. J. Agric. Res.* **3**, 364–376.
14. Andrews, E. D., and Stephenson, B. J. (1966). *N.Z. J. Agric. Res.* **9**, 491–507.
15. Andrews, E. D., Stephenson, B. J., Anderson, J. P., and Faithful, W. C. (1958). *N.Z. J. Agric. Res.* **1**, 125–139.
16. Andrews, E. D., Stephenson, B. J., Isaacs, C. E., and Register, R. H. (1966). *N.Z. Vet. J.* **14**, 191–196.
17. Anonymous (1968). *Nutr. Rev.* **26**, 173–175.
18. Archibald, J. G. (1947). *J. Dairy Sci.* **30**, 293–297.
19. Askew, H. O. (1946). *N.Z. J. Sci. Technol. Sect. A* **28**, 37–43.
20. Askew, H. O., and Watson, J. (1943). *N.Z. J. Sci. Technol. Sect. A* **25**, 81–85.
21. Aston, B. C. (1924). *N.Z. J. Agric.* **28**, 215–238.
22. Aston, B. C. (1924). *N.Z. J. Agric.* **28**, 381–390.
23. Aston, B. C. (1924). *N.Z. J. Agric.* **29**, 14–17.
24. Aston, B. C. (1924). *N.Z. J. Agric.* **29**, 84–91.
25. Babior, B. M., and Krouwer, J. S. (1979). *CRC Crit. Rev. Biochem.* **6**, 35–102.
26. Beck, W. S., Flavin, M., and Ochoa, S. (1957). *J. Biol. Chem.* **229**, 997–1010.
27. Becker, D. E., and Smith, S. E. (1951). *J. Anim. Sci.* **10**, 226–271.
28. Becker, D. E., Smith, S. E., and Loosli, J. K. (1949). *Science* **110**, 71–72.
29. Becker, G., Huebers, H., and Rummel, W. (1979). *Blut* **38**, 397–406.
30. Bigger, G. W., Elliott, J. M., and Rickard, T. R. (1976). *J. Anim. Sci.* **43**, 1077–1081.
31. Blaylock, B. A., and Stadtman, T. C. (1963). *Biochem. Biophys. Res. Commun.* **11**, 34–38.
32. Blokhima, R. I. (1970). *In* "Trace Element Metabolism in Animals" (C.F. Mills, ed.), Vol. 1, pp. 426–432. Livingstone, Edinburgh.
33. Bowie, E. A., and Hurley, P. J. (1975). *Aust. N.Z. J. Med.* **5**, 306–314.
34. Braude, R., Free, A. A., Page, J. E., and Smith, E. L. (1953). *Br. J. Nutr.* **3**, 289–292.
35. Brown, F., and Cuthbertson, W. F. J. (1956). *Nature (London)* **177**, 188.
36. Burch, R. E., Williams, R. V., and Sullivan, J. F. (1973). *Am. J. Clin. Nutr.* **26**, 403–408.
37. Burke, D. H., Brooks, J. C., and Treml, S. B. (1983). *J. Pharm. Sci.* **72**, 824–826.
38. Cikrt, M., and Tichy, M. (1981). *J. Hyg. Epidemiol. Microbiol. Immunol.* **25**, 364–368.
39. Clark, R. G., Cornforth, I. S., Jones, B. A. H., McKnight, L. J., and Oliver, J. (1978). *N.Z. Vet. J.* **26**, 316.
40. Comar, C. L. (1948). *Nucleonics* **3**, 30–42.
41. Conrad, C. H., Brooks, W. W., Ingwall, J. S., and Bing, O. H. L. (1984). *J. Mol. Cell. Cardiol.* **16**, 345–354.
42. Consolazio, C. F., Nelson, R. A., Matoush, L. O., Hughes, R. C., and Urone, P. (1964). *U.S. Army Med. Res. Nutr. Lab. Rep.* **284.**
43. Culvenor, C. C., DalBon, R., and Smith, L. W. (1964). *Aust. J. Chem.* **17**, 1301–1304.
44. Darken, M. A. (1953). *Bot. Rev.* **19**, 99–130.
45. Davis, G. B. (1974). *N.Z. Vet. J.* **22**, 39–42.
46. Dawbarn, M. C., and Hine, D. C. (1955). *Aust. J. Exp. Biol. Med. Sci.* **33**, 335–348.
47. Dawbarn, M. C., Hine, D. C., and Smith, J. (1957). *Aust. J. Exp. Biol. Med. Sci.* **35**, 273–276.
48. Dewey, D. W., Lee, H. J., and Marston, H. R. (1958). *Nature (London)* **181**, 1367–1371.

49. Dickman, S. R. (1977). *J. Mol. Evol.* **10,** 251–260.
50. Dickson, J., and Bond, M. P. (1974). *Aust. Vet. J.* **50,** 236–237.
51. Dinusson, W. E., Klosterman, E. W., Lasley, E. L., and Buchanan, M. L. (1953). *J. Anim. Sci.* **12,** 623–627.
52. Dryden, L. P., Hartman, A. M., Bryant, M. P., Robinson, J. M., and Moore, L. A. (1962). *Nature (London)* **195,** 201–202.
53. Duncan, W. R. H., Morrison, E. R., and Garton, G. A. (1981). *Br. J. Nutr.* **46,** 337–344.
54. Edwin, E. E. (1977). *Vet. Rec.* **101,** 393–394.
55. Elliot, J. M., Kay, R. N. B., and Goodall, E. D. (1971). *Life Sci.* **10,** 647–654.
56. Engel, R. W., Price, N. O., and Miller, R. F. (1967). *J. Nutr.* **92,** 197–204.
57. Everist, S. L. (1974). "Poisonous Plants of Australia," pp. 241–243. Angus & Robertson, Sydney.
58. Filmer, J. F. (1933). *Aust. Vet. J.* **9,** 163–179.
59. Filmer, J. F. (1941). *N.Z. J. Agric.* **63,** 287–290.
60. Filmer, J. F., and Underwood, E. J. (1934). *Aust. Vet. J.* **10,** 83–87.
61. Filmer, J. F., and Underwood, E. J. (1937). *Aust. Vet. J.* **13,** 57–64.
62. Flavin, M., and Ochoa, S. (1957). *J. Biol. Chem.* **229,** 965–979.
63. Ford, J. E., and Hutner, S. H. (1955). *Vitam. Horm.* **13,** 101–136.
64. Forth, W., and Rummel, R. (1971). *In* "Intestinal Absorption of Metal Ions, Trace Elements and Radionuclides" (S. C. Skoryna and D. Waldron-Edwards, eds.), pp. 173–191. Pergamon, Oxford.
65. Fox, G. E., Magrum, L. J., Balch, W. E., Wolfe, R. S., and Woese, C. R. (1977). *Proc. Natl. Acad. Sci. U.S.A.* **74,** 4537–4541.
66. Fraser, F. J. (1972). *Stain Technol.* **17,** 147–154.
67. Gallagher, C. H., Koch, J. H., Moore, R. M., and Steel, J. D. (1964). *Nature (London)* **204,** 542–545.
68. Gardner, F. H. (1953). *J. Lab. Clin. Med.* **41,** 56–63.
69. Gawthorne, J. M. (1969). *Aust. J. Exp. Biol. Med. Sci.* **47,** 311–317.
70. Gawthorne, J. M. (1970). *Aust. J. Exp. Biol. Med. Sci.* **48,** 285–292.
71. Gawthorne, J. M. (1970). *Aust. J. Exp. Biol. Med. Sci.* **48,** 293–300.
72. Gawthorne, J. M. (1968). *Aust. J. Biol. Sci.* **21,** 789–794.
73. Gawthorne, J. M., Somers, M., and Woodliff, H. J. (1966). *Aust. J. Exp. Biol. Med. Sci.* **44,** 585–588.
74. Gawthorne, J. M., and Smith, R. M. (1974). *Biochem. J.* **142,** 119–126.
75. Gawthorne, J. M., and Stokstad, E. L. R. (1971). *Proc. Soc. Exp. Biol. Med.* **136,** 42–46.
76. Gebgardt, A. G., Kucheras, R. V., Laska, D. V., and Vogrin, A. G. (1970). *Microbiology* **39,** 380–384.
77. Gessert, C. F., Berman, D. T., Kastelic, J., Bentley, O. G., and Phillips, P. H. (1952). *J. Dairy Sci.* **35,** 693–698.
78. Gleason, F. K., and Wood, J. M. (1976). *Science* **192,** 1343–1344.
79. Gould, S. J. (1980). "The Panda's Thumb," pp. 181–188. Penguin, London.
80. Grant, W. C., and Root, W. S. (1952). *Physiol. Rev.* **32,** 449–498.
81. Grimmett, R. E. R., and Shorland, F. B. (1934). *Trans. R. Soc. N.Z.* **64,** 191–213.
82. Grinvalsky, H. T., and Fitch, D. M. (1969). *Ann. N.Y. Acad. Sci.* **156,** 544–565.
83. Gurnani, S., Mistry, S. P., and Johnson, B. C. (1960). *Biochim. Biophys. Acta* **38,** 187–188.
84. Halbrook, E. R., Cords, F., Winter, A. R., and Sutton, T. S. (1950). *J. Nutr.* **41,** 555–563.
85. Hamilton, H. G. (1956). *South. Med. J.* **49,** 1056–1059.
86. Hannam, R. J., Judson, G. J., Reuter, D. J., McLaren, L. D., and McFarlane, J. D. (1980). *Aust. J. Agric. Res.* **31,** 347–355.
87. Harp, M. J., and Scoular, F. I. (1952). *J. Nutr.* **47,** 67–72.

88. Hart, I. I., and Andrews, E. D. (1959). *Nature (London)* **184**, 1242.
89. Hartley, W. J. (1978). *In* "Effects of Poisonous plants on Livestock" (R. F. Keeler, K. R. Van Kampen, and L. F. James, eds.), pp. 391–393. Academic Press, New York.
90. Hartman, A. M., and Dryden, L. P. (1952). *Arch. Biochem. Biophys.* **40**, 310–313.
91. Hartroft, W. S. (1954). *Ann. N.Y. Acad. Sci.* **57**, 633–645.
92. Hedrich, M. F., Elliot, J. M., and Lowe, J. E. (1973). *J. Nutr.* **103**, 1646–1651.
93. Hendlin, D., and Rudger, M. L. (1950). *Science* **111**, 541–542.
94. Hine, D. C., and Dawbarn, M. C. (1954). *Aust. J. Exp. Biol. Med. Sci.* **32**, 641–652.
95. Hoekstra, W. G., Pope, A. L., and Phillips, P. H. (1952). *J. Nutr.* **48**, 421–430.
96. Hoekstra, W. G., Pope, A. L., and Phillips, P. H. (1952). *J. Nutr.* **48**, 431–441.
97. Holm-Hansen, O., Gerloff, G. C., and Skoog, F. (1954). *Physiol. Plant.* **1**, 665–675.
98. Houk, A. E. H., Thomas, A. W., and Sherman, H. C. (1946). *J. Nutr.* **31**, 609–620.
99. Hubbard, D. M., Creech, F. M., and Cholak, J. (1966). *Arch. Environ. Health* **13**, 190–194.
100. Ibbotson, R. N., Allen, S. H., and Gurney, C. W. (1970). *Aust. J. Exp. Biol. Med. Sci.* **48**, 161–169.
101. International Commission on Radiological Protection. (1975). No. 23. "Report of the Task Group on Reference Man," pp. 300–301. Pergamon, Oxford.
102. IUPAC-IUB Commission on Biochemical Nomenclature (1974). *Biochemistry* **13**, 1555–1560.
103. Iyengar, G. V., Kasperek, K., Feinendegen, L. E., Wang, Y. X., and Weese, H. (1982). *Sci. Total Environ.* **24**, 267–274.
104. Iyengar, G. V., Kollmer, W. E., and Bowen, H. J. M. (1978). "The Elemental Composition of Human Tissues and Body Fluids." Verlag Chemie, Weinheim.
105. Kasperek, K., Kiem, J., Iyengar, G. V., and Feinendegen, L. E. (1981). *Sci. Total Environ.* **17**, 133–143.
106. Katsuoka, Y., Beckman, B., George, W. J., and Fisher, J. W. (1983). *Am. J. Physiol.* **244**, F129–F133.
107. Kenealy, W., and Zeikus, J. G. (1981). *J. Bacteriol.* **146**, 133–140.
108. Kennedy, A., Dornan, J. D., and King, R. (1981). *Lancet* **1**, 412–414.
109. Kent, N. L., and McCance, R. A. (1941). *Biochem. J.* **35**, 877–883.
110. Kercher, C. J., and Smith, S. E. (1955). *J. Anim. Sci.* **14**, 458–464.
111. Kercher, C. J., and Smith, S. E. (1955). *J. Anim. Sci.* **14**, 878–884.
112. Kercher, C. J., and Smith, S. E. (1956). *J. Anim. Sci.* **15**, 550–558.
112a. Knott, P., Algar, B., Zervas, G., and Telfer, S. B. (1985). *In* "Trace Elements in Man and Animals" (C. F. Mills, I. Bremner, and J. K. Chesters, eds.), Vol. 5, pp. 708–713. Commonwealth Agricultural Bureaux, Slough, England.
113. Koch, B. A., and Smith, S. E. (1951). *J. Anim. Sci.* **10**, 1017–1021.
114. Koch, H. J., Smith, E. R., Shimp, N. F., and Connor, J. (1956). *Cancer* **9**, 499–511.
115. Kondo, H., Kolhouse, J. F., and Allen, R. H. (1980). *Proc. Natl. Acad. Sci. U.S.A.* **77**, 817–821.
116. Kovalsky, V. V. (1970). *In* "Trace Element Metabolism in Animals" (C.F. Mills, ed.), Vol. 1, pp. 385–397. Livingstone, Edinburgh.
117. Kriss, J. P., Carnes, W. H., and Gross, R. T. (1955). *J. Am. Med. Assoc.* **157**, 117–121.
118. Krzycki, S., and Zeikus, J. G. (1980). *Curr. Microbiol.* **3**, 243–245.
119. Kutzbach, C., Galloway, E., and Stokstad, E. L. R. (1967). *Proc. Soc. Exp. Biol. Med.* **124**, 801–805.
120. Lassiter, C. A., Ward, G. M., Huffman, C. F., Duncan, C. W., and Webster, H. D. (1953). *J. Dairy Sci.* **36**, 997–1005.
121. Lau, K.-S., Gottlieb, C., Wasserman, L. R., and Herbert, V. (1965). *Blood* **26**, 202–214.
122. Lee, C. C., and Wolterink, L. F. (1955). *Am. J. Physiol.* **183**, 173–177.

123. Lee, C. C., and Wolterink, L. F. (1955). *Poult. Sci.* **34**, 764–776.
124. Lee, H. J. (1950). *Aust. Vet. J.* **26**, 152–159.
125. Lee, H. J. (1974). *In* "Final report of the Division of Nutritional Biochemistry," pp. 11–14, CSIRO, Adelaide.
126. Lee, H. J., and Kuchel, R. E. (1953). *Aust. J. Agric. Res.* **4**, 88–99.
127. Lee, H. J., Kuchel, R. E., Good, B. F., and Trowbridge, R. F. (1957). *Aust. J. Agric. Res.* **8**, 494–501.
128. Lee, H. J., Kuchel, R. E., Good, B. F., and Trowbridge, R. F. (1957). *Aust. J. Agric. Res.* **8**, 502–511.
129. Lee, H. J., Kuchel, R. E., and Trowbridge, R. F. (1956). *Aust. J. Agric. Res.* **7**, 333–344.
130. Lee, H. J., and Marston, H. R. (1969). *Aust. J. Agric. Res.* **20**, 905–918.
131. Lee, H. J., and Marston, H. R. (1969). *Aust. J. Agric. Res.* **20**, 1109–1116.
132. LeGoff, J. M. (1930). *J. Pharmacol. Exp. Ther.* **38**, 1–9.
133. Lezius, A. G., and Barker, H. A. (1965). *Biochemistry* **4**, 510–518.
134. Lines, E. W. (1935). *J. Counc. Sci. Ind. Res. (Aust.)* **8**, 117–119.
135. Linnell, J. C., and Matthews, D. M. (1984). *Clin. Sci.* **66**, 113–121.
136. Looney, J. W., Gille, G., Preston, R. L., Graham, E. R., and Pfander, W. H. (1976). *J. Anim. Sci.* **42**, 693–698.
137. McAdam, P. A., and O'Dell, G. D. (1982). *J. Dairy Sci.* **65**, 1219–1226.
138. McCance, R. A., and Widdowson, E. M. (1944). *Annu. Rev. Biochem.* **13**, 315–346.
139. McDonald, I. W. (1942). *Aust. Vet. J.* **18**, 182–189.
140. McDonald, I. W. (1946). *Aust. Vet. J.* **22**, 91–94.
141. MacLaren, A. P. C., Johnston, W. G., and Voss, R. C. (1964). *Vet. Rec.* **76**, 1148–1149.
142. MacLaughlin, M. F., Rice, D. A., and Taylor, S. M. (1984). *Vet. Rec.* **115**, 325.
143. McNaught, K. J. (1938). *N.Z. J. Sci. Technol. Sect. A* **20**, 14–30.
144. McNaught, K. J. (1948). *N.Z. J. Sci. Technol. Sec. A* **30**, 26–48.
145. MacPherson, A., and Moon, F. E. (1974). *In* "Trace Element Metabolism in Animals" (G. W. Hoekstra, J. W. Suttie, H. E. Ganther, and W. Mertz, eds.), Vol. 2, pp. 624–627. Univ. Park Press, Baltimore.
146. MacPherson, A., Moon, F. E., and Voss, R. C. (1973). *Br. Vet. J.* **129**, 414–426.
147. MacPherson, A., Moon, F. E., and Voss, R. C. (1976). *Br. Vet. J.* **132**, 294–308.
148. Marston, H. R. (1935). *J. Counc. Sci. Ind. Res. (Aust.)* **8**, 111–116.
149. Marston, H. R. (1952). *Physiol. Rev.* **32**, 66–121.
150. Marston, H. R. (1970). *Br. J. Nutr.* **24**, 615–633.
151. Marston, H. R., Allen, S. H., and Smith, R. M. (1961). *Nature (London)* **190**, 1085–1091.
152. Marston, H. R., Allen, S. H., and Smith, R. M. (1972). *Br. J. Nutr.* **27**, 147–157.
153. Marston, H. R., and Lee, H. J. (1952). *Nature (London)* **170**, 791.
154. Marston, H. R., Lee, H. J., and McDonald, I. W. (1948). *J. Agric. Sci.* **38**, 222–228.
155. Marston, H. R., Thomas, R. C., Murnane, D., Lines, E. W., McDonald, I. W., Moore, H. O., and Bull, L. B. (1938). *Commonw. Counc. Sci. Ind. Res. (Aust.), Bull.* **113**.
156. Martinovich, D. (1974). *Proc. N.Z. Vet. Assoc. Sheep Soc. 4th Semin.* pp. 99–101.
157. Mason, R. W., and McKay, R. (1983). *Aust. Vet. J.* **60**, 219–220.
158. Millar, K. R., and Andrews, E. D. (1964). *N.Z. Vet. J.* **12**, 9–12.
159. Millar, K. R., and Lorentz, P. P. (1979). *N.Z. Vet. J.* **27**, 90–92.
160. Mitchell, P. J., McOrist, S., Thomas, K. W., and McCausland, I. P. (1982). *Aust. Vet. J.* **58**, 181–184.
161. Monroe, R. A., Sauberlich, H. E., Comar, C. L., and Hood, S. L. (1952). *Proc. Soc. Exp. Biol. Med.* **80**, 250–257.
162. Morris, J. G., and Gartner, R. J. W. (1967). *J. Agric. Sci.* **68**, 1–9.
163. Muerling, S., and Plantin, L.-O. (1981). *Acta Chir. Scand.* **147**, 481–485.

164. Murthy, G. K., Rhea, U., and Peeler, J. T. (1971). *Environ. Sci. Technol.* **5**, 436–442.
165. National Research Council (1975,1976,1978,1980). "Mineral Tolerance of Domestic Animals." National Academy of Sciences—National Research Council, Washington, D.C.
166. O'Halloran, M. W., and Skerman, K. D. (1961). *Br. J. Nutr.* **15**, 99–108.
167. Onkelinx, C. (1976). *Toxicol. Appl. Pharmacol.* **38**, 425–438.
168. Oram, R. N. (1970). *Proc. Int. Grassl. Congr. 11th, 1969* pp. 785–788.
169. Orr, J. B., and Holm, A. (1931). "Mineral Content of Pastures." 6th Rep. Econ. Advis. Counc., Great Britain.
170. Ozanne, P. G., Greenwood, E. A., and Shaw, T. C. (1963). *Aust. J. Agric. Res.* **14**, 39–50.
171. Paley, K. R., and Sussman, E. S. (1963). *Metab. Clin. Exp.* **12**, 975–982.
172. Pearson, P. B., Struglia, L., and Lindahl, I. L. (1953). *J. Anim. Sci.* **12**, 213–218.
173. Peters, J. P., and Elliott, J. M. (1983). *J. Dairy Sci.* **66**, 1917–1925.
174. Phillipson, A. T., and Mitchell, R. L. (1952). *Br. J. Nutr.* **6**, 176–189.
175. Pol, A., van der Drift, C., and Vogels, D. G. (1982). *Biochem. Biophys. Res. Commun.* **108**, 731–737.
176. Pollack, S., George, J. N., Reba, R. C., Kaufman, R. M., and Crosby, W. H. (1965). *J. Clin. Invest.* **44**, 1470–1473.
177. Porter, J. W. (1953). *Proc. Nutr. Soc.* **12**, 106–114.
178. Poston, J. M. (1976). *Science* **195**, 301–302.
179. Poston, J. M. (1978). *Phytochemistry* **17**, 401–402.
180. Powrie, J. K. (1960). *Aust. J. Sci.* **23**, 198–199.
181. Provasoli, L. (1958). *Annu. Rev. Microbiol.* **12**, 279–308.
182. Rajagopalan, K. (1976). *J. Appl. Bacteriol.* **40**, 111–114.
183. Raun, N. S., Stables, G. L., Pope, L. S., Harper, O. F., Waller, G. R., Renbarger, R., and Tillman, A. D. (1968). *J. Anim. Sci.* **27**, 1695–1702.
184. Rendig, V. V., Cooper, D. W., Dunbar, J. R., Lawrence, C. M., Clawson, W. J., Bushnell, R. B., and McComb, E. A. (1976). *Calif. Agric.* **30**, 8–10.
185. Richards, R. B., and Harrison, M. R. (1981). *Aust. Vet. J.* **57**, 565–568.
186. Rickard, T. R., and Elliot, J. M. (1978). *J. Anim. Sci.* **46**, 304–308.
187. Robey, J. S., Veazey, P. M., and Crawford, J. D. (1956). *N. Engl. J. Med.* **255**, 955–957.
188. Roginski, E. E., and Mertz, W. (1977). *J. Nutr.* **107**, 1537–1542.
189. Rossiter, R. C., Curnow, D. H., and Underwood, E. J. (1948). *J. Aust. Inst. Agric. Sci.* **14**, 9–14.
190. Rothery, P., Bell, J. M., and Spinks, J. W. T. (1953). *J. Nutr.* **49**, 173–181.
190a. Ruelke, O. C., and McCall, J. T. (1961). *Agron. J.* **53**, 406–407.
191. Russel, A. J. F., and Whitelaw, A. (1983). *Vet. Rec.* **112**, 418.
192. Russell, F. C., and Duncan, D. (1956). *Commonw. Bur. Anim. Nutr. Tech. Commun.* No 15.
193. Sandusky, G. E., Henk, W. G., and Roberts, E. D. (1981). *Toxicol. Appl. Pharmacol.* **61**, 89–98.
194. Schroeder, H. A., Nason, A. P., and Tipton, I. H. (1967). *J. Chronic Dis.* **20**, 869–890.
195. Sheehan, R. G. (1974). *Proc. Soc. Exp. Biol. Med.* **146**, 993–996.
196. Sheline, G. E., Chaikoff, I. L., and Montgomery, M. L. (1946). *Am. J. Physiol.* **145**, 285–290.
197. Sherlock, S. (1968). "Diseases of the Liver and Biliary System," 4th Ed., pp. 387–394. Blackwell, Oxford.
198. Simpson, B. H., Jolly, R. D., and Thomas, S. H. M. (1969). *N.Z. Vet. J.* **17**, 240–244.
199. Sinclair, J. F., Sinclair, P. R., Healey, J. F., Smith, E. L., and Bonkowsky, H. L. (1982). *Biochem. J.* **204**, 103–109.
200. Skerman, K. D., Sutherland, A. K., O'Halloran, M. W., Bourke, J. M., and Munday, B. L. (1959). *Am. J. Vet. Res.* **20**, 977–984.

201. Smith, E. L. (1965). "Vitamin B$_{12}$," p. 84. Methuen, London.
202. Smith, R. M., and Marston, H. R. (1970). *Br. J. Nutr.* **24,** 857–877.
203. Smith, R. M., and Marston, H. R. (1970). *Br. J. Nutr.* **24,** 879–891.
204. Smith, R. M., and Marston, H. R. (1971). *Br. J. Nutr.* **26,** 41–53.
205. Smith, R. M., and Monty, K. J. (1959). *Biochem. Biophys. Res. Commun.* **1,** 105–109.
206. Smith, R. M., and Osborne-White, W. S. (1973). *Biochem. J.* **136,** 279–293.
207. Smith, R. M., Osborne-White, W. S., and Gawthorne, J. M. (1974). *Biochem. J.* **142,** 105–117.
208. Smith, R. M., Osborne-White, W. S., and Russell, G. R. (1969). *Biochem. J.* **112,** 703–707.
209. Smith, S. E., Becker, D. E., Loosli, J. K., and Beeson, K. C. (1950). *J. Anim. Sci.* **9,** 221–230.
210. Smith, S. E., Koch, B. A., and Turk, K. L. (1951). *J. Nutr.* **44,** 455–464.
211. Smith, S. E., and Loosli, J. K. (1957). *J. Dairy Sci.* **40,** 1215–1227.
212. Somers, M. (1969). *Aust. J. Exp. Biol. Med. Sci.* **47,** 219–225.
213. Somers, M., and Gawthorne, J. M. (1969). *Aust. J. Exp. Biol. Med. Sci.* **47,** 227–233.
214. Southcott, W. H. (1956). *Aust. Vet. J.* **32,** 225–228.
215. Speijers, G. J. A., Krajnc, E. I., Berkvens, J. M., and van Logten, M. J. (1982). *Food Chem. Toxicol.* **30,** 311–314.
216. Stadtman, T. C. (1960). *J. Bacteriol.* **79,** 904–905.
217. Stebbings, R. St. J., and Lewis, G. (1983). *Vet. Rec.* **112,** 328.
218. Stenberg, T. (1983). *Acta Odontol. Scand.* **41,** 143–148.
219. Sutherland, R. J., Cordes, D. O., and Carthew, G. C. (1979). *N.Z. Vet. J.* **27,** 227–232.
220. Sutton, A. L., and Elliot, J. M. (1972). *J. Nutr.* **102,** 1341–1346.
221. Taylor, C. D., and Wolfe, R. S. (1974). *J. Biol. Chem.* **249,** 4879–4885.
222. Telfer, S. B., Zervas, G., and Knott, P. (1983). UK Patent Application 2116424A.
223. Tevetoglu, F. (1956). *J. Pediatr.* **49,** 46–55.
224. Thiers, R. E., Williams, J. F., and Yoe, J. H. (1955). *Anal. Chem.* **27,** 1725–1731.
225. Thompson, J. F., and Ellis, G. H. (1947). *J. Nutr.* **34,** 121–127.
226. Thomson, A. B. R., Valberg, L. S., and Sinclair, D. G. (1971). *J. Clin. Invest.* **50,** 2384–2394.
227. Thorndike, J., and Beck, W. S. (1984). *Blood* **64,** 91–98.
228. Tipton, I. H., and Cook, M. J. (1963). *Health Phys.* **9,** 103–145.
229. Tipton, I. H., Stewart, P. L., and Martin, P. G. (1966). *Health Phys.* **12,** 1683–1689.
230. Tosic, J., and Mitchell, R. L. (1948). *Nature (London)* **162,** 502–504.
231. Toskes, P. P., Smith, G. W., and Conrad, M. E. (1973). *Am. J. Clin. Nutr.* **26,** 435–437.
232. Ulvund, M. J., and Overas, J. (1980). *N.Z. Vet. J.* **28,** 19.
233. Underwood, E. J. (1934). *Aust. Vet. J.* **10,** 87–92.
234. Underwood, E. J., and Elvehjem, C. A. (1938). *J. Biol. Chem.* **124,** 419–424.
235. Underwood, E. J., and Filmer, J. F. (1935). *Aust. Vet. J.* **11,** 84–92.
236. Underwood, E. J., and Harvey, R. J. (1938). *Aust. Vet. J.* **14,** 183–189.
237. Unverferth, D. V., Fertel, R. H., Thomas, J., Leier, C. V., Croskery, R., Hunsaker, R., Miller, M., Gibb, L., and Hamlin, R. (1984). *Cardiovasc. Res.* **18,** 44–50.
238. Valberg, L. S. (1971). *In* "Intestinal Absorption of Metal Ions, Trace Elements and Radionuclides" (S. C. Skoryna and D. Waldron-Edward, eds.), p. 257 *et seq.* Pergamon, Oxford.
239. Valberg, L. S., Ludwig, J., and Olatunbosun, D. (1969). *Gastroenterology* **56,** 241–251.
240. van der Meijden, P., van der Lest, C., van der Drift, C., and Vogels, G. D. (1984). *Biochem. Biophys. Res. Commun.* **118,** 760–766.
241. Versiek, J., Hoste, J., Barbier, F., Steyaert, H., de Rudder, J., and Hilde, M. (1978). *Clin. Chem.* **24,** 303–308.
242. Waltner, K., and Waltner, K. (1929). *Klin. Wochenschr.* **8,** 313.

243. Whitelaw, A., and Russel, A. J. F. (1979). *Vet. Rec.* **104,** 8–11.
244. Wolff, H. (1950). *Klin. Wochenschr.* **28,** 280.
245. Wood, J. M., Moura, I., Moura, J. J. G., Santos, M. H., Xavier, A. V., LeGall, J., and Scandellari, M. (1982). *Science* **216,** 303–305.
246. Yamagata, N., Kurioka, W., and Shimizu, T. (1963). *J. Radiat. Res.* **4,** 8–15.

6

Manganese

LUCILLE S. HURLEY and CARL L. KEEN
Department of Nutrition
University of California
Davis, California

I. MANGANESE IN ANIMAL TISSUES AND FLUIDS

A. Total Content and Distribution

Manganese (Mn), though widely distributed in the biosphere, occurs in only trace amounts in animal tissues. The body of a normal 70-kg man is estimated to contain a total of 10–20 mg manganese (52,257). This relatively small amount of manganese is distributed widely throughout the tissues and fluids, without notable concentration in any particular location and with comparatively little variation among organs or species, or with age, relative to some other trace elements (257). Similarly, in contrast to several of the other essential trace elements, the fetus does not normally accumulate liver manganese before birth, and values are actually lower in fetal liver than in adult liver (204). Manganese tends to be higher in tissues rich in mitochondria and is more concentrated in the mitochondria than in the cytoplasm or other organelles of the cell (193,279). The pigmented portions of the eye are not exceptionally rich in manganese, as they are in zinc and copper, although the retina is richer in this metal than most body tissues. The pigmented melanin-containing parts of the conjunctiva are higher in manganese than the nonpigmented parts (56).

Typical concentrations of manganese in several tissues of humans, cattle, and rats based on papers cited in this chapter and data from the authors' laboratories are shown in Table I. It is apparent that the bones, liver, and kidney normally carry higher manganese concentrations than do other organs, and that the muscles are among the lowest in this element of the tissues of the body. The levels of manganese in the bones can be raised or lowered by substantially varying the manganese intakes of the animal. This has been demonstrated in the newborn calf (129), rat (178), rabbit (69), pig (146), and chick (29,190,294). Mathers and Hill (190) reported a skeletal manganese concentration of 2.5 ppm on the dry, fat-free basis for pullets fed a low-manganese diet from 18 weeks of age to after 6–7 months of egg production, compared with 7.8 ppm for similar birds fed a high-manganese diet for this period. The skeletal manganese amounted to about 25% of total-body manganese but did not constitute an important mobilizable store of the element. Black and co-workers (29) have shown that for chickens, the increase in bone manganese concentration with increasing dietary manganese levels is linear for diets containing between 90 and 3000 ppm manganese. The relative increase in bone (tibia) manganese concentration with increasing dietary levels was much higher than that found for kidney, liver, plasma, or muscle, and these authors suggested that changes in bone manganese levels could be used as a bioassay of dietary manganese bioavailability for the chick. Most of the bone manganese is believed to be deposited in the inorganic portion of the bone, but a small proportion is associated with the organic matrix (84). How the distribution of manganese in bone is affected by age or by a deficiency or excess of the element is not known.

Table I
Concentrations of Manganese in Adult
Animal Tissue

Tissue	Manganese concentration (ppm, fresh basis)		
	Human	Bovine	Rat
Bone (long)	2.55	1.50	2.00
Liver	1.40	2.50	2.40
Pancreas	1.21	1.58	1.35
Testes	0.15	1.00	1.33
Intestine	1.10	1.80	1.00
Kidney	1.20	1.30	0.72
Brain	0.32	0.23	0.35
Heart	0.23	0.33	0.40
Spleen	0.22	0.20	0.24
Lung	0.27	0.25	0.11
Muscle	0.06	0.80	0.07
Blood	0.01	0.02	0.012
Plasma	0.001	0.005	0.001

Howes and Dyer (129) doubled the manganese concentration in the marrow-free radius of calves at 1 week of age by supplementing their milk diet with manganese at 14.5 ppm from birth. This treatment had little influence on the level of manganese in the calves' muscle or hair but strikingly increased the levels in their liver. For example, the liver of 7-day-old unsupplemented calves fed a 13-ppm manganese diet contained 6.36 ± 0.6 μg manganese per gram of dry matter, compared with 614.67 ± 195.89 for 7-day-old calves from similar dams that were manganese-supplemented during the first week of life. It was also apparent from this study that newborn calves preferentially store manganese in their livers when this element is added to the diet of their mothers. Higher liver manganese in the newborn of manganese-supplemented mothers has similarly been demonstrated in rats and mice (39). On the other hand, reserve manganese stores do not normally occur in the livers of newborn rats, rabbits, guinea pigs, cattle, or humans (41,184,266,289). The lack of fetal storage of manganese is not surprising, for in contrast to zinc, copper, and iron, there does not appear to be a manganese "storage" protein. In addition, the emergence of many of the manganese enzymes, such as arginase, pyruvate carboxylase, and superoxide dismutase (SOD), occurs during postnatal life.

Human livers from healthy individuals of all ages contain about 6–8 ppm manganese (dry basis) (41,282), with appreciable individual variation but very little variation from one part of the liver to another (233). Typical manganese concentrations of normal sheep and cattle livers are 8–10 ppm (dry basis). (284,289), with similar levels in the livers of laying hens (190). The concentration of manganese in human rib cartilage has been reported to decrease with age, with cartilage collected from children under 1 year of age having levels up to 10 times higher than cartilage collected from adults (256).

The partition between free and bound manganese in liver has been receiving attention. Using electron paramagnetic resonance, Mn^{2+} concentration of rat hepatocyte cells has been estimated at 35 nmol/ml cell water (10). It has been suggested that changes in intracellular free Mn^{2+} may be an important mechanism of cellular metabolic control, in a manner analogous to that of free Mg^{2+} and Ca^{2+} (10,301).

B. Manganese in Hair, Wool, and Feathers

The concentration of manganese in mammalian hair varies with the species, individual, season, color, and less certainly with the manganese status of the diet. Van Koetsveld (289) found the hair of healthy adult cows to range from 8 to 15 ppm (mean 12), and those showing signs of manganese deficiency to be below 8 ppm. At very high intakes of manganese, concentrations as high as 80 ppm were observed. Meyer and Engelbartz (208) found the hair of 351 cattle to range from 3.9 to 49.9 ppm manganese (mean 15.8), and O'Mary et al. (224)

observed a range of 6–104 ppm manganese in Hereford cattle and calves. Red and black hair from cattle and from humans is consistently higher in manganese than white or gray hair (36,52,224). Groppel and Anke (103) reported that the manganese level in the hair of mature goats receiving a low-manganese diet averaged 3.5 ppm, compared with 11.1 ppm in comparable goats receiving adequate manganese. These workers maintained that the manganese level in the hair reflects the manganese dietary supply better than any other part of the body studied. No such relationship was observed in two later studies with cattle. Thus Howes and Dyer (129) found no significant differences in the hair manganese levels of 7-day-old calves that had received a manganese supplement from birth and those that had not, and Hartmans (112) found no difference between identical twin cattle that had received either a diet containing 21 ppm manganese or one containing 130 ppm manganese. This worker did not find a significantly higher level of hair manganese in these cattle when at pasture than when receiving hay. The mean levels were 6.0 and 1.8 ppm, and 6.7 and 2.7 ppm for the low-manganese and the high-manganese groups, respectively. Marked seasonal changes in hair manganese levels have also been reported in the Alaskan moose *Alces gigas* (81). Remarkably low levels, <1 ppm, were observed in the late winter–early spring samples, with levels >8 ppm appearing in the autumn samples. The extent to which these changes are a reflection of varying dietary manganese intakes remains to be determined. Mehnert has argued that for cattle, hair manganese values are not useful in assessing manganese status, as environmental contamination is too great (202,203); however, on a herd basis values that are either very high or very low may be useful indications of manganese nutrition.

It has been suggested by Collipp *et al.* (51) that human infants receiving formulas high in manganese have higher hair manganese levels than breast-fed infants. This report has raised some concern as children with learning disabilities have been reported to have significantly higher hair manganese levels than controls (21,235), leading Collipp to suggest that some of these children may be suffering from manganese toxicity (see Section VII).

The manganese levels in wool and feathers vary significantly with dietary manganese intakes. Lassiter and Morton (171) reported a mean of 6.1 ppm in the wool of lambs fed a low-manganese diet for 22 weeks, compared with 18.7 ppm in the wool of control lambs. Mathers and Hill (190) found the skin and feathers of pullets fed a low-manganese diet for several months to average 1.2 ppm manganese, as opposed to 11.4 ppm in comparable birds fed a high-manganese diet.

C. Manganese in Blood

Widely varying values have appeared for the manganese levels in normal human whole blood. In a study of 102 individuals in England a mean of 6.88 ±

0.86×10^{-2} µg/ml was reported (109). This is very close to 69 µg/liter. Bowen (37) found normal human blood to average 24 ± 8 µg manganese per liter, divided fairly equally between cells and plasma. Cotzias and co-workers (55,229), in two separate studies, obtained much lower values and higher concentrations in the red cells than in the serum. In the first study 8.44 ± 2.73 µg/liter manganese was reported for whole blood and only 0.59 ± 0.18 µg/liter for plasma. In the second the following mean concentrations were obtained: whole blood, 9.84 ± 0.4; serum, 1.42 ± 0.2; and red cells, 23.57 ± 1.2 µg/liter. Similar low values for whole blood and serum were reported by others (80,236,271,290).

Hatano and co-workers (114) have reported erythrocyte manganese levels of about 100 µg/liter in Japanese infants <1 month of age compared to levels of about 35 µg/liter in Japanese adults. These investigators suggested that the relatively high erythrocyte manganese levels in the Japanese population studied may be due to a higher dietary intake of manganese in Japan compared to several other countries. The pronounced developmental change in erythrocyte manganese levels observed by Hatano *et al.* is as yet unexplained, although it may in part be related to an immature biliary excretion of this element in the infant.

The manganese in human serum is selectively and almost totally bound in the trivalent form by a β_1-globulin (83). In rat serum the globulin that binds manganese has been suggested to be transferrin (151). In erythrocytes a firmly bound manganese compound exists which is probably a manganese porphyrin (34,110).

Elevated whole-blood and serum levels of manganese have been reported to occur under various conditions. Increases in serum levels up to twice normal have been reported to occur due to congestive heart failure, infections, psychosis, myocardial infarctions, and Alzheimer-like diseases (72,128,273). Increases in whole-blood manganese levels have been reported to occur with excess manganese intake (55), rheumatoid arthritis (57), and iron deficiency (206). Low levels of serum manganese have been reported in children suffering from the inborn errors of metabolism, maple syrup urine disease and phenylketonuria (143). Whole-blood-manganese has been reported to be lower than normal in some epileptics (228). While it is not known for humans if whole-blood manganese reflects body manganese stores, in the rat it has been shown that low blood manganese is reflective of low tissue manganese levels (156).

Levels of 20 µg manganese per liter have been reported for the blood of calves (115), with lower values for bovine blood in other studies (27,30). Bovine blood plasma contains approximately 5 µg manganese per liter even at widely varying manganese intakes (248). Comparable low values have also been obtained for avian blood (31). The manganese concentration in the blood plasma of pullets increases markedly with the onset of egg-laying (122). At 19 weeks of age the levels were 30–48 µg/liter, and at 25 weeks the levels had risen to 85–91 µg/liter. The influence of estrogen activity on plasma manganese levels is further apparent from the studies of Panic and co-workers (226). The radioactivity in the plasma 4 hr after the intramuscular injection of ^{54}Mn was shown to be about 15 and 70 times higher in laying hens and estrogen-treated immature pullets than in controls.

Greger et al. (100) have reported that adolescent girls have significantly higher blood manganese levels than males, supporting the observation that estrogens increase blood manganese levels, although as this population group is often marginal in iron status, the higher blood manganese observed could also be due to the iron–manganese interaction.

D. Manganese in Milk

Investigations in several countries have shown normal mature cow's milk to contain 20–50 μg manganese per liter, with concentrations of 100–160 μg/liter in colostrum (8,9,144,164,250). The level of manganese in milk responds rapidly to changes in dietary manganese intakes. Archibald and Lindquist (9) increased the milk manganese concentration 2- to 4-fold by feeding manganese sulfate to cows in amounts equivalent to 10 or 13 g/day manganese. The provision of high-manganese feeds or feeding supplements providing only 3 g/day manganese produced smaller increases (160). Manganese levels in mature sheep and goat milk have been reported to be between 20 and 50 μg/liter (182). In contrast to these species, human mature milk has been reported to contain manganese in the range of 4–15 μg/liter, although values as high as 70 μg/liter have been found by some investigators (40,98,155,199). Stastny et al. (271) have reported that manganese intakes of human milk-fed infants are linearly correlated with the serum manganese concentration at 3 months of age (see below, Section V,E).

E. Manganese in the Avian Egg

The manganese content of eggs varies widely with the manganese level of the diet. In two early studies, raising the manganese level in the hen's diet from 13 to 1000 ppm increased the amount in the yolk from 4 to 33 μg (87), and supplementing a manganese-deficient diet with 40 ppm manganese increased the concentration in the whole egg from 0.5 to 0.9 ppm (dry basis) (185). Hill and Mather (123) reported that eggs from pullets fed a low-manganese diet contained 4–5 μg manganese, while those from pullets fed a normal-manganese diet contained 10–15 μg. In one study, hens fed a diet containing 81 ppm manganese produced eggs containing 28.6 ± 0.78 μg manganese (62). The manganese concentration of the yolk is four to five times that of the white. Eggshell quality is best when diets contain at least 70 ppm manganese (126).

II. MANGANESE METABOLISM

A. Absorption and Bioavailability

Studies on the absorption of Mn have been carried out primarily with the isotope manganese-54 (^{54}Mn), a γ emitter with a half-life of 312 days. In the early work by Greenberg and co-workers (99), it was estimated than only 3–4% of an orally administered dose of manganese is absorbed in adult rats. However, in young rats the absorption of manganese is on the order of 20% (155). In cattle, approximately 1–4% of dietary manganese is absorbed, irrespective of dietary concentration (248,273,288). For the adult human, it has been reported that approximately 3–4% of dietary manganese is absorbed, and it is thought that the percentage of manganese absorbed from a meal is, to a large extent, independent of the amount of manganese in the diet, or the body burden of the element.

Absorption of manganese apparently occurs equally well throughout the length of the small intestine (280). Garcia-Aranda *et al.* (88) have reported that for the rat intestinal manganese absorption is a rapidly saturable process, with manganese absorption declining linearly with time when the element is perfused into the intestine at concentrations between 0.0125 and 0.10 mM. These investigators also reported that the absorption of manganese could be enhanced by low molecular weight ligands such a L-histidine and citrate.

The percentage of manganese absorbed from a meal can be influenced by other dietary factors. Considerable work has been done on the interaction between iron and manganese. Matrone and co-workers (192) demonstrated that high levels of dietary manganese could depress hemoglobin formation in the pig, presumably by decreasing iron absorption. Reduced iron absorption with high dietary manganese has also been reported in cattle (125). Conversely, in chickens, high levels of dietary iron can accentuate the severity of perosis, probably by decreasing manganese absorption (300). Thomson *et al.* (280,281) have provided data that the interaction between manganese and iron is at the level of absorption. These investigators reported that manganese was absorbed by a two-step mechanism involving initial uptake from the lumen and then transfer across the mucosal cells. The two kinetic processes operate simultaneously, with manganese competing with iron and cobalt for common binding sites in both processes (280,281). In this way one of the metals exerts an inhibitory effect on the absorption of the others. Thomson *et al.* (280) showed that the addition of iron competitively inhibited manganese absorption in iron-deficient rats. In patients with varying iron stores subjected to duodenal perfusion with manganese, the rate of manganese absorption was found to be increased in iron deficiency and the enhanced manganese absorption to be inhibited by iron (205). It is apparent that the absorption mechanisms of manganese and iron show many similarities— similarities that are not shared by their excretion processes.

The absorption and retention of manganese from foods low in iron, such as milk, is relatively high. Using mice as models, it has been shown that if milk is supplemented with iron, both the absorption and retention of manganese is reduced (104,157). Since this effect of iron on manganese absorption from milk occurs at a level of iron supplementation similar to that used in infant formulas, it will be important to determine in future studies if an unwanted side effect of iron supplementation is suboptimal manganese nutrition.

Manganese availability is further affected by excess dietary calcium. In birds, the effect of high dietary levels of calcium phosphate in aggravating manganese deficiency is believed to be due to a reduction in soluble manganese through adsorption by solid mineral (61,252,300). Chemical forms as diverse and varying in solubility as oxide, carbonate, sulfate, and chloride were shown some years ago to be equally valuable as sources of manganese in poultry rations (252). Differences in manganese availability among various inorganic manganese sources have now been demonstrated using leg abnormality scores and bone manganese levels as the response criteria (294). The carbonate ore (rhodochrosite) and silicate ore (rhodomite) are relatively unavailable (87,252). The fecal excretion of parenterally administered [54]Mn is higher and liver retention much lower in rats fed a 1.0% than in those given a 0.6% calcium diet (172). It appears that calcium can influence manganese metabolism by affecting retention of absorbed manganese as well as by affecting its absorption.

Ethanol feeding has been shown to increase hepatic manganese in humans, monkeys, and rats (18,66,159). This is apparently due to a significant effect (at least 2-fold increase) on manganese absorption, mediated by ethanol in the gut (25).

Manganese metabolism is influenced by the estrogenic hormones of the ovary and by the adrenal cortical hormones. The profound effect of estrogen on plasma manganese levels in poultry has already been mentioned (226). The administration of glucocorticoid hormones markedly affects [54]Mn distribution in the mouse. There is a shift in the partition within the body from the liver to the carcass (130). Stimulation of the animal's own adrenal cortices with ACTH results in similar changes (52,131). However, adrenalectomy did not alter the manganese concentrations in the liver and diaphragm, except in animals receiving high manganese intakes when the levels increased.

Increased absorption of manganese has been reported during pregnancy in sows (166), also supporting an effect of estrogens on absorption of the element. Coccidiosis infection in chickens has also been reported to increase manganese absorption (268); the mechanisms underlying this effect are not known.

In addition to the influence of other dietary factors on manganese absorption, the molecular localization of the element in the diet may have a considerable impact on its bioavailability. For example, the bioavailability of manganese from human milk is considerably higher than that from cow's milk, a difference which is presumably due in part to a large difference in the localization of manganese in the two fluids. Thus in human milk, the major part of manganese is found in whey, with only a small amount of the element in the casein fraction, while in cow's milk most of the manganese is in the casein fraction (183). The protein source of the diet can also affect manganese absorption. In chicks the bioavailability of manganese from soybean meal and cottonseed meal is >70%, while from rapeseed, average manganese availability is only 50% (222). In humans the absorption of manganese from meat and fish is higher than that from legumes (240).

B. Transport and Tissue Distribution

Manganese from the gastrointestinal tract entering into the portal circulation may either remain free or rapidly become bound to α_2-macroglobulin before traversing the liver, where it is removed nearly quantitatively, although some of the manganese bound to α_2-macroglobulin may enter the systemic circulation, become oxidized to the manganic state, and become bound to transferrin (248). Transferrin-bound manganese may be taken up by extrahepatic tissue (92). Manganese flux in plasma is extremely rapid; for humans it has been reported that only 1% of intravenously injected ^{54}Mn is still present in blood after 10 min (54). Injected ^{54}Mn is cleared from the bloodstream in three phases (35). The first and fastest of these is identical with a clearance rate of other small ions, suggesting the normal transcapillary movement; the second can be identified with the entrance of manganese into the mitochondria of the tissues; the third and slowest component could indicate the rate of nuclear accumulation of the metal. These interpretations are supported by studies demonstrating early and preferential accumulation of ^{54}Mn in the mitochondria-rich organs of the body (150), localization of manganese in the mitochondria of the cell, and high mitochondrial and low nuclear manganese turnover rates (52). The kinetic patterns for blood clearance and liver uptake of manganese are similar, indicating that the two manganese pools, blood manganese and liver mitochondrial manganese, rapidly enter into equilibrium. The uptake and release of manganese by mitochondria may be via a Ca^{2+} carrier (145). Hepatic release of manganese can be increased by glucocorticoid or ACTH injections; however, the physiological significance of this effect of the hormones is not clear at the present time (131).

The retention of ingested ^{54}Mn by adult humans has been estimated at 10% 14 days after feeding an oral dose, and 16% in the premature infant (205). The higher retention in the premature infant may be the result of increased manganese absorption by the immature gut, immaturity of the excretory pathways for the element, and/or a higher requirement for manganese due to tissue synthesis and emergence of manganese enzymes. Kirchgessner *et al.* (166), using everted gut sacs from rats of varying ages, have shown that manganese absorption decreases with age. Miller *et al.* (211) have shown that manganese excretion can be virtually absent in the neonatal period due to low bile output, the principal route of manganese excretion (see below).

C. Excretion

Absorbed manganese is almost totally excreted via the intestinal wall by several routes. These routes are interdependent and combine to provide the body with an efficient homeostatic mechanism regulating the level of manganese in the tissues (28,229). Under ordinary conditions the bile flow is the main route of excretion. A suggested mechanism for the movement of manganese into bile is that manganese taken up by the liver and not shunted to extrahepatic tissue is rapidly incorporated into the mitochondrial and lysosomal compartments; manganese complexes from the lysosomal compartment are subsequently transferred to the bile. In rats this process is very rapid, with intravenously injected ^{54}Mn appearing in bile within an hour of injection (274). Over 50% of manganese injected intravenously can be recovered in the feces within 24 hr and 17% additional within the second 24-hr period. By 3 days most of the injected dose of ^{54}Mn will be eliminated through the biliary route (168). There is an overall bile–plasma concentration ratio >150 for manganese. It has been estimated that about two-thirds of this overall gradient is due to the gradient from plasma to bile, and one-third from liver to bile. As plasma and liver contain ligands with a higher affinity for manganese than bile, the transport of manganese into bile must be active (168). It has been estimated by Cikrt (49) that about 35% of excreted manganese undergoes enterohepatic circulation. The manganese concentration of the bile fluid can be increased 10-fold or more by the administration of large amounts of manganese to the animal (243).

Excretion of manganese also occurs via the pancreatic juice (44). When the hepatic (biliary) route is blocked, or when overloading with manganese occurs, pancretic excretion increases (229). Manganese excretion also takes place in the duodenum, jejunum, and to a smaller extent into the terminal ileum (28). These must be regarded as auxiliary routes. Very little manganese is excreted in the urine, even when injected or added to the diet (162,194,247). A marked rise in urinary manganese can be achieved by the administration of chelating agents (45,194). The remarkable capacity of the bovine liver to remove and presumably excrete excess absorbed manganese has been demonstrated by Sansom *et al.* (248,249).

Loading the body with stable manganese, but not with other elements, rapidly elutes manganese from the body and redistributes it within the tissues (39,53). The turnover of parenterally administered ^{54}Mn has been directly related to the level of stable manganese in the diet of mice over a wide range (39). An inverse relationship between dietary manganese and the percentage of ^{54}Mn taken up by the tissues of chicks has also been demonstrated (261). These findings support the concept that variable excretion rather than variable absorption is the regulator of manganese homeostasis. However, Lassiter and co-workers (163,170) have provided evidence that dietary manganese level has a greater effect on absorption than on endogenous excretion, and that both variable excretion and absorption play important roles in manganese homeostasis. For example, 3 days after a single oral dose, total-body ^{54}Mn retention in low-manganese calves was nine times greater (18.2 versus 2.2%) than in manganese-supplemented calves. In an experiment with rats given a single oral ^{54}Mn dose, liver ^{54}Mn was 15 times higher after 4 hr in those fed a diet containing 4 ppm manganese than in those fed 1000 ppm manganese. Such a great difference indicates a major effect on absorption. Increased absorption of manganese under conditions of low manganese intakes and decreased absorption at higher manganese intakes, in a manner reminiscent of iron absorption in iron deficiency, is equally apparent from experiments of Howes and Dyer (129) with calves.

III. BIOCHEMISTRY OF MANGANESE

A. Metal-Activated Reactions

Like other essential trace elements, manganese can function both as an enzyme activator and as a constituent of metalloenzymes. For manganese-activated reactions, the metal can act either by binding to the substrate (such as ATP) or to the protein directly, with induction of subsequent conformational changes. While the number of manganese metalloenzymes is limited, the enzymes that can be activated by manganese are numerous. They include hydrolases, kinases, decarboxylases, and transferases (103). Many of these metal activations are nonspecific, so that manganese may be partly replaced by other metal ions, particularly magnesium. Even the manganese metalloenzyme pyruvate carboxylase is subject to partial substitution of manganese by magnesium, with only minor alteration in the catalytic properties of the enzyme. It is important to point out that manganese activation of an enzyme *in vitro* does not prove a role for the element *in vivo*. In addition, a problem with those enzymes not specifically activated by manganese is the difficulty of correlating pathological defects seen with manganese deficiency and specific biochemical lesions. Reviews discussing this topic and listing manganese-activated enzymes have been published (158,197).

One exception to the nonspecific activation of enzymes by manganese is the manganese-specific activation of glycosyltransferases. Other divalent cations will not activate this enzyme, and some of the pathological effects of manganese deficiency can be ascribed to a low activity of this enzyme class (see below, Section IV,A). Another example of a manganese-activated enzyme that can be affected by dietary manganese deficiency is phosphoenolpyruvate carboxykinase(see Section IV,F). This manganese-activated enzyme is unusual in that Fe^{2+}, not Mg^{2+}, is the principal competing cation (186). A third example of an enzyme which may be specifically activated by manganese is xylosyltransferase (200). This enzyme, isolated from chick cartilage, catalyzes the transfer of xylose in the formation of the linkage of proteins and polysaccharides. Cartilage isolated from manganese-deficient chicks has been reported to be xylose poor.

B. Manganese Metalloenzymes

In contrast to the manganese-activated enzymes, there are few manganese metalloenzymes. Manganese-containing enzymes include arginase (EC 3.5.3.1; MW 120,000), pyruvate carboxylase (EC 6.4.1.1; MW 500,000) and manganese-superoxide dismutase (EC 1.15.1.1; MW 80,000). Arginase, the cytosolic enzyme responsible for urea formation, contains 4 mol of Mn^{2+} per mole of enzyme (124). The activity of arginase is affected by diet, with liver levels in rats fed manganese-deficient diets being less than 50% that of controls (165). Despite this dramatic influence of manganese deficiency on arginase activity the potential functional significance of this biochemical effect has not been studied.

Pyruvate carboxylase isolated from chicken liver mitochondria (260) contains 4 mol Mn(II) per mole of enzyme. The following reaction has been proposed for this enzyme (209): enzyme-biotin + CO_2 + pyruvate enzyme-biotin + oxaloacetate. Scrutton et al. (259) found that magnesium replaced manganese as the bound metal in pyruvate carboxylase from manganese-deficient chicks. The relative content of the two metals was related to the severity of the deficiency. It seems, therefore, that the chicken can adapt to manganese deficiency with respect to pyruvate carboxylase by substituting magnesium for manganese. In contrast, the manganese-deficient rat may have low liver pyruvate carboxylase activity levels under some conditions (see below, Section IV,F).

The SOD enzymes, which catalyze the disproportionation of O_2^- to H_2O_2 + O_2, are metalloenzymes containing copper, zinc, iron, or manganese, or in some cases combinations of these (101). The SOD isolated from E. coli has a molecular weight of 40,000, is composed of two subunits of equal size, and contains one Mn^{2+} per molecule (153). The manganese SOD from chicken liver mitochondria is similar to the corresponding bacterial enzyme but is twice as large and contains four subunits instead of two (101). In contrast to pyruvate carboxylase and arginase, the manganese in resting SOD is trivalent. The catalytic cycle of this enzyme involves reduction and then reoxidation of the metal center during successive encounters with O_2^-. The activity of manganese SOD in tissues of manganese-deficient animals can be significantly lower than in control animals. The functional significance of this is discussed in Section IV,D.

C. Manganese and Brain Function

Manganese is essential for normal brain function. In rats a deficiency of manganese is manifested in part by increased susceptibility to convulsions and by electroencephalographic recordings that are similar to those of epileptics (139). The significance of this abnormality is underscored by the work of Papavasiliou et al. (228), which has shown that whole-blood manganese levels are lower in epileptics with frequent seizures than in controls. As whole-blood manganese can be reflective of low soft tissue manganese concentrations (156), the work by Papavasiliou et al. suggests that suboptimal manganese status may be a problem in some epileptics, either as a cause or an effect. While the biochemical lesions occurring in the brain with manganese deficiency have not been identified, it is obvious that this should be an area of future research. Manganese toxicity can also have severe effect on brain function (see Section VII).

In the human brain the highest concentrations of manganese are found in the pineal gland, olfactory bulb, median eminence of the hypothalamus, and basal ganglia (19,33). Manganese in the brain accumulates mainly in pigmented structures, such as melanocytes of the substantia nigra. It has been suggested that if too much manganese is taken up by these structures, the metabolism of dopamine may be affected by manganese-induced free-radical damage (227).

The biological functions of manganese in the brain are not well understood. Wedler et al. (295) have suggested that manganese functions as the metal constituent of glutamine synthase. Yip and Dain (306) have suggested a role for manganese in the activation of UDPgalactose : GM_2 ganglioside galactosyltransferase, and Tagliamonte et al. (275) have provided evidence than manganese is the activator for brain catechol o-methyltransferase. In addition, manganese may have a role in the regulation of adenylate cyclase activity and hence cyclic AMP levels in the brain (293). None of the above enzyme systems has been studied under conditions of dietary manganese deficiency, thus at this time it is not known if a disturbance in these pathways may explain some of the neurological lesions seen with manganese deficiency.

A direct interaction between manganese and biogenic amines via complex formation has also been suggested by Papavasiliou (227), who proposed that by such an interaction manganese may be important in the binding, transport, and storage of catechols and ethanolamines.

IV. MANGANESE DEFICIENCY AND RELATION TO BIOCHEMICAL FUNCTION

Manganese deficiency has been demonstrated in mice, rats, rabbits, guinea pigs, pigs, poultry, sheep, goats, and cattle, occurs naturally on certain diets composed of normal feeds fed to pigs and poultry, and has been reported in humans in association with a vitamin K deficiency (63). The main manifestations of manganese deficiency, namely impaired growth, skeletal abnormalities, disturbed or depressed reproductive function, ataxia of the newborn, and defects in lipid and carbohydrate metabolism, are displayed in all species studied, but their actual expression varies with the degree and duration of the deficiency and with the developmental stage during which the deficiency occurs. Many of the gross effects of manganese deficiency on the skeleton can now be explained in terms of its effect on mucopolysaccharide synthesis, as discussed in the following section.

Hemoglobin levels do not appear to be significantly affected by lack of manganese (267,291). The growth inhibition of manganese deficiency results from both reduced food consumption and impaired efficiency of food use (38), but severe inappetence is not a conspicuous feature of manganese deficiency, as it is in zinc and cobalt deficiencies.

A. Manganese and Bone Growth

The skeletal abnormalities of manganese deficiency are particularly characterized in rats, rabbits, and guinea pigs by disproportionate and retarded bone growth, with shortening and bowing of the forelegs (22,263,267) and defective development of the skull and the otoliths of the inner ear during gestation (134,141,264). In pigs the skeletal abnormalities are characterized by lameness and enlarged hock joints with crooked and shortened legs (210,216); in calves by difficulty in standing (129); in sheep by joint pains with poor locomotion and balance (171); and in goats by tarsal joint excrescences, leg deformities, and ataxia (103). Manganese deficiency in chicks, poults, and ducklings is manifested as the disease perosis or "slipped tendon" (299), and in chick embryos by chondrodystrophy (185). Perosis is characterized by enlargement and malformation of the tibiometatarsal joint, twisting and bending of the tibia, thickening and shortening of the long bones, and slipping of the gastrocnemius tendon from its condyles. With increasing severity of the condition the chicks are reluctant to move, walk on their hocks, and soon die. Nutritional chondrodystrophy is characterized by shortened and thickened legs and wings, "parrot beak" resulting from a disproportionate shortening of the lower mandible, globular contour of the head, and high mortality. A similar defect in bone development occurs in the offspring of manganese-deficient rats. Severe shortening of the radius, ulna, tibia, and fibula is evident in these young at birth (136). In addition, marked epiphyseal dysplasia at the proximal end of the tibia and shortening and doming of the skull with anomalous ossification of the inner ear develops in such animals (140,141). The skulls of deficient young are shorter, wider, and higher than those of controls.

Impairment of the calcification process per se is not a primary causal factor in the bone abnormalities of manganese deficiency just described. Neither the volume of the bone nor the gross composition of the bone ash is reduced, although the length, density, and breaking strength are lowered (4,46,267). In addition, studies of calcium metabolism in manganese-deficient animals demonstrated that calcium metabolism itself was normal (137). Retarded skeletal maturation and chondrodystrophic growth is due to inhibition in endochondral osteogenesis at the epiphyseal cartilages (85,216,304). Deficient bones appear normal to X-ray examination and silver nitrate staining (46,86), despite some reduction in bone ash content, and differ distinctly from those of calcium-, phosphorus-, vitamin D-, or copper-deficient animals (46,175,267).

In view of these findings, attention was turned to a possible involvement of manganese in the synthesis of the organic matrix of cartilage. Leach and Muenster (176) discovered that radiosulfate uptake is lowered in the cartilage of the manganese-deficient chick, and that the total concentration of hexosamines and hexuronic acid is reduced in this tissue. A less pronounced reduction was observed in the hexosamine content of other tissues. Chondroitin sulfate is the mucopolysaccharide most severely affected by manganese deficiency. Subsequently Everson and associates (76,264,283) demonstrated a reduction in the concentration of acid mucopolysaccharides (AMPS) in rib and epiphyseal cartilage and in the otoliths of manganese-deficient newborn guinea pigs, and Hurley *et al.* (138) showed that manganese deficiency markedly reduced the *in vitro* rate of incorporation of radiosulfate into cartilage matrix in fetal rat tibia. The same group studied histologically mucopolysaccharide synthesis in the developing inner ear of manganese-deficient and pallid mutant mice (265). Manganese deficiency or the pallid gene caused reduced incorporation of ^{35}S in the macular cells, the formation of nonmetachromatic, variably PAS-positive matrix which did not contain ^{35}S, and failure of otolithic calcification. Hidiroglou and co-workers (119) have shown that newborn lambs born to manganese-deficient ewes had lower than normal concentrations of glucosamine and galactosamine in their epiphyseal cartilage.

The impairment in mucopolysaccharide synthesis associated with manganese deficiency has been related to the activation of glycosyltransferases by this element (175). These enzymes are important in polysaccharide and glycoprotein synthesis, and manganese is usually the most effective of the metal ions required for their activity. The critical sites of manganese function in chondroitin sulfate synthesis have been identified by Leach and co-workers (177) as the two enzyme systems: (1) polymerase enzyme, which is responsible for the polymerization of UDP-*N*-acetylgalactosamine to UDPglucuronic acid to form the polysaccharide, and (2) galactotransferase, an enzyme which incorporates galactose from UDPgalactose into the galactose-galactose-xylose trisaccharide which serves as the linkage between the polysaccharide and the protein associated with it. Since mucopolysaccharides are vital structural components of cartilage, these findings provide a likely biochemical explanation of the skeletal defects associated with manganese deficiency.

It is also probable that the effect of manganese on mucopolysaccharide synthesis explains some of the deficiency defects observed in laying hens, such as reduced egg production and poor shell formation (47,107), and chondrodystrophy in chick embryos (185). A reduction in the hexosamine content of the shell matrix has been associated with poor shell formation in laying hens (180). Of further interest is the finding that several types of chondrodystrophy of genetic origin are characterized by a reduced limb mucopolysaccharide content (191).

B. Manganese and Neonatal Ataxia

Ataxia in the offspring of manganese-deficient animals was first observed in the chick by Caskey and Norris (48) and in rats by Shils and McCollum (263). Hurley and Everson and their co-workers (77,134,135) later showed that manganese deficiency during pregnancy in rats and guinea pigs can produce an irreversible congenital defect in the young characterized by ataxia and loss of equilibrium, and often also by head retraction and tremors, increased susceptibility to stimuli, and delayed development of the body-righting reflexes. The defect responsible for these disturbances arises relatively late in gestation—in the rat between the fourteenth and eighteenth days. There is a critical need for manganese on the fifteenth and sixteenth days, because manganese supplementation begun at that time results in an improvement of survival time and of the incidence of ataxia in the young, whereas supplementation begun on the eighteenth day is ineffective in preventing ataxia (135) (Table II). A structural defect in the inner ear then emerged as the causative factor in the ataxia and postural defects of

Table II

Effect of Manganese Supplementation at Various Times during Gestation[a]

Initiation of supplementation (day of gestation)[b]	No. of litters	Young born		Survival to 28 days	
		Total	Per litter	Live young	Ataxic
7–12	14	105	7.5	53	0
14	6	42	7.0	87	0
15	8	60	7.5	36	48[c]
16	8	54	6.8	44	46[c]
18	8	65	8.1	26	100

[a] From Hurley et al. (135).
[b] Day of finding sperm considered first day of gestation.
[c] Mild.

manganese-deficient rats and guinea pigs (134,141,264). These disorders were shown to arise from impaired vestibular function, itself a reflection of a specific effect of lack of manganese on cartilage mucopolysaccharide synthesis and hence bone development of the skull, particularly the otoliths. Deficient animals do not exhibit normal otolith development in the utricular and saccular maculae as a consequence of impaired mucopolysaccharide synthesis (265).

A specific congenital ataxia in mice resulting from defective development of the otoliths, and caused by the presence of a mutant gene affecting coat color (pallid), was shown by Erway *et al.* (72) to be completely prevented by high levels of manganese supplementation during pregnancy. The congenital ataxia resulting from maternal dietary deficiency of manganese thus appears to be a phenocopy of the condition produced by this mutant gene. The prevention of the congenital genetic ataxia by manganese supplementation of the maternal diet showed that a genetic defect could be prevented by nutritional manipulation of the prenatal diet.

In adult pallid mice, manganese concentration of bone, but not of liver or kidney, was lower than normal, and slow transport of manganese, L-dopa, and L-tryptophan was observed (58). The relationship of manganese to the abnormal melanin formation [in retina and inner ear (278) as well as skin] of pallid mice and otolith development is not understood; it has been speculated that the alteration of melanin formation may reduce intracellular manganese concentration, producing localized manganese deficiency. The mutant gene screwneck, in pastel mink, is analogous to pallid. Defective development of otoliths and abnormal postural reflexes can be prevented by dietary manganese supplementation of pregnant screwneck females (73).

C. Manganese and Reproductive Function

Defective ovulation, testicular degeneration, and infant mortality were observed in the earliest studies demonstrating the essentiality of manganese in the diet of rats (161,225,292). In the female three stages of manganese deficiency can be recognized. In the least severe stage the animals give birth to viable young, some or all of which exhibit ataxia. In the second, more severe stage the young are born dead or die shortly after birth. In the third, acute stage of deficiency, estrous cycles are absent or irregular, the animals will not mate, and sterility results. A delay in the opening of the vaginal orifice may also occur (38). A similar impairment of reproductive performance occurs in hens, since lowered egg production and decreased hatchability can occur even in the absence of perosis and chondrodystrophy (11,253). The severely manganese-deficient male rat and rabbit exhibit sterility and absence of libido, associated with seminal tubular degeneration, lack of spermatozoa, and accumulation of degenerating cells in the epididymis (38,267). In guinea pigs omission of manganese from the maternal diet increases the proportion of young born dead or delivered prematurely and reduces litter size (77).

The feeding of low-manganese rations to cows (6,27,129) and goats (6) causes depressed or delayed estrus and conception, as well as increased abortion and stillbirths and lowered birth weights. The biochemical lesions underlying the reproductive dysfunctions have not been delineated. Doisy (63) has suggested that they may result from defective steroid synthesis, but this hypothesis has yet to be tested. It is of interest that the corpus luteum is reported to contain a high concentration of manganese (120).

Several claims that manganese supplements improve the fertility of dairy cows in certain areas have also been made (118,121,297,303). However, experiments by Hartmans (113) with identical cattle twins fed from the age of 1 or 2 months on low-manganese rations (16–21 ppm, dry basis)—that is, low in comparison with most field pasture levels—revealed no differences in fertility compared with controls receiving supplementary manganese. It seems, therefore, that unidentified environmental factors that limit manganese absorption or utilization must be operating where manganese responses have been obtained.

D. Manganese and Ultrastructural Abnormalities

Abnormalities in cell function and ultrastructure, particularly involving the mitochondria, occur in manganese deficiency. Oxidative phosphorylation was studied in isolated liver mitochondria of adult manganese-deficient mice and rats (142). In both species, ratios of ATP formed to oxygen consumed (P/O) were normal but oxygen uptake was reduced. Electron microscopy revealed ultrastructure abnormalities of the mitochondria, including elongation and reorientation of cristae (142). Subsequently Bell and Hurley (24) confirmed and extended these findings. A dietary deficiency of manganese was found to cause changes in the ultrastructural parameters of liver, pancreas, kidney, and heart in aged mice. All the manganese-deficient tissues observed revealed alterations in the integrity of their cell membranes. In addition, the endoplasmic reticulum was swollen and irregular mitochondria were found with elongated stacked cristae in liver, heart, and kidney cells, and there was an overabundance of lipid in liver parenchymal and kidney tubule cells. Further work showed normal ultrastructure in liver from manganese-deficient rats up to 60 days of age, but at 9 months of age, mitochondria were abnormal. Vacuoles were present in the matrix, and inner and outer mitochondrial membranes were separated (309).

Dietary manganese deficiency in the rat resulted in lower than normal activity of liver manganese SOD from 20 to 60 days of age (310) and higher than normal levels of mitochondrial lipid peroxidation by 60 days of age (310). These observations support the hypothesis that mitochondrial damage results from manganese deficiency because of depressed manganese SOD activity leading to increased lipid peroxidation in the membranes. Thus, the mitochondrial damage apparent at 9 months of age may result, at least in part, from the low manganese SOD activity occurring earlier. There is also *in vitro* evidence that manganese is important for mitochondrial integrity. Manganous ions together with ATP, ADP, or AMP, can reverse or prevent swelling of mitochondria (82,187,238), and such swelling has been correlated with formation of lipid peroxides (20,132).

Conditions that cause production of superoxide radicals, such as exposure to hyperbaric oxygen (272), ozone, or ethanol, result in increased activity of manganese SOD in rats and monkeys (66,67,159). Manganese-deficient rats were more susceptible to ethanol toxicity than normal rats (308). Pasquier *et al.* (230) have reported that polymorphonuclear leukocytes of human adults with rheumatoid arthritis had less manganese SOD activity than did those of healthy volunteers.

E. Manganese and Lipid Metabolism

An association between manganese and choline in metabolism has been recognized for some years (5,75,148). Liver and bone fat in manganese-deficient rats are reduced by both manganese and choline supplements (5), and a manganese-deficient diet can produce excessive fat deposition in pigs (237) and mice (24). Both nutrients are needed for complete protection against perosis in chicks (148). Rats fed a choline-deficient diet for 25 days exhibited lower hepatic manganese levels than those of controls (152). The authors suggest that this is due to reduced intestinal transport of the metal. Furthermore, the changes in liver ultrastructure that arise in choline deficiency (42) are very similar to those that have been observed in manganese deficiency (24) (see above). Deficiencies of manganese and choline both appear to affect membrane integrity. These nutrients may therefore be, as Bell and Hurley (24) have suggested, "linked in a common pathway to establish the normal structure of the mitochondrial and cellular membranes."

These authors also examined the histochemical enzyme changes in cephalic tissues of manganese-deficient fetal mice. No differences in enzyme activities were found in the inner ear tissues between the manganese-deficient, pallid, and control fetuses, but in the cephalic epidermis of manganese-deficient fetuses, particularly the soft keratin layer, cytochrome oxidase and choline esterase activities and the false-positive yellow reaction for alkaline phosphatase were reduced, and there was a reduced concentration of choline lipids.

Manganese may also be related to mitochondrial structure through effects on lipid peroxidation. Liver mitochondria of manganese-deficient rats and mice had higher rates of lipid peroxidation than did those from controls; increased lipid peroxidation was correlated with lower activity of manganese SOD in manganese-deficient animals (310) (see above).

Manganese also has a role in the biogenesis of cholesterol. Curran (60) has shown that the addition of manganese to rat liver *in vitro* markedly increases cholesterol synthesis, and Olson (223) has defined two sites between acetate and mevalonate that require manganese. In addition, farnesyl pyrophosphate synthase is known to require Mn^{2+}, and the two have been found to act together to add one five-carbon unit to geranyl pyrophosphate to make farnesyl pyrophosphate (26). Lack of farnesyl pyrophosphate could inhibit production of squalene, thus limiting the formation of important precursors of cholesterol. Roby *et al.* (244) have reported that manganese-deficient male weanling Sprague–Dawley rats had lower levels of cholesterol in plasma and liver than controls, and a similar finding of hypocholesterolemia has been reported in young men fed a manganese-deficient diet (302). However, a pronounced effect of manganese deficiency on cholesterol metabolism was not found in estrogen-treated chicks, laying hens, Wistar rats, or rats of the genetically hypercholesterolemic RICO strain (169).

There appears to be a strain difference in response to manganese deficiency in rats as well as mice (68); in addition to the difference in cholesterol levels, Sprague–Dawley rats are more susceptible than Wistar rats to other effects of manganese deficiency as well, for example, neonatal mortality and congenital ataxia (68). However, fatty acid synthesis was depressed in Wistar rats by manganese deficiency (169).

F.   Manganese and Carbohydrate Metabolism

Defects in carbohydrate metabolism, other than those known to arise from the impairment of the activities of glycosyltransferases already described, have been observed in manganese-deficient guinea pigs and rats. Everson and Shrader (78) found that newborn guinea pigs severely affected with manganese deficiency exhibited aplasia or marked hypoplasia of all the cellular components of the pancreas. Smaller numbers of islet cells, containing fewer and less intensely granulated β cells, were apparent than in the islets of control guinea pigs. Young adult manganese-deficient guinea pigs also had subnormal numbers of pancreatic islets, with less intensely granulated β cells and more α cells than manganese-supplemented controls. When glucose was administered orally or intravenously the manganese-deficient guinea pigs revealed a decreased capacity to utilize glucose and displayed a diabeticlike curve in response to glucose loading. Manganese supplementation completely reversed the reduced glucose utilization.

Baly *et al.* (13) have observed that the adult manganese-deficient offspring of manganese-deficient rats also showed diabeticlike glucose tolerance curves, while at the same time plasma insulin levels were not commensurate with their high plasma glucose levels. Pancreatic manganese and insulin concentrations were low. Insulin output of perfused pancreas from these deficient rats was considerably lower than in controls. Further experiments with perfused pancreas indicated that insulin synthesis as well as release was impaired by manganese deficiency (14), indicating a role for manganese in insulin biosynthesis. Pancreatic glucagon release was not affected by manganese deficiency (14).

It is also possible that the effect of manganese deficiency on insulin production occurs through the destruction of β cells in the pancreas. Diabetogenic agents such as alloxan have been postulated to act through increased formation of superoxide radicals (50). As the activity of manganese SOD in pancreatic islet cells is low (compared to that of other tissues) (97), a reduction in its activity by manganese deficiency could make the pancreas particularly susceptible to free-radical damage.

It has been suggested that insulin may regulate the utilization of glucose in the synthesis of mucopolysaccharides (254), and decreased concentrations of stainable mucopolysaccharides have been demonstrated in the skin of young rats born to diabetic mothers (307). Sand rats (*Psammonys obesus*), whose natural diet is high in manganese, developed an insulin-resistant diabetes when given a commercial rat feed. The condition was reversed on reintroduction of the manganese-rich natural food (262). Unfortunately, supplementation with manganese alone was not attempted. Failla and co-workers (79) have reported that liver manganese concentrations can be significantly higher in streptozotocin-diabetic animals than in controls. This increase is due in part to an increase in the level of arginase. Increased tissue manganese levels have also been reported in the pregnant streptozotocin-diabetic rat and her fetuses (71,286).

Manganese deficiency may also influence carbohydrate metabolism through effects on phosphoenolpyruvate carboxykinase, for which manganese is a cofactor, or pyruvate carboxylase, a manganese metalloenzyme (see above). Baly *et al.* (15) found that pyruvate carboxylase activity of liver in adult manganese-deficient rats in the fed state was normal, but if they were fasted for 48 hr, pyruvate carboxylase specific activity was lower than in controls. In chickens, however, Scrutton *et al.* (259) found that magnesium replaced manganese as the bound metal in pyruvate carboxylase.

Both pyruvate carboxylase and phosphoenolpyruvate carboxykinase are key enzymes in gluconeogenesis, necessary for the conversion of lactate or pyruvate to oxaloacetate, phosphoenolpyruvate, and glucose. As the neonate must accommodate to its new nutrient supply, gluconeogenesis is important for glucose homeostasis in the newborn. In the rat, increases in the activities of pyruvate carboxylase and phosphoenolpyruvate carboxykinase occur during the first week of postnatal life (231). Offspring of manganese-deficient rats showed lower plasma glucose levels in the neonatal period than controls (15). While the relationship of phosphoenolpyruvate carboxykinase and pyruvate carboxylase to glucose levels in the manganese-deficient pups was not clear, it is possible that impairment of gluconeogenesis may be a factor in the high neonatal mortality that occurs at the same period in manganese-deficient offspring.

A direct effect of manganese on gluconeogenesis has also been reported. In perfused rat liver producing glucose from lactate, Mangnall et al. (189) found that adding alanine caused a shift to alanine utilization; when manganese was added to the system, both substrates were used for gluconeogenesis, with a net increase in glucose output. With isolated hepatocytes from fasted rats, glucagon and epinephrine produced an additive increase in gluconeogenesis, but only in the presence of added manganese (245). Rognstad (245) also found that addition of manganese increased the specific activity of glucose formed from lactate and $NaH^{14}CO_3$, and speculated that this could be due to an increased exchange reaction of either pyruvate carboxylase or phosphoenolpyruvate carboxykinase by stimulation from manganese.

Acute manganese toxicity can also have a strong effect on carbohydrate metabolism. Rats injected with 10–40 mg manganese per kilogram body weight show a rapid rise in blood glucose, and sharp decline in blood insulin levels. The effect of injected manganese occurred in both fed and fasted rats, suggesting that part of the response is due to increased gluconeogenesis (16,154).

Manganese may also be involved in carbohydrate metabolism as it relates to lectin chemistry. Lectins (phytohemagglutinins) are known to contain metals, and in some cases there is evidence for a manganese requirement for activity (179). Most lectins contain manganese in combination with calcium; removal of these cations abolishes carbohydrate-binding properties for some lectins but not others (181). The physiological role of manganese-containing lectins in mammalian systems is not well characterized. Grinnell (102) has suggested that manganese may be essential for fibronectin-dependent cell adhesion.

G. Manganese and Immune Function

There is some evidence that manganese plays a role in immunological function. In rats fed a diet marginally deficient in manganese, IgG agglutinins and the 19S fraction of γ-globulin were low, but the hemolysin titer was elevated after intraperitoneal injection with sheep red blood cells (195). Interaction of manganese-containing salts with neutrophils and macrophages has also been demonstrated (12,239), possibly through interactions with the plasma membrane of cells employed in the immune response. Manganese may also act as a cofactor in an enzyme possibly associated with the stimulation of macrophage spreading induced by glass-bound antigen–antibody complexes (43,239).

Manganese is a component of the concanavalin A (Con A) molecule, and binding of manganese may be a prelude to the binding of calcium and the subsequent interaction between Con A and cell surface carbohydrates, which must occur before it can influence cellular processes. There may also be an interaction between manganese and calcium; calcium ions are involved in many aspects of lymphocyte activation, and manganese may interact with calcium at these steps (90).

V. MANGANESE REQUIREMENTS

The minimum dietary requirements of manganese vary with the species and genetic strain of animal, the stage of the life cycle, the chemical form in which the element is ingested, the composition of the rest of the diet, and the criteria of adequacy employed.

A. Laboratory Species

Mice, rats, and rabbits are unable to grow normally on milk diets containing 0.1–0.2 ppm manganese (dry basis), or on purified diets containing 0.2–0.3 ppm manganese (263,267,291). A level of 1 ppm manganese appears to be adequate for growth but is inadequate for normal fetal development in mice or rats (24,141). The genetic constitution of mice has been shown to affect their response to dietary deficiencies of manganese, suggesting that at low or borderline levels of intake of this element the responses of individuals may vary greatly depending in part on their genetic background (133). Rats also show strain differences in response to dietary manganese deficiency. Of two widely used strains, Sprague–Dawley rats were much more susceptible to the effects of manganese deficiency on fetal development than were Wistar rats (68). The response of different mouse and rat strains to prenatal manganese deficiency with regard to otolith development is shown in Table III to illustrate the above comments. Sand rats, whose natural diet is high in manganese, developed an insulin-resistant diabetes when fed a commercial feed containing less manganese (262). Holtkamp and Hill (127) concluded that 40 ppm manganese of the dry diet is optimum for rats. A level of 45 ppm manganese was found to be fully adequate for growth and normal reproduction in mice and rats in the studies of Hurley and co-workers (24,141). High intakes of phosphorus relative to calcium increase the manganese requirements of rats (291).

Yamato (305) has estimated that for total parenteral nutrition in rats, manganese at the level of 9 μg/kg/day was necessary in order to maintain tissue concentrations in the absence of loss of gastrointestinal fluid. In that case, however, 16 μg/kg/day were needed.

Table III
Effect of Manganese Nutrition during Pregnancy on Otolith Development of Progeny[a]

Strain	Dietary manganese (ppm)	First litter young	
		No. of litters	Otolith (% normal) score
Mouse			
Hybrid	3	20	33.8
	45	17	99.6
C57BL/10J	3	6	31.1
	45	5	100.0
DBA/2J	3	9	4.4
	45	7	100.0
SEC/REJ	3	14	5.5
	45	9	98.3
C57BL/10J-pa/pa	3	9	0.0
	45	14	5.8
	1500	6	93.6
Rat			
Sprague-Dawley	3	8	8.3
	45	9	96.1
Wistar	3	8	74.0
	45	7	98.8

[a] Adapted from Dungan et al. (68) and Hurley and Bell (133).

B. Pigs

The requirements of pigs for satisfactory reproduction are substantially higher than those needed for body growth. In fact satisfactory growth in young pigs has been reported with diets supplying only 0.5 or 1.0–1.5 ppm manganese (146,237). However, marked tissue manganese depletion was observed, and when such diets were fed throughout gestation and lactation, skeletal abnormalities and impaired reproduction became apparent. All these manifestations of manganese deficiency were prevented by supplemental manganese at a level of 40 ppm, but whether 40 ppm manganese is necessary or whether some lower level is adequate for all functional purposes still cannot be answered with certainty.

C. Sheep, Goats, and Cattle

The minimum dietary manganese requirements of sheep, goats, and cattle cannot be given with any precision. Hawkins et al. (115) maintain that the minimum requirement of young calves for growth is probably not more than 1 ppm, and that it is increased by high dietary intakes of calcium and phosphorus. Bentley and Phillips (27) state that 10 ppm manganese is adequate for growth in heifers but is marginal for optimal reproductive performance. Rojas and co-workers (246) concluded that the manganese requirements of Hereford cows for maximum fertility are in excess of 16 ppm of the dry diet. Howes and Dyer (129) found no differences in growth or weight gain in 2-year-old Hereford heifers fed diets containing 13 ppm manganese, 14 ppm manganese + 1.5% calcium, and 21 ppm manganese for several months, but calves born to heifers fed the 13 and 14 ppm manganese diets were weak and had difficulty in standing. These find-ings suggested that a level of about 20 ppm manganese is adequate for growth and satisfactory reproductive performance. This finds support from the experi-ments of Hartmans (112), who fed practical rations containing 16 and 21 ppm manganese for 2.5–3.5 years without any clinical evidence of manganese defi-ciency or significant improvement from manganese supplementation of identical twin heifers.

Female goats fed rations containing 20 ppm manganese in the first year and 6 ppm manganese in the second grew as well as those consuming the same rations supplemented with manganese at 100 ppm (7), but the former animals exhibited greatly impaired reproductive performance. The U.S. National Research Council lists the requirement for manganese as 20–40 ppm for sheep (219), 1–10 ppm for growing and finishing cattle, 20 ppm for dry pregnant cows (220), and 40 ppm for dairy cattle (221).

D. Poultry

The minimum dietary manganese requirements of poultry for the growth of chicks and for normal egg production and hatchability approximate 40 ppm under normal dietary conditions. A total intake of 50 ppm is recommended to provide a margin of safety and to cope with variations in calcium and phosphorus intakes. For optimal reproductive performance in turkey hens it has been shown that an ordinary diet containing 3.22% calcium and 0.78% inorganic phosphorus needs to be supplemented with between 54 and 108 ppm manganese (11).

Excess dietary calcium and phosphorus increases manganese requirements by reducing manganese availability (253). For example, 64% of the chicks fed a ration containing 3.2% calcium, 1.6% phosphorus, and 37 ppm manganese developed perosis, whereas no perosis was observed when a diet was fed which contained the same level of manganese but only 1.2% calcium and 0.9% phos-phorus. Freedom from perosis was achieved either by omitting the bone meal, without additional manganese, or by retaining the bone meal and increasing the manganese levels to 62 and 145 ppm (253).

The heavier breeds of poultry have slightly higher manganese requirements than the lighter ones. Egg production does not impose particularly high extra manganese demands on the hen, so that the dietary manganese requirement does not increase greatly with a higher rate of egg production.

The higher manganese requirements of birds compared with mammals arise mainly from lower absorption from the gut. The special importance of manganese in poultry nutrition stems largely from this fact. A second factor is the low-manganese content of corn (maize) compared with other cereal grains, as discussed below (Section VI,B). Where corn is the main dietary component manganese supplementation becomes essential, even when calcium and phosphorus intakes are not unduly high.

E. Human

Estimations of the human requirement for manganese are based primarily on balance studies. Although metabolic balance studies of manganese are difficult to carry out, data from several sources suggest that manganese intakes of 0.035–0.070 mg/kg of body weight per day would result in a balance of manganese intake with excretion (174,199,255). Engel et al. (70) concluded that 1.25 mg/day manganese is the requirement for girls between 6 and 10 years of age, but adolescent girls were in negative balance with 3 mg/day manganese (100). In young women, intakes of 2.5–3.2 mg/day resulted in balance (198), but some apparently healthy subjects had a pattern of manganese intake of <2 mg/day (62). In adult males, on the other hand, the work of Spencer et al. (269) suggested that the manganese requirement may be >2 mg/day, and Rao and Rao (241) reported that 3.7 mg/day were required for manganese balance in Indian men.

Very little is known of the manganese requirements of infants. Widdowson (298) reported negative manganese balances in neonates, and Sampson et al. (247), who studied manganese balance in 16 infants after cardiac surgery, also found negative balances before the age of 4–5 months. However, as these authors did not include controls not surgically treated in their study, it is not clear whether postsurgical effects played a role. On the other hand, Zlotkin (311) found that young infants under total parenteral nutrition retained all the manganese given. This latter observation is reminiscent of the report by Miller et al. (211) of complete retention of all manganese injected during the first 17 days of life in the mouse.

Daily manganese intakes from human dietaries vary greatly with the amounts of unrefined cereals, nuts, leafy vegetables, and tea consumed (see below). The average daily intakes of two adults in which 40–50% of the calories came from white flour were 2.2–2.7 mg manganese, whereas the corresponding intakes for two individuals in which the same proportion of calories came from 92% extraction flour were 8.5–8.8 mg manganese (162). The latter figures are close to the

7 mg/day manganese calculated some years ago for adults consuming a typical English winter diet, although in this case no less than 3.3 mg of the total was estimated to come from tea (213). The total mean daily manganese intakes for nine college women in the United States was 3.7 mg (217), and four young adult women in New Zealand had a mean daily intake of 2.48–3.15 mg manganese (198). Of these intakes, 0.05–0.46 mg/day manganese were retained. Similar average total manganese intakes from adult diets in the United Kingdom of 2.7 ± 0.8 mg/day (108) and from Japanese adult diets of 2.7–2.9 mg/day (214) have been reported. Similarly, the average manganese intakes of women in Canada and other parts of North America fell between 2 and 4 mg/day (93,167,207,270).

Dietary manganese supply during the first 6 months of life varies depending on the age of the infant, the type of milk given, and the quantity of solid foods consumed. McLeod and Robinson (199) reported that manganese intake in New Zealand infants varied from 2.5 to 7.5 μg/day/kg body weight, and that at 3 months, breast-fed infants consumed 14 μg/day from human milk reported to contain 15 μg/liter. Stastney et al. (271), however, found in eight American infants 3 months of age that breast milk contained 3.5 μg/liter manganese and provided a manganese intake of 0.42 μg/kg. The reasons for this large difference in results is unknown but may be related to differences in the analytical methods.

Estimates of requirements of manganese for patients given total parenteral nutrition are 1.0 mg/day for adults, and 0.002–0.01 mg/kg/day for children (3).

VI. SOURCES OF MANGANESE

A. Manganese in Human Foods and Dietaries

The common foods in human dietaries are highly variable in manganese concentration. Many years ago Peterson and Skinner (234) listed 12 major food groups in descending order of their manganese content on the fresh basis. These groups were nuts, whole cereals, and dried fruits; roots, tubers, and stalks; fruits, nonleafy vegetables, animal tissues, and fluids; poultry and poultry products; and fish and seafoods. The average concentrations ranged from 20–23 ppm for the first three groups to as low as 0.2–0.5 ppm for the last three. The leafy vegetables would, of course, rank much higher if expressed on a dry-weight basis. Essentially similar but highly variable levels of manganese have been reported for fresh vegetables obtained from different parts of the United States (128), and in a later study of the mineral content of a very wide range of foods and beverages used in hospital menus there (95). Schroeder et al. (257) confirmed that the highest manganese levels occur normally in nuts and whole cereals,

Table IV

Manganese Content of Foods and Beverages[a]

Food group	Manganese content (mg/kg)	
	Mean	Range
Cereals	6.77	2.4–14.0
Meat	0.59	0.35–1.1
Fish	0.85	0.45–1.3
Milk	<0.10	—
Fats	<0.20	—
Root vegetables	1.34	0.5–2.2
Other vegetables	1.51	1.0–2.2
Fruit and sugars	1.67	0.4–3.6
Beverages		
Beer	0.1	0.1–0.5
Wine		
White	8.8	0.3–23.4
Red	14.4	5.3–19.5
Coffee (instant)	20	13–39
Tea	610	350–900

[a] Adapted from Wenlock et al. (296).

variable amounts in vegetables, and low concentrations in meat, fish, and dairy products. Tea was found to be exceptionally rich in manganese (217,257). Similar findings were reported from a later study of English foods (296) (Table IV).

The wide range for manganese in cereal grains and their products is due partly to plant species differences and partly to the efficiency with which milling separates the manganese-rich from the manganese-poor parts of the grain. In one U.S. study whole wheat containing 31 ppm manganese yielded 160 ppm in the germ, 119 ppm in the bran, and only 5 ppm in the white flour (253). In another such study (312), common hard wheat averaging 38 ppm manganese yielded patent flour containing 4.5 ppm manganese, and common soft wheat averaging 35 ppm manganese yielded patent flour containing 4.8 ppm manganese.

B. Manganese in Animal Feeds and Forages

The manganese levels in pastures and forages are extremely variable. This variation reflects substantial species differences and soil and fertilizer effects. Thus Beeson et al. (23), in an investigation of 17 grass species grown together on a sandy loam soil and sampled at the same time, found the Mn concentrations to range from 96 to 815 ppm (dry basis). Mitchell (212) reports mean levels of 58 and 140 ppm manganese for red clover and ryegrass, respectively, when grown on a granitic, unlimed soil. Heavy liming reduced the levels in the red clover to 40 ppm manganese and in the ryegrass to 120 ppm manganese. In a study of mixed pastures in New Zealand the manganese levels were found to range from 140 to 200 ppm (dry basis), but levels above 400 ppm manganese were reported to occur in some areas (96).

The cereal grains and their by-products also vary greatly in manganese concentration, mainly due to inherent species differences. Typical concentrations in Australian-grown feeds are wheat, 40; bran, 120; barley, 25; oats, 50; maize (corn) 8; and sorghum, 16 ppm manganese (dry basis) (89,196,285). Soybean meal, an important protein supplement for poultry, contains 30–40 ppm manganese (196,253). It is apparent that poultry rations based on corn, and to a lesser extent sorghum and barley, are deficient in manganese unless supplemented with this element or with manganese-rich feeds such as wheaten bran or middlings, in which the manganese of the wheat grain is concentrated. Rations based on wheat or oats, by contrast, are likely to be adequate in manganese unless the diet contains calcium and phosphorus in excess. Diets containing appreciable quantities of the high-protein seeds of *Lupinus albus* are even more likely to be adequate in manganese. Gladstones and Drover (94) obtained the remarkably high concentrations of 817–3397 ppm manganese for this species, or 10–15 times those of other lupin species growing on the same sites.

Protein supplements of animal origin are usually low or very low in manganese. Levels of 5–15 ppm manganese are common in such feeds as dried skim milk or buttermilk, fish meal, and meat meal (196).

VII. MANGANESE TOXICITY

Although manganese can produce toxic effects if taken into the body in excessive amounts, it is often considered to be among the least toxic of the trace elements to birds and mammals. Hens tolerate 1000 ppm in the diet without ill effects (87). Chicks have been reported to tolerate diets up to 3000 ppm manganese with no ill effects (29), but 4800 ppm is highly toxic (117). Growing pigs are less tolerant, since 500 ppm manganese retards growth and depresses appetite (105). Male calves (130 days old) fed diets containing 1000 ppm manganese for 18 days showed no signs of toxicosis (125), while depressed feed intake and lowered body weight were observed in calves fed a low-manganese ration supplemented with manganese sulfate at levels of 2460 and 4920 ppm manganese. When supplemented at a level of 820 ppm manganese, no effect on growth or appetite was apparent (59). The adverse effects of excess manganese on growth were shown to be mainly a reflection of depressed appetite. Sheep grazing pastures containing 140–200 ppm manganese (dry basis), supplemented with 250 or 500 mg/day manganese as the sulfate exhibited a significant depression in growth rate from both levels of manganese compared with untreated controls, together with decreased heart and plasma iron levels in one experiment (96). These findings indicate that an overall intake of 400 ppm of dietary manganese can be toxic to growing sheep, despite the fact that the pastures grazed were reported to contain 1100–2200 ppm iron. In support of this idea, Grace (96) has reported that sheep grazing pastures containing 500 ppm manganese show less weight gain than sheep consuming lower manganese grasses.

A relationship between manganese, iron, and hemoglobin formation has been demonstrated in lambs (111,192), pigs, rats, and rabbits (192,242). Hemoglobin regeneration was retarded and serum iron depressed in anemic lambs fed diets containing 1000 or 2000 ppm manganese. In normal lambs higher manganese levels up to 5000 ppm produced similar effects and also brought about decreased concentrations of iron in the liver, kidney, and spleen. The depressing effects were overcome by a dietary supplement of 400 ppm iron (192) and undoubtedly arise as a consequence of a mutual antagonism between manganese and iron at the absorptive level, as discussed in Section II. Calves fed diets containing 1000 ppm manganese were found to have reduced liver iron compared to controls receiving diets with 55 ppm manganese, but blood parameters were unaffected (125); however, as this feeding study was continued for only 18 days, the period of time may have been too short for anemia to develop.

The postnatal growth of rats is unaffected by dietary manganese intakes as high as 1000–2000 ppm, provided dietary iron is adequate (87). However if dietary iron is low (20 ppm), dietary manganese intakes of 1100 ppm can retard growth (242). Offspring of rats fed diets containing adequate iron and 3550 ppm manganese showed depressed weight gain but normal survival, while offspring of rats fed a low-iron diet with 3550 ppm manganese showed both low weight gain and increased mortality (242). Reproductive dysfunction with long-term feeding of excess manganese (≥ 1050 ppm) has been reported in male and female rats (173).

While the levels of manganese in the diet needed for signs of overt toxicity are quite high, Murthy *et al.* (215) have reported that daily intake of manganese in drinking water at a level of 1 ng/ml impaired learning ability and memory consolidation of young rats. Magour *et al.* (188), in studying this phenomenon, found that young rats given water containing 55 μg/ml manganese for 3 weeks had a significant reduction in brain RNA and protein synthesis. This effect was not present in animals given the diet for 4 weeks, suggesting a developmental component to the toxicity of excess manganese, with young animals being most sensitive. Ali and co-workers (1,2) have suggested that the neurotoxicity of dietary manganese can be amplified by dietary protein deficiency; the mechanisms underlying this effect of protein are not known.

In humans, incidents of oral manganese poisoning are quite rare. Kawamura *et al.* (149) described a series of cases of manganese toxicity in individuals who consumed well water contaminated with manganese. Banta and Markesbury (17) reported a single case of manganese toxicity in an individual who consumed high levels of mineral supplements for several years. Taylor and Price (276) have reported one case of acute manganese toxicity in a patient receiving a dialysate contaminated with manganese; this patient developed pancreatitis presumably as a result of the excess manganese.

In contrast to the few reports of oral manganese poisoning in humans, there is an extensive literature on manganese toxicity in humans who chronically inhale large amounts of airborne manganese, a situation which can occur in manganese mines, steel mills, and some chemical industries. Excess manganese enters the body mainly as oxide dust via the lungs and also via the gastrointestinal tract from the contaminated environment (52). The lungs apparently act as a depot from which the manganese is continuously absorbed. Manganese poisoning is characterized by a severe psychiatric disorder (locura manganica) resembling schizophrenia, followed by a permanently crippling neurological disorder clinically similar to Parkinson's disease. Comparative studies of a population of "healthy" manganese miners and patients suffering from chronic manganese poisoning revealed faster losses of injected ^{54}Mn from the whole body and from an area representing the liver, and higher tissue manganese concentration in the former group (54,201) than in those suffering from chronic manganese poisoning. The presence of elevated tissue manganese levels is thus not necessary for the continuance of the neurological manifestations of the disease, and metal chelation therapy is unlikely to secure their remission (54,158).

The neurotoxicity resulting from manganese has been receiving increased attention because of the use of manganese-containing compounds in gasoline as a replacement for lead (277). In addition, it has been reported that children living in urban areas can have elevated blood manganese levels (147).

The biochemical lesions underlying manganese-induced neurotoxicity have not been delineated. Several investigators have suggested that the initial lesion may be in the metabolism of dopamine and serotonin (32,91) and/or increased neural tissue lipid peroxidation (65). Detailed discussions of manganese toxicity and the central nervous system have been published (64,158,227).

REFERENCES

1. Ali, M. M., Murthy, R. C., Saxena, D. K., and Chandra, S. V. (1983). *Neurobehav. Toxicol. Teratol.* **5**, 385.
2. Ali, M. M., Murthy, R. C., Saxena, D. K., Srivastava, R. S., and Chandra, S. V. (1983). *Neurobehav. Toxicol. Teratol.* **5**, 377.
3. AMA guidelines for essential trace element preparations for parenteral use. (1979). *J. Am. Med. Assoc.* **241**, 2051.
4. Amdur, M. O., Norris, L. C., and Heuser, G. F. (1945). *Proc. Soc. Exp. Biol. Med.* **59**, 254.

5. Amdur, M. O., Norris, L. C., and Heuser, G. F. (1946). *J. Biol. Chem.* **164**, 783.
6. Anke, M., Groppel, B., Reissig, W., Ludke, H., Grun, M., and Dittrich, G. (1973). *Arch. Tierernaehr.* **23**, 197.
7. Anke, M., and Groppel, B. (1970). *In* "Trace Element Metabolism in Animals" (C. F. Mills, ed.), p. 133. Livingstone, Edinburgh.
8. Antila, P., and Antila, V. (1971). *Suom. Kemistil. B* **44**, cited by M. Heikonen, Publ. No. 1, Valio Lab., Helsinki, 1973.
9. Archibald, J. G., and Lindquist, J. G. (1943). *J. Dairy Sci.* **26**, 325.
10. Ash, D. E., and Schramm, V. L. (1982). *J. Biol. Chem.* **257**, 9261.
11. Atkinson, R. L., Bradley, J. W., Couch, J. R., and Quinsenberry, J. H. (1967). *Poult. Sci.* **46**, 472.
12. Attramadal, A. (1969). *J. Nutr. Period. Res.* **4**, 281.
13. Baly, D. L., Curry, D. L., Keen, C. L., and Hurley, L. S. (1984). *J. Nutr.* **114**, 1438.
14. Baly, D. L., Curry, D. L., Keen, C. L., and Hurley, L. S. (1985) *Endocrinology,* **116**, 1734.
15. Baly, D. L., Keen, C. L., Curry, D. L., and Hurley, L. S. (1985) *In Proc. Int. Symp. Trace Elem. Metab. Man Anim., 5th (TEMA 5)* (C. F. Mills, I. Bremner, and J. K. Chesters, eds.), p. 254. Commonwealth Agric. Bureaux, Farnham Royal, U.K.
16. Baly, D. L., Lonnerdal, B., and Keen, C. L. (1985). *Toxicol. Lett.,* **25**, 95.
17. Banta, R. G., and Markesbery, W. R. (1977). *Neurology* **27**, 213.
18. Barak, A. J., Keefer, R. C., and Tuma, D. J. (1971). *Nutr. Rep. Int.* **3**, 243.
19. Barbeau, A., Inoue, N., and Cloutier, T. (1975). *In* "Advances in Neurology" (R. Elridge and S. Fahn, eds.), p. 339. Raven, New York.
20. Barber, M. M., and Bernheim, F. (1967). *Adv. Gerontol. Res.* **2**, 355.
21. Barlow, P. J., Kapel, M. (1979). *Annu. Meet. Trace Minerals Health Sci., 2nd, Boston.*
22. Barnes, L. L., Sperling, G., and Maynard, L. A. (1941). *Proc. Soc. Exp. Biol. Med.* **46**, 562.
23. Beeson, K. C., Gray, L. S., and Adams, M. G. (1947). *J. Am. Soc. Agron.* **39**, 356.
24. Bell, L. T., and Hurley, L. S. (1973). *Lab. Invest.* **29**, 723.
25. Bell, L. T., and Hurley, L. S. (1974). *Proc. Soc. Exp. Biol. Med.* **145**, 1321.
26. Benedict, C. R., Kett, J., and Porter, J. W. (1965). *Arch. Biochem. Biophys.* **110**, 611.
27. Bentley, O. G., and Phillips, P. H. (1951). *J. Dairy Sci.* **34**, 396.
28. Bertinchamps, A. J., Miller, S. T., and Cotzias, G. C. (1966). *Am. J. Physiol.* **211**, 217.
29. Black, J. R., Ammerman, C. B., Henry, P. R., and Miles, R. D. (1984). *Nutr. Rep. Int.* **29**, 807.
30. Blakemore, F., Nicholson, J. A., and Stewart, J. (1937). *Vet. Rec.* **49**, 415.
31. Bolton, W. (1955). *Br. J. Nutr.* **9**, 170.
32. Bonilla, E. (1980). *Neurobehav. Toxicol.* **2**, 37.
33. Bonilla, E., Salazar, E., Villasmil, J., and Villalobos, R. (1982). *Neurochem. Res.* **7**, 221.
34. Borg, D. C., and Cotzias, G. C. (1958). *Nature (London)* **182**, 1677.
35. Borg, D. C., and Cotzias, G. C. (1958). *J. Clin. Invest.* **37**, 1269.
36. Bosch, S., Van der Grift, J., and Hartmans, J. (1966). *Versl. Landouwk. Onderz.* **666**.
37. Bowen, H. J. M. (1956). *J. Nucl. Energy* **3**, 18.
38. Boyer, P. D., Shaw, J. H., and Phillips, P. H. (1942). *J. Biol. Chem.* **143**, 417.
39. Britton, A. A., and Cotzias, G. C. (1966). *Am. J. Physiol.* **211**, 203.
40. Brock, A., and Wolff, L. K. (1935). *Acta Brevia Neerl. Physiol. Parmacol. Microbiol.* **5**, 80.
41. Bruckman, G., and Zondek, S. G. (1933). *Biochem. J.* **33**, 1845.
42. Bruni, C., and Hegsted, D. M. (1970). *Am. J. Pathol.* **61**, 413.
43. Bryant, R. E. (1969). *Proc. Soc. Exp. Biol. Med.* **130**, 975.
44. Burnett, W. T., Bigelon, R. R., Kimbol, A. W., and Sheppard, C. W. (1952). *Am. J. Physiol.* **168**, 520.
45. Cantilena, L. R., and Klaassen, C. D. (1982). *Toxicol. Appl. Pharmacol.* **63**, 344.
46. Caskey, C. D., Gallup, W. D., and Norris, L. C. (1939). *J. Nutr.* **17**, 407.

47. Caskey, C. D., and Norris, L. C. (1938). *Poult. Sci.* **17**, 433.
48. Caskey, C. D., and Norris, L. C. (1940). *Proc. Soc. Exp. Biol. Med.* **44**, 332.
49. Cikrt, M. (1973). *Arch. Toxciol.* **31**, 51.
50. Cohen, G., and Heikkila, R. E. (1974). *J. Biol. Chem.* **249**, 2447.
51. Collipp, P. J., Chen, S. Y., and Maitinsky, S. (1983). *Ann. Nutr. Metab.* **27**, 488.
52. Cotzias, G. C. (1958). *Physiol. Rev.* **38**, 503.
53. Cotzias, G. C., and Greenough, J. J. (1958). *J. Clin. Invest.* **37**, 1298.
54. Cotzias, G. C., Horiuchi, K., Fuenzalida, S., and Mena, I. (1968). *Neurology* **18**, 376.
55. Cotzias, G. C., Miller, S. T., and Edwards, J. (1966). *J. Lab. Clin. Med.* **67**, 836.
56. Cotzias, G. C., Papavasiliou, P. S., and Miller, S. T. (1964). *Nature (London)* **201**, 1228.
57. Cotzias, G. C., Papavasiliou, P. S., Hughes, E. R., Tang, L., and Borg, D. C. (1968). *J. Clin. Invest.* **47**, 992.
58. Cotzias, G. C., Tang, L. C., Miller, S. T., Sladic-Simic, D., and Hurley, L. S. (1972). *Science* **176**, 410.
59. Cunningham, G. N., Wise, M. B., and Barrick, E. R. (1966). *J. Anim. Sci.* **25**, 532.
60. Curran, G. L. (1954). *J. Biol. Chem.* **210**, 765.
61. Davies, W. T., and Nightingale, R. (1975). *Br. J. Nutr.* **34**, 243.
62. Dewar, W. A., Teague, P. W., and Downie, J. N. (1974). *Br. Poult. Sci.* **15**, 119.
63. Doisy, E. A., Jr. (1974). In "Trace Element Metabolism in Animals" (W. G. Hoekstra, J. W. Suttie, H. E. Ganther, and W. Mertz, eds.), p. 664. Univ. Park Press, Baltimore.
64. Donaldson, J., Crammer, J. M., and Graham, D. G. (1984). *Neurotoxicology* **5**, 1.
65. Donaldson, J., Labella, F. J., and Gesser, D. (1980). *Neurotoxicology* **2**, 53.
66. Dreosti, I. E., Manuel, S. J., and Buckley, R. A. (1982). *Br. J. Nutr.* **48**, 205.
67. Dubick, M. A., and Keen, C. L. (1983). *Toxicol. Lett.* **17**, 355.
68. Dungan, D. D., Zidenberg-Cherr, S., Keen, C. L., Lonnerdal, B., and Hurley, L. S. (1984). *Fed. Proc., Fed. Am. Soc. Exp. Biol.* **43**, 1054.
69. Ellis, G. H., Smith, S. E., and Gates, E. M. (1947). *J. Nutr.* **34**, 21.
70. Engel, R. W., Price, N. O., and Miller, R. F. (1967). *J. Nutr.* **92**, 197.
71. Eriksson, V. J. (1984). *J. Nutr.* **114**, 477.
72. Erway, L., Hurley, L. S., and Fraser, A. (1966). *Science* **152**, 1766.
73. Erway, L. C., and Mitchell, S. E. (1973). *J. Hered.* **64**, 111.
74. Erway, L. C., and Purichia, N. A. (1974). In "Trace Element Metabolism in Animals" (W. G. Hoekstra, J. W. Suttie, H. E. Ganther, and W. Mertz, eds.), p. 249. Univ. Park Press, Baltimore.
75. Evans, R. J., Rhian, M., and Draper, C. I. (1943). *Poult. Sci.* **22**, 88.
76. Everson, G. J., DeRafols, W., and Hurley, L. S. (1964). *Fed. Proc., Fed. Am. Soc. Exp. Biol.* **23**, 448.
77. Everson. G. J., Hurley, L. S., and Geiger, J. F. (1959). *J. Nutr.* **68**, 49.
78. Everson, G. J., Shrader, R. E. (1968). *J. Nutr.* **94**, 89; Shrader, R. E., and Everson, G. J., *J. Nutr.* **94**, 269.
79. Failla, M. L., Caperna, T. J., and Dougherty, J. M. (1985) *Proc. Int. Symp. Trace Elem. Metab. Man Anim., 5th (TEMA 5)*, (C. F. Mills, I. Bremner, and J. K. Chesters, eds.), p. 254. Commonwealth Agric. Bureaux, Farnham Royal, U.K.
80. Fernandez, A. A., Sobel, C., and Jacobs, S. L. (1962). *Anal. Chem.* **35**, 1721.
81. Flynn, A., and Franzmann, A. W. (1974). In "Trace Element Metabolism in Animals" (W. C. Hoekstra, J. W. Suttie, H. E. Ganther, and W. Mertz, eds.), p. 444. Univ. Park Press, Baltimore.
82. Fonnesu, A., and Davies, R. E. (1956). *Biochem. J.* **64**, 769.
83. Foradori, A. C., Bertinchamps, A., Gulebon, J. M., and Cotzias, G. C. (1967). *J. Gen. Physiol.* **50**, 2255.
84. Fore, H., and Morton, R. A. (1952). *Biochem. J.* **51**, 594, 598, 600, 603.

85. Frost, G., Asling, C. W., and Nelson, M. M. (1959). *Anat. Rec.* **134**, 37.
86. Gallup, W. D., and Norris, L. C. (1938). *Science* **87**, 18.
87. Gallup, W. D., and Norris, L. C. (1939). *Poult. Sci.* **18**, 76, 83.
88. Garcia, J. A., Wapnir, R. A., and Lifshitz, F. (1983). *J. Nutr.* **113**, 2601.
89. Gartner, R. J. W., and Twist, J. (1968). *Aust. J. Exp. Agric. Anim. Husb.* **8**, 210.
90. Gershwin, M. E., Beach, R., and Hurley, L. S. (1985). "Nutrition and Immunity." Academic Press, Orlando.
91. Gianutsos, G., and Murray, M. I. (1982). *Neurotoxicology* **3**, 75.
92. Gibbons, R. A., Dixson, S. N., Hallis, K., Russell, A. M., Sanson, B. F., and Symonds, H. W. (1976). *Biochim. Biophys. Acta* **444**, 1.
93. Gibson, R. S., and Scytnes, C. A. (1982). *Br. J. Nutr.* **48**, 241.
94. Gladstones, J. S., and Drover, D. P. (1962). *Aust. J. Exp. Agric. Anim. Husb.* **2**, 46.
95. Gormican, A. (1970). *J. Am. Diet. Assoc.* **56**, 397.
96. Grace, N. D. (1973). *N.Z. J. Agric. Res.* **16**, 177.
97. Grankuist, K., Marklund, S. L., and Taljedal, J-B. (1981). *Biochem. J.* **199**, 393.
98. Grebennikov, E. P., Soroka, V. R., and Sabadash, E. V. (1964). *Fed. Proc., Fed. Am. Soc. Exp. Biol.* **23** (Transl. Suppl.), T461.
99. Greenberg, D. M., Copp, H. D., and Cuthbertson, E. M. (1943). *J. Biol. Chem.* **147**, 749.
100. Greger, J. L., Balingen, P., Abernathy, R. P., Bennett, O. A., and Peterson, T. (1978). *Am. J. Clin. Nutr.* **31**, 117.
101. Gregory, E. M., and Fridovich, I. (1974). *In* "Trace Element Metabolism in Animals" (W. G. Hoekstra, J. W. Suttie, H. E. Granther, and W. Mertz, eds.), p. 486. Univ. Park Press, Baltimore.
102. Grinnell, F. (1984). *J. Cell Sci.* **65**, 61.
103. Groppel, B., and Anke, M. (1971). *Arch. Veterinaermed.* **25**, 779.
104. Gruden, N. (1977). *Nutr. Metab.* **21**, 305.
105. Grummer, R. H., Bentley, O. G., Phillips, P. H., and Bohstedt, G. (1950). *J. Anim. Sci.* **9**, 170.
106. Guthrie, B. E., and Robinson, M. F. (1977). *Br. J. Nutr.* **38**, 55.
107. Gutowska, M. S., and Parkhurst, R. T. (1942). *Poult. Sci.* **21**, 277.
108. Hamilton, E. I., and Minski, M. J. (1972/1973). *Sci. Total Environ.* **1**, 375.
109. Hamilton, E. I., Minski, M. J., and Cleary, J. J. (1972/1973). *Sci. Total Environ.* **1**, 341.
110. Hancock, R. G. V., and Fritze, K. (1973). *Bioinorg. Chem.* **3**, 77.
111. Hartman, R. H., Matrone, G., and Wise, G. H. (1955). *J. Nutr.* **57**, 429.
112. Hartmans, J. (1974). *In* "Trace Element Metabolism in Animals" (W. G. Hoekstra, J. W. Suttie, H. E. Granther, and W. Mertz, eds), p. 261. Univ. Park Press, Baltimore.
113. Hartmans, J. (1972). *Landwirtsch, Forsch.* **27**, 1.
114. Hatano, S., Nishi, Y., and Usui, T. (1983). *Am. J. Clin. Nutr.* **37**, 457.
115. Hawkins, G. E., Wise, G. H., Matrone, G., and Waugh, R. K. (1955). *J. Dairy Sci.* **38**, 536.
116. Hedge, B., Griffith, G. C., and Butt, E. M. (1961). *Proc. Soc. Exp. Biol. Med.* **107**, 734.
117. Heller, V. H., and Penquite, R. (1937). *Poult. Sci.* **16**, 243.
118. Hidiroglou, M. (1979). *Can. J. Anim. Sci.* **59**, 217.
119. Hidiroglou, M., Williams, C. J., Siddiqui, I. R., and Khan, S. V. (1979). *Am. J. Vet. Res.* **40**, 1375.
120. Hidiroglou, M., and Shearer, D. A. (1976). *Can. J. Comp. Med.* **40**, 306.
121. Hignett, S. L. (1956). *Proc. Int. Congr. Anim. Reprod.* **3**, 116.
122. Hill, R. (1974). *In* "Trace Element Metabolism in Animals" (W. G. Hoekstra, J. W. Suttie, H. E. Granther, and W. Mertz, eds.), p. 632. Univ. Park Press, Baltimore.
123. Hill, R., and Mather, J. S. (1968). *Br. J. Nutr.* **22**, 625.
124. Hirsch-Kolb, H., Kolb, H. J., and Greenberg, D. M. (1971). *J. Biol. Chem.* **246**, 395.

125. Ho, S. Y., Miller, W. J., Gentry, R. P., Neathery, M. W., and Blackmon, D. M. (1984). *J. Dairy Sci.* **67**, 1489.
126. Holden, D. P., and Huntley, D. M. (1978). *Poultry Sci.* **57**, 1629.
127. Holtkamp, D. E., and Hill, R. M. (1950). *J. Nutr.* **41**, 307.
128. Hopkins, H., and Eisen, J. (1959). *J. Agric. Food Chem.* **7**, 633.
129. Howes, A. D., and Dyer, I. A., (1971). *J. Anim. Sci.* **32**, 141.
130. Hughes, E. R., and Cotzias, G. C. (1961). *Am. J. Physiol.* **201**, 1061.
131. Hughes, E. R., Miller, S. T., and Cotzias, G. C. (1966). *Am. J. Physiol.* **211**, 207.
132. Hunter, E. F., Gibick, J. M., Hoffsten, P. E., Weinstein, J., and Scott, H. (1963). *J. Biol. Chem.* **238**, 828.
133. Hurley, L. S., and Bell, L. T. (1974). *J. Nutr.* **104**, 133.
134. Hurley, L. S., and Everson, G. J. (1959). *Proc. Soc. Exp. Biol. Med.* **102**, 360.
135. Hurley, L. S., Everson, G. J., and Geiger, J. F. (1958). *J. Nutr.* **66**, 309; (1959) **67**, 445.
136. Hurley, L. S., Everson, G. J., Wooten, E., and Asling, C. W. (1961). *J. Nutr.* **74**, 274.
137. Hurley, L. S., Gowan, J., and Milhaud, G. (1969). *Proc. Soc. Exp. Biol. Med.* **130**, 856.
138. Hurley, L. S., Gowan, J., and Shrader, R. (1968). *In* "Les tissues calcifiés. V. Symposium Européen," p. 101. Soc. d'Enseigment Superieur, Paris.
139. Hurley, L. S., Wooley, D. E., Rosenthal, F., and Timiras, P. S. (1963). *Am. J. Physiol.* **204**, 493.
140. Hurley, L. S., Wooten, E., and Everson, G. J. (1961). *J. Nutr.* **74**, 282.
141. Hurley, L. S., Wooten, E., Everson, G. J., and Asling, C. W. (1960). *J. Nutr.* **71**, 15.
142. Hurley, L. S., Theriault, L. T., and Dreosti, I. E. (1970). *Science* **170** 1316.
143. Hurry, V. J., and Gibson, R. S. (1982). *Biol. Trace Elem. Res.* **4**, 157.
144. Jaulmes, P., and Hamelle, G. (1971). *Ann. Nutr. Aliment.* **25**, B133.
145. Jeng, A. Y., and Shamoo, A. E. (1980). *J. Biol. Chem.* **255**, 6897.
146. Johnson, S. R. (1943). *J. Anim. Sci.* **2**, 14.
147. Joselow, M. M., Tobias, E., Koehler, R., Coleman, S., Bodgen, J., and Gause, D. (1978). *Am. J. Public Health* **68**, 557.
148. Jukes, T. H. (1940). *J. Biol. Chem.* **134**, 789; *J. Nutr.* **20**, 445.
149. Kawamura, R., Ikuta, H., Fukozumi, S., Yamada, R., Tsubuki, S., Kodama, T., and Kurata, S. (1941). *Kisasato Arch. Exp. Med.* **18**, 145.
150. Kato, M. (1963). *Q. J. Exp. Physiol. Cogn. Med. Sci.* **48**, 355.
151. Keefer, R. C., Barak, A. J., and Boyett, J. D. (1970). *Biochim. Biophys. Acta* **221**, 390.
152. Keefer, R. C., Tuma, D. J., and Barak, A. J. (1973). *Am. J. Clin. Nutr.* **26**, 409.
153. Keele, B. B., McCord, J. M., and Fridovich, I. (1970). *J. Biol. Chem.* **245**, 6176.
154. Keen, C. L., Baly, D. L., and Lonnerdal, B. (1984). *Biol. Trace Elem. Res.,* **6**, 309.
155. Keen, C. L., Bell, J., and Lonnerdal, B. (1986). *J. Nutr.* **116**, 395.
156. Keen, C. L., Clegg, M. S., Lonnerdal, B., and Hurley, L. S. (1983). *N. Engl. J. Med.* **308**, 1230.
157. Keen, C. L., Frannson, G.-B., and Lonnerdal, B. L. (1984). *J. Pediatr. Gastroenterol. Nutr.* **3**, 290.
158. Keen, C. L., Lonnerdal, B., and Hurley, L. S. (1985). *In* "Biochemistry of the Elements" (E. Frieden, ed.). Plenum, New York.
159. Keen, C. L., Tamura, T., Lonnerdal, B., Hurley, L. S., and Halsted, C. H. (1983). *Am. J. Clin. Nutr.* **35**, 836.
160. Kemmerer, A. R., and Todd, W. R. (1931). *J. Biol. Chem.* **94**, 317.
161. Kemmerer, A. R., Elvehjem, C. A., and Hart, E. B. (1931). *J. Biol. Chem.* **92**, 623.
162. Kent, N. L., and McCance, R. A. (1941). *Biochem. J.* **35**, 877.
163. King, B. D., Lassiter, J. W., Neatheny, M. N., Miller, W. J., and Gentry, R. P. (1980). *J. Anim. Sci.* **50**, 452.
164. Kirchgessner, M. (1955). *Mangelkrankheiten* **6**, 61, 105.

165. Kirchgessner, M., and Heiske, D. (1978). *Int. Z. Vitam. Ernaehrungsforsch.* **48**, 75.
166. Kirchgessner, M., Schwarz, F. J., and Roth-Maier, D. A. (1981). *In* "Trace Element Metabolism in Man and Animals (TEMA-4)" (J. McC. Howell, J. M. Gawthorne, and C. L. White, eds.), p. 85. Australian Acad. of Sciences, Canberra.
167. Kirkpatrick, D. C., and Coffin, D. E. (1974). *J. Inst. Can. Sci. Technol. Aliment.* **7**, 56.
168. Klassen, C. D., and Watkins, J. B. (1984). *Pharmacol. Rev.* **36**, 1.
169. Klimis-Tavantzis, D. J., Leach, R. M., and Kris-Etherton, P. M. (1983). *J. Nutr.* **113**, 328.
170. Lassiter, J. W., Miller, W. J., Neathery, M. W., Gentry, R. P., Abrams, E., Carter, J. C., Jr., and Stake, P. E. (1974). *In* "Trace Element Metabolism in Animals" (W. G. Hoekstra, J. W. Suttie, H. E. Granther, and W. Mertz, eds.), Vol. 2, p. 557. Univ. Park Press, Baltimore.
171. Lassiter, J. W., and Morton, J. D. (1968). *J. Anim. Sci.* **27**, 776.
172. Lassiter, J. W., Morton, J. D., and Miller, W. J. (1970). *In* "Trace Element Metabolism in Animals" (C. F. Mills, ed.), Vol. 1, p. 130. Livingstone, Edinburgh.
173. Laskey, J. W., Rehnberg, G. L., Hein, J. F., and Carter, S. D. (1982). *J. Toxicol. Environ. Health* **9**, 677.
174. Lawson, M. S., Clayton, B. E., Delves, H. T., and Mitchell, J. D. (1977). *Arch. Dis. Childhood* **52**, 62.
175. Leach, R. M., Jr. (1971). *Fed. Proc., Fed. Am. Soc. Exp. Biol.* **30**, 991.
176. Leach, R. M., Jr., and Muenster, A. M. (1962). *J. Nutr.* **78**, 51.
177. Leach, R. M., Jr., Muenster, A. M., and Wien, E. M. (1969). *Arch. Biochem. Biophys.* **133**, 22.
178. Lindow, C. W., Petersen, W. H., and Steenbock, H. (1929). *J. Biol. Chem.* **84**, 419.
179. Lis, H., and Sharon, N. (1981). *In* "The Biochemistry of Plants" (A. Marcus, ed.), Vol. 6., p. 371. Academic Press, New York.
180. Longstaff, M., and Hill, R. (1972). *Br. Poult. Sci.* **13**, 377.
181. Lonnerdal, B., Borrebaeck, C. A. K., Etzler, M. E., and Errson, B. (1983). *Biochem. Biophys. Res. Comm.* **115**, 1069.
182. Lonnerdal, B., Keen, C. L., and Hurley, L. S. (1981). *Annu. Rev. Nutr.* **1**, 149.
183. Lonnerdal, B., Keen, C. L., and Hurley, L. S. (1985). *Am. J. Clin. Nutr.*, **41**, 550.
184. Lorenzen, E. J., and Smith, S. E. (1947). *J. Nutr.* **33**, 143.
185. Lyons, M., and Insko, W. M. (1937). *Ky. Agric. Exp. Stn. Bull.* **371**.
186. MacDowald, M. J., Bentle, L. A., and Lardy, H. A. (1978). *J. Biol. Chem.* **253**, 116.
187. MacFarlane, M. G., and Spencer, A. G. (1953). *Biochem. J.* **54**, 569.
188. Magour, S., Mäser, H., and Steffen, I. (1983). *Acta Pharmacol. Toxicol.* **53**, 88.
189. Mangnall, D., Giddings, A. E. B., and Clark, R. G. (1976). *Int. J. Biochem.* **7**, 293.
190. Mathers, J. R., and Hill, R. (1968). *Br. J. Nutr.* **22**, 635.
191. Mathews, M. B. (1967). *Nature (London)* **213**, 215.
192. Matrone, G., Hartman, R. H., and Clawson, A. J. (1959). *J. Nutr.* **67**, 309.
193. Maynard, L. S., and Cotzias, G. C. (1955). *J. Biol. Chem.* **214**, 489.
194. Maynard, L. S., and Fink, S. (1956). *J. Clin. Invest.* **35**, 83.
195. McCoy, J. H., Kenney, M. A., and Gillmam, B. (1979). *Nutr. Rep. Int.* **19**, 165.
196. McDonald, M. W., Humphries, C., Short, C. C., Smith, L., and Solvyns, A. (1969). *Proc. Aust. Poult. Sci. Conv.*, p. 223.
197. McEuen, A. R. (1981). *In* "Inorganic Chemistry" (H. A. O. Hill, ed.), p. 249. Royal Society of Chemistry, Burlington House, London.
198. McLeod, B. E., and Robinson, M. F. (1972). *Br. J. Nutr.* **27**, 221.
199. McLeod, B. E., and Robinson, M. F. (1972). *Br. J. Nutr.* **27**, 229.
200. McNatt, M. L., Fisher, F. M., Elders, M. J., Kilgore, B. S., Smith, W. G., and Hughes, E. R. (1976). *Biochem. J.* **160**, 211.
201. Mena, I., Marin, O., Fuenzalida, S., and Cotzias, G. C. (1967). *Neurology* **17**, 128.
202. Mehnert, E. (1984). *Arch. Exp. Vet. Med. Leipzig* **38**, 16.

203. Mehnert, E., and Hudec, R. (1984). *Arch. Exp. Vet. Med. Leipzig* **38**, 1.
204. Meinel, B., Bode, J. C., Koenig, W., and Richter, F. W. (1979). *Biol. Neonate* **36**, 225.
205. Mena, I. (1981). *In* "Disorders of Mineral Metabolism" (F. Bronner and J. W. Coburn, eds.), p. 233. Academic Press, New York.
206. Mena, I., Horiuchi, K., Burke, K., and Cotzias, G. C. (1969). *Neurology* **19**, 1000.
207. Meranger, J. C., and Smith, D. C. (1972). *Can. J. Public Health.* **63**, 53.
208. Meyer, H., and Engelbartz, T. (1960). *Dtsch. Tieraerztl. Wochenschr.* **67**, 124.
209. Mildvan, A. S., Scrutton, M. C., and Utter, M. F. (1966). *J. Biol. Chem.* **241**, 3488.
210. Miller, R. C., Keith, T. B., McCarty, M. A., and Thorp, W. T. S. (1940). *Proc. Soc. Exp. Biol. Med.* **45**, 50.
211. Miller, S. T., Cotzias, G. C., and Evert, H. A. (1975). *Am. J. Physiol.* **229**, 1080.
212. Mitchell, R. L. (1957). *Research (London)* **10**, 357.
213. Monier-Williams, G. W. (1949). "Trace Elements in Foods." Chapman & Hall, London.
214. Murakami, Y., Suzuki, Y., and Yamagata, T. (1965). *J. Radiat. Res.* **6**, 105.
215. Murthy, R. C., Lal, S., Saxena, D. K., Shukla, G. S., Mohd, M., and Chandra, V. (1981). *Chem.-Biol. Interact.* **37**, 299.
216. Neher, G. M., Doyle, L. P., Thrasher, D. M., and Plumlee, M. P. (1956). *Am. J. Vet. Res.* **17**, 121.
217. North, B. B., Leichsenring, J. M., and Norris, L. M. (1960). *J. Nutr.* **72**, 217.
218. Nozdryukina, L. R., Grinkevich, N. I., and Gribovskaya, I. F. (1973). *Trace Subst. Environ. Health-7, Proc. Univ. Mo. Annu. Conf., 7th* p. 353.
219. NRC Nutrient Requirements of Domestic Animals (1975). No. 5: Nutrient Requirements of Sheep. 5th revised Ed. National Academy of Sciences-National Research Council, Washington, D.C.
220. NRC Nutrient Requirements of Domestic Animals (1976). No. 4: Nutrient Requirements of Cattle. 5th revised Ed. National Academy of Sciences-National Research Council, Washington, D.C.
221. NRC Nutrient Requirements of Domestic Animals (1978). No. 3: Nutrient Requirements of Dairy Cattle. 5th revised Ed. National Academy of Sciences-National Research Council, Washington, D.C.
222. Nwokolo, E., and Bragg, D. B. (1980). *Poultry Sci.* **59**, 155.
223. Olson, J. A. (1965). *Rev. Physiol. Biochem. Exp. Pharmacol.* **56**, 173.
224. O'Mary, C. C., Butts, W. T., Reynolds, R. A., and Bell, M. C. (1969). *J. Anim. Sci.* **28**, 268.
225. Orent, E. R., and McCollum, E. V. (1931). *J. Biol. Chem.* **92**, 661.
226. Panic, B., Bezbradica, L. J., Nedeljkov, N., and Istwani, A. G. (1974). *In* "Trace Element Metabolism in Animals" (W. G. Hoekstra, J. W. Suttie, H. E. Granther, and W. Mertz, eds.), Vol. 2, p. 635. Univ. Park Press, Baltimore.
227. Papavasiliou, P. S. (1981). *In* "Electrolytes and Neuropsychiatric Disorders" (P. A. Alexander, ed.), p. 187. SP Medical and Scientific.
228. Papavasiliou, P. S., Kutt, H., Miller, S. T., Rosal, V., Wang, Y. Y., and Aronson, R. B. (1979). *Neurology* **29**, 1466.
229. Papavasiliou, P. S., Miller, S. T., and Cotzias, G. C. (1966). *Am. J. Physiol.* **211**, 211.
230. Pasquier, C., Mach, P. S., Raichvarg, D., Sarfati, G., Amor, B., and Delbarre, F. (1984). *Inflammation* **8**, 27.
231. Patel, M. S., van Lelyveld, P., and Hanson, R. W. (1982). *In* "The Biochemical Development of the Fetus and Neonate" (C. T. Jones, ed.), pp. 553–572. Elsevier, New York.
232. Paynter, D. I. (1980). *J. Nutr.* **110**, 437.
233. Perry, H. M., Jr., Perry, E. F., Purifoy, J. E., and Erlanger, J. N. (1973). *Trace Subst. Environ. Health-7, Proc. Univ. Mo. Annu. Conf., 7th* p. 281.
234. Peterson, W. H., and Skinner, J. T. (1931). *J. Nutr.* **4**, 419.

235. Pihl, R. D., and Parkes, M. (1977). *Science* **198**, 204.
236. Pleban, P. A., and Pearson, K. H. (1979). *Clin. Chem.* **25**, 1915.
237. Plumlee, M. P., Thrasher, D. M., Beeson, W., Andrews, F. N., and Parker, H. E. (1954). *J. Anim. Sci.* **14**, 996; (1956) **15**, 352.
238. Price, C. H., Fonnesu, A., and Davies, R. E. (1955). *Biochem. J.* **64**, 754.
239. Rabinovitch, M., and Destefano, M. J. (1973). *Exp. Cell Res.* **79**, 423.
240. Rao, C. N., and Rao, B. S. N. (1980). *Nutr. Metab.* **24**, 244.
241. Rao, C. N., and Rao, B. S. N. (1982). *Nutr. Rep. Int.* **26**, 1113.
242. Rehnberg, G. L., Hein, J. F., Carter, S. D., Linko, R. J., and Laskey, J. W. (1982). *J. Toxicol. Environ. Health* **9**, 175.
243. Reiman, C. K., and Minot, A. S. (1920). *J. Biol. Chem.* **45**, 133.
244. Roby, M. J., Vann, K. L., Freeland-Graves, J. H., and Shorey, R. L. (1982). *Fed. Proc., Fed. Am. Soc. Exp. Biol.* **41**, 786.
245. Rognstad, R. (1981). *J. Biol. Chem.* **256**, 1608.
246. Rojas, M. A., Dyer, I. A., and Cassat, W. A. (1965). *J. Anim. Sci.* **24**, 664.
247. Sampson, B., Barlow, G. B., and Wilkinson, A. W. (1983). *Pediatr. Res.* **17**, 263.
248. Sansom, B. F., Gibbons R. A., Dixon, S. N., Russell, A. M., and Symonds, H. W. (1976). *In* "Nuclear Techniques in Animal Production and Health." IAEA, Vienna.
249. Sansom, B. F., Symonds, H. W., and Vagg, M. J. (1978). *Res. Vet. Sci.* **24**, 366.
250. Sato, M., and Murata, K. (1932). *J. Dairy Sci.* **15**, 461.
251. Schafer, D. F., Stephenson, D. V., Barak, A. J., and Sorrell, M. F. (1974). *J. Nutr.* **104**, 101.
252. Schaible, P. J., and Bandemer, S. L. (1942). *Poult. Sci.* **21**, 8.
253. Schaible, P. J., Bandemer, S. L., and Davison, J. A. (1938). *Mich. Agric. Exp. Stn. Tech. Bull.* **159**.
254. Schiller, S., and Dorfman, A. (1957). *J. Biol. Chem.* **227**, 625.
255. Schlage, C., and Wortberg, B. (1972). *Acta Paediatr. Scand.* **61**, 648.
256. Schor, R. A., Prussin, S. G., Jewett, D. L., Phil, D., Ludowieg, J. J., and Bhatnagar, R. S. (1973). *Clin. Orthop. Relat. Res.* **93**, 246.
257. Schroeder, H. A., Balassa, J. J., and Tipton, I. H. (1966). *J. Chronic Dis.* **19**, 545.
258. Scrutton, M. C. (1971). *Biochemistry* **10**, 3897.
359. Scrutton, M. C., Griminger, P., and Wallace, J. C. (1972). *J. Biol. Chem.* **247**, 3305.
260. Scrutton, M. C., Utter, M. F., and Mildvan, A. S. (1966). *J. Biol. Chem.* **241**, 3480.
261. Settle, E. A., Mraz, F. R., Douglas, C. R., and Bletner, J. K. (1969). *J. Nutr.* **97**, 141.
262. Shani, J., Ahronson, Z., Sulman, F. G., Mertz, W., Frenkel, A., and Kraicer, P. F. (1972). *Isr. J. Med. Sci.* **8**, 757.
263. Shils, M. E., and McCollum, E. V. (1943). *J. Nutr.* **26**, 1.
264. Shrader, R. E., and Everson, G. J. (1967). *J. Nutr.* **91**, 453.
265. Shrader, R. E., Erway, L. C., and Hurley, L. S. (1973). *Teratology* **8**, 257.
266. Smith, S. E., and Ellis, G. H. (1947). *J. Nutr.* **34**, 33.
267. Smith, S. E., Medlicott, M., and Ellis, G. H. (1944). *Arch. Biochem.* **4**, 281.
268. Southern, L. L., and Baker, D. H. (1983). *J. Nutr.* **113**, 172.
269. Spencer, H., Asmussen, C. R., Haltzman, R. B., and Kramer, L. (1979). *Am. J. Clin. Nutr.* **32**, 1867.
270. Srivastava, V. S., and Nadeau, M. H. (1978). *Nutr. Rep. Int.* **18**, 325.
271. Stastny, D., Vogel, R. S., and Picciano, M. F. (1984). *Am. J. Clin. Nutr.* **39**, 872.
272. Stevens, J. B., and Autor, A. P. (1980). *Fed. Proc., Fed. Am. Soc. Exp. Biol.* **39**, 3138.
273. Sullivan, J. F., Blotcky, A. J., Jetton, M. M., Hahn, H. K., and Burch, R. E. (1979). *J. Nutr.* **109**, 1432.
274. Suzuki, H., and Wada, O. (1981). *Environ. Res.* **26**, 521.

275. Tagliamonte, E., Tagliamonte, P., and Gessa, G. L. (1970). *J. Neurochem.* **17**, 733.
276. Taylor, P. A., and Price, J. D. E. (1982). *Can. Med. Assoc. J.* **125**, 503.
277. Ter Harr, G. L., Griffing, M. E., Brandt, M., Oberding, D. G., and Kapron, M. (1975). *J. Air Pollut. Control Assoc.* **25**, 858.
278. Theriault, L., and Hurley, L. S. (1970). *Dev. Biol.* **23**, 261.
279. Theirs, R. E., and Vallee, B. L. (1957). *J. Biol. Chem.* **226**, 911.
280. Thomson, A. B. R., Olatunbosun, D., and Valberg, L. S. (1971). *J. Lab. Clin. Med.* **78**, 642.
281. Thomson, A. B. R., and Valberg, L. S. (1971). *Am. J. Physiol.* **223**, 1327.
282. Tipton, I. H., and Cook, M. J. (1956). *Health Phys.* **9**, 103.
283. Tsai, H., and Everson, G. J. (1967). *J. Nutr.* **91**, 447.
284. Underwood, E. J., and Curnow, D. H. (1944). Unpublished data.
285. Underwood, E. J., Robinson, T. J., and Curnow, D. H. (1947). *J. Agric. West. Aust.* **24**, 259.
286. Uriu, J. Y., Keen, C. L., Reaven, G. M., and Stern, J. S. (1984). *Teratology* **29**, 61A.
287. Vallee, B. L., and Coleman, J. E. (1964). *Compr. Biochem.* **12**, 165.
288. Van Bruwaene, R., Gerber, G. B., Kirchmann, R., Colard, J., and Van Kerkom, J. (1984). *Health Phys.* **46**, 1069.
289. Van Koetsveld, E. E. (1958). *Tijdshr. Diergeneesk.* **83E**, 229.
290. Versieck, J., Barbier, F., Speecke, A., and Hoste, J. (1974). *Acta Endocrinol.* **76**, 783.
291. Wachtel, L. W., Elvehjem, C. A., and Hart, E. B. (1943). *Am. J. Physiol.* **140**, 72.
292. Waddell, J., Steenbock, H., and Hart, E. B. (1931). *J. Nutr.* **4**, 53.
293. Walton, K. G., and Baldessarini, R. J. (1976). *J. Neurochem.* **27**, 557.
294. Watson, L. T., Ammerman, C. B., Miller, S. M., and Harms, R. H. (1971). *Poult. Sci.* **50**, 1693.
295. Wedler, F. C., Denmen, R. B., and Roby, W. G. (1982). *Biochemistry* **24**, 6389.
296. Wenlock, R. W., Buss, D. H., and Dixon, E. J. (1979). *Br. J. Nutr.* **41**, 253.
297. Werner, A., and Anke, M. (1960). *Arch. Tierernaehr.* **10**, 142.
298. Widdowson, E. M. (1969). *In* "Mineral Metabolism in Paediatrics" (D. Barltrop and W. G. Burland, eds.), p. 85. Blackwell, Oxford.
299. Wilgus, H. S., Jr., Norris, L. C., and Heuser, G. F. (1937). *Science* **84**, 252; *J. Nutr.* **14**, 155 (1937).
300. Wilgus, H. R., Jr., and Patton, A. R. (1939). *J. Nutr.* **18**, 35.
301. Williams, R. J. P. (1982). *FEBS Lett.* **140**, 3.
302. Willis, R. A., Peng, C.-J., Freeland-Graves, J., Shorey, R., and Bales, C. (1984). *Fed. Proc., Fed. Am. Soc. Exp. Biol.* **43**, 1053.
303. Wilson, J. G. (1966). *Vet. Rec.* **64**, 621; **79**, 562 (1966).
304. Wolbach, S. B., and Hegsted, D. M. (1953). *Arch. Pathol.* **56**, 437.
305. Yamato, H. (1982). *Tokushima J. Exp. Med.* **29**, 21.
306. Yip, G. B., and Dain, J. A. (1970). *Biochim. Biophys. Acta* **206**, 252.
307. Zhuk, V. P. (1964). *Bull. Exp. Biol. Med. (Engl. transl.)* **54**, 1271.
308. Zidenberg-Cherr, S., Hurley, L. S., Lonnerdal, B., and Keen, C. L. (1985). *J. Nutr.* **115**, 460.
309. Zidenberg-Cherr, S., Keen, C. L., and Hurley, L. S. (1985) *Biol. Trace Elem. Res.* **7**, 31.
310. Zidenberg-Cherr, S., Keen, C. L., Lonnderdal, B., and Hurley, L. S. (1983). *J. Nutr.* **113**, 2498.
311. Zlotkin, S. (1985). *In* "Nestles 8th Nutrition Workshop on Trace Elements" (R. K. Chandra, ed.), p. 175. Raven Press, New York.
312. Zook, E. G., Greene, F. E., and Morris, E. R. (1970). *Cereal Chem.* **47**, 720.

7

Chromium

RICHARD A. ANDERSON

U.S. Department of Agriculture
Agricultural Research Service
Beltsville Human Nutrition Research Center
Beltsville, Maryland

I. CHROMIUM IN ANIMAL TISSUES AND FLUIDS

Chromium (Cr) is widely distributed throughout the human body in low concentrations without special concentration in any known tissue or organ. There have been few definitive studies on the chromium concentration of human or animal tissues. Problems with analysis and extreme problems with contamination during collection of samples have hampered the determination of verifiable values. Analytical methods are presently adequate to determine chromium content of tissues, but comprehensive studies have not been completed. Therefore, values for chromium content of tissues should be viewed with caution.

The chromium concentration of tissues and organs except the lungs appears to decline with age. Chromium concentrations of tissues from human stillborns and infants are reported to be higher than those of adults. These levels decline rapidly in the first decade of life in the heart, lung, aorta, and spleen, while in the liver and kidney the neonatal concentrations are maintained until the second decade, when a decline occurs (94,103). Following the first decade of life, chromium concentration in the lungs increases due likely to inhalation of pollutants (41). Substantial variations in human liver and kidney chromium levels have been observed in different geographic regions (94), presumably as a reflection of regional differences in environmental chromium intakes, but this has not been verified. In general, tissue chromium concentrations tend to be lower in areas of the world where the incidence of maturity-onset diabetes and atherosclerosis is high. For example, chromium in the aorta is significantly lower in samples from the United States than in samples from other countries where the incidence of atherosclerosis is lower (95). However, values obtained using newer methods and increased awareness of chromium contamination indicate that chromium concentration of blood samples from different geographic regions is similar (see below). Insulin-dependent diabetes also affects the chromium content of human tissues. Chromium content of the liver of diabetics (72) is reported to be lower than that of controls. In contrast, Eatough et al. (29) found no difference in the chromium concentration of the liver, pancreas, and spleen between diabetic and nondiabetic Pima Indians. However, chromium concentrations of individual subjects varied greatly in that study, and the nondiabetic group was 20 years younger and may have included subjects predisposed to diabetes.

Exposure to chromium also affects chromium content of tissues. Teroka (102) studied the distribution of chromium in autopsy samples obtained from seven workers exposed to chromium and other heavy metals and five control subjects. Although absolute concentrations of chromium varied extensively among individuals, the greatest concentrations of chromium were found in the hilar lymph nodes and lungs, followed by spleen, liver, kidney, and heart, which all contained approximately the same level of chromium. The distribution pattern among the samples tested was similar for the workers exposed to chromium and the nonexposed workers, but the absolute chromium concentrations of tissues from exposed workers was higher. For example, chromium concentrations of hilar lymph nodes from control subjects was almost 300-fold lower than that of two chromium plating workers. Similarly, chromium content of tissues from rats given higher levels of chromium in the food or drinking water is also increased (106).

Reported serum and plasma chromium concentrations of normal subjects have varied more than 5000-fold over the period since the early 1950s (Table I). Values tend to decrease due mainly to improved analytical instrumentation and procedures, as well as sufficient precautions to avoid contamination. Serum chromium values <0.5 μg/liter for human serum from normal individuals have been determined using five separate independent methods, and have been verified in several countries (14,107). Mean serum chromium for normal individuals is usually 0.01–0.3 μg/liter (12,14,52,108) and is reported to be increased by strenuous exercise (12), a glucose challenge (36), and supplemental chromium (14). Serum chromium levels are reported to decline in pregnancy (26) and during acute infectious illness (84,85).

The reported values for daily urinary excretion of chromium have decreased as the sensitivity of the methods and the awareness of the problems associated with

Table I
Reported Chromium Concentrations in Human
Serum or Plasma[a]

Mean	Year
28–1000	1948–1970
13	1971
7	1971
28	1971
10	1972
5.1	1972
9.3	1972
5.7	1973
4.7	1973
1.6	1974
150	1974
0.73	1974
1.3	1976
1.7	1978
0.16	1978
6.1	1978
0.14	1978
0.20	1979
2.0	1980
1.7	1980
0.43	1980
0.43	1983
0.13	1984
0.12	1984
0.13	1985

[a] From Anderson *et al.* (14).

chromium contamination have increased. Values reported for daily urinary chromium excretion prior to 1970 were usually >10 μg/day; they decreased to 3–10 μg from 1970 to 1978, and are usually <1 μg from 1978 to the present (Table II). Urinary chromium excretion does not appear to be related to concentrations of blood glucose, insulin, lipids, or any other known clinical variables (10), but does appear to be a meaningful indicator of chromium intake when dietary intake is >40 μg/day (15). Urinary chromium excretion at normal daily chromium intakes <40 μg was constant, due likely to the inverse relationship between chromium absorption and chromium intake (see below). Urinary chromium excretion is increased in diabetes (52), glucose loading (8), strenuous running (9,12), and physical trauma (17). However, urinary excretion does not appear to be a meaningful indicator of nutritional status.

Table II
Reported 24-Hr Urinary Chromium
Excretion of Human Subjects[a]

Cr excretion (μg/day)	Year
150	1966
115	1969
18	1964
8.4	1971
3.1	1975
7.2	1975
3.6	1977
2.7	1977
0.8	1978
11.3	1978
6.6	1978
4.3	1980
0.16	1982
0.22	1984

[a] Adapted from Anderson *et al.* (10).

The analysis of chromium in hair has been suggested as a method of assessing chromium nutritional status. The advantages of hair chromium determinations are as follows:

1. Analytical methods are less difficult because hair chromium concentrations are relatively high compared to serum or urine.
2. Hair chromium concentrations are not subject to rapid fluctuations because of diet or other variables and therefore would more likely reflect long-term nutritional status.
3. Sample collection is noninvasive.
4. Samples can be stored at room temperature (16).

However, methods for hair chromium analysis still have not been verified. For example, four different wash procedures of a pooled hair sample led to four different values for hair chromium concentration (56).

Hair chromium concentrations appear to decrease with age, similar to body tissues except the lungs (41). The fetus accumulates chromium, especially in the last months before birth; premature infants have lower hair chromium than full-term babies. The hair chromium concentration of a full-term baby may be more than double that of the mother and decreases to near adult concentrations within 3 years (39,41).

Pregnancy and time between pregnancies also affects hair chromium concentrations. Nulliparous women usually have higher hair chromium than parous women. However, hair chromium concentration does not appear to decrease further with increasing parity and actually increases significantly between pregnancies separated by ≥4 years (64). Hair chromium concentration appears to be decreased by diabetes and arteriosclerosis. Insulin-dependent diabetic children have lower hair chromium levels than control children (42), and adult female diabetics also appear to have lower hair chromium levels than control females. However, hair chromium concentration of adult male diabetics was similar to that of male controls (90). A greater percentage of individuals with arteriosclerotic heart disease had hair chromium concentrations in the lowest quartile compared to age-matched controls, and mean hair chromium concentration of individuals with arteriosclerosis was lower than that of controls (24).

Normal cow's milk is reported to contain 5–15 ng of chromium per milliliter (53,55), and cow's colostrum 5-fold higher levels. Breast milk has been reported to contain similar concentrations of chromium to those reported for cow's milk (41,96); however, subsequent reports indicate that breast milk contains approximately 0.3–0.4 ng of chromium per milliliter (21,57).

Chromium concentration of chicken egg samples from Canada ranged from 50 to 150 ng/g on a fresh weight basis (54). Chromium concentration of egg white from turkey eggs ranged from 1 to 2 ng/g wet weight, and that of the yolk was 11–17 ng/g (Anderson, Bryden, and Polansky, unpublished observation). Finnish workers also reported higher levels of chromium in the yolk compared to the white (55). Chromium content of other foods is discussed in Section IV.

II. CHROMIUM METABOLISM

Inorganic chromium compounds are poorly absorbed in animals and humans to the extent of 0.4–3% or less, regardless of dose and dietary chromium status (10,28,48,69,86,109). There is some evidence that the natural complexes in the diet are better available than simple chromium salts (66). Chromium absorption is rapid, with substantial absorption of labeled chromium within 15 min (86), and in humans urinary chromium excretion is elevated within 2 hr of chromium ingestion (10). Chromium absorption by human subjects is inversely related to dietary intake at dietary levels found routinely in normal self-selected diets in the United States (<40 μg/day) (15). At a daily dietary intake of 10 μg, chromium absorption was ~2%, and with increasing chromium intake to 40 μg, chromium absorption decreased to 0.5%. At dietary intakes >40 μg/day, chromium absorption appears constant at ~0.4%.

Little is yet known of the site or mechanism of chromium absorption. In the rat the midsection of the small intestine appears to be the most diffusible segment for chromium, followed by the ileum and duodenum (23). Interaction with other cations and anions is also beginning to shed some light on the mechanism of absorption. Hahn and Evans (40) showed that in zinc-deficient rats, chromium-51 (^{51}Cr) absorption and intestinal content were increased and that this increase was prevented by oral zinc administration. It was further shown that chromium inhibited ^{65}Zn absorption in zinc-deficient rats and decreased the intestinal content of this isotope. These findings suggest that chromium and zinc may be metabolized by a common pathway in the intestine—a suggestion that finds support from the demonstration of the presence of the two metals in the same mucosal supernatant fraction and their similar behavior on an anion exchange column. Chromium and iron may also share a common gastrointestinal transport mechanism, since iron-deficient animals appear to absorb more chromium than iron-supplemented controls (48). Oral administration of iron to the deficient animals inhibited chromium absorption. Chromium absorption also appears to be influenced by vanadium (45). High intakes of chromium prevent the growth depression and mortality of chicks associated with feeding high levels of vanadate. Vanadate inhibits the uptake of chromate by respiring mitochondria, and similarly, chromate inhibits the uptake of vanadate.

The presence of other anions constitutes another factor that can affect chromium absorption. For example, Chen et al. (23) found that oxalate significantly increased and phytate significantly decreased trivalent chromium transport through the rat intestine, both in vitro and in vivo, whereas two other chelating agents, citrate and EDTA, showed no significant effects.

Hexavalent chromium is better absorbed than trivalent chromium. Mackenzie et al. (62) observed 3- to 5-fold greater blood radioactivity levels following intestinal administration of hexavalent ^{51}Cr than after trivalent ^{51}Cr; Donaldson and Barreras (28) obtained greater intestinal chromium uptake from $Na_2^{51}CrO_4$ than from $^{51}CrCl_3$ in humans and in rats; and Mullor et al. (74) found the uptake of Cr^{6+} to be double that of Cr^{3+} in the organs of rats given 400 μg $^{51}Cr^{6+}$ or $^{51}Cr^{3+}$ per day orally for several months. Chromium retention of rats given 5 ppm of sodium chromate (Cr^{6+}) was approximately 10-fold higher than when the drinking water contained chromium chloride (Cr^{3+}) (86). When the drinking water contained carrier-free sodium chromate, 10-fold less radioactive chromate was retained, and severalfold less labeled Cr^{3+} retained by rats consuming carrier-free chromium chloride. When Cr^{6+} was administered, ~50% of the labeled chromium in the whole blood was present in the cells, but when Cr^{3+} was administered, <5% of the counts in the blood were bound to the cells. It is known that Cr^{6+} and not Cr^{3+} is incorporated into red blood cells; therefore, the ingested Cr^{6+} was not converted to the +3 state prior to absorption. Chromium absorption was approximately twofold higher when the radioactive chromium lost in the urine within hours of administration was included with the total chromium retained. Threfore, in rats, significant quantities of absorbed chromium are excreted within hours of oral intake, absorption of Cr^{6+} may be severalfold higher than that of Cr^{3+}, and Cr^{6+} may be absorbed in the +6 valence state.

Absorbed anionic hexavalent chromium readily passes through the membrane of the red cells and becomes bound to the globin fraction of the hemoglobin. Cationic trivalent chromium cannot pass this membrane; it combines with the α-globulin of the plasma and, in physiological quantities, is transported to the tissues bound to transferrin (siderophilin) (48). Transferrin has two binding sites that have different affinities for iron, depending on pH. At lower levels of iron saturation, iron and chromium preferentially occupy site A and B, respectively. However, at higher iron concentrations, iron and chromium compete for binding mainly to site B. This may explain why patients with hemochromatosis, with >50% saturation of transferrin by iron, retain less chromium than iron-depleted patients or normal subjects (91). Since diabetes is a frequent complication associated with hemochromatosis, the effect of high iron saturation on the transport of chromium may play a role (16). In addition to binding to transferrin, at higher than physiological concentrations, chromium binds nonspecifically to several plasma proteins (48).

Tissue uptake of chromium is rapid and the plasma is cleared of a dose of ^{51}Cr within a few days of administration (46,109). Whole-body radioactivity disappears much more slowly and can be expressed by at least three components, with half-lives of 0.5, 6, and 83 days, respectively (48). Hopkins (46) injected ^{51}Cr chloride into rats at levels of 0.01 and 0.10 μg chromium per 100 g body weight and found little difference in blood clearance, tissue distribution, or excretion due to dose level, previous diet, or sex. The bones, spleen, testes, and epididymides retained more ^{51}Cr after 4 days than the heart, lungs, pancreas, or brain. The fact that various tissues retain chromium much longer than the plasma suggests that there is no equilibrium between tissue stores and circulating chromium, and therefore that plasma chromium levels may not be good indicators of body chromium status. Tissue uptake of chromate is markedly affected by age in mice (110). When older mice were used, the concentration of intraperitoneally injected ^{51}Cr in liver, stomach, epididymal fat pads, thymus, kidneys, and especially the testes declined to almost half the values observed in young animals. The chromium entering the tissues is distributed among the subcellular fractions in unusual proportions. Edwards and co-workers (3) found 49% to be concentrated in the nuclear fraction, 23% in the supernatant, and the remainder divided equally between the mitochondria and the microsomes. A high concentration of chromium in nucleic acids has long been known (111) and the hypothesis proposed that chromium and other transition metals may play a role in nucleic acid metabolism (see Section III,D).

III. CHROMIUM DEFICIENCY AND FUNCTIONS

Glucose intolerance is usually one of the first signs of chromium deficiency followed by additional abnormalities in glucose and lipid metabolism, and nerve disorders (Table III). Most of the signs of chromium deficiency listed in Table III have been observed in both experimental animals and humans. Signs that have not been conclusively demonstrated in humans include those not readily differentiated using human subjects, such as impaired growth and decreased longevity. Although it may be a factor, chromium has not yet been shown to affect sperm count and fertility in humans.

A. Carbohydrate Metabolism

In 1957 Schwarz and Mertz (98) observed impaired glucose tolerance in rats fed certain diets and postulated that the condition was due to a deficiency of a new dietary agent, designated the glucose tolerance factor (GTF). The active component was subsequently shown to be trivalent chromium, and various trivalent chromium compounds administered at dose levels of 20–50 μg chromium per 100 g body weight were found to be fully effective in restoring glucose tolerance (99). Fractionation studies of brewer's yeast yielded fractions with much greater biological activity than inorganic chromium, suggesting the existence of a chromium-containing complex or GTF (20). Efforts to purify this factor have led to the detection of nicotinic acid, glycine, glutamic acid, and

Table III
Signs and Symptoms of Chromium Deficiency[a]

Function	Animal
Impaired glucose tolerance	Human, rat, mouse, squirrel monkey, guinea pig
Elevated circulating insulin	Human, rat
Glycosuria	Human, rat
Fasting hyperglycemia	Human, rat, mouse
Impaired growth	Rat, mouse, turkey
Decreased longevity	Rat, mouse
Increased incidence of aortic plaques	Rabbit, rat, mouse
Elevated serum cholesterol and triglycerides	Human, rat, mouse
Neuropathy	Human
Encephalopathy	Human
Reduction in aortic intimal plaque areas	Rabbit
Corneal lesions	Rat, squirrel monkey
Decreased fertility and sperm count	Rat

[a] Adapted from Anderson (4).

cysteine, as well as chromium, in purified fractions (68). Synthetic complexes of these ligands with chromium display insulin-potentiating activity, but the specific activity of the synthetic chromium complexes are several hundred-fold lower than that of the naturally occurring insulin-potentiating chromium complexes (5). The exact structures of the insulin-potentiating native and synthetic chromium complexes have not been determined.

Other systems involving carbohydrate metabolism, including glucose uptake by the isolated rat lens (31), glucose utilization for lipogenesis and carbon dioxide production (70), and glycogen formation from glucose (89), respond to chromium plus insulin, with little or no response in the absence of the hormone. Under more severe chromium deficiency conditions, a syndrome resembling diabetes mellitus, with fasting hyperglycemia and glycosuria, has been observed in rats and mice and shown to be reversed when 2 or 5 ppm chromium was supplied in the drinking water (71,92). These findings are consistent with the hypothesis that a decreased sensitivity of peripheral tissue to insulin is the primary biochemical lesion in chromium deficiency. In chromium-deficient systems, the response to doses of insulin *in vitro* and *in vivo* is significantly inferior, and higher doses of the hormone are required to elicit metabolic responses similar to those of chromium-sufficient controls, whereas in the absence of insulin these responses in chromium-deficient animals or tissues do not differ significantly from those of controls (70,71,89,92). Chromium increases or potentiates insulin activity, but it is not a substitute for insulin. In the presence of specific organic forms of chromium, much lower levels of insulin are required to elicit similar biological responses.

Exogenous insulin requirements of a female patient receiving total parenteral nutrition (TPN) decreased from 45 units per day to zero following chromium supplementation (51). The patient had been on TPN for 3 years and had severe diabetic-like symptoms including glucose intolerance, inability to utilize glucose for energy, neuropathy, high fatty acid levels, low respiratory quotient, and abnormalities of nitrogen metabolism. Addition of 250 μg of chromium as chromium chloride restored intravenous glucose tolerance and respiratory quotient to normal. The subject was then placed on a daily maintenance dose of 20 μg, and the symptoms mentioned were eliminated. Similar results were reported for a female patient who had been receiving TPN for 5 months following complete bowel resection (33).

Since chromium is present in most TPN solutions as a contaminant, its concentration varies depending on the lot, manufacturer, type of solutions, and other unknown variables. Chromium concentration is usually higher in protein or protein hydrolysates than synthetic amino acid mixtures, fat, carbohydrate, or electrolyte solutions (17,32,43,100). Therefore, the amount of chromium a patient may receive fluctuates widely, and chromium intake of subjects on long-term TPN should be monitored.

The American Medical Association (3) recommends daily supplementation of TPN solutions with 10–15 μg of chromium for stable adult patients and 20 μg/day for stable adults with intestinal fluid losses. No definite amount is recommended for patients in acute catabolic states.

Improvements in patients or subjects following chromium supplementation are not limited to subjects on TPN. A severe impairment of glucose tolerance accompanying kwashiorkor or protein-calorie malnutrition (PCM) of infants was shown to improve following chromium supplementation in some subjects (38,49) but not in others (22). For example, glucose removal rates of six malnourished infants from Jordan improved from 0.6 to 2.9% per minute within 18 hr of supplementation with 250 μg of chromium as chromium chloride; similar results were observed for malnourished infants from Nigeria following chromium supplementation. In 9 of 14 Turkish infants suffering from marasmic PCM, the glucose removal rates showed a similar striking response to 50 μg of chromium as chromium chloride, and the effect of a single dose of chromium continued during the period of observation of 8–40 days (38). However, malnourished children from an area of Egypt where the drinking water was higher in chromium showed no effect of chromium supplementation on fasting blood glucose or glucose tolerance (22). In addition to the drinking water, many foods consumed by these children were high in chromium. Plasma chromium of these children was also high relative to levels considered normal at the time.

Adults have also been shown to respond to supplemental chromium, although adults usually respond within several days or weeks while infants usually respond within hours. Glinsmann and Mertz (36) obtained significant improvement in glucose tolerance in four of six maturity-onset diabetics following daily administration of 180–1000 μg chromium as chromium chloride for periods of 7–13 weeks. Shorter periods of chromium supplementation did not improve glucose utilization. In another study, 50% of a group of similar subjects treated with 150 μg chromium daily for 6 months displayed a marked improvement in glucose tolerance (47). Many elderly subjects over the age of 70 exhibit impaired glucose tolerance (27). Levine et al. (61) reported improvements in the glucose tolerance of 4 of 10 elderly individuals treated with 150 μg chromium for 4 months.

Sixty days of chromium supplementation (500 μg/day) of 12 maturity-onset diabetics resulted in significantly lower glucose and insulin after a glucose challenge (75). In a separate study, there was a consistent drop in fasting blood glucose in 13 diabetic patients taking 600 μg of chromium per day as chromium chloride. Insulin requirements of 4 of the 13 controls increased during the 3-month duration of the study, while the exogenous insulin requirement of 5 of the 13 patients on chromium decreased (73). Anderson et al. (11) reported that response to chromium varies with degree of glucose intolerance. The study design was double-blind crossover, with each test period lasting 3 months.

Following a glucose tolerance test (1 g glucose per kilogram of body weight), 20 of the 76 subjects had 90-min glucose values >100 mg/dl, glucose tolerance of 18 of these 20 subjects improved following daily chromium supplementation of 200 μg chromium as chromium chloride, glucose tolerance of one subject was unchanged, and that of the remaining subject continued to decline during both the placebo and chromium supplementation periods. Glucose tolerance of subjects with near-optimal glucose tolerance was not altered following chromium supplementation. These subjects with near-optimal glucose tolerance presumably are receiving sufficient dietary chromium. Since chromium is a nutrient and not a therapeutic drug, only individuals not receiving sufficient chromium should respond, as was observed in the above-mentioned study.

In the study by Anderson et al. (11), the 90-min glucose of subjects with 90-min glucose values less than fasting values was increased following supplemental chromium, suggesting that hypoglycemics (people with low blood sugar) may also be helped by supplemental chromium. In a follow-up study, eight hypoglycemic female patients were supplemented with chromium to define the effects of supplemental chromium on hypoglycemia. The study design was double-blind crossover, and the chromium supplement was 200 μg daily as described for the previous study. Serum samples were obtained at fasting and 30, 60, 90, 120, 180, 240, and 300 min following a glucose load. The area of the glucose tolerance curve (glucose concentration versus time) less than fasting, which includes the hypoglycemic values, was compared during the placebo and chromium supplementation periods. There was a significant improvement in hypoglycemic glucose values following chromium supplementation (13). Immunoreactive serum insulin also tended to decrease and insulin binding to red blood cells improved in seven of eight subjects. Therefore, supplemental chromium appears not only to lead to a decrease in blood sugar of subjects with elevated serum glucose, but also an increase in glucose values of subjects with low blood glucose and no detectable changes in subjects with near-optimal glucose tolerance.

However, not all studies have shown improvements in glucose tolerance following chromium supplementation (87,101,104,112). It is apparent that many individuals are ingesting insufficient chromium to maintain normal glucose utilization; however, not all glucose intolerance is due to chromium deficiency. Chromium is a nutrient and not a therapeutic agent; one would not expect to observe changes following chromium supplementation in persons who already had adequate chromium nutriture.

B. Lipid Metabolism

Evidence linking marginal chromium intakes with abnormal lipid metabolism and ultimately artherosclerosis is accumulating from both animal and human studies. Animal studies indicate that with increasing age, rats fed a low-chromium diet have increased serum cholesterol, aortic lipids, and plaque formations (93). The addition of chromium to a low-chromium diet suppressed serum cholesterol levels in rats and inhibited the tendency of cholesterol levels to increase with age. Addition of 1 or 5 µg chromium per milliliter in the drinking water depressed serum cholesterol of male and female rats, respectively (93,96). Chromium, in conjunction with insulin, increases glucose uptake and incorporation of glucose carbons into epididymal fat (70). Examination of the aortas of rats at the end of their natural lives revealed a significantly lower incidence of spontaneous plaques in the chromium-fed animals than in the chromium-deficient animals, 2% versus 19%, respectively. There were also lower amounts of stainable lipids in the aorta of chromium-supplemented animals (93). In rabbits fed a high-cholesterol diet to induce atherosclerotic plaques, subsequent intraperitoneal administration of 20 µg potassium chromate per day led to a reduction in the size of aortic plaques and a decrease in aortic cholesterol concentration (1,2).

Humans who died of coronary artery disease were reported to have lower chromium concentration in aortic tissue compared to subjects dying from accidents, although the chromium concentration of other tissues analyzed was similar (95). Subjects with coronary artery disease have also been reported to have lower serum chromium than subjects without symptoms of disease (77). In that study decreased serum chromium correlated highly ($p < 0.01$) with appearance of coronary artery disease, whereas elevated serum triacylglycerol correlated less significantly ($p < 0.05$), and there was no correlation between other risk factors, such as serum cholesterol, blood pressure, or body weight.

Chromium supplementation (200 µg daily, 5 days a week for 12 weeks) of 12 adult men resulted in significant decreases in serum triglycerides and increases in high-density lipoprotein cholesterol compared to placebo-treated subjects (88). Chromium supplementation (600 µg chromium daily as chromium chloride) of patients being treated for diabetes also resulted in a significant increase in high-density lipoprotein cholesterol as well as a decrease in total cholesterol (73).

The effects of chromium on serum lipids are difficult to predict. Subjects with the highest levels of lipids (e.g., cholesterol or triglycerides) usually improve the most (27). Elevated serum lipids, like elevated glucose levels, are due to many factors other than dietary chromium; therefore, improved chromium nutrition would only lead to improvements in individuals whose elevated serum lipids are due to marginal dietary chromium.

C. Protein Synthesis

Rats fed diets deficient in chromium and protein have an impaired capacity to incorporate several amino acids into heart proteins (89). Slightly improved incorporation was achieved with insulin alone, which was significantly enhanced by

Cr^{3+} supplementation. The amino acids affected by chromium were glycine, serine, and methionine. No such effect of chromium was observed with lysine, phenylalanine, and a mixture of 10 other amino acids.

Insulin also stimulated the *in vivo* cell transport of an amino acid analog, α-aminoisobutyric acid, to a greater degree in rats fed a low-protein, chromium-supplemented diet than it did in chromium-deficient controls (89). The claim that chromium acts as a cofactor for insulin can therefore also be applied to two insulin-responsive steps in amino acid metabolism which are independent of the action of insulin on glucose utilization.

D. Nucleic Acid Metabolism

Chromium is also postulated to be involved in maintaining the structural integrity of nucleic acids. The interaction between chromium and nucleic acids was demonstrated by Herrman and Speck (44), who reported that treatment of tissues with chromates and dichromates greatly reduced the amount of nucleic acids extractable with trichloroacetic acid. This effect was specific for chromates and was not observed with other compounds tested. Chromates were apparently being reduced to Cr^{3+} and subsequently forming a complex with the tissues as evidenced by the greenish color characteristic of trivalent chromium complexes. The interaction of chromium with nucleic acids was verified by Wacker and Vallee (111), who observed that beef liver fractions that are high in nucleic acids were also high in chromium. The bond between chromium and nucleic acids is strong, since precipitation of beef liver ribonucleic acids (RNA) six times from EDTA solutions did not reduce the amount of chromium associated with RNA; the concentration of all other metals measured decreased dramatically. Chromium also protects RNA against heat denaturation, indicating that chromium may be involved in maintaining the tertiary structure of nucleic acids (34). Sperm cells which are rich in nucleic acids are also affected by low levels of dietary chromium in rats (7). Chromium accumulates in the nuclei; therefore, it is likely that nuclear trivalent chromium may alter or regulate gene function. Findings that chromium enhances RNA synthesis both *in vitro* (81) and *in vivo* (82) support this postulate. Hepatic RNA synthesis induced by partial hepatectomy was enhanced by intraperitoneal dosing with trivalent chromium without changes in nucleotide pool size for at least 12 hr. Of the RNA species newly synthesized, the synthesis of nuclear RNA was enhanced predominantly and its synthesis followed normal processing into ribosomal RNA. When the nucleoli of regenerating livers from rats dosed with chromium were fractionated, the nucleic acid-enhancing activity was associated with a protein of ~70,000 MW that was postulated to bind 5–6 gram atoms of chromium per mole. Therefore, chromium accumulated in the nucleoli may participate in nucleolar gene expression (83).

E. Chromium and Stress

During different forms of stress, such as strenuous exercise, physical trauma, infection, and intense heat or cold, glucose metabolism is greatly altered. Factors that alter glucose metabolism often also alter chromium metabolism. In experimental animals, stress induced by a low-protein diet, controlled exercise, acute blood loss, or infection aggravated the symptoms of depressed growth and survival caused by low chromium diets (67). Exercise training of rats also increases tissue chromium (105).

Anderson *et al.* (9) found that the increased glucose utilization associated with running 6 miles caused a nearly 5-fold increase in urinary chromium concentration of urine samples taken 2 hr following a 6-mile run, and total urinary daily chromium losses were approximately 2-fold greater on a run day compared to the following nonrun day. Serum chromium was also elevated immediately following running and was still elevated 2 hr following running (12). Severely traumatized patients also excrete severalfold more chromium than control subjects (17). However, in the trauma patients, some of the additional chromium excreted in the urine was due to high levels of chromium present in the blood components administered in the treatment of the injuries. Even when this was accounted for, trauma still appeared to cause substantial chromium losses.

F. Growth and Longevity

On a diet of rye, skim milk, and corn oil with added vitamins and zinc, copper, manganese, cobalt, and molybdenum in the water, male mice and rats receiving 2 or 5 ppm Cr^{3+} in the drinking water grew significantly better than controls (97). This effect was associated with decreased mortality and greater longevity. The median age of male mice at death was 99 days longer when they were fed chromium. No such differences in longevity due to chromium were observed in female mice and rats (97). Supplemental chromium had no effect on the incidence of tumors but appeared to protect female rats against lung infection. Mertz and Roginski (67) have shown that raising rats in plastic cages on a low-protein, low-chromium diet results in a moderate depression of growth which can be alleviated by chromium supplementation.

IV. CHROMIUM SOURCES AND REQUIREMENTS

The reported dietary intakes of chromium for the past two decades are listed in Table IV. Most of the earlier studies listed also determined the intake of several elements, in addition to chromium, and unless conditions for sample collection and analysis are optimized for chromium, reported values would be expected to

Table IV
Reported Dietary Chromium Intakes[a]

Chromium intake (μg/day)	Country	Diet	Year
78	United States	College	1962
102	United States	Hospital	1963
130–140	Japan	Self-selected	1965
52	United States	Nursing home	1968
200–290	United States	Nutrition diet study	1969
170	Japan	Self-selected	1969
150	India	Self-selected	1969
231	United States	School lunch	1971
123	United States	Hospital	1971
282	Canada	Self-selected (representative)	1972
39–190	New Zealand	Self-selected	1973
149	Italy	Self-selected	1977
77	United States	College	1977
62	Federal Republic of Germany	Self-selected	1977
190	Sweden	Self-selected	1979
62, 89	United States	Nutrition diet study	1979
29	Finland	Self-selected (calculated)	1980
240	Belgium	Hospital and self-selected	1983
56	Canada	Self-selected	1984
24.5	England	Self-selected (elderly subjects)	1984
28	United States	Self-selected	1985

[a] From Anderson and Kozlovsky (15).

be too high. For example, if conditions are not optimized for chromium and stainless-steel blender blades are used during the homogenization of the samples, erroneously high values for dietary chromium intake would be observed. Chromium leaching from stainless steel into foods and beverages is well documented (6,58,80). Determination of other elements in addition to chromium using neutron activation analysis, under conditions not optimized for chromium, may also lead to erroneously high results (108). To obtain meaningful values for dietary chromium intake, conditions must be optimized for chromium during the collection, homogenization, and analysis steps, and suitable standard reference materials must be used to verify the accuracy of the results.

Appreciable losses of chromium occur in the refining and processing of certain foods. The recovery of chromium in white flour was only 35–44% of that of the parent wheat products (114). Chromium content of unrefined sugar was only 60% of parent molasses, and brown and white sugar contained only 24 and 8%, respectively, of the chromium content of the unrefined product (65). The high intake of refined sugar in typical U.S. diets (~120 g/day per person) not only contributes virtually no chromium but could lead to a loss of body chromium through the chromium-depleting action of refined sugars (10,59).

There are presently no reliable data bases from which to calculate dietary chromium intake from diet records. Therefore chromium intake must be measured. Studies from Finland, England, Canada and the United States all report that the daily dietary chromium intake is <60 μg (15,19,35,55,57). In the study from the United States (15) there were ~15 μg of chromium per 1000 kcal; therefore, ~3000 calories would need to be eaten to obtain the minimum suggested safe and adequate intake of chromium of 50 μg (76). Intake of chromium was highly correlated with intake of potassium, fat, saturated fat, and sodium, and less significantly with oleic acid, phosphorus, vitamin B_6, copper, protein, and total carbohydrate (15).

V. CHROMIUM TOXICITY

Hexavalent chromium is much more toxic than the trivalent form. In fact trivalent chromium has a low order of toxicity, and a wide margin of safety exists between the amounts ordinarily ingested and those likely to induce deleterious effects. Cats tolerate 1000 mg/day and rats showed no adverse effects from 100 mg/kg diet (113). Lifetime exposure to 5 mg/liter of Cr^{6+} in the drinking water induced no toxic effects in rats and mice, and exposure of mice for three generations to chromium oxide at levels up to 20 ppm of the diet had no measureable effect on mortality, morbidity, growth, or fertility (50).

Chronic exposure to chromate dust has been correlated with increased incidence of lung cancer (18), and oral administration of 50 ppm of chromate has been associated with growth depression and liver and kidney damage in experimental animals (63). Chromium exposure is important to the general public because it is a common skin sensitizer in allergic eczema (79). Dermatitis among homemakers may be related to chromium in detergents and bleaches. Likewise, the high incidence of allergies among construction workers may, in part, be related to the presence of chromium in cement (60). However, chromium toxicity appears to be a very limited problem, while marginal dietary chromium intakes are widespread.

REFERENCES

1. Abraham, A. S., Sonnenblick, M., Eini, M., Shemash, O., and Batt, A. P. (1980). *Am. J. Clin. Nutr.* **33**, 2294.
2. Abraham, A. S., Sonnenblick, M., and Eini, M. (1982). *Atherosclerosis* **42**, 185.
3. American Medical Association, Department of Foods and Nutrition (1979). *J. Am. Med. Assoc.* **241**, 2051.

4. Anderson, R. A. (1981). *Sci. Total Environ.* **17**, 13.
5. Anderson, R. A., Brantner, J. H., and Polansky, M. M. (1978). *J. Agric. Food Chem.* **26**, 1219.
6. Anderson, R. A., and Bryden, N. A. (1983). *J. Agric. Food Chem.* **31**, 308.
7. Anderson, R. A., and Polansky, M. M. (1981). *Biol. Trace Elem. Res.* **3**, 1.
8. Anderson, R. A., Polansky, M. M., Bryden, N. A., Roginski, E. E., Patterson, K. Y., Veillon, C., and Glinsmann, W. (1982). *Am. J. Clin. Nutr.* **36**, 1184.
9. Anderson, R. A., Polansky, M. M., Bryden, N. A., Roginski, E. E., Patterson, K. Y., and Reamer, D. C. (1982). *Diabetes* **31**, 212.
10. Anderson, R. A., Polansky, M. M., Bryden, N. A., Patterson, K. Y., Veillon, C., and Glinsmann, W. (1983). *J. Nutr.* **113**, 276.
11. Anderson, R. A., Polansky, M. M., Bryden, N. A., Roginski, E. E., Mertz, W., and Glinsmann, W. (1983). *Metabolism* **32**, 894.
12. Anderson, R. A., Polansky, M. M., and Bryden, N. A. (1984). *Biol. Trace Elem. Res.* **6**, 327.
13. Anderson, R. A., Polansky, M. M., Bryden, N. A., Bhathena, S. J., and Canary, J. J. (1987). *Metabolism* **36**, 351.
14. Anderson, R. A., Bryden, N. A., and Polansky, M. M. (1985). *Am. J. Clin. Nutr.*, **41**, 571.
15. Anderson, R. A., and Kozlovsky, A. S. (1985). *Am. J. Clin. Nutr.*, **41**, 1177.
16. Borel, J. S., and Anderson, R. A. (1984). *In* "Biochemistry of the Essential Ultratrace Elements" (E. Frieden, ed.), pp. 175–199. Plenum, New York.
17. Borel, J. S., Majerus, T. C., Polansky, M. M., Moser, P. B., and Anderson, R. A. (1984). *Biol. Trace Elem. Res.* **6**, 317.
18. Brinton, H. P., Fraiser, E. S., and Koven, A. L. (1952). *Public Health Rep.* **67**, 835.
19. Bunker, W., Lawson, M. S., Delves, H. T., and Clayton, B. E. (1984). *Am. J. Clin. Nutr.* **39**, 797.
20. Burkeholder, J. N., and Mertz, W. (1966). *Fed. Proc., Fed. Am. Soc. Exp. Biol.* **25**, 759.
21. Casey, C., and Hambidge, K. M. (1984). *Br. J. Nutr.* **52**, 73.
22. Carter, J. P., Kattab, A., Abd-al-Hadi, K. A., Davis, J. T., Gholmy, A. E., and Patwardhan, V. N. (1968). *Am. J. Clin. Nutr.* **21**, 195.
23. Chen, N. S. C., Tsai, A., and Dyer, I. A. (1973). *J. Nutr.* **103**, 1182.
24. Cote, M., Munan, L., Gagne-Billon, M., Kelley, A., Di Pietro, O., and Shapcott, D. (1979). *In* "Chromium in Nutrition and Metabolism" (D. Shapcott and J. Huber, eds.), pp. 223–228. Elsevier, Amsterdam.
25. Curran, G. L. (1954). *J. Biol. Chem.* **210**, 765.
26. Davidson, I. W. F., and Burt, R. L. (1973). *Am. J. Obstet.* **116**, 601.
27. Doisy, R. J., Streeten, D. H. P., Freiberg, J. M., and Schneider, A. J. (1976). *In* "Trace Elements in Human Health and Disease" (A. S. Prasad, ed.), Vol. 2, pp. 79–104. Academic Press, New York.
28. Donaldson, R. M., and Barreras, R. F. (1966). *J. Lab. Clin. Med.* **68**, 484.
29. Eatough, D. J., Hansen, L. O., Starr, S. E., Astin, M. S., Larsen, S. B., Izatt, R. M., and Christensen, J. J. (1978). *In* "Trace Element Metabolism in Man and Animals" (M. Kirchgessner, ed.), Vol. III, pp. 259–263. Institute fur Ernährungspysiologie, Freising-Weihenstephan.
30. Edwards, C., Olson, K. B., Heggen, G., and Glenn, J. (1961). *Proc. Soc. Exp. Biol. Med.* **107**, 94.
31. Farkas, T. G., and Robertson, S. L. (1965). *Exp. Eye Res.* **4**, 124.
32. Fell, G. S., Halls, D., and Shenkin, A. (1979). *In* "Chromium in Nutrition and Metabolism" (D. Shapcott and J. Hubert, eds.), pp. 105–172. Elsevier, Amsterdam.
33. Freund, H., Atamian, S., and Fisher, J. E. (1979). *J. Am. Med. Assoc.* **249**, 496.

34. Fuwa, K., Wacker, W. E. C., Druyan, R., Bartholomay, A. F., and Vallee, B. L. (1960). *Proc. Natl. Acad. Sci. U.S.A.* **46,** 1298.
35. Gibson, R. D., and Scythes, C. A. (1984). *J. Biol. Trace Elem. Res.* **6,** 105.
36. Glinsmann, W. H., and Mertz, W. (1966). *Metab. Clin. Exp.* **15,** 510.
37. Glinsmann, W. H., Feldman, F. J., and Mertz, W. (1966). *Science* **152,** 1243.
38. Gurson, C. T., and Saner, G. (1971). *Am. J. Clin. Nutr.* **24,** 1313.
39. Gurson, C. T., Saner, G., Mertz, W., Wolf, W. R., and Sokucu, S. (1975). *Nutr. Rep. Int.* **12,** 9.
40. Hahn, C. J., and Evans, G. W. (1975). *Am. J. Physiol.* **228,** 1020.
41. Hambidge, K. M. (1971). *In* "Newer Trace Elements in Nutrition" (W. Mertz and W. E. Cornatzer, eds.), p. 169. Dekker, New York.
42. Hambidge, K. M., Rodgerson, D. O., and O'Brien, D. (1968). *Diabetes* **17,** 517.
43. Hauer, E. C., and Kaminski, M. V. (1978). *Am. J. Clin. Nutr.* **31,** 264.
44. Hermann, H., and Speck, L. B. (1954). *Science* **119,** 221.
45. Hill, C. H. (1975). *In* "Trace Elements in Human Disease" (A. S. Prasad, ed.), Vol. 2, pp. 281–300. Academic Press, New York.
46. Hopkins, L. L., Jr. (1965). *Am. J. Physiol.* **209,** 731.
47. Hopkins, L. L., Jr., and Price, M. G. (1968). *Proc. West. Hemisphere Congr., 2nd* **II,** 40.
48. Hopkins, L. L., Jr., and Schwarz, K. (1964). *Biochim. Biophys. Acta* **90,** 484.
49. Hopkins, L. L., Jr., Ransome-Kuti, O., and Majaj, A. S. (1968). *Am. J. Clin. Nutr.* **21,** 203.
50. Hutcheson, D. P., Gray, D. H., Venugopal, B., and Luckey, T. D. (1975). *J. Nutr.* **105,** 670.
51. Jeejeebhoy, K. N., Shu, R., Marliss, E. B., Greenburg, G. R., and Bruce-Robertson, A. (1975). *Clin. Res.* **23,** 636A.
52. Kayne, F. J., Komar, G., Laboda, H., and Vanderlinde, R. E. (1978). *Clin. Chem.* **24,** 2151.
53. Kirchgessner, M. (1959). *Z. Tierphysiol. Tierernaehr. Futtermittelkd.* **14,** 270, 278.
54. Kirkpatrick, D. C., and Coffin, D. E. (1975). *J. Sci. Food Agric.* **26,** 99.
55. Koivistoinen, P. (ed.) (1980). Mineral element composition of Finnish foods: N, K, Ca, Mg, P, S, Fe, Cu, Mn, Zn, Mo, Co, Ni, Cr, F, Se, Si, Rb, Al, B, Br, Hg, As, Cd, Pd, and Ash. *Acta Agric. Scand. (Suppl.)* **22.**
56. Kumpulainen, J. T., Salmela, S., Vuori, E., and Lehto, J. (1982). *Anal. Chim. Acta* **138,** 361.
57. Kumpulainen, J., Vuori, E., Makinen, S., and Kara, R. (1980). *Br. J. Nutr.* **44,** 257.
58. Kumpulainen, J. T., Wolf, W. R., Veillon, C., and Mertz, W. (1979). *J. Agric. Food Chem.* **27,** 490.
59. Kozlovsky, A. S., Moser, P., Reiser, S., and Anderson, R. A., (1986) *Metabolism* **35,** 515.
60. Leonard, A., and Lauwerys, R. R. (1980). *Mutat. Res.* **76,** 227.
61. Levine, R. A., Streeten, D. H. P., and Doisy, R. J. (1968). *Metab. Clin. Exp.* **17,** 114.
62. Mackenzie, R. D., Anwar, R., Byerrum, R. U., and Hoppert, C. (1959). *Arch. Biochem.* **79,** 200.
63. Mackenzie, R. D., Byerrum, R., Decker, C. F., Hoppert, C. A., and Langham, R. (1958). *AMA Arch. Ind. Health* **18,** 232.
64. Mahalko, J. R., and Bennion, M. (1976). *Am. J. Clin. Nutr.* **29,** 1069.
65. Masironi, R., Wolf, W., and Mertz, W. (1973). *Bull. W.H.O.* **49,** 322.
66. Mertz, W. (1969). *Physiol, Rev.* **49,** 163.
67. Mertz, W., and Roginski, E. E. (1969). *J. Nutr.* **97.** 531.
68. Mertz, W., Toepfer, E. W., Roginski, E. E., and Polansky, M. M. (1974). *Fed. Proc., Fed. Am. Soc. Exp. Biol.* **33,** 2275.
69. Mertz, W., Roginski, E. E., and Reba, R. C. (1965). *Am. J. Physiol.* **209,** 489.
70. Mertz, W., Roginski, E. E., and Schwarz, K. (1961). *J. Biol. Chem.* **236,** 318.

71. Mertz, W., Roginski, E. E., and Schroeder, H. A. (1965). *J. Nutr.* **86,** 107.
72. Morgan, J. M. (1972). *Metab. Clin. Exp.* **21,** 313.
73. Mossop, R. T. (1983). *Centr. Afr. J. Med.* **29,** 80.
74. Mullor, J. B., Vigil, J., Bielsa, L. B., Imaz, F., and Prat, J. C. (1972). *Rev. Fac. Ing. Quim. Univ. Nac. Litoral* **39,** 73.
75. Nath, R., Minocha, J., Lyall, V., Sunder, S., Kumar, V., Kapoor, S., and Dhar, K. L. (1970). *In* "Chromium in Nutrition and Metabolism" (D. Shapcott and J. Hubert, eds.), p. 213. Elsevier, Amsterdam.
76. National Research Council (1980). Recommended Dietary Allowances. National Academy of Sciences, Washington, D.C.
77. Newman, H. A. I., Leighton, R. F., Lanese, R. R., and Freedland, N. A. (1978). *Clin. Chem.* **24,** 541.
78. Niedermeier, W., and Griggs, J. H. (1971). *J. Chronic Dis.* **23,** 527.
79. Norseth, T. (1981). *Environ. Health Perspect.* **40,** 121.
80. Offenbacher, E. G., and Pi-Sunyer, F. X. (1983). *J. Agric. Food Chem.* **31,** 308.
81. Okada, S., Taniyama, M., and Ohba, H. J. (1982). *Inorg. Biochem.* **17,** 41.
82. Okada, S., Susuki, M., and Ohba, H. (1983). *J. Inorg. Biochej.* **19,** 95.
83. Okada, S., Tsukada, H., and Ohba, H. (1984). *J. Inorg. Biochem.* **21,** 113.
84. Pekarek, R. S., Hauer, E. C., Bayfield, E. J., Wannemacher, R. W., and Beisel, W. R. (1975). *Diabetes* **24,** 350.
85. Pekarek, R. S., Hauer, E. C., Wannemacher, R. W., and Beisel, W. R. (1974). *Anal. Biochem.* **59,** 283.
86. Polansky, M. M., and Anderson, R. A. (1983). *Fed. Proc., Fed. Am. Soc. Exp. Biol.* **42,** 925.
87. Rabinowitz, M. B., Gonick, H. C., Levin, S. R., and Davidson, M. B. (1983). *Diabetes Care* **6,** 319.
88. Riales, R., and Albrink, M. J. (1981). *Am. J. Clin. Nutr.* **34,** 2670.
89. Roginski, E. E., and Mertz, W. (1969). *J. Nutr.* **97,** 525.
90. Rosson, J. W., Foster, K. J., Walton, R. J., Monro, P. P., Taylor, T. G., and Alberti, K. G. M. (1979). *Clin. Chim. Acta* **93,** 299.
91. Sargent, T., III, Lim, T. H., and Jenson, R. L. (1979). *Metabolism* **28,** 70.
92. Schroeder, H. A. (1966). *J. Nutr.* **88,** 439.
93. Schroeder, H. A., and Balassa, J. J. (1965). *Am. J. Physiol.* **209,** 433.
94. Schroeder, H. A., Balassa, J. J., and Tipton, I. H. (1962). *J. Chronic Dis.* **15,** 941.
95. Schroeder, H. A., Balassa, J. J., and Tipton, I. H. (1970). *J. Chronic Dis.* **23,** 123.
96. Schroeder, H. A., Vinton, W. H., and Balassa, J. J. (1962). *Proc. Soc. Exp. Biol. Med.* **109,** 859.
97. Schroeder, H. A., Vinton, W. H., and Balassa, J. J. (1963). *J. Nutr.* **80,** 39, 48.
98. Schwarz, K., and Mertz, W. (1957). *Arch. Biochem. Biophys.* **72,** 515.
99. Schwarz, K., and Mertz, W. (1959). *Arch. Biochem. Biophys.* **85,** 292.
100. Seeling, W., Ahnefeld, F. W., Grunert, A., Kienle, K. H., and Swobodnik, M. (1979). *In* "Chromium in Nutrition and Metabolism" (D. Shapcott and J. Hubert, eds.), p. 95. Elsevier, Amsterdam.
101. Sherman, L., Glendon, J. A., Brech, W. J., Klomber, G. H., and Gordon, E. S. (1968). *Metabolism* **17,** 439.
102. Teroka, H. (1981). *Arch. Environ. Health* **36,** 155.
103. Tipton, I. H. (1960). *In* "Metal-Binding in Medicine" (M. J. Seven, ed.), p. 27. Lippincott, Philadelphia.
104. Uusitupa, M. I. J., Kumpulainen, J. T., Voutilainen, E., Hersio, K., Sarlund, H., Pyorala, K. P., Koivistoinen, P. E., and Lehto, J. T. (1983). *Am. J. Clin. Nutr.* **38,** 404.

105. Vallerand, A. L., Cuerrier, J., Shapcott, D., Vallerand, R. J., and Gardiner, P. F. (1984). *Am. J. Clin. Nutr.* **39,** 402.
106. Verch, R. L., Chu, R., Wallach, S., Peabody, R. A., Jain, R., and Hannan, E. (1983). *Nutr. Rep. Intl.* **27,** 531.
107. Veillon, C., Patterson, K. Y., and Bryden, N. A. (1984). *Anal. Chim. Acta,* **164,** 67.
108. Versieck, J., Hoste, J., Barbier, F., Steyaert, H., De Rudder, J., and Michels, H. (1978). *Clin. Chem.* **24,** 303.
109. Visek, W. J., Whitney, I. B., Kuhn, U. S. G., and Comar, C. I. (1963). *Proc. Soc. Exp. Biol. Med.* **84,** 610.
110. Vittorio, P. V., and Wright, E. W. (1963). *Can. J. Biochem. Biophys.* **41,** 1349.
111. Wacker, W. E. C., and Vallee, B. L. (1959). *J. Biol. Chem.* **234,** 3257.
112. Wise, A. (1978). *J. Am. Med. Assoc.* **240,** 2045.
113. World Health Organization (1973). W.H.O., Tech. Rep. Ser. 532.
114. Zook, E. G., Greene, F. E., and Morris, E. R. (1970). *Cereal Chem.* **47,** 720.

8

Nickel

FORREST H. NIELSEN

United States Department of Agriculture
Agricultural Research Service
Grand Forks Human Nutrition Research Center
Grand Forks, North Dakota

I. NICKEL IN ANIMAL TISSUES AND FLUIDS

Nickel (Ni) occurs in low concentrations in all animal tissues and fluids that have been examined by sufficiently sensitive and reliable analytical methods. The nickel is distributed throughout the body without remarkable concentration in any tissue or organ and apparently does not accumulate with age in any human organ except possibly the lungs (167,193). Very low normal levels were found in human tissues by Sunderman *et al.* (187) and Nomoto (132). Their findings were combined by Sunderman (184) to give the following means (in ng/g fresh tissue): lung, 85 ± 65; kidney, 10.5 ± 4.1; liver, 8.2 ± 2.3; heart, 6.4 ± 1.6; and bone 333 ± 147. Most reports indicate slightly higher tissue nickel concentrations than these. Some reported mean values are given in Table I. Age apparently affects the nickel content of some tissues. Schneider *et al.* (166) found that in the developing embryo the nickel concentration decreased from 3.57 μg/g dry weight at 3–4 months to 1.65 μg/g at 6–7 months. Casey and Robinson (21) found that the nickel content in fetal kidney decreased from 0.86 μg/g dry weight at 22–25 weeks to 0.36 μg/g at term. Schneider *et al.* (166) also found that ribs, liver, and kidneys of newborns to 1-year-old children contained more nickel than 1- to 90-year humans. Sex might affect the nickel content of some tissues. Schneider *et al.* (166) found more nickel in ribs from males than fe-

Table I
Reported Mean Nickel Concentrations in Human Tissue[a]

Tissue	Nickel Concentration (ng/g) Fresh	Dry	Tissue	Nickel Concentration (ng/g) Fresh	Dry
Artery, healthy			Liver	80	647
basilar	150			8.2	
Artery,			Liver (fetal)		630
atheromatous	170		Lung	230	531
Bladder, urinary		101		160	380
Bone	230	1760			85
		370	Lymph node (hilar)	810	
		333	Muscle	100	240
Cerebrum	50	381	Placenta	30	
Hair, pubic	700		Prostate		320
Hair, scalp	1700		Skin	100	
Heart	6.4	690	Testis	549	
		362	Tonsils (palatine)	135	
Intestine	130		Trachea	90	
Kidney	100	683			
	10.5				

[a] From Refs. 10a,14,21,32,132,166,169a,181,183,184,194.

males. There are several reports (31,43,77,168,179) that indicate hair contains between 0.6 and 4.0 μg/g nickel, and that hair from females contains more nickel than hair from males. In contrast, Nechay and Sunderman (117) found a much lower nickel content in hair (~0.2 μg/g), and found no difference between male and female hair. Scheiner et al. (160) reported that the sex of a guinea pig did not affect the hair nickel content (~4 μg/g). Occupational exposure to nickel can markedly elevate the nickel content of hair (13,179). For example, Bencko et al. (13), found mean hair nickel concentrations (μg/g) of 330 ± 409 in nickel smelter workers, 44 ± 31 in cobalt smelter workers, 2.4 ± 0.1 in welders, and 0.9 ± 0.3 in controls.

Table II shows that if animals are not fed extremely high or low levels of nickel, their tissue nickel concentrations are similar to humans—generally <1 μg/g (dry weight) (7,8,9,64,65,83,164,176). Nickel content of animal organs is markedly affected by age or nickel exposure. For example, Spears et al. (176) found that the dry kidney nickel concentration was 0.15 μg/g in pigs fed nickel at 0.16 μg/g diet, and 1.187 μg/g in pigs fed nickel at 25 μg/g diet. They also found that when dietary nickel was 5 μg/g, the dry kidney nickel concentration was 0.645 μg/g in 21-day-old pigs and 0.218 μg in 49-day-old pigs.

Serum nickel varies among species but comparatively little within species if nickel exposure is not excessively altered. Sunderman *et al.* (187) gave the following mean serum concentrations and range of values for various species (in µg/liter): humans, 2.6 (1.7–4.4); dogs, 2.7 (1.8–4.2); rats, 2.7 (0.9–4.1); goats, 3.5 (2.7–4.4); cats, 3.7 (1.5–6.4); guinea pigs, 4.1 (2.4–7.1); pigs 5.0 (4.2–5.6); rabbits, 9.3 (6.5–14.0); and Maine lobsters, 12.4 (8.3–20.1). Others have found that serum nickel content (µg/liter) normally ranges between 1.5 and 5.0 for humans (25,27,179,184,200), 4 and 5 for pigs (176), and 3.5 and 8.7 for rats (130). Because less sensitive and reliable methods were used, most earlier reports that human serum contains 12–40 µg nickel per liter were probably incorrect. However, under appropriate conditions, serum can contain those levels of nickel and more. Solomons *et al.* (171), and Christensen and Lagesson (25), found >100 µg nickel per liter in serum or plasma a few hours after the ingestion of 5 mg of nickel as a soluble salt. Animals ingesting high dietary nickel also have elevated serum or plasma nickel. Serum nickel was 22 µg/liter in pigs fed 25 mg nickel per kilogram diet for 21 days (176). Plasma nickel levels as high as 67 µg/liter were found in rats fed 50 mg nickel per kilogram diet for 8 weeks (130).

Abnormal serum nickel concentrations occur in response to some pathological conditions. Serum nickel is elevated to about twice the normal 12–36 hr after the onset of symptoms in patients with acute myocardial infarction (34,108,135). For example, McNeely *et al.* (108) found a mean of 5.2 ± 2.8 µg nickel per liter

Table II

Reported Mean Nickel Concentrations in Selected Animal Tissues[a]

Tissue	Nickel concentrations[b] (ng/g)			
	Goat	Minipig	Pig	Rat
Amniotic fluid	—	—	—	250
Bone	907	603	—	120
Fetus	—	—	—	42
Heart	591	351	174	—
Kidney	1209	1176	645	357
Liver	1115	564	127	120
Lung	764	304	236	—
Milk	288	110	230	1088
Muscle	526	259	—	—
Placenta	—	—	—	65
Spleen	576	—	180	749
Uterus	—	—	—	270

[a] From Refs. 7,8,81–83,164,165.
[b] Goat, minipig, and pig are all on dry basis, except milk of minipig and pig, which are on fresh basis. All rat tissues are on fresh basis.

in 33 patients 13–36 hr after acute myocardial infarction, compared with a mean of 2.6 ± 0.8 µg/liter in 47 healthy adults. Abnormally high serum nickel concentrations also have been found in patients with acute stroke and severe burns (>25% body surface) (108,135), and in rats with experimental nephritis, severe burns, or hemorrhagic shock (11,50,182). Significantly diminished mean nickel concentrations were found in patients with hepatic cirrhosis (1.6 ± 0.8 µg/liter) and psoriasis (1.5–0.7 µg/liter) (41,108).

Most likely, the cause for the changes in serum nickel during certain pathological disorders is a redistribution of nickel throughout the body. This redistribution is indicated by the following reports. In rats with hemorrhagic shock, the nickel content was elevated in both liver and serum, depressed in heart and kidney, and unchanged in spleen (59). In pigs dead from myocardial infarction, the nickel content in the infarcted heart region was lower than that of intact heart tissue (143). In humans with chronic renal failure, the nickel content was depressed in the skeleton, unchanged in liver and cardiac muscle, and elevated in lungs and testis (181). In patients at the time of mastectomy, the nickel concentration was higher in neoplastic than histologically normal breast tissue (150a).

Reported values for other body fluids of animals and humans include the following. The normal nickel values (µg/liter) are apparently near 4.8 for human whole blood (27,184) and 2.2 for human parotid saliva (22). Reported mean nickel concentrations in human milk are 20 ng/liter (19), 17 ng/liter (51), and about 120 ng/g (dry weight) (149). The nickel content of sweat is surprisingly high. Horak and Sunderman (71) found a mean concentration of 49 µg/liter in the sweat from the arms of five healthy men during sauna bathing. Cohn and Emmett (30) found 55 µg nickel per liter of total body wash-down sweat. However, the nickel content of sweat can be variable. Cohn and Emmett (30) found occlusive arm-bag sweat contained a mean of 293 µg/liter nickel. Christensen et al. (27) found that the nickel content of trunk sweat was generally lower (2–20 µg/liter) than hand sweat (3–115 µg/liter). Regardless of the variation, all the analyses to date indicate that sweat contains much higher nickel levels than normal blood serum.

II. NICKEL METABOLISM

Most ingested nickel remains unabsorbed by the gastrointestinal tract and is excreted in the feces. Limited studies suggest that typically <10% of ingested nickel is absorbed. However, a higher percentage may be absorbed in an iron-deficient (122,146) or gravid state (82,88,177). Kirchgessner et al. (82) found that pigs absorbed >19% of nickel ingested from day 21 of gravidity until delivery. The kidney is the primary route of excretion for absorbed nickel. Some absorbed nickel may be excreted via the bile; as nickel has been found in the bile of rats and rabbits injected with nickel-63 ($^{63}Ni^{2+}$) (141,170). The high nickel content of sweat, pointing to active nickel secretion by the sweat glands, was mentioned in Section I. It is obvious that under conditions of excessive sweating, dermal losses of nickel could be relatively high.

Tedeschi and Sunderman (189) found that normal dogs excreted 90% of ingested nickel in the feces and 10% in the urine, with no significant retention in the body. Sayato et al. (159) found that rats excreted about 90% of an oral dose of ^{63}Ni in the feces. Low efficiency of absorption and predominant excretion of ingested nickel in the feces are also apparent from several human studies. In nickel balance studies carried out on 10 Russian males (131) ingesting 289 ± 23 μg/day nickel, fecal excretion averaged 258 ± 23 μg/day. Analyses of urine from persons not exposed to unusual amounts of nickel indicate that urinary nickel is <5 μg/liter, or 5 μg/day. For example, Mikac-Devic et al. (111) found 2.7 ± 1.6 μg nickel per liter of urine from 19 healthy adults living in central Connecticut; and Andersen et al. (5) found 4.45 ± 1.9 μg nickel per liter of urine from 15 healthy adults in Norway. Myron et al. (115) found that with nine different diets prepared in North Dakota, nickel intake averaged 165 ± 11 μg/day. Horak and Sunderman (71) found that the fecal excretion of nickel in 10 healthy adults was 259 ± 126 μg/day. Thus, normal ingestion and fecal excretion of nickel is about 100 times greater than normal urinary excretion of nickel.

Data for ruminants and other farm species are meager. O'Dell and co-workers (140) reported that calves excreted >20 times as much nickel in feces as in the urine when consuming a normal ration. When this ration was supplemented with 62.5, 250, and 1000 ppm nickel as nickelous carbonate, the animals excreted only 2.7, 1.9, and 4.3% of the respective total nickel excretion in the urine. Spears et al. (175) found that by 72 hr postingestion of the tracer $^{63}Ni^{2+}$, lambs had excreted about 65% of the dose in the feces, and 2% in the urine.

Becker et al. (12) suggested that the transport of nickel across the mucosal epithelium is an energy-driven process rather than simple diffusion, and that nickel ions use the iron transport system located in the proximal part of the small intestine. Once in the blood, the transport of nickel is accomplished by serum albumin and by ultrafilterable serum amino acid ligands. As indicated in a review by Sunderman (185), albumin is the principal Ni^{2+}-binding protein in human, bovine, rabbit, and rat sera. Nickel apparently is bound to albumin by a square planar ring formed by the terminal amino group, the first two peptide nitrogen atoms at the N-terminus, and the imidazole nitrogen of the histidine residue located at the third position from the N-terminus. Findings of Glennon and Sarkar (60) indicate that the side-chain carboxyl group of an aspartic acid residue also is involved in the binding, and that the nickel–albumin complex could be either square planar or square pyramidal. Canine and porcine albumins, which contain tyrosine instead of histidine at the third position, have less affinity for Ni^{2+} than albumins from other species (61,185).

The identities of the ultrafilterable binding ligands that play an important role in the extracellular transport of nickel have not been clearly established. When added to human serum at physiological pH *in vitro*, Ni^{2+} was bound mainly to L-histidine via a coordination with the imidazole nitrogen (101). Asato *et al.* (10) found five distinct ^{63}Ni complexes in serum ultrafiltrates from rabbits that had received an intravenous injection of $^{63}NiCl_2$. Preliminary findings suggested that $^{63}Ni^{2+}$ may be complexed with cysteine, histidine, and aspartic acid, either singly or as mixed-ligand species (10,185). Computer approaches have predicted that the predominant interaction with naturally occurring low-molecular-weight ligands would occur with histidine and cysteine (75).

Two other proteins in serum that might influence nickel transport or metabolism are histidine-rich glycoprotein (HRG) and nickeloplasmin. Equilibrium dialysis and immunoadsorbent chromatography showed that HRG was capable of binding significant amounts of Ni^{2+} in serum under conditions similar to those found physiologically, for example, in the presence of histidine and albumin (66). The nickel-containing macroglobulin, nickeloplasmin, has been found in human and rabbit sera (133,134). Nomoto (133) found 43% of the total serum nickel in humans in the form of nickeloplasmin. Characteristics of nickeloplasmin include an estimated molecular weight of 7.0×10^5, nickel content of 0.90 gram atom per mole, and esterolytic activity. Sunderman (185) noted that a 9.5S α_1-glycoprotein that strongly binds Ni^{2+} has been isolated from human serum, and thus suggested that nickeloplasmin might represent a complex of the 9.5S α_1-glycoprotein with serum α_1-macroglobulin. Unfortunately, there is no clear indication of the physiological significance or function of nickeloplasmin. The nickel in nickeloplasmin is not readily exchangeable with $^{63}Ni^{2+}$ *in vivo* or *in vitro* (35).

The kinetics of $^{63}Ni^{2+}$ metabolism in rodents apparently fits a two-component model (141). A summary of the tissue retention and clearance of $^{63}Ni^{2+}$ administered by all routes of entry has been given by Kasprzak and Sunderman (76). This summary shows that kidney retains significant levels of nickel shortly after $^{63}Ni^{2+}$ is given. The retention probably reflects the role of the kidney in nickel excretion. The level of $^{63}Ni^{2+}$ in kidney falls quickly over time. After 6 days, the lung apparently has the highest affinity for nickel (68). Studies with $^{63}Ni^{2+}$ show that nickel readily passes through the placenta. Jacobsen *et al.* (74) found that, 2 days after administration, embryonic tissue retained greater amounts of parenterally administered nickel than did maternal tissue. Also, amniotic fluid retains relatively high amounts of orally administered nickel (82). The level of nickel in the fetus does not fall quickly after parenteral administration to the dam, thus suggesting greater retention and/or inhibited clearance by the fetus (74,100).

After it was established that the kidney has a major role in nickel metabolism, attempts were made to ascertain the form of nickel in kidney and urine. Sunderman *et al.* (186) found that renal cytosol of rats, 0.5–4 hr after intravenous injection of ^{63}NiCl$_2$, contained five macromolecular ^{63}Ni constituents in addition to low-molecular-weight components. Based on high-performance size-exclusion chromatography, the apparent molecular weights of the five macromolecules were 168,000, 84,000, 51,000, 24,000, and 10,000 (188). Abdulwajid and Sarkar (1) found that nickel in kidney was primarily bound to a glycoprotein with a molecular weight of about 15,000–16,000. The glycoprotein contained 10% carbohydrate (high mannose with galactose/glucose and glucosamine) and a protein moiety with a molecular weight of about 12,000. The protein moiety contained high amounts of glycine and low amounts of cysteine and tyrosine, thus indicating the protein was not metallothionein. Sarkar (158a) indicated that the glycoprotein composition was similar to that of renal basement protein. Urine also contained this nickel-binding glycoprotein. Subsequently, Templeton and Sarkar (190a) found that the fractionation of kidney low-molecular-weight complexes yielded two fractions which bound nickel. Fraction I was a carbohydrate-rich material with a composition suggesting an origin from the basement membrane glycosaminoglycan, heparan sulfate. Fraction II was a single 36-amino acid peptide, rich in aspartic acid and glutamic acid, but devoid of cystine. Sayato *et al.* (159) found that in rat urine collected after oral and intravenous administration of ^{63}NiCl$_2$, the ^{63}Ni metabolite behaved like a complex of ^{63}Ni and creatine phosphate.

III. NICKEL DEFICIENCY AND FUNCTIONS

A. Signs of Nickel Deficiency

The first description of possible signs of nickel deprivation appeared in 1970 (see Ref. 119). However, those findings, and others that followed shortly thereafter, were obtained under conditions that produced suboptimal growth in the experimental animals (119). Also, some reported signs of nickel deprivation appeared inconsistent. During the years 1975–1983, diets and environments that apparently allowed for optimal growth and survival of experimental animals were perfected for the study of nickel nutrition. As a result, what were thought to be clear signs of nickel deprivation were described for six animal species: chick, cow, goat, minipig, rat, and sheep. Unfortunately, the described signs probably will have to be redefined because later studies indicate that many of the reported signs of nickel deprivation may have been misinterpreted and might be manifestations of pharmacological actions of nickel (127). That is, a high dietary level of nickel alleviated an abnormality caused by something other than a nutritional deficiency of nickel, or was causing a change in a parameter that was not necessarily subnormal. Thus, in the following descriptions of signs of nickel deprivation, those obtained from studies in which the nickel-deprived animals were compared to controls fed high levels of supplemental nickel (i.e., 5–20 µg/g diet) should be accepted with caution, especially if dietary iron was low. Nickel can partially alleviate many manifestations of iron deficiency apparently by pharmacological mechanisms.

The suggestion that some of the reported signs of nickel deprivation are misinterpreted manifestations of a pharmacological action of nickel does not detract from the conclusion that nickel is an essential nutrient. The following signs appear representative of nickel deficiency. Iron utilization is impaired. As a consequence, the trace element profile of both the femur and liver changes. In the femur, the concentrations of calcium and manganese are depressed, and the concentrations of copper and zinc are elevated. If the nickel deficiency is severe, hematopoiesis is depressed, especially in marginally iron-adequate animals. Other possible signs for specific species follow.

1. Chickens

Nielsen et al. (124) reported that the signs of nickel deprivation in chicks included depressed hematocrit, depressed oxidative ability of the liver in the presence of α-glycerophosphate, and ultrastructural abnormalities in the liver. Among the liver abnormalities were dilated cisternal lumens of the rough endoplasmic reticulum. Also, the ribosomes were more irregular in their spacing on the rough endoplasmic reticulum and large amounts of membrane appeared devoid of ribosomes. In the studies of Nielsen et al. (124), the controls were fed diets containing relatively low amounts of nickel (<3 µg/g). The diet may have contained marginal iron (50 µg/g).

2. Cows

Signs of nickel deprivation have been found only for cattle fed low or marginal levels of protein. These signs include depressed ruminal urease, serum urea, serum nitrogen, and growth (173). The nickel-deprived animals were fed diets containing 310–400 ng/g nickel; the controls were fed the same diet supplemented with 5 µg/g nickel.

3. Goats

Anke and co-workers (8,9,70) have described the signs of nickel deprivation for goats. The signs included depressed growth, elevated perinatal mortality, unthriftiness characterized by a rough coat and dermatitis, depressed levels of calcium in the skeleton and of zinc in rib and liver, depressed triglycerides, β-

lipoproteins, glutamate-oxaloacetate transaminase activity and glutamate dehy-drogenase activity in serum, and an elevated level of α-lipoprotein in serum. Lactating nickel-deprived goats also showed depressed hematocrit, hemoglobin, and iron content in the liver. Because nickel-deprived goats absorbed less of an oral ^{65}Zn dose and exhibited abnormalities similar to those of zinc-deficient goats, Anke and co-workers (9,70) suggested that nickel deficiency disturbs zinc metabolism. In adult goats, the nickel content was lower in bone, kidney, cere-brum, liver, and heart of nickel-deprived goats (dietary nickel: 137 ng/g) than of control goats (dietary nickel: 4.36 μg/g).

4. Pigs

Anke et al. (6) described the signs of nickel deprivation in minipigs as de-pressed growth, delayed estrus, elevated perinatal mortality, unthriftiness char-acterized by scaly and crusty skin, and depressed levels of calcium in the skeleton and of zinc in liver, hair, rib, and brain. The nickel-deprived minipigs were fed diets containing nickel at 100 ng/g and compared with minipigs fed a dietary nickel supplement of 10 μg/g. Spears et al. (176) fed diets containing 0.12–0.16, 5, or 25 μg nickel per gram to day-old regular pigs and obtained findings suggesting 0.12–0.16 μg/g of diet is adequate for growth of neonatal pigs. However, because some pigs fed the higher levels of nickel had higher liver and/or lung copper, iron, nickel, and zinc contents, Spears et al. (176) thought that high dietary nickel may improve the iron and zinc status of the young pig.

5. Rats

Two independent laboratories have described signs of nickel deprivation for the rat. However, for both, the iron status of the rats probably had a major influence on the extent and severity of the signs.

In one study by Nielsen et al. (125), successive generations of rats were fed a low-nickel diet throughout fetal, neonatal, and adult life and compared to simi-larly treated rats supplemented with 3 μg nickel per gram diet. Signs of nickel deprivation included elevated perinatal mortality, unthriftiness characterized by a rough coat and/or uneven hair development in pups, pale livers, and ultrastruc-tural changes in the liver. The most obvious effect of nickel deprivation on liver ultrastructure was a reduced amount of rough endoplasmic reticulum which appeared disorganized in that the normal "stacking" of the cisternae was par-tially or totally absent. Nickel deprivation appeared to depress both growth and hematocrits of rats which were fed apparently adequate dietary iron, but these signs were not consistently significant. Subsequent studies by Nielsen and co-workers (121,129) showed that, when relatively unavailable ferric sulfate at marginally adequate levels was the dietary source of iron, depressed growth and hematopoiesis were consistently found in nickel-deprived offspring or in weanling rats fed a low-nickel diet for 9–11 weeks. However, as mentioned previously, these apparent signs of nickel deficiency are clouded by the suggestion (127) that high levels of dietary nickel (5, 20, or 50 μg/g) fed to nickel-adequate controls may have been acting pharmacologically to alleviate manifestations of iron deficiency. Nonetheless, based on their latest study, Nielsen *et al.* (127) suggested that the signs of nickel deprivation in rats include an impairment of iron utilization. As a consequence, the trace element profile of both femur and liver changes. In the femur, the concentrations of calcium and manganese are depressed and the concentrations of copper and iron are elevated. In the liver, the concentrations of copper and zinc are elevated.

In a series of reports summarized in 1980, Kirchgessner and Schnegg (85) described the following nickel deprivation signs for offspring of nickel-deprived dams. (Supplemented controls were fed 20 μg nickel per gram diet.) At age 30 days, nickel-deprived pups exhibited significantly depressed growth, hematocrit, hemoglobin, and erythrocyte counts; those signs were still evident, but less marked, at age 50 days. At age 50 days, depressed hematopoiesis was not evident in the nickel-deprived offspring fed 100 instead of 50 μg iron per gram diet (163). Thus, like those of Nielsen (see above), the findings with regard to hematopoiesis were inconsistent. Most likely, the iron status of the rats also had a major influence on the extent and severity of the apparent signs of nickel deprivation in the studies of Kirchgessner and Schnegg (85). Although the initial form of iron used in their studies was ferrous sulfate, the treatment of the diet may have rendered some of the iron unavailable for absorption. The ferrous sulfate was put into distilled water, added to a moist diet mixture, and heated to 50°C to remove water (162). Heating of food apparently reduces the availability of iron for absorption (112).

Kirchgessner and Schnegg (85) reported that other nickel deprivation signs in young pups fed 15 ng nickel and 50 μg iron per gram diet, and compared to pups fed 20 μg/g diet were as follows:

1. At age 30 days, the activities of the liver enzymes malate dehydrogenase, glucose-6-phosphate dehydrogenase, isocitrate dehydrogenase, lactate dehydrogenase, glutamate-oxaloacetate transaminase, and glutamate-pyruvate transaminase were depressed, whereas the activities of alkaline phosphatase and creatine kinase were elevated. At age 50 days, nickel deprivation did not significantly depress liver malate dehydrogenase, or glucose-6-phosphate dehydrogenase activity (84).

2. At age 30 days, the activities of the kidney enzymes glutamate dehydrogenase, glutamate-oxaloacetate transaminase, and glutamate-pyruvate transaminase were depressed.

3. The levels of urea, adenosine triphosphate (ATP), and glucose in serum were reduced.

4. The levels of triglycerides, glucose, and glycogen in liver were reduced.
5. Iron absorption was impaired.
6. Levels of iron, copper, and zinc were depressed in the liver, kidney, and spleen.
7. Levels of calcium and phosphorus were depressed and the level of magnesium was elevated in femur (79).
8. The nickel content of muscle, hair, bone, and kidney was reduced.
9. Activities of proteinase and leucine amylamidase increased, and the activity of α-amylase decreased in the pancreas. Kirchgessner *et al.* (82) suggested that the reduction in the digestion of starch by α-amylase might be indirectly responsible for the observed large depressions in the activities of hepatic enzymes and in the concentrations of hepatic metabolites in nickel-deprived rats.

6. *Sheep*

Spears and co-workers (173–175) have described nickel deprivation signs for sheep. When compared to nickel-supplemented lambs (5 μg/g diet), nickel-deprived (~30 ng/g diet) lambs exhibited depressed erythrocyte counts, ruminal urease activity, and hepatic total lipids, cholesterol, and copper. Iron content was elevated in liver, spleen, lung, and brain. Other possible (borderline or inconsistent significant change) signs of nickel deprivation in lambs were depressed growth, serum total protein, albumin, α_2-globulin and γ-globulin, serum calcium, phosphorus, and alanine transaminase activity, liver oxygen uptake, and kidney, lung, and liver retention of $^{63}Ni^{2+}$ at 72 hr after oral dosing.

B. Nickel Function

To date, there is no firmly established biological function of nickel in humans or animals. However, some findings indicate that nickel functions either as a cofactor or structural component in specific metalloenzymes, or as a bioligand cofactor facilitating the intestinal absorption of the ferric ion.

1. *Enzyme Cofactor or Structural Component*

Development of the hypothesis that nickel functions as an enzyme cofactor or structural component has been stimulated by the discovery of several nickel-containing enzymes in plants and microorganisms (see Table III). Urease from jack bean was the first natural nickel metalloenzyme discovered (39,40). Subsequently, ureases from several other plants and microorganisms were identified as nickel metalloenzymes (62,90,144,172). Jack bean urease (EC 3.5.1.5) contains stoichiometric amounts of nickel, 2.00 ± 0.12 gram atoms per mole of 96,600-dalton subunits, or about 12 nickel atoms per enzyme molecule (39,40). The

Table III
Nickel Metalloenzymes

Enzyme	Species from which isolated	Reference	Properties of nickel in enzyme
Urease (EC 3.5.1.5)	Jack bean	39,40	Tightly bound at the active site; apparently in an octahedral configuration; 2.00 ± 0.12 gram atoms per mole of 96,000-dalton subunit
	Soybean	90,144	
	Rice, tobacco	144	
	Lemna paucicostata	62	
	Ruminal	172	
Factor F_{430} (component of methyl coenzyme M methylreductase)	Methanogenic bacteria	36,47 48,191	2 gram atoms per mole of enzyme: F_{430} apparently is a nickel tetrapyrrole
Hydrogenase			
Soluble	*Alcaligenes eutrophus*	55,56	1–2 gram atoms per mole of enzyme
Particulate	*Alcaligenes eutrophus*	55,56	
Periplasmic	*Desulfovibrio gigas*	18,97, 113	About 1 gram atom per mole of 89,500-dalton enzyme; apparently present in a redox-sensitive form—Ni(III)
One-electron reducing (viologen)	*Methanobacterium thermoautotrophicum*	63,91	1 gram atom per mole of enzyme; apparently present in a redox-sensitive form—Ni(III)—in both enzymes
Two-electron reducing (coenzyme factor 420)			
Membrane	*Vibrio succinogenes*	195	1 gram atom per mole of enzyme with a molecular weight of 100,000
Hydrogenase I	*Desulfovibrio desulfuricans*	93	0.6 gram atom per mole of enzyme with a molecular weight of 77,600
Hydrogenase II	*Desulfovibrio desulfuricans*	93	0.6 gram atom per mole of enzyme with a molecular weight of 75,500
Carbon monoxide dehydrogenase	*Clostridium thermoaceticum*	38,42, 63,148	Enzyme with a molecular weight of 440,000 and composed of three each of two different subunits $(\alpha\beta)_3$; 2 gram atoms of nickel per mole of dimer; at active site Ni(III) may be present in a porphyrinlike factor
	Acetobacterium woodii	37	

nickel is tightly bound at the active site in an octahedral configuration (39). Binding of the substrate urea to a nickel ion in the enzyme molecule is an integral part of the mechanism in the hydrolysis reaction catalyzed by urease. Jack bean urease was the first enzyme to be crystallized. After 50 years, nickel was identified in the structure of urease.

More than 50% of the nickel taken up by methanogenic bacteria is incorporated into a low-molecular-weight compound which apparently contains a nickel tetrapyrrol structure and is called factor F_{430} (36,191). The tetrapyrrol is unusual because it contains a six-membered ring which introduces a "pucker" into the planar configuration, thus making it possible for the firm binding of the nickel atom (198a). Factor F_{430} apparently is a component of methyl coenzyme M methylreductase which reduces CH_3-S-CoM to methane and HS-CoM (47,48). The role of nickel tetrapyrrol F_{430} has not been defined, but it is believed to be involved in methyl group reduction.

In bacteria that derive energy from the conversion of H_2 to CH_4 (methanogenic bacteria) or H_2O (Knallgas bacteria), the hydrogenase involved contains nickel, or requires nickel for its synthesis (55,56,95,142,191). Also, nickel is required for derepression of hydrogenase activity, and apparently protein synthesis is necessary for the participation of nickel in hydrogenase expression of the free-living rhizobia, *Rhizobium japonicum* (90). In the hydrogenases of *Methanobacterium bryantii*, *Methanobacterium thermoautotrophicum*, *Chromatrium vinosum*, and the sulfate-reducing bacterium, *Desulfovibrio gigas*, a substantial amount of the nickel is in the Ni^{3+} state (3,91,95,97). It has been suggested (3,91,190) that redox-sensitive nickel probably is the binding site for the substrate H_2, even though the hydrogenases from these microorganisms contain iron–sulfur centers.

In acetogenic bacteria, the energy source is the conversion of carbon dioxide, or carbon monoxide, and hydrogen to acetate and water. The last step of this conversion, the reductive carboxylation of methyl tetrahydrofolate to acetate, is catalyzed by a multienzyme complex. This complex has a moiety with carbon monoxide dehydrogenase activity, that is, the ability to convert CO to CO_2. The synthesis of the moiety requires nickel (38,42,147,148,191). Furthermore, analyses of the multienzyme complex indicate that the moiety is a iron–sulfur protein with a nickel-containing prosthetic group (147,148).

Nickel can activate many enzymes *in vitro*, but a role as a specific cofactor for any animal enzyme has not been shown (120). However, there are findings that suggest nickel may be important for at least three enzymes. Calcineurin, a Ca^{2+}- and calmodulin-dependent phosphoprotein phosphatase, was found to be markedly activated by nickel ions (24,78). King and Huang (78) found that the activation was not reversed by high concentrations of chelators, thus indicating that nickel tightly binds to the enzyme, and that nickel may have a role in regulating the substrate specificity of phosphoprotein phosphatases. Fishelson and co-workers (52,53) found that Ni^{2+} could replace, and was more efficient than, Mg^{2+} in the formation of the two C3 convertases of the complement system: C3b,Bb of the alternative pathway, and C4b,2a of the classical pathway. Furthermore, up to nine times more factor B was specifically bound to C3b-bearing sheep erythrocytes (EC3b) in the presence of nickel than in the presence of magnesium under identical conditions. To form one effective hemolytic site per EC3b cell with nickel, three times less factor B, 12 times less factor D, and 66 times fewer metal ions were required than when using magnesium. These findings suggest that nickel has a role in the complement pathway.

Pullarkat et al. (145) presented evidence suggesting that ATP and Ni^{2+} are part of a reaction that converts phosphatidic acid to pyrophosphatidic acid, which is a precursor of phosphatidylserine biosynthesis by rat brain microsomes. In isotope studies, the specific activity of the ATP,Ni^{2+}-dependent phosphatidyl-serine was increased >2-fold during active myelination.

2. Bioligand Cofactor Facilitating Fe^{3+} Absorption

The hypothesis that nickel may function as a bioligand cofactor facilitating the intestinal absorption of Fe^{3+} was supported by findings showing that, depending on the form of dietary iron, nickel interacts with iron in the rat. Nickel synergistically interacted with iron to affect hematopoiesis in rats fed dietary iron as ferric sulfate only, but not as a mixture of ferric and ferrous sulfates (120,129). Furthermore, when only ferric sulfate was supplemented to the diet, liver content of iron was depressed in nickel-deprived rats (126). On the other hand, when a ferric–ferrous mixture was supplemented to the diet, nickel deprivation elevated the liver content of iron (126).

The mechanism through which nickel possibly enhances Fe^{3+} absorption is unclear. Because Fe^{3+} tends not to stay in a soluble form in the duodenum, it needs to be complexed or converted to the more soluble Fe^{2+} form for efficient absorption (107). According to May et al. (107), only ligands, such as porphyrinlike molecules, that form high-spin complexes and thereby increase the electrode potential, stabilize Fe^{2+} over Fe^{3+}. Most other bioligands lower the electrode potential and thus enhance the stability of the Fe^{3+} state. Therefore, the preferred chelated state of iron in vivo is probably Fe^{3+}, and the reduction to Fe^{2+} occurs spontaneously only in the presence of high local concentrations of a reducing metabolite, or under the influence of special enzyme mechanisms. The idea that nickel might act in an enzyme mechanism that converts Fe^{3+} to Fe^{2+} for absorption is attractive because of the finding of redox-sensitive nickel in enzymes of microorganisms (see above). However, the possibility that nickel promotes the absorption of Fe^{3+} per se by enhancing its complexation to a molecule that can be absorbed cannot be overlooked. Such a molecule could be similar to the nickel tetrapyrrole molecule found in certain microorganisms.

3. Other Physiological Effects of Nickel

Rubanyi *et al.* (154) found that the concentration of nickel was elevated 20-fold in the serum of women immediately after delivery of an infant, but before delivery of the placenta. Sixty minutes after parturition, the serum concentration of nickel was similar to healthy control subjects. Based on this finding, and the finding that nickel has oxytocic action on rat uterine strips, Rubanyi and Balogh (152) suggested that nickel may support the separation of the placenta and/or contribute to the prevention of atonic bleeding in the postpartum period.

Other findings suggesting a functional role for nickel include (1) restoration of the porcine lymphocyte response to a mitogen that was inhibited by EDTA (199); (2) *in vitro* polymerization of purified goat brain tubulin was stimulated at low Ni^{2+} concentrations, and inhibited at high Ni^{2+} concentration ($>10^{-4}$ M) (151); (3) the finding of sufficient nickel in the cytoplasmic membrane (0.16% of the dry weight) and sheath of the archaebacterium *Methanospirillum hungatei* to suggest that nickel had an important structural role (178); (4) stabilization of RNA (58) and DNA (45) against thermal denaturation, suggesting a structural role in these macromolecules; (5) inhibition of prolactin secretion by nickel, thus suggesting that nickel plays a role in lactation at the pituitary level (94). On the other hand, another study (29) showed that a subcutaneous injection of nickel promoted high circulating prolactin levels lasting from 1 to 4 days, apparently mediated through the neuroendocrine system. Regardless of the contrast, these findings suggest nickel can affect prolactin metabolism.

IV. NICKEL REQUIREMENTS AND SOURCES

A. Nickel Requirements

The minimum amounts of nickel required by animals to maintain health, based on the addition of graded increments to a known deficient diet in the conventional manner, are not yet known. Nielsen *et al.* (124) reported that 50 ng of nickel per gram satisfied the dietary nickel requirement of chicks, and Kirchgessner and co-workers (82) reported that rats had a similar requirement for growth. Subsequently, Kirchgessner and co-workers (80,86) obtained findings that suggested an even higher nickel requirement for rats. Although growth was not affected, rats fed 60 or 150 ng nickel per gram diet exhibited lower activities of the enzymes pancreatic α-amylase, hepatic glucose-6-phosphate dehydrogenase, and hepatic lactate dehydrogenase than rats fed 20 μg nickel per gram diet—thus suggesting a requirement higher than 150 ng/g diet (86). Kirchgessner *et al.* (80) also found that the iron concentrations in liver, serum, femur, spleen, and muscle increased when dietary nickel was increased from 60 to 10,000 or 15,000 ng/g.

Thus, they suggested that the nickel requirement of growing rats is much higher than the 50–100 ng/g which prevents the appearance of clinical signs of deficiency. Anke and co-workers (6,7) indicated that minipigs and goats must have a nickel requirement >100 ng/g diet, because upon feeding those species a diet containing that level of nickel, deprivation signs appeared. Spears and Hatfield (172) found that a dietary nickel supplement of 5 μg/g markedly increased ruminal urease activity in lambs fed a corn-based diet containing 320 ng nickel per gram and 9.5% crude protein. Similar findings were obtained with cattle (173). Reidel *et al.* (150) found a nickel supplement of 5 μg/g diet did not increase growth, nickel content in organs, or the activity of several enzymes in cattle fed a diet containing 500 ng nickel per gram. Considering the preceding, it seems prudent to accept the suggestions of Anke *et al.* (7) that ruminants have a dietary nickel requirement in the range of 300 to 350 ng/g, probably because some rumen bacteria use nickel as part of their urease enzyme, and that monogastric animals have a dietary nickel requirement of <200 ng/g. Calculated from data for chicks, Nielsen (123) suggested that the dietary nickel requirement of humans would be near 35 μg daily.

B. Nickel in Human Foods and Dietaries

The nickel content of foods has attracted increased attention since nickel was described as nutritionally essential, and because oral nickel was found to produce a positive skin reaction in some nickel-sensitive individuals. Schlettwein-Gsell and Mommsen-Straub (161) adequately summarized the nickel content of foods reported up to 1971. Some of the more extensive and reliable listings of the nickel content in foods have been summarized by Ellen *et al.* (49), Thomas *et al.* (192), Brun (16), and Anke *et al.* (7). Casey (20) reported the nickel content of infant milk foods and supplements. Foods that generally contain high concentrations (>0.3 μg/g) of nickel include nuts (49,57), leguminous seeds (7,49,192), shellfish (73), cacao products (49), and hydrogenated solid shortenings (116) (see Table IV). Grains (49,198), cured meats (49,89), and vegetables (49,192) are generally of intermediate (0.1–0.3 μg/g) nickel content. Foods of animal origin, such as fish (73), milk (20), and eggs (49), are generally low (<0.1 μg/g) in nickel. The canning of some fruits and vegetables apparently increases their nickel content. For example, Thomas *et al.* (192) found that fresh tomatoes, plums, and rhubarb contained 0.09, 0.16, and 0.15 μg nickel per gram, respectively, whereas the canned products contained 0.49, 0.36, and 0.49 μg/g, respectively. Canned pineapple contained 0.85 μg nickel per gram.

Total dietary nickel intakes vary greatly with the amount and proportion of foods of animal (nickel-low) and plant origin (nickel-high) consumed, and with the amounts of refined and processed foods in the diet. However, several reports indicate that nickel intake probably is in the range of 150–700 μg/day.

Table IV
High Nickel Content Foods

Food	Nickel content (μg/g) (dry weight)[a]	(fresh weight)[b]	Food	Nickel content (μg/g) (fresh weight)[b,c,d]
Nuts			Cacao products	
Almond	1.6	1.3	Cacao powder	9.8
Black walnut	4.8	—	Bittersweet chocolate	2.6
Brazil nut	5.8	—	Milk chocolate	1.2
Butter nut	4.3	—	Leguminous seeds	
Cashew	5.0	5.1	Red kidney beans	0.45
English walnut	1.1	3.6	Field peas	0.32
Filbert	1.8	1.6	Broad beans, frozen	0.55
Hickory nut	9.8	—	Peas, frozen	0.35
Pecan	1.6	—	Beans, frozen	0.35
Pistachio	1.1	0.8	Other	
Peanuts	—	1.6	Spinach	0.39
Coconut	2.1	—	Asparagus, frozen	0.42
			Shortening, solid	0.592–2.772

[a] From Furr *et al.* (57).
[b] From Ellen *et al.* (49).
[c] From Thomas *et al.* (192).
[d] From Nash *et al.* (116).

Schroeder *et al.* (167) reported that the dietary intake of nickel by adults in the United States ranged from 300 to 600 μg/day. Murthy *et al.* (114) found that institutionalized children, age 9–12, from 28 U.S. cities consumed diets that supplied 288–696 μg/day nickel (mean of 451 μg/day). Myron *et al.* (115) measured the nickel content in nine diets prepared in university or hospital kitchens in North Dakota. They found that the diets would supply 165 ± 11 μg nickel per day or 75 ± 10 μg/1000 calories. Several other reports indicate that the dietary intake of nickel is somewhere near 300–500 μg/day (67,71,131).

C. Nickel in Animal Feeds and Pastures

The nickel content has been reported for hundreds of substances, many of which could be used for animal feeds (see Ref. 136). Most animal foods, because they are plant based, contain relatively high levels of nickel. Common pasture plants contain 0.5–3.5 μg nickel per gram dry matter (7,103,157,169). Reported nickel content (in μg/g) of some feed grains and protein sources include wheat, 0.08–0.3 (7,103,198); corn, 0.20 (7); oats, 0.71–2.09 (7,103); linseed cake,

5.24 (103); soybean cake, 7.91 (103); and sunflower cake, 7.78 (103).Based on nickel concentrations found in animal products (see above), milk products and meat meals used as protein supplements apparently are poor nickel sources. Nonetheless, fish protein concentrate was found to contain 0.7–2.8 μg/g nickel (96). The high concentration of nickel found in many animal feeds indicate that a nickel deficiency is unlikely to arise in animals under natural conditions.

V. NICKEL TOXICITY

A. Oral Nickel Toxicity Signs

The preceding sections show that there are mechanisms for the homeostatic regulation of nickel. Thus, it is not surprising that life-threatening toxicity of nickel through oral intake is low, ranking with such elements as zinc, chromium, and manganese. Nickel salts exert their toxic action mainly by gastrointestinal irritation and not by inherent toxicity. Generally, $\geq$250 μg of nickel per gram of diet are required to produce signs of nickel toxicity in rats, mice, chicks, dogs, cows, rabbits, pigs, ducks, and monkeys (118) (see below). Furthermore, when nickel toxicity occurs, many of the apparent signs are the result of reduced food intake (196,197), partially caused by reduced palatability (139).

Reported signs of oral nickel toxicity are numerous. Kirchgessner et al. (87) found that, when piglets were fed 0, 125, 250, 375, or 500 μg nickel per gram diet, only the two higher levels depressed growth. Those levels also induced coarse, shaggy hair, diarrhea, and dark-colored feces. Kirchgessner et al. (87) also reported that rats fed 1000 μg nickel per gram diet exhibited both diminished food intake and a cessation of growth. After 14 days, the rats exhibited elevated erythrocyte counts, hematocrits, hemoglobin, serum protein and liver protein, urea, glutamate dehydrogenase, iron, zinc, copper, and nickel, and depressed liver glucose-6-phosphate dehydrogenase. Depressed iodine-binding activity of the thyroid may be another sign of nickel toxicity (46).

Ambrose et al. (4) found that dogs fed 2500 μg nickel per gram diet vomited and salivated excessively. After return to the control diet, followed by a gradual increase of dietary nickel to 2500 μg/g, there were no acute problems. After 2 years, the dogs exhibited a moderately depressed growth, mild anemia, granulocytic hyperplastic bone marrow, elevated urine volume, and severe lung lesions. Dogs fed 100 or 1000 μg nickel per gram diet apparently performed similarly to dogs fed no supplemental nickel.

Mallard ducks and ducklings, except for a black tarry feces and elevated tissue nickel, were not markedly affected when fed 800 μg nickel per gram diet for 90 days (17,44). Ducklings fed 1200 μg nickel per gram diet began to tremble and show signs of paresis after 14 days. They grew subnormally, and had a 71% mortality rate by 60 days (17).

Experiments with chicks indicate a lower nickel tolerance in this species (98,197). Ling and Leach (98) found that chicks fed $\geq$300 μg nickel per gram of diet exhibited depressed growth. Mortality and anemia were exhibited by chicks fed 1100 μg/g diet. Weber and Reid (197) found that chicks fed >700 μg/g diet exhibited depressed food intake, impairment of energy metabolism, and a marked reduction in nitrogen retention. However, when these chicks, or those fed 1100 μg/g diet, were compared to pair-fed controls, only nitrogen retention was depressed.

The health, feed consumption, milk production, and milk composition of dairy cows were unaffected by dietary supplements of nickel carbonate at 50 and 250 μg/g nickel (137). Dairy calves fed nickel at 62.5 and 250 μg/g diet were similarly unaffected, but at 1000 μg/g diet, feed intake was greatly depressed and nitrogen retention significantly lowered (138).

Subsequent findings suggest that oral nickel, in not particularly high doses, can adversely affect the health of humans. Nickel dermatitis is a relatively common form of nickel toxicity in humans. Several surveys have shown that the incidence of sensitivity to nickel is between 4 and 13% (118). Nickel dermatitis had been thought to be caused mainly by the percutaneous absorption of nickel. However, Christensen and Moller (26) presented evidence suggesting that the ingestion of small amounts of nickel may be of greater importance in maintaining hand eczema than external contacts. Cronin *et al.* (33) observed that an oral dose of 0.6 mg of nickel as nickel sulfate produced a positive reaction in some nickel-sensitive individuals. That dose is only 12 times as high as the human daily requirement postulated from animal studies.

B. Nickel Toxicity in Parenteral Nutrition

Nickel is a ubiquitous element. The nickel content in purified proteins, amino acids, and minerals often is near 0.1–1.0 μg/g. Thus, not surprisingly, Leach and Sunderman (96a) found albumin replacement solutions from six different sources contained relatively high nickel concentrations; they ranged from 11 to 222 μg nickel/liter. For those concerned with parenteral nutrition, nickel toxicity may be a greater problem than deficiency. Recent findings suggest that the infusion of solutions highly contaminated with nickel could cause adverse reactions including undesirable changes in cardiac and uterine function, and allergic reactions (see above). Furthermore, other signs of parenteral nickel toxicity observed in animals might well be expected to occur in humans given parenteral nickel.

Rubanyi and co-workers (153,155,156) found that the addition of 60 μg nickel per liter to the fluid used in the perfusion of isolated rat hearts increased coronary artery resistance, reduced myocardial contractility, and induced ultrastructural myocardial damage. They (92,156) also found that an intravenous injection of nickel at 4.7 μg/kg body weight in anesthetized dogs elevated coronary artery resistance by 46%. These kinds of cardiac changes might be hazardous to patients with acute myocardial infarction.

Lu *et al.* (99) found that the intraperitoneal injection of nickel chloride caused teratology in mice. They found also that the kinetics of nickel chloride in fetal tissues were different from those in maternal tissues, and suggested that anomalies caused by elevated nickel levels could occur in the embryo without recognizable adverse affects in maternal mice (100). Another concern during pregnancy is the effect of nickel on rat uterine muscle strips *in vitro* (152). The tendency of the fetus to retain nickel and the oxytocic effect of nickel suggest that elevated levels of nickel in the blood are not desirable during pregnancy.

Other toxicity signs of parenterally administered nickel include (1) suppression of antibody synthesis in rats (109), (2) elevation of α-amino acid levels in the urine and plasma of rats (106), and (3) hyperglycemia in rodents and fowl (28,54,72).

C. Factors Influencing Nickel Toxicity

The manifestations of high dietary nickel can be affected by the dietary levels of several nutrients. Furthermore, the deficiency of some other nutrients can be exacerbated by high dietary nickel.

1. Iron

Blalock and Hill (15) reported that iron-deficient chicks were more susceptible to nickel toxicity, as judged by growth depression, than were iron-adequate chicks. An adverse effect of nickel on iron metabolism was shown by Nielsen *et al.* (127). In rats, 20 μg nickel per gram diet, perhaps through both physiological and pharmacological mechanisms, partially alleviated the marginal iron deficiency (15 μg iron per gram diet) signs such as depressed hematocrit, hemoglobin, and liver iron content, and elevated heart weight–body weight ratio, and liver copper and zinc content. In contrast, when a 100 μg nickel per gram diet was fed, the beneficial effect of nickel was aborted, and iron metabolism was impaired. No adverse effects were seen when 100 μg nickel per gram diet was fed to rats consuming 100 μg iron per gram diet. Nielsen *et al.* (129) also found that as little as 20 μg nickel per gram diet apparently exacerbated some signs of *severe* iron deficiency. Thus, the dietary level at which nickel stops pharmacologically alleviating iron deprivation and begins to toxicologically impair iron metabolism depends on the severity of iron deprivation. Most of the interaction between nickel and iron probably occurs during absorption because, as described in Section II, both Ni^{2+} and Fe^{2+} apparently use the same transport system. A competitive interaction between nickel and iron is possible because they both can form the same type of complex. Nickel ions can possess outer orbital bonding, in which the coordination number is 6, and form octahedral complexes. Both ions of iron (Fe^{2+}, Fe^{3+}) can have a coordination number of 6 and can form octahedral complexes regardless of whether there is inner or outer orbital bonding. Thus, the same orbitals could be involved in bonding both nickel and iron when they form outer orbital complexes. However, some of the interaction might be the result of nickel affecting certain metabolic pathways involving iron. For example, the depression in hematopoiesis might be due, in part, to nickel enhancing the activity of heme oxygenase, an enzyme that degrades heme, or nickel inhibiting the activity of δ-aminolevulinate synthase, the initial and rate-limiting enzyme in the heme-biosynthetic pathway (102).

2. Copper

Cuprous and cupric ions have a preferred coordination number of 4 and form tetrahedral and square coplanar complexes, respectively. Nickel can possess inner orbital bonding, in which the preferred coordination number is 4, and form either tetrahedral or square coplanar complexes. Thus, if nickel is present in biological systems with a coordination number of 4, nickel and copper could have similar chemical parameters and a nickel–copper interaction probably would be competitive. Sarkar (158) supported this suggestion by reporting that copper and nickel demonstrates similar binding components in serum, occupying the same specific transport site of albumin with a similar symmetry. Further support for this suggestion are the results from factorial experiments that show an antagonistic interaction between nickel and copper (128,173a). Thus, it is not surprising that, in copper-deficient animals, relatively low levels of nickel can have detrimental effects. Nielsen and co-workers (128,130) found that in rats deficient in copper, and significantly but not severely anemic, the copper deficiency signs of elevated heart weight and plasma cholesterol, and depressed hemoglobin were exacerbated by nickel supplementation. The effect was greater when dietary nickel was 50 μg/g rather than 5 μg/g. Spears and Hatfield (173a) found that the copper deficiency signs of depressed growth and hemoglobin were exacerbated by nickel supplementation. The effect on growth seemed greater when dietary nickel was 225 μg/g rather than 15 or 30 μg/g.

3. Zinc

Zinc status can also influence the level at which nickel begins to affect an animal adversely. Mathur and co-workers (104,105) found that high dietary nickel exacerbated zinc deficiency signs including depressed growth, feed efficiency, and feed intake. High dietary zinc also enhanced the negative effects of high dietary nickel on growth, feed efficiency, and feed intake.

4. Other Substances

In addition to trace elements, dietary vitamin C, protein, and ionophores could possibly influence nickel toxicity. Chatterjee *et al.* (23) found that high dietary vitamin C alleviated nickel toxicity in rats. Hill (69) found that increasing dietary protein from 10% to 30% decreased the toxicity of 400 or 800 μg nickel per gram diet fed to chicks. Starnes *et al.* (180) reported that the ionophores monensin and lasalocid (33 μg/g diet) depressed nickel-dependent bacterial urease in steers by 66% and 28%, respectively, after 33 days. This suggests that ionophores affect the transport of nickel across membranes, and thus may affect nickel toxicity.

D. Clinical Tests for Nickel Toxicity

The most convenient and apparently reliable evaluation of nutrition status is the measurement of serum and urine nickel. Values above the reported normal range (>7.5 μg/liter) would be indicative of an excess intake of nickel. Evaluation of nickel status in this manner is supported by reports that show urine and serum nickel are quickly and markedly elevated after the ingestion of a soluble nickel salt (27,110,130,171,179), and are also elevated in persons with occupational exposure to nickel (2,179,184).

E. Toxicity of Nickel Carbonyl and Nickel Subsulfide

The metabolism and toxicity of the carcinogen nickel carbonyl differs markedly from that of Ni^{2+}. Nickel carbonyl is highly volatile and is absorbed readily through the lungs. The inhalation route is most important in respect to nickel carbonyl toxicity. Also, the metabolism and toxicity of relatively insoluble nickel compounds such as Ni_3S^2 (usually administered intramuscularly to cause cancer, or intrarenally to cause erythropoiesis) differ from Ni^{2+}. Sunderman (185) has written a review of the toxicity of nickel carbonyl and relatively insoluble nickel compounds. These compounds have limited nutritional interest.

REFERENCES

1. Abdulwajid, A. W., and Sarkar, B. (1983). *Fed. Proc., Fed. Am. Soc. Exp. Biol.* **42,** 819; *Proc. Natl. Acad. Sci. U.S.A.* **80,** 4509–4512.
2. Adams, D. B., Brown, S. S., Sunderman, F. W., Jr., and Zachariasen, H. (1978). *Clin. Chem.* **24,** 862–867.
3. Albracht, S. P. J., Albrecht-Ellmer, K. J., Schmedding, D. J. M., and Slater, E. C. (1982). *Biochim. Biophys. Acta* **681,** 330–334.

4. Ambrose, A. M., Larson, P. S., Borzelleca, J. F., and Hennigar, G. R., Jr. (1976). *J. Food Sci. Technol.* **13**, 181–187.

5. Andersen, I., Torjussen, W., and Zachariasen, H. (1978). *Clin. Chem.* **24**, 1198–1202.

6. Anke, M., Grün, M., Dittrich, G., Groppel, B., and Hennig, A. (1980). *In* "3rd Spurenelement-Symposium, Nickel" (M. Anke, H.-J. Schneider, and Chr. Brückner, eds.), pp. 715–718. Freidrich-Schiller Univ., Jena.

7. Anke, M., Grün, M., Groppel, B., and Kronemann, H. (1983). *In* "Biological Aspects of Metals and Metal-Related Diseases" (B. Sarkar, ed.), pp. 89–105. Raven, New York.

8. Anke, M., Grün, M., and Kronemann, H. (1980). *In* "3rd Spurenelement-Symposium, Nickel" (M. Anke, H.-J. Schneider, and Chr. Brückner, eds.), pp. 237–244. Friedrich-Schiller Univ., Jena.

9. Anke, M., Kronemann, H., Groppel, B., Hennig, A., Meissner, D., and Schneider, H.-J. (1980). *In* "3rd Spurenelement-Symposium, Nickel" (M. Anke, H.-J. Schneider, and Chr. Brückner, eds.), pp. 3–10. Friedrich-Schiller Univ., Jena.

10. Asato, N., Van Soestbergen, M., and Sunderman, F. W., Jr. (1975). *Clin. Chem.* **21**, 521–527.

10a. Badica, T., Brazdes, L., and Popescu, I. (1984). *Nucl. Instrum. Methods Phys. Res. B3* **231**, 385–387.

11. Balogh, I., Szabo, K., and Gergely, A. (1983). *In* "4th Spurenelement-Symposium" (M. Anke, W. Bauman, H. Bräunlich, and Chr. Brückner, eds.), pp. 66–72. Friedrich-Schiller Univ., Jena.

12. Becker, G., Dörstelmann, U., Frommberger, U., and Forth, W. (1980). *In* "3rd Spurenelement-Symposium, Nickel" (M. Anke, H.-J. Schnieder, and Chr. Brückner, eds.), pp. 79–85. Friedrich-Schiller Univ., Jena.

13. Bencko, V., Geist, T., Arbetová, D., and Dharmadikari, D. M. (1983). *In* "4th Spurenelement-Symposium" (M. Anke, W. Bauman, H. Bräunlich, and Chr. Brückner, eds.), pp. 355–361. Friedrich-Schiller Univ., Jena.

14. Bernstein, D. M., Kneip, T. J., Kleinman, M. T., Riddick, R., and Eisenbud, M. (1974). *In* "Trace Substances in Environmental Health—8" (D. D. Hemphill, ed.), pp. 329–334. Univ. of Missouri Press, Columbia.

15. Blalock, T. L., and Hill, C. H. (1985). *In* "Health Effects and Interactions of Essential and Toxic Elements" (M. Abdulla, B. M. Nair, and R. K. Chandra, eds.), pp. S648–S654. Nutrition Research, Supplement I, Pergamon Press, New York.

16. Brun, R. (1979). *Dermatologica* **159**, 365–370.

17. Cain, B. W., and Pafford, E. A. (1981). *Arch. Environ. Contam. Toxicol.* **10**, 737–745.

18. Cammack, R., Patil, D., Aguirre, R., and Hatchikian, E. C. (1982). *FEBS Lett.* **142**, 289–292.

19. Casey, C. E. (1976). *Proc. Univ. Otago Med. School* 54, 7–8.

20. Casey, C. E. (1977). *N.Z. Med. J.* **85**, 275–278.

21. Casey, C. E., and Robinson, M. F. (1978). *Br. J. Nutr.* **39**, 639–646.

22. Catalanatto, F. A., and Sunderman, F. W., Jr. (1977). *Ann. Clin. Lab. Sci.* **7**, 146–151.

23. Chatterjee, K., Chakarborty, D., Majumdar, K., Bhattacharyya, A., and Chatterjee, G. C. (1979). *Int. J. Vitam. Nutr. Res.* **49**, 264–275.

24. Chernoff, J., and Li, H.-C. (1984). *Fed. Proc., Fed. Am. Soc. Exp. Biol.* **43**, 766.

25. Christensen, O. B., and Lagesson, V. (1981). *Ann. Clin. Lab. Sci.* **11**, 119–125.

26. Christensen, O. B., and Möller, H. (1975). *Contact Dermatitis* **1**, 136–141.

27. Christensen, O. B., Möller, H., Andrasko, L., and Lagesson, V. (1979). *Contact Dermatitis* **5**, 312–316.

28. Clary, J. J. (1975). *Toxicol. Appl. Pharmacol.* **31**, 55–65.

29. Clemons, G. K., and Garcia, J. F. (1981). *Toxicol. Appl. Pharmacol.* **61**, 343–348.

30. Cohn, J. R., and Emmett, E. A. (1978). *Ann. Clin. Lab. Sci.* **8**, 270–275.
31. Creason, J. P., Hinners, T. A., Bumgarner, J. E., and Pinkerton, C. (1975). *Clin. Chem.* **21**, 603–612.
32. Creason, J. P., Svendsgaard, D., Bumgarner, J., Pinkerton, C., and Hinners, T. (1976). *In* "Trace Substances in Environmental Health—10" (D. D. Hemphill, ed.), pp. 53–62. Univ. of Missouri Press, Columbia.
33. Cronin, E., DiMichiel, A. D., and Brown, S. S. (1980). *In* "Nickel Toxicology" (S. S. Brown, and F. W. Sunderman, Jr., eds.), pp. 149–152. Academic Press, New York.
34. D'Alonzo, C. A., and Pell. S. (1963). *Arch. Environ. Health* **6**, 381–385.
35. Decsy, M. I., and Sunderman, F. W., Jr. (1974). *Bioinorg. Chem.* **3**, 95–105.
36. Diekert, G., Konheiser, U., Piechulla, K., and Thauer, R. K. (1981). *J. Bacteriol.* **148**, 459–464.
37. Diekert, G., and Ritter, M. (1982). *J. Bacteriol.* **151**, 1043–1045.
38. Diekert, G., and Thauer, R. K. (1980). *FEMS Microbiol. Lett.* **7**, 187–189.
39. Dixon, N. E., Blakely, R. L., and Zerner, B. (1980). *Can. J. Biochem.* **58**, 481–488.
40. Dixon, N. E., Gazzola, C., Asher, C. J., Lee, D. S. W., Blakely, R. L., and Zerner, B. (1980). *Can. J. Biochem.* **58**, 474–480.
41. Donadini, A., Pazzaglia, A., Desirello, G., Minoia, C., and Colli, M. (1980). *Acta Vitaminol. Enzymol.* **2**, 9–16.
42. Drake, H. L., Hu, S.-I., and Wood, H. G. (1980). *J. Biol. Chem.* **255**, 7174–7180.
43. Eads, E. A., and Lambdin, C. E. (1973). *Environ. Res.* **6**, 247–252.
44. Eastin, W. C., Jr., and O'Shea, T. J. (1981). *J. Toxicol. Environ. Health* **7**, 883–892.
45. Eichhorn, G. L. (1962). *Nature (London)* **194**, 474–475.
46. Eliseev, I. N. (1975). *Gig. Sanit.* 7–9.
47. Ellefson, W. L., and Whitman, W. B. (1982). *Basic Life Sci.* **19**, 403–414.
48. Ellefson, W. L., Whitman, W. B., and Wolfe, R. S. (1982). *Biochemistry* **79**, 3707–3710.
49. Ellen, G., Bosch-Tibbesma, G., and Douma, F. F. (1978). *Z. Lebensm. Unters. Forsch.* **166**, 145–147.
50. Farago, M., Szabó, K., Gergely, A., Balogh, I., and Rubányi, G. (1983). *In* "4th Spurenelement-Symposium" (M. Anke, W. Bauman, H. Bräunlich, and Chr. Brückner, eds.), pp. 73–80. Friedrich-Schiller Univ., Jena.
51. Feely, R. M., Eitenmiller, R. R., Jones, J. B., Jr., and Barnhart, H. (1983). *Fed. Proc., Fed. Am. Soc. Exp. Biol.* **42**, 921.
52. Fishelson, Z., and Müller-Eberhard, J. (1982). *J. Immunol.* **129**, 2603–2607.
53. Fishelson, Z., Pangburn, M. K., and Müller-Eberhard, H. J. (1983). *J. Biol. Chem.* **258**, 7411–7415.
54. Freeman, B. M., and Langslow, D. R. (1973). *Comp. Biochem. Physiol.* **47A**, 427–436.
55. Friedrich, B., Heine, E., Finck, A., and Friedrich, C. G. (1981). *J. Bacteriol.* **145**, 1144–1149.
56. Friedrich, C. G., Schneider, K., and Friedrich, B. (1982). *J. Bacteriol.* **152**, 42–48.
57. Furr, A. K., MacDaniels, L. H., St. John, L. E., Jr., Gutenmann, W. H., Pakkala, I. S., and Lisk, D. J. (1979). *Bull. Environ. Contam. Toxicol.* **21**, 392–396.
58. Fuwa, K., Wacker, W. E. C., Druyan, R., Batholomay, A. F., and Vallee, B. L. (1960). *Proc. Natl. Acad. Sci. U.S.A.* **46**, 1298–1307.
59. Gergely, A., Rubányi, G., Bakos, M., Gaál, O., and Kovách, A. (1980). *In* "3rd Spurenelement-Symposium, Nickel" (M. Anke, H.-J. Schneider, and Chr. Brückner, eds.), pp. 143–147. Friedrich-Schiller Univ., Jena.
60. Glennon, J. D., and Sarkar, B. (1982). *Biochem. J.* **203**, 15–23.
61. Glennon, J. D., and Sarkar, B. (1982). *Biochem. J.* **203**, 25–31.
62. Gordon, W. R., Schwemner, S. S., and Hillman, W. S. (1978). *Planta* **140**, 265–268.

63. Graf, E.-G., and Thauer, R. K. (1981). *FEBS Lett.* **136,** 165–169.
64. Groppel, B., Anke, M., Reidel, E., and Grün, M. (1980). *In* "3rd Spurenelement-Symposium, Nickel" (M. Anke, H.-J. Schneider, and Chr. Brückner, eds.), pp. 269–276. Friedrich-Schiller Univ., Jena.
65. Grun, M., Anke, M., Regius, A., and Szentmihályi, S. (1980). *In* "3rd Spurenelement-Symposium, Nickel" (M. Anke, H.-J. Schneider, and Chr. Brückner, eds.), pp. 245–251. Friedrich-Schiller Univ., Jena.
66. Guthans, S. L., and Morgan, W. T. (1982). *Arch. Biochem. Biophys.* **218,** 320–328.
67. Hamilton, E. I., and Minski, M. J. (1972–1973). *Sci. Total Environ.* **1,** 375–394.
68. Herlant-Peers, M.-C., Hildebrand, H. F., and Biserte, G. (1982). *Zentralbl. Bakteriol. Hyg., I. Abt. Orig. B* **176,** 368–382.
69. Hill, C. H. (1979). *J. Nutr.* **109,** 501–507.
70. Hoffman, G., Anke, M., Groppel, B., Gruhn, K., and Faust, H. (1983). *In* "4th Spurenelement-Symposium" (M. Anke, W. Bauman, H. Bräunlich, and Chr. Brückner, eds.), pp. 29–33. Friedrich-Schiller Univ., Jena.
71. Horak, E., and Sunderman, F. W., Jr. (1973). *Clin. Chem.* **19,** 429–430.
72. Horak, E., Zygowicz, E. R., Tarabishy, R., Mitchell, J. M., and Sunderman, F. W., Jr. (1978). *Ann. Clin. Lab. Sci.* **8,** 476–482.
73. Ikebe, K., and Tanaka, R. (1979). *Bull. Environ. Contam. Toxicol.* **21,** 526–532.
74. Jacobsen, N., Alfheim, I., and Jonsen, J. (1978). *Res. Commun. Chem. Pathol. Pharmacol.* **20,** 571–584.
75. Jones, D. C., May, P. M., and Williams, D. R. (1980). *In* "Nickel Toxicology" (S. S. Brown and F. W. Sunderman, Jr., eds.), pp. 73–76. Academic Press, New York.
76. Kasprzak, K. S., and Sunderman, F. W., Jr. (1979). *Pure Appl. Chem.* **51,** 1375–1389.
77. Katz, S. A., Bowen, H. J. M., Comaish, J. S., and Samitz, M. H. (1975). *Br. J. Dermatol.* **92,** 187–190.
78. King, M. M., and Huang, C. Y. (1983). *Biochem. Biophys. Res. Commun.* **114,** 955–961.
79. Kirchgessner, M., Perth, J., and Schnegg, A. (1980). *Arch. Tierernaehr.* **30,** 805–810.
80. Kirchgessner, M., Reichlmayr-Lais, A., and Maier, R. (1984). *In* "Trace Element Metabolism in Man and Animals—5 (TEMA-5)" (C. F. Mills, I. Bremner, and J. K. Chesters, eds.), pp. 147–151. Commonwealth Agricultural Bureaux, Farnham Royal, U.K.
81. Kirchgessner, M., Roth-Maier, D. A., Grassman, E., and Mader, H. (1982). *Arch. Tierernaehr.* **32,** 853–858.
82. Kirchgessner, M., Roth-Maier, D. A., and Schnegg, A. (1981). *In* "Trace Element Metabolism in Man and Animals (TEMA-4)" (J. McC. Howell, J. M. Gawthorne, and C. L. White, eds.), pp. 621–624. Australian Academy of Science, Canberra.
83. Kirchgessner, M., Roth-Maier, D. A., and Schnegg, A. (1982). *Res. Exp. Med. (Berlin)* **180,** 247–254.
84. Kirchgessner, M., and Schnegg, A. (1976). *Bioinorg. Chem.* **6,** 155–161.
85. Kirchgessner, M., and Schnegg, A. (1980). *In* "Nickel in the Environment" (J. O. Nriagu, ed.), pp. 635–652. Wiley, New York.
86. Kirchgessner, M., and Schnegg, A. (1981). *Ann. Nutr. Metab.* **25,** 307–310.
87. Kirchgessner, M., Schnegg, A., and Roth, F. X. (1980). *In* "3rd Spurenelement-Symposium, Nickel" (M. Anke, H.-J. Schneider, and Chr. Brückner, eds.), pp. 309–313. Friedrich-Schiller Univ., Jena.
88. Kirchgessner, M., Spoerl, R., and Roth-Maier, D. A. (1980). *Z. Tierphysiol. Tierernaehr. Futtermittelkd.* **44,** 98–111.
89. Kirkpatrick, D. C., and Coffin, D. E. (1975). *J. Sci. Food Agric.* **26,** 43–46.
90. Klucas, R. V., Hanus, F. J., Russell, S. A., and Evans, H. J. (1983). *Proc. Natl. Acad. Sci. U.S.A.* **80,** 2253–2257.

91. Kojima, N., Fox, J. A., Hau Singer, R. P., Daniels, L., Orme-Johnson, W. H., and Walsh, C. (1983). *Proc. Natl. Acad. Sci., U.S.A.* **80**, 378–382.

92. Kovach, A. G. B., Rubányi, G., Ligeti, L., and Koller, A. (1980). *In* "Nickel Toxicology" (S. S. Brown, and F. W. Sunderman, Jr., eds.), pp. 137–140. Academic Press, New York.

93. Krüger, H.-J., Huynh, B. H., Ljungdahl, P. O., Xavier, A. V., DerVartanian, D. V., Moura, I., Peck, H. D., Jr., Teixeira, M., Moura, J. J. G., and LeGall, J. (1982). *J. Biol. Chem.* **257**, 14620–14623.

94. LaBella, F. S., Dular, R., Lemon, P., Vivian, S., and Queen, G. (1973). *Nature (London)* **245**, 330–332.

95. Lancaster, J. R., Jr. (1982). *Science* **216**, 1324–1325.

96. Langmyhr, F. J., and Orre, S. (1980). *Anal. Chim. Acta* **118**, 307–311.

96a. Leach, C. H., Jr., and Sunderman, F. W., Jr. (1985). *N. Engl. J. Med.* **313**, 1232.

97. LeGall, J., Ljungdahl, P. O., Moura, I., Peck, H. D., Jr., Xavier, A. V., Moura, J. J. G., Teixeira, M., Huynh, B. H., and DerVartanian, D. V. (1982). *Biochem. Biophys. Res. Commun.* **106**, 610–616.

98. Ling, J. R., and Leach, R. M., Jr. (1979). *Poult. Sci.* **58**, 591–596.

99. Lu, C.-C., Matsumoto, N., and Iijima, S. (1979). *Teratology* **19**, 137–142.

100. Lu, C.-C., Matsumoto, N., and Iijima, S. (1981). *Toxicol. Appl. Pharmacol.* **59**, 409–413.

101. Lucassen, M., and Sarkar, B. (1979). *J. Toxicol. Environ. Health* **5**, 897–905.

102. Maines, M. D., and Kappas, A. (1977). *In* "Clinical Chemistry and Chemical Toxicology of Metals" (S. S. Brown, ed.), pp. 75–81. Elsevier, Amsterdam.

103. Malaiškaité, B. S., and Radišouskas, J. G. (1980). *In* "3rd Spurenelement-Symposium, Nickel" (M. Anke, H.-J. Schneider, and Chr. Brückner, eds.), pp. 253–258. Friedrich-Schiller Univ., Jena.

104. Mathur, A. K. (1984). *In* "Trace Element Metabolism in Man and Animals—5 Abstracts," p. 221. Aberdeen, Scotland.

105. Mathur, A. K., Reichlmayr-Lais, A., and Kirchgessner, M. (1982). *Z. Tierphysiol. Tierernaehr. Futtermittelkd.* **47**, 101–109.

106. Mathur, A. K., and Tandon, S. K. (1981). *Toxicol. Lett.* **9**, 211–214.

107. May, P. M., Williams, D. R., and Linder, P. W. (1978). *In* "Iron in Model and Natural Compounds" (H. Sigel, ed.). *(Metal Ions Biol. Syst.* **7**, 29–76). Dekker, New York.

108. McNeely, M. D., Sunderman, F. W., Jr., Nechay, M. W., and Levine, H. (1971). *Clin. Chem.* **17**, 1123–1128.

109. Mehrmofakham, S., and Treagan, L. (1981). *Biol. Trace Eleme. Res.* **3**, 7–11.

110. Menne, T., Mikkelsen, H. I., and Solgaard, P. (1978). *Contact Dermatitis* **4**, 106–108.

111. Mikac-Dević, D., Sunderman, F. W., Jr., and Nomoto, S. (1977). *Clin. Chem.* **23**, 948–956.

112. Miller, J. (1983). *Nutr. Res.* **3**, 351–359.

113. Moura, J. J. G., Moura, I., Huynh, B. H., Krüger, H.-J., Teixeira, M., DuVarney, R. C., DerVartanian, D. V., Xavier, A. V., Peck, H. D., Jr., and LeGall, J. (1982). *Biochem. Biophys. Res. Commun.* **108**, 1388–1393.

114. Murthy, G. K., Rhea, U. S., and Peeler, J. T. (1973). *Environ. Sci. Technol.* **7**, 1042–1045.

115. Myron, D. R., Zimmerman, T. J., Shuler, T. R., Klevay, L. M., Lee, D. E., and Nielsen, F. H. (1978). *Am. J. Clin. Nutr.* **31**, 527–531.

116. Nash, A. M., Mounts, T. L., and Kwolek, W. F. (1983). *J. Am. Oil Chem. Soc.* **60**, 811–814.

117. Nechay, M., and Sunderman, F. W., Jr. (1973). *Ann. Clin. Lab. Sci.* **3**, 30–35.

118. Nielsen, F. H. (1977). *Adv. Mod. Toxicol.* **2**, 129–146.

119. Nielsen, F. H. (1980). *Adv. Nutr. Res.* **3**, 157–172.

120. Nielsen, F. H. (1980). *In* "Inorganic Chemistry in Biology and Medicine, ACS Symposium Series 140" (A. E. Martell, ed.), pp. 23–42. American Chemical Society, Washington, D.C.

121. Nielsen, F. H. (1980). *J. Nutr.* **110**, 965–973.
122. Nielsen, F. H. (1983). *In* "4th Spurenelement-Symposium" (M. Anke, W. Bauman, H. Bräunlich, and Chr. Brückner, eds.), pp. 11–18. Friedrich-Schiller Univ., Jena.
123. Nielsen, F. H. (1984). *Annu. Rev. Nutr.* **4**, 21–41.
124. Nielsen, F. H., Myron, D. R., Givand, S. H., and Ollerich, D. A. (1975). *J. Nutr.* **105**, 1607–1619.
125. Nielsen, F. H., Myron, D. R., Givand, S. H., Zimmerman, T. J., and Ollerich, D. A. (1975). *J. Nutr.* **105**, 1620–1630.
126. Nielsen, F. H., and Shuler, T. R. (1981). *Biol. Trace Elem. Res.* **3**, 245–256.
127. Nielsen, F. H., Shuler, T. R., McLeod, T. G., and Zimmerman, T. J. (1984). *J. Nutr.* **114**, 1280–1288.
128. Nielsen, F. H., and Zimmerman, T. J. (1981). *Biol. Trace Elem. Res.* **3**, 83–98.
129. Nielsen, F. H., Zimmerman, T. J., Collings, M. E., and Myron, D. R. (1979). *J. Nutr.* **109**, 1623–1632.
130. Nielsen, F. H., Zimmerman, T. J., and Shuler, T. R. (1982). *Biol. Trace Elem. Res.* **4**, 125–143.
131. Nodiya, P. I. (1972). *Gig. Sanit.* **37**, 108–109.
132. Nomoto, S. (1974). *Shinshu Igaku Zasshi* **22**, 39–44.
133. Nomoto, S. (1980). *In* "Nickel Toxicology" (S. S. Brown, and F. W. Sunderman, Jr., eds.), pp. 89–90. Academic Press, New York.
134. Nomoto, S., McNeely, M. D., and Sunderman, F. W., Jr. (1971). *Biochemistry* **10**, 1647–1651.
135. Nozdryukhina, L. R., Grinkevich, N. I., and Gribovskaya, I. F. (1973). *In* "Trace Substances in Environmental Health—7" (D. D. Hemphill, ed.), pp. 353–357. Univ. of Missouri Press, Columbia.
136. Nriagu, J. O. ed. (1980). "Nickel in the Environment." Wiley, New York.
137. O'Dell, G. D., Miller, W. J., King, W. A., Ellers, J. C., and Jurecek, H. (1970). *J. Dairy Sci.* **53**, 1545–1548.
138. O'Dell, G. D., Miller, W. J., King, W. A., Moore, S. L., and Blackmon, D. M. (1970). *J. Nutr.* **100**, 1447–1454.
139. O'Dell, G. D., Miller, W. J., Moore, S. L., and King, W. A. (1970). *J. Dairy Sci.* **53**, 1266–1269.
140. O'Dell, G. D., Miller, W. J., Moore, S. L., King, W. A., Ellers, J. C., and Jurecek, H. (1971). *J. Anim. Sci.* **32**, 769–772.
141. Onkelinx, C., Becker, J., and Sunderman, F. W., Jr. (1974). *Trace Elem. Metab. Anim.* **2**, 560–563.
142. Partridge, C. D. P., and Yates, M. G. (1982). *Biochem. J.* **204**, 339–344.
143. Peters, H.-J., Köhler, H., Duck, H.-J., Günther, K. R., and Pankau, H. (1981). *In* "Mengen-und Spurenelemente, Arbeitstagung" (M. Anke and H. J. Schneider, eds.), pp. 197–200. Karl-Marx-Univ., Leipzig.
144. Polacco, J. C. (1977). *Plant Sci. Lett.* **10**, 249–255.
145. Pullarkat, R. K., Sbaschnig-Agler, M., and Reha, H. (1981). *Biochim. Biophys. Acta* **663**, 117–123.
146. Ragan, H. A. (1978). "Pacific Northwest Lab. Annual Report—1977, DOE Assist. Secre. Environ." PNL-2500—Pt 1, 1.9–1.10.
147. Ragsdale, S. W., Clark, J. E., Ljungdahl, L. G., Lundie, L. L., and Drake, H. L. (1983). *J. Biol. Chem.* **258**, 2364–2369.
148. Ragsdale, S. W., Ljungdahl, L. G., and DerVartanian, D. V. (1982). *Biochem. Biophys. Res. Commun.* **108**, 658–663.
149. Rica, C. C., and Kirkbright, G. F. (1982). *Sci. Total Environ.* **22**, 193–201.

150. Riedel, E., Anke, M., Schwarz, S., Regius, A., Szilagyi, M., Löhnert, H.-J., Flachowsky, G., Zenker, G., and Glós, S. (1980). *In* "3rd Spurenelement-Symposium, Nickel" (M. Anke, H.-J. Schneider, and Chr. Brückner, eds.), pp. 55–61. Friedrich-Schiller Univ., Jena.

150a. Rizk, S. L., and Sky-Peck, H. H. (1984). *Cancer Research* **44**, 5390–5394.

151. Roychaudhury, S., Banerjee, A., and Bhattacharyya, B. (1982). *Biochim. Biophys. Acta* **707**, 46–49.

152. Rubányi, G., and Balogh, I. (1982). *Am. J. Obstet. Gynecol.* **142**, 1016–1020.

153. Rubányi, G., Balogh, I., Somogyi, E., Kovách, A. G. B., and Sótony, P. (1980). *J. Mol. Cell. Cardiol.* **12**, 609–618.

154. Rubányi, G., Birtalan, I., Gergely, A., and Kovách, A. G. B. (1982). *Am. J. Obstet. Gynecol.* **143**, 167–169.

155. Rubányi, G., and Kovách, A. G. B. (1980). *In* "3rd Spurenelment-Symposium, Nickel" (M. Anke, H.-J. Schneider, and Chr. Brückner, eds.), pp. 111–115. Friedrich-Schiller Univ., Jena.

156. Rubányi, G., Ligeti, L., and Koller, A. (1981). *J. Mol. Cell. Cardiol.* **13**, 1023–1026.

157. Sapek, A., and Sapek, B. (1980). *In* "3rd Spurenelement-Symposium, Nickel" (M. Anke, H.-J. Schneider, and Chr. Brückner, eds.), pp. 215–220. Friedrich-Schiller Univ., Jena.

158. Sarkar, B. (1983). *Chem. Scripta* **21**, 101–108.

158a. Sarkar, B. (1985). *In* "Health Effects and Interactions of Essential and Toxic Elements" (M. Abdulla, B. M. Nair, and R. K. Chandra, eds.), pp. S489–S498. Nutrition Research, Supplement I, Pergamon, New York.

159. Sayato, Y., Nakamuro, K., Matsui, S., and Tanimura, A. (1981). *J. Pharmacobio.-Dyn.* **4**, S-73.

160. Scheiner, D. M., Katz, S. A., and Samitz, M. H. (1976). *Environ. Res.* **12**, 355–357.

161. Schlettwein-Gsell, D., and Mommsen-Straub, S. (1971). *Int. J. Vitam. Nutr. Res.* **41**, 429–437.

162. Schnegg, A. (1975). Ph.D. thesis., Univ. Munich, Weihenstephen.

163. Schnegg, A., and Kirchgessner, M. (1975). *Nutr. Metab.* **19**, 268–278.

164. Schnegg, A., and Kirchgessner, M. (1977). *Zentralbl. Vet. Med. A* **24**, 394–401.

165. Schnegg, A., and Kirchgessner, M. (1978). *In* "Trace Element Metabolism in Man and Animals—3" (M. Kirchgessner, ed.), pp. 236–243. Tech. Univ. Munich, Freising-Weihenstephen.

166. Schneider, H. J., Anke, M., and Klinger, G. (1980). *In* "3rd Spurenelement-Symposium, Nickel" (M. Anke, H.-J. Schneider, and Chr. Brückner, eds.), pp. 277–283. Freidrich-Schiller Univ., Jena.

167. Schroeder, H. A., Balassa, J. J., and Tipton, I. H. (1961). *J. Chronic Dis.* **15**, 51–65.

168. Schroeder, H. A., and Nason, A. P. (1969). *J. Invest. Dermatol.* **53**, 71–78.

169. Seay, W. A., and DeNumbrum, L. E. (1958). *Agron. J.* **50**, 237–240.

169a. Shand, C. A., Aggett, P. J., and Ure, A. M. (1984). *In* "Trace Element Metabolism in Man and Animals-5 (TEMA-5)" (C. F. Mills, I. Bremner, and J. K. Chesters, eds.), pp. 642–645. Commonwealth Agricultural Bureaux, Farnham Royal, U.K.

170. Smith, J. C., and Hackley, B. (1968). *J. Nutr.* **95**, 541–546.

171. Solomons, N. W., Viteri, F., Shuler, T. R., and Nielsen, F. H. (1982). *J. Nutr.* **112**, 39–50.

172. Spears, J. W., and Hatfield, E. E. (1978). *J. Anim. Sci.* **47**, 1345–1350.

173. Spears, J. W., and Hatfield, E. E. (1980). *In* "3rd Spurenelement-Symposium, Nickel" (M. Anke, H.-J. Schneider, and Chr. Brückner, eds.), pp. 47–53. Friedrich-Schiller Univ., Jena.

173a. Spears, J. W., and Hatfield, E. E. (1985). *Biol. Trace Elem. Res.* **7**, 181–193.

174. Spears, J. W., Hatfield, E. E., and Fahey, G. C., Jr. (1978). *Nutr. Rep. Int.* **18**, 621–629.

175. Spears, J. W., Hatfield, E. E., Forbes, R. M., and Koenig, S. E. (1978). *J. Nutr.* **108**, 313–320.

176. Spears, J. W., Jones, E. E., Samsell, L. J., and Armstrong, W. D. (1983/1984). *Fed. Proc., Fed. Am. Soc. Exp. Biol.* **42,** 819; *J. Nutr.,* **114,** 845–853.
177. Spoerl, R., and Kirchgessner, M. (1977). *Z. Tierphysiol. Tierernaehr. Futtermittelkd.* **38,** 205–210.
178. Sprott, G. D., Shaw, K. M., and Jarrell, K. F. (1983). *J. Biol. Chem.* **258,** 4026–4031.
179. Spruit, D., and Bongaarts, P. J. M. (1977). *Dermatologica* **154,** 291–300.
180. Starnes, S. R., Spears, J. W., Froetschel, M. A., and Croom, W. J., Jr. (1984). *J. Nutr.* **114,** 518–525.
181. Stein, G., Schneider, H. J., and Stadie, G. (1980). *In* "3rd Spurenelement-Symposium, Nickel" (M. Anke, H.-J. Schneider, and Chr. Brückner, eds.), pp. 293–299. Friedrich-Schiller Univ., Jena.
182. Strizhkova, S. N. (1975). *Patol. Friziol. Eksp. Ter.* **1975,** 72–74.
183. Sumino, K., Hayakawa, K., Shibata, T., and Kitamura, S. (1975). *Arch. Environ. Health* **30,** 487–494.
184. Sunderman, F. W., Jr. (1980). *Pure Appl. Chem.* **52,** 527–544.
185. Sunderman, F. W., Jr. (1981). *In* "Disorders of Mineral Metabolism, Vol. 1" (C. F. Bronner, and J. W. Coburn, eds.), pp. 201–232. Academic Press, New York.
186. Sunderman, F. W., Jr., Costa, E. R., Fraser, C., Hui, G., Levine, J. J., and Tse, T. P. H. (1981). *Ann. Clin. Lab. Sci.* **11,** 488–496.
187. Sunderman, F. W., Jr., Decsy, M. J., and McNeely, M. D. (1972). *Ann. N.Y. Acad. Sci.* **199,** 300–312.
188. Sunderman, F. W., Jr., Mangold, B. L. K., Wong, S. H. Y., Shen, S. K., Reid, M. C., and Jansson, I. (1983). *Res. Commun. Chem. Pathol. Pharmacol.* **39,** 477–492.
189. Tedeschi, R. E., and Sunderman, F. W. (1957). *AMA Arch. Ind. Health* **16,** 486–488.
190. Teixeira, M., Moura, I., Xavier, A. V., DerVartanian, D. V., LeGall, Jr., Peck, H. D., Jr., Huynh, B. H., and Moura, J. J. G. (1983). *Eur. J. Biochem.* **130,** 481–484.
190a. Templeton, D. M., and Sarkar, B. (1985). *Fed. Proc., Fed. Am. Soc. Exp. Biol.* **44,** 497.
191. Thauer, R. K., Diekert, G., and Schonheit, P. (1980). *Trends Biochem. Sci.* **5,** 304–306.
192. Thomas, B., Roughan, J. A., and Watters, E. D. (1974). *J. Sci. Food Agric.* **25,** 771–776.
193. Tipton, I. H., and Cook, M. J. (1963). *Health Phys.* **9,** 103–145.
194. Torjussen, W., Andersen, I., and Zachariasen, H. (1977). *Clin. Chem.* **23,** 1018–1022.
195. Unden, G., Böcher, R., Knecht, J., and Kröger, A. (1982). *FEBS Lett.* **145,** 230–234.
196. Weber, C. W., and Reid, B. L. (1968). *J. Nutr.* **95,** 612–616.
197. Weber, C. W., and Reid, B. L. (1969). *J. Anim. Sci.* **28,** 620–623.
198. Welch, R. M., and Cary, E. E. (1975). *Agric. Food Chem.* **23,** 479–482.
198a. Wolfe, R. S. (1985). *Trends Biol. Sci.* **10,** 396–399.
199. Yang, W. C., and Schultz, R. D. (1982). *Biol. Trace Elem. Res.* **4,** 175–182.
200. Zachariasen, H., Anderson, I., Kostol, C., and Barton, R. (1975). *Clin. Chem.* **21,** 562–567.

9

Vanadium

FORREST H. NIELSEN

United States Department of Agriculture
Agricultural Research Service
Grand Forks Human Nutrition Research Center
Grand Forks, North Dakota

I. VANADIUM IN ANIMAL TISSUES AND FLUIDS

Older data on the vanadium content of animal tissues are meager and discordant, presumably as a consequence of analytical difficulties. Recent findings, however, have made it clear that vanadium is widely distributed in very low concentrations in most animals.

Extremely high vanadium concentrations have long been known to occur in the blood of ascidian worms (44,75,187). Specialized blood cells called vanadocytes accumulate vanadate via a specific transport system in the plasma membrane (51), reduce transported V^{5+} to V^{4+} (in some species to V^{3+}), and store the reduced vanadium at concentrations up to 0.15 M within special vacuoles that also contain a high concentration of a reducing agent called tunichrome (101). Contrary to initial beliefs, vanadium apparently exists in the vanadocyte as a small complex, not as a vanadium–protein complex (62). Hemovanadin, once thought to be the functional form of vanadium *in vivo*, apparently was an artifactual finding. It is not known if vanadium performs any vital function in the ascidian worm. The vanadocyte does not act as an oxygen carrier (102). The hypothesis that vanadium has an oxidation–reduction function in these organisms, perhaps in the manufacture of the tunic material that forms an extracellular sheath, is currently being investigated.

Early analyses of vertebrate tissues for vanadium were limited by the lack of sensitive methods. However, even today the true vanadium content of some tissues and fluids is uncertain because of the disparity of the analytical findings when using various methods with high sensitivity. For example, using neutron activation techniques, Cornelis et al. (36) found human serum vanadium to be in the range of 0.016 to 0.939 μg/liter; Simonoff et al. (164) found a range of 0.26 to 1.30 μg/liter. Stroop et al. (170) used flameless atomic absorption techniques to find human serum vanadium values in the range of 2.22 to 3.94 μg/liter. Perhaps some of the variation can also be explained by the fact that vanadium exposure markedly influences tissue vanadium content (15). Regardless of the problems, recent reports indicate that the vanadium content of vertebrate tissues and fluids is remarkably low. Hamilton et al. (67) reported the following values with standard error, determined by spark source mass spectrometry, for adult human tissues: brain, 0.03 ± 0.008; liver, 0.04 ± 0.01; lung, 0.10 ± 0.02; lymph nodes, 0.40 ± 0.2; muscle, 0.01 ± 0.003; and testis, 0.20 ± 0.08 μg vanadium per gram wet weight. Using the same methodology, Shand et al. (161) found 0.33 μg of vanadium per gram dry fetal human liver (range 0.02–1.96 μg/g); Curzon and Crocker (42) found vanadium present in human dental enamel at concentrations near 20 ng/g. Yukawa et al. (194) reported the following values with standard deviations, determined by neutron activation analysis and instrumental semiconductor γ-ray spectrometry, for human tissues: aorta, 0.33 ± 0.13; brain, 0.13 ± 0.11; heart, 0.14 ± 0.09; kidney, 0.11 ± 0.06; liver, 0.11 ± 0.08; lung, 0.13 ± 0.08; muscle, 0.11 ± 0.07; pancreas, 0.13 ± 0.05; and spleen, 0.08 ± 0.08 μg vanadium per gram wet weight. Others, using neutron activation or photometry techniques (21,35,177,182), found much lower vanadium concentrations in human tissues including fat and muscle, 0.55; heart, 1.1; kidney, 3; liver, 7.5; lung, 2.1; placenta, 9; and thyroid, 3.1 ng vanadium per gram wet weight. Gibson and Dewolfe (61) reported that concentrations of vanadium in hair are low at birth, near 40 ng/g, and increase with age, to about 400 ng/g at 6 months. Other reported mean concentrations of vanadium in scalp hair of healthy adults range from 25 to 90 ng/g (91,113,149,186). Marumo et al. (103) found extremely low concentrations of vanadium in hair (433 ± 15 pg/g). Neutron activation analysis was used to show that human colostrum, transitional and mature milk generally contained less than 1.0 ng vanadium per gram dry weight (94).

Occupational exposure or pathological conditions can alter the vanadium concentration in blood and organs of humans. When compared to healthy or normal controls, the vanadium concentration was higher in the hair of welders (91) and lower in the hair of multiple sclerosis patients (186). Marumo et al. (103) found the vanadium concentration in hair was higher in nondialyzed and hemodialyzed patients with chronic renal failure than in healthy controls or hemofiltered patients with chronic renal failure. Although hair vanadium apparently was not affected by cancer (113), the vanadium concentration of neoplastic human breast tissue (1.34 ± 0.76 μg/g dry weight) was higher than histologically normal breast tissue 0.78 ± 0.46 μg/g dry weight) from the same patient (145).

Recent analyses using neutron activation analysis (152) or atomic absorption spectrometry (15,53,68) indicate that most rat tissues contain less than 10 ng vanadium per gram wet weight, and many sheep tissues contain less than 200 ng vanadium per gram dry weight. However, tissue vanadium is markedly elevated in animals fed high dietary vanadium. In rats, liver vanadium increased from 10 to 55 ng/g wet weight when dietary vanadium was increased from 0.1 to 25 μg/g (15). In sheep, bone vanadium increased from 220 to 3320 ng/g dry weight when dietary vanadium was increased from 10 to 200 μg/g (68). Rat tissue vanadium is affected also by age. Edel al. (53) found that the vanadium concentration decreased in kidney, liver, lung, and spleen, and increased in fat and bone, in rats between ages 21 and 115 days.

II. VANADIUM METABOLISM

A. Absorption

Limited information exists about vanadium metabolism in animals at physiological levels. Very small amounts (<0.2–0.8 μg/liter) are normally excreted in the urine of man (21,178). Urinary vanadium is greatly increased upon vanadium exposure (178) or when vanadium salts are administered orally at toxic or subtoxic levels (15,69). Nonetheless, the very low concentrations of vanadium normally in urine compared with the estimated dietary intake and fecal level of vanadium indicates that ≥1% of vanadium ingested is absorbed (21), but more evidence is needed to confirm this. In one study, only 0.1–1.0% of the vanadium in 100 mg of soluble diammonium oxytartarovanadate was found to be absorbed from the human gut (40). Hansard et al. (69) estimated that only 0.13–0.75% (mean of 0.34%) of ingested vanadium as ammonium metavanadate was absorbed from the sheep gut. On the other hand, two studies with rats indicated much greater vanadium absorption from the gut. Wiegmann et al. (191) fasted rats overnight then gavaged them with 5 μmol of Na_3VO_4 in 1.0 ml 0.9% NaCl containing 1 μC [48]V. After 4 days, 86.6 ± 2.4% of the administered [48]V was recovered in the feces and urine. Only 69.1 ± 1.8% of the dose was recovered in the feces; recovery from feces increased to 85.7 ± 1.5% if 1.0 ml of a suspension of $Al(OH)_3$ was administered simultaneously with the vanadium. In either case, the findings indicated an absorption of greater than 10%. Bogden et al. (15) found that rats retained 39.7% ± 18.5%, and excreted in the feces 59.1% ± 18.8%, of vanadium ingested as sodium metavanadate supplemented at 5 or 25 ppm in a casein–sucrose–dextrin based diet. Dietary composition and vanadium form probably affect the percentage of ingested vanadium absorbed from the gut. Regardless, the rat studies suggest caution in assuming that ingested vanadium will always be poorly absorbed from the gastrointestinal tract.

B. Excretion

Most ingested vanadium is excreted via the feces; the feces probably contain mostly unabsorbed vanadium. Based upon studies in which vanadium is administered parenterally, urine is the major excretory route for absorbed vanadium, and bone apparently is a major sink for retained vanadium.

In timed distribution studies of intravenously injected [48]V in rats, no significant differences were seen in the rate or amount of uptake of the three oxidation states of vanadium (87). Liver, kidney, spleen, and testis accumulated [48]V for 1.5 to 4 hr and retained most of this radioactivity up to 96 hr, at which time other major organs (bones were not examined) had retained less (87,150). At 96 hr, 30 to 46% of the [48]V had been excreted in the urine and 9 to 10% in the feces. Evidence was obtained that the marked soft tissue retention of [48]V was caused by its movement into the nuclear and mitochondrial fractions of the cells. Vanadium in the blood was mainly present in the plasma and disappeared from the bloodstream with different rates corresponding to the clearance of three plasma components (150). Initially, only a small amount of vanadium in the plasma was associated with proteins, but after 96 hr it was present only as a vanadium–transferrin bicomplex. In urine, [48]V was mainly associated with high-molecular-weight components 72 hr after [48]V administration.

Vanadium administered subcutaneously (147), intramuscularly (140), intraperitoneally (134,147,191), or intratracheally (134,140,147) is generally metabolized as when administered intravenously. For example, 5 days after intraperitoneal administration of [48]V-labeled vanadate to rats, 41% of the dose appeared in the urine, 8.3% in the feces (191). Another study showed that 66% of the [48]V intramuscularly injected as $VOCl_2$ was eliminated in the urine within 24 hours (140). These studies also showed that the kidney retained the highest percentage of the dose 24 hr after radioactive vanadium was given by means other than oral ingestion. As the length of time subsequent to injection increased, bone and, in some cases, spleen and liver became major sites of vanadium retention (133,137,147). For example, Roschin et al. (147) found that 30 min after intraperitoneally injecting rats with [48]V, the kidney retained 7.2%, and the bone retained 2.1%, of the dose; at 48 hr, the kidney retained 1.6%, and the bone retained 3.45% of the dose. In mice 1 day after an intraperitoneal injection of [48]$VOCl_3$, the highest amount of [48]V was found in the kidney; 4 to 12 days after injection, bone, spleen and liver contained markedly higher levels of [48]V than kidney (137).

C. Intermediary Metabolism

The rich aqueous chemistry of vanadium suggests that its metabolism is complex. Both the anion vanadate (VO_3^-) and the cation vanadyl (VO^{2+}) can complex with molecules of physiological significance. It is not known whether either of these forms is of greater importance in vanadium metabolism or biological function *in vivo*. Evidence to date suggests that the binding of the vanadyl ion to iron-containing proteins is important in vanadium metabolism. The paramagnetic vanadyl ion has been used as a sensitive electron paramagnetic resonance probe in the study of metal binding mechanisms of metalloproteins such as serotransferrin (31,70), lactoferrin (25), conalbumin (28), and apoferritin (30). Regardless of the oxidation state administered to animals, vanadium apparently is converted into vanadyl–transferrin and vanadyl–ferritin complexes in plasma and body fluids (71,87,153). Sabbioni and Marafante (152) found that 1 day after intravenous administration of $^{48}VO^{2+}$, 29% of ^{48}V incorporated in rat liver cytosol existed as a vanadium-low-molecular-weight complex ($<$5000 mol wt). By day 9, however, they found the low-molecular-weight complex had disappeared and vanadium was present only as vanadyl–ferritin (15%) and vanadyl–transferrin (85%) in rat liver cytosol. Nine days after the administration of vanadium, partial purification of heart myoglobin, liver mitochondrial and microsomal cytochromes b, b_5, and c, ferriporphyrin, and red blood cell hemoglobin showed no significant incorporation of ^{48}V into these proteins (151). These findings suggest that non-heme iron-metalloproteins are involved more than iron hemoproteins in the metabolism of vanadium *in vivo*. It remains to be determined whether ferritin is a storage vehicle for vanadium as well as for iron in the liver and whether vanadyl–transferrin can transfer vanadium to cells through the transferrin receptor. *In vitro* studies suggest vanadium enters cells as vanadate through phosphate or other anion transport systems. Erythrocytes take up vanadate via the Band 3 protein (24), the normal function of which is to catalyze Cl^-/HCO_3^- exchange. In tunicate blood cells an anion transport system exists that is specific for phosphate and vanadate (51). *In vitro* studies also show that, within erythrocytes (100) and fat cells (46), vanadate is slowly but quantitatively reduced to vanadyl by glutathione. In erythrocytes, vanadyl binds to hemoglobin. In adipocytes, vanadyl complexes with glutathionine.

III. VANADIUM DEFICIENCY AND FUNCTIONS

A. Signs of Deficiency

Vanadium has long been suspected to have a biological function. In 1963, Schroeder *et al.* (155) reviewed the early studies of vanadium essentiality and concluded that, although vanadium behaves like an essential trace metal, final proof of essentiality for mammals was still lacking. Between 1971 and 1974 a number of findings reported by four different research groups led many to conclude that vanadium is an essential nutrient. However, close examination of the findings from the vanadium deprivation studies revealed that they were confusing and inconsistent. For example, in studies of rats, findings were as follows. In 1971, Strasia (169) reported that rats fed less than 100 ng of vanadium per gram of diet exhibited slower growth, higher plasma and bone iron, and higher hematocrits than controls fed 0.5 μg of vanadium per gram of diet. Williams (193), however, could not duplicate those findings in the same laboratory under similar conditions. Schwarz and Milne (158) reported that a vanadium supplement of 0.25–0.50 μg/g of diet gave a positive growth response in suboptimally growing rats fed a purified diet with an unknown vanadium content. On the other hand, Hopkins and Mohr (85) reported that the only effect of vanadium deprivation on rats was impaired reproductive performance (decreased fertility and increased perinatal mortality) that became apparent only in the fourth generation.

Deprivation studies with chicks also gave inconsistent findings. Hopkins and Mohr (86) found that vanadium-deprived chicks exhibited significantly depressed wing- and tail-feather development, depressed plasma cholesterol and triglycerides at age 28 days, and elevated plasma cholesterol at age 49 days. Nielsen and Ollerich (129) reported that vanadium deprivation depressed growth, elevated hematocrits and plasma cholesterol, and adversely affected bone development in chicks.

Recently, Anke and co-workers (7) studied vanadium-deprivation in goats. They found that deprived goats aborted their young at a high rate and produced less milk. Also during lactation, the mortality rate was much higher in vanadium-deprived than vanadium-supplemented mothers and kids.

Nielsen (125,126) became concerned about the inconsistency of the effect of vanadium deprivation on chicks and rats, and attempted to establish a reproducible set of signs of vanadium deprivation in these species. In 16 experiments, in which chicks were fed several diets of different composition, vanadium deprivation adversely affected growth, feathering, hematocrit, plasma cholesterol, bone morphology, and the amounts of lipid, phospholipid, and cholesterol in liver. In several experiments with rats, vanadium deprivation adversely affected perinatal survival, growth, physical appearance, hematocrit, plasma cholesterol, and lipids and phospholipids in liver. Unfortunately, throughout all experiments, no variable was consistently affected by vanadium deprivation.

Recent findings suggested that the reported differences between vanadium-deprived and vanadium-supplemented animals were the consequence of high vanadium supplements (0.5–3.0 μg/g diet) that resulted in pharmacologic-type changes (128). The dose of available vanadium given to supplemented animals probably ranged from 10 to 100 times that normally found in a diet under natural conditions. Vanadium is a relatively toxic element and, therefore, it would not be surprising that a few micrograms per gram of diet would exert a pharmacologic effect *in vivo*, especially if the nutritional status of the organism were suboptimal.

High dietary vanadium can influence the metabolism of a number of nutrients, and vice versa. Hill found that the toxicity of vanadate for the chick was inversely related to the chloride (83) and protein (79) content of the diet. He also found that high dietary chromium (77), ascorbic acid (78), ferrous iron (81), copper (82), and mercury (82) alleviated vanadium toxicity in chicks. Iron deficiency made chicks more susceptible to vanadium toxicity (13).

Nutritional studies by others also indicate that diet composition can affect the response of rats to nontoxic levels of dietary vanadium. For example, Nielsen and co-workers (130,163) found that when dietary cystine was 4.65 mg/g and iron was moderately deficient, hematocrits were lower in rats fed 2–5 ng vanadium per gram diet than in rats fed 1 μg vanadium per gram diet. On the other hand, dietary vanadium did not affect hematocrits in rats fed adequate iron or in rats fed a moderately iron-deficient diet containing 10.15 mg/g cystine. Subsequent studies suggested that vanadium might affect iron metabolism through an effect on copper because dietary vanadium was found to affect copper metabolism (127). Moreover, the direction and extent of the effect were markedly influenced by the sulfur-containing amino acid content of the diet. For example, when the diet contained marginal copper (1 μg/g), 18.34 mg/g methionine, and 1.68 mg/g cystine, plasma ceruloplasmin was lower in vanadium-deprived than vanadium-supplemented rats. Increasing dietary cystine substantially elevated ceruloplasmin in vanadium-deprived, but not vanadium-supplemented, rats fed marginal copper. As a result, plasma ceruloplasmin was higher in vanadium-deprived than vanadium-supplemented rats when they were fed the diet containing marginal copper, 14.28 mg/g cystine, and 5.74 mg/g methionine. Regardless of dietary sulfur amino acids, plasma cholesterol was higher in vanadium-deprived than vanadium-supplemented rats fed deficient or marginal copper. With luxuriant copper, the opposite was found if dietary methionine was low.

Many of the early studies were done with animals fed unbalanced diets, thus resulting in suboptimal performance such as depressed growth. Moreover the diets used in early studies had widely varying contents of protein, sulfur-containing amino acids, ascorbic acid, iron, copper, and perhaps other nutrients that affect vanadium metabolism. Under these conditions, if the relatively high vanadium supplements were acting pharmacologically, the response to dietary vanadium might be expected to be quite variable. In other words, the nature of the difference between vanadium-deprived and vanadium-supplemented animals would depend on the unbalanced condition that is being affected by vanadium, and the extent of response would depend on whether some dietary components were present in concentrations that would blunt the action of vanadium.

The following will show some of the variation in treatments in early studies. Strasia (169), who found that vanadium affected iron metabolism in rats, used a diet that contained only 20 μg/g iron, no ascorbic acid, and 269 g/kg vitamin-free casein. Williams (193), who found that vanadium deprivation did not affect the growing rat, used a diet that contained 35 μg/g iron, 900 mg/kg ascorbic acid, and 175 g/kg vitamin-free casein. Hopkins and Mohr (85,86) used a diet that contained a high ratio of methionine to cystine (11.44 mg and 0.6 mg/g, respectively). The diet also contained a high level of arginine (20 mg/g), an amino acid that can affect methionine metabolism. In the study of Schwarz and Milne (158), rats gained only 0.8–1.5 g/day, much less than the expected 3–5 g/day. The basal diet fed to the rats was not well described, but apparently was deficient in riboflavin (114). In experiments done by Nielsen et al. (125,126,129), iron, methionine, arginine, cystine, copper, and ascorbic acid were all fed at variable and, perhaps, nonoptimal levels.

In summary, there has been no demonstration to date that vanadium deficiency reproducibly and consistently impairs a biological function in any animal. Moreover, findings described below indicate that vanadium is a very active substance from a pharmacologic point of view. Thus, a definitive nutritional experiment establishing vanadium essentiality may be difficult. The discovery of a specific physiological role for vanadium may be necessary to establish it as essential.

B. Vanadium Function

Failure to define the conditions that induce reproducible deficiency in animals has hampered the establishment of vanadium essentiality. Kustin and Macara (98) stated that three types of *in vivo* behavior can be predicted for vanadium from a consideration of its aqueous chemistry. First, as vanadate, it will compete with phosphate at the active sites of phosphate-transport proteins, phosphohydrolases and phosphotransferases. Second, as vanadyl, it will compete with other transition metal ions for binding sites on metalloproteins and for small ligands such as ATP, and third, it will participate in redox reactions within the cell, particularly with relatively small molecules that can reduce vanadate nonenzymatically, such as glutathione. With such a range of possibilities, it is not surprising that, in spite of the uncertainty about essentiality, there have been numerous suggestions as to possible biochemical and physiological functions of vanadium.

1. (Na,K)-ATPase and the Sodium Pump

Although Rifkin (143) first found that vanadium inhibits (Na,K)-ATPase, interest in the inhibition was negligible until Cantley et al. (23) identified vanadate as the impurity in a commercial preparation of horse muscle ATP that interfered with the *in vitro* studies of that enzyme. (Na,K)-ATPase plays a key role in the maintenance of membrane potential and sodium gradient across the plasma membrane. Thus, it was hypothesized that *in vivo* vanadium might function as a physiological regulator of sodium pump activity (99,100). Vanadate inhibits (Na,K)-ATPase by binding to the ATP hydrolysis site, but reduction of vanadate to vandyl reverses that inhibition (22,100). The finding that the predominant form of vanadium in tissue is vanadyl has caused some to dismiss the possibility that vanadium regulates sodium pump activity (98). Others still allow for the possibility that vanadium plays a regulatory role, especially in organs, such as the kidney, which often contains significant amounts of vanadium (142). Such a role would be supported by the finding of an *in vivo* mechanism whereby vanadium in tissue is converted from vanadyl to vanadate.

It has been proposed that a presumably genetic defect in manic depressive psychosis inhibits the production of new pump sites in response to elevated cell sodium. Therefore, manic depressives would be unduly vulnerable to factors affecting the sodium pump, one of which is vanadium (122). Evidence supporting the suggestion that vanadium is a factor involved in the causation of manic-depressive illness include the following. Mean plasma vanadium concentrations were higher in manic-depressive patients than in normal controls (49). Significant negative correlations were found between plasma vanadium concentration and the ratio of Na-K-Mg ATPase to Mg-ATPase in two manic-depressive subjects, but not in normal subjects (49). Drugs that have a therapeutic value in manic-depressive illness are very effective in the reduction of vanadate to vanadyl (121). This suggests that the drugs were effective by optimizing (Na,K)-ATPase activity through the reduction in the body content of the inhibitor vanadate. This explanation is probably too simplistic because the (Na,K)-ATPase most likely involved in manic-depressive illness is that found in brain microsomes; the activity of this (Na,K)-ATPase apparently is effectively inhibited by both vanadate and vanadyl (172).

2. Regulation of Phosphoryl Transfer Enzymes

Vanadium is present in tissues at concentrations that might inhibit phosphoryl transfer enzymes *in vivo*, possibly by forming a transition state analog (trigonal bipyramidal species) of phosphate in the phosphoryl-enzyme intermediate, and such inhibition could reflect a regulation function for vanadium in addition to, or other than, inhibition of (Na,K)-ATPase. Phosphohydrolases and phosphotransferases inhibited by vanadate *in vitro* include glucose-6-phosphatase (165), alkaline phosphatase (159), acid phosphatase (181), 2,3-biphosphoglycerate-dependent phosphoglycerate mutase (27), and phosphoglucomutase (27). The inhibitory action of vanadium on phosphoryl transfer enzymes is quite selective.

It is this selectivity that prompts the belief that vanadium has a regulatory role *in vivo*. For example, micromolar concentrations of vanadate strongly inhibit the membrane-dependent dephosphorylation of histones containing phosphotyrosine but do not inhibit the dephosphorylation of histones containing phosphoserine and phosphothreonine (173). Also, most phosphoryl transfer enzymes other than the ATPases probably are not inhibited by physiological concentrations of vanadium (34). The number of ATPases inhibited by vanadium is quite extensive, and as listed by Kustin and Macara (98), include plasma membrane Ca^{2+}-ATPase, sarcoplasmic reticulum Ca^{2+}-ATPase, fungal H^+-ATPase, gastric mucosa and colon epithelium H^+,K^+-ATPase, myosin ATPase, and dynein-1-ATPase.

3. Adenyl Cyclase

Vanadium stimulates the synthesis of cyclic AMP in a variety of cell membranes through the activation of adenylate cyclase (29,43,156). Vanadium apparently activates the adenylate cyclase complex via the guanine nucleotide regulatory protein called the G unit or G protein (43,95). When GTP is bound to the G unit, adenylate cyclase is activated; when GDP is bound, it is not. The data of Krawietz *et al.* (95) support the suggestion that vanadium acts by promoting an association of an otherwise inactive G unit with the catalytic unit of the adenylate cyclase complex to cause increased activity. Thus, the idea that vanadium as vanadate is bound to enzyme·GDP to form a stable, ternary enzyme·GDP·vanadate complex, so that the G protein behaves as though GTP were bound and adenylate cyclase is activated (98), seems reasonable. The elevation of cyclic AMP may explain why vanadium stimulates cyclic AMP-dependent protein kinase activity in rat liver (29).

However, vanadium having a physiological role in regulating adenylate cyclase activity is questionable because the effective concentration needed to affect the enzyme is relatively high—much higher than that required for ATPase inhibition. Also, like the hypotheses of a regulatory role for vanadium in phosphoryl transfer, adenylate cyclase regulation would need a mechanism whereby vanadyl is converted to vanadate *in vivo*.

4. NADH Oxidation

The suggestion that vanadium has a regulatory role for some enzymes would be enhanced if there were a specific mechanism whereby vanadyl is converted to vanadate *in vivo*. If such a mechanism exists it may involve the oxidation of reduced pyridine nucleotides. Erdmann *et al.* (54) reported that cardiac and erythrocyte cell membranes contain a NADH-vanadate oxidoreductase which reversibly converts vanadate to vanadyl. Similarly, rat liver and sugar beet microsomal membranes were found to contain vanadate-dependent systems, enzymatic in nature, for the oxidation of NADPH and NADH (18,37). On the other hand, Vyskocil *et al.* (185) reported findings that cast some doubt on the presence of a membrane NADH-vanadate-oxidoreductase and suggested that, within the cell, the oxidation of NADH by vanadate proceeds by forming a complex without any additional support from a specific enzyme. This suggestion was supported by findings from other studies (183).

Another possibility for the *in vivo* oxidation of vanadyl to vanadate is cytochrome-*c* oxidase; Crane (38) found that this enzyme isolated from mitochondrial membrane oxidized the vanadyl ion *in vitro*.

Only further research will resolve the question of whether vanadyl is converted to vanadate in sufficient local quantity so that it affects a target enzyme under physiological conditions.

5. Protein Kinase

Vanadium can enhance DNA synthesis (92,106) and can mimic and potentiate the effects of growth factors such as insulin (see below) and epidermal growth factor (26). The action of these growth factors are thought to be partly regulated by protein kinase activity with a specificity for tyrosine residues (93). Brown and Gordon (19) found that vanadate increased pp60^{v-src} kinase activity which has specificity toward tyrosine residues. With the stimulation of activity, they observed a concomitant increase in the phosphotyrosine content, without a change in phosphoserine content, in the kinase enzyme. They speculated that the phosphorylation of tyrosine in the amino terminal fragment of the enzyme represents a regulatory function which may be modulated by vanadate; this speculation indicates a cellular function for vanadium. Gentleman *et al.* (59) found that vanadium enhanced ^{32}P incorporation into phosphotyrosine proteins in Nakano mouse lens epithelial cells. The enhancement was apparently the result of both enhanced protein tyrosine kinase activity and inhibited protein phosphotyrosine phosphatase activity. These findings also suggest that vanadium might have a cellular function which regulates tyrosine phosphorylation.

6. Vanadyl in Enzyme Action and Iron Metabolism

Other than a possible regulatory role, vanadium as the vanadyl cation might have a catalytic function or be a cofactor for some enzmye. As mentioned previously, vanadyl can bind to proteins of biological importance. In all of the proteins studied so far, the ligands that chelate the VO^{2+} are imidazole and carboxylic groups (98). When vanadyl replaces other metals, the metals replaced include Zn^{2+}, Cu^{2+}, and Fe^{3+}, so it is possible that vanadyl has a role similar to these cations. Further support for this role is that VO^{2+} can replace active site Zn^{2+} and retain activity of carboxypeptidase A (47).

A possible vanadium-dependent enzyme is a peroxidase isolated from *As-*

cophyllum nodosum (184). This peroxidase was completely inactivated by dialysis in pH 3.8 citrate–phosphate buffer containing EDTA, and slowly reactivated by vanadium (+5) in suitable buffers.

If vanadyl does have a catalytic function, there is a strong possibility it is associated with iron metabolism. This possibility is supported by the following findings. Early studies of the effect of vanadium on iron metabolism showed (118) that when rats, made anemic by feeding only whole milk, were supplemented with 0.5 mg iron daily, without vanadium, blood regeneration was complete in 6 weeks. However, addition of 0.05 mg vanadium daily with the iron accelerated the recovery, and hemoglobin became normal in 2 to 3 weeks. Also, a blood-regenerative response from the vanadium supplement occurred in rats fed milk diet with an ineffective dose of iron. In unicellular green algae, vanadium overcame a limited iron deficiency (108). In algae, vanadium enhances chlorophyll formation apparently by increasing the synthesis of porphyrin precursor δ-aminolevulinic acid through stimulation of the pyridoxalphosphate-dependent transamination of 4,5-dioxovaleric acid to δ-aminolevulinic acid (107). In nonenzymatic model reactions, vanadyl forms a Schiff base complex with pyridoxalphosphate and 4,5-dioxovaleric acid (109). Wilhelm and Wild (192) reported that vanadium elevates chlorophyll content, dry weight production, and photosynthetic activity in *Chlorella fusca* grown autotrophically. The vanadium effect depended on the iron-complexing status of the nutrient medium. Under conditions of facilitated iron uptake the vanadium effect on photosynthesis disappeared. Vanadium also enhanced chlorophyll formation and iron metabolism in tomato plants, and increased the Hill reaction activity of isolated tomato chloroplasts (8). Finally, Sakurai *et al.* (154) found that vanadyl, but not vanadate, markedly enhanced the affinity of oxygen for hemoglobin and myoglobin *in vitro*.

7. Hormone Interactions

There is a possibility that vanadium might be involved in the expression of some endocrine function. In rats, vanadium metabolism is disturbed by endocrine deficiency induced by hypophysectomy or thyroidectomy–parathyroidectomy (138,139). Hypophysectomized rats retain elevated amounts of injected [48]V in most of their organs. Normal or near normal retention of [48]V in most organs occurred upon growth hormone or thyroxine replacement. One exception was the testis. On the other hand, the gonadotrophins LH and FSH affected [48]V retention by the testis. LH and FSH reduced the [48]V concentration, but because they also increased testicular weight, they increased the total retention of [48]V in the testis. Another study suggesting that vanadium can affect endocrine function involved the cell-free activation of progesterone receptor from the avian oviduct (131). Vanadate inhibited the rate of transformation of receptor–hormone complex to the activated form. Further research is needed to see if these findings have any bearing on the reported impaired reproductive physiology in vanadium-deprived goats (85) and rats (7).

The potent inhibition of (Na,K)-ATPase *in vitro* can be reversed by catecholamines (89). This reversal apparently involves the complexation between the catecholamine and vanadate, followed by oxidation of the catecholamine and reduction of vanadate. In the complexation between vanadate and catecholamines, two adjacent hydroxyl groups on the phenyl ring are required, and the ethylamine side chain may enhance the affinity of the catechol for vanadate. Whether this binding and reduction of vanadate by catecholamines is of any physiologic significance is unclear. It has been suggested that this process, along with the reduction of vanadate by ascorbic acid, protects synaptosomal membrane ATPase from inhibition by vanadate (1). Most likely, the affinity between catecholamines and vanadate was involved in the findings that vanadate enhanced the release of noradrenaline evoked from rabbit isolated perfused pulmonary artery by electrical stimulation, and that noradrenaline inhibited this enhancement (180).

The relationship between vanadium and insulin is discussed in the following section.

8. Glucose Metabolism

Vanadium may affect glucose metabolism by altering or mimicking the action of insulin. Tolman *et al.* (179) found that vanadium stimulated glucose oxidation and transport in adipocytes and glycogen synthesis in liver and diaphragm, and inhibited hepatic gluconeogenesis and intestinal glucose transport. Moreover, they found that glucose oxidation was further stimulated by the addition of vanadate or vanadyl to a submaximally effective concentration of insulin in isolated rat adipocytes. Hori and Aka (88) found that vanadium enhanced the stimulatory effect of insulin on DNA synthesis in cultured tissue. Vanadate by itself caused only a slight stimulation of DNA synthesis, and inhibited Li^+-stimulated DNA synthesis, in mouse mammary gland cultured in a serum-free, chemically defined medium. Subsequent to these *in vitro* studies, it was found that pharmacologic levels of dietary vanadium improved oral glucose tolerance in guinea pigs (52). Vanadate was also found to control high blood glucose and to prevent the decline in cardiac performance in rats made diabetic with streptozotocin (76). Despite low serum insulin, the blood glucose level of diabetic rats fed vanadium as sodium orthovanadate (0.6 to 0.8 mg/ml) in the drinking water was not different from that of nondiabetic controls.

Vanadium seems to act like insulin by altering membrane function for ion transport processes. Tamura *et al.* (174) found that vanadate stimulated the phosphorylation of tryosine residues on the 95,000-dalton subunit of the insulin receptor both in intact adipocytes and in a solubilized insulin receptor fraction. Their findings suggest that vanadium has a beneficial effect on glucose tolerance, or in diabetes, because it enhances the phosphorylation of the insulin receptor by stimulating a kinase reaction in a manner similar to insulin. Other possible mechanisms of action for vanadium, either as vanadate or vanadyl, acting like insulin, have been proposed.

For example, Clausen *et al.* (33) found that vanadate induced an increase in cytoplasmic Ca^{2+} level that activates the glucose transport system in muscle cells, adipocytes, and a number of other cell types. Clausen *et al.* (33) hypothesized that the increased cytoplasmic Ca^{2+} was brought about by an inhibitory effect of vanadate on the Ca^{2+}-sensitive ATPase of the sarcoplasmic or endoplasmic reticulum. On the other hand, Zemkova *et al.* (195) reported findings indicating that vanadyl hyperpolarized muscle fiber plasma membrane, presumably by increasing intracellular K^+ through triggering the release of the potassium associated with cellular proteins. Glucose transport was preceded by hyperpolarization. Other hypotheses state that vanadyl acts as a second messenger playing an important role in controlling protein phosphorylation (162), and that vanadate couples insulin receptors by oxidation of sulfhydryl residues to form a disulfide bridge (98).

Vanadium may also affect glucose metabolism by altering the activity of the multifunctional enzyme glucose-6-phosphatase. Vanadate was found to be a potent inhibitor of this enzyme (165). On the other hand, vanadate up to a concentration of 65 μM has no effect on either rat hepatic glucokinase or hexokinase. This selective inhibition of vanadate on cellular glucose phosphorylation/dephosphorylation mechanisms might be physiologically important in modifying blood glucose levels or maintaining glucose homeostasis.

9. Lipid Metabolism

Based on the effects of vanadium on glucose metabolism, hormone expression, and enzyme action, it is not surprising to find that vanadium affects lipid metabolism. An inhibition of cholesterol synthesis by vanadium was observed *in vivo* in humans and animals. Vanadium was used at pharmacologic levels (41). This inhibition was accompanied by decreased plasma phospholipid and cholesterol levels and by reduced aortic cholesterol concentrations. In older individuals and in patients with hypercholesterolemia or ischemic heart disease no such effect from vanadium was apparent (41,50,155,167), while in older rats the inhibition could be demonstrated *in vitro* but not *in vivo*. The site of the inhibition by vanadium apparently was the microsomal enzyme system referred to as squalene synthetase. In contrast to the lowering of plasma phospholipids and cholesterol findings, Hafez and Kratzer (64) found that vanadium fed as ammonium vanadate to young chicks at 100 ppm increased liver and plasma total lipid and cholesterol levels and plasma cholesterol turnover rate. Also, rabbits intoxicated chronically with 0.15 mg vanadium per kilogram developed atheroscelerosis (132). Because vanadium affects copper metabolism (127) the high levels of vanadium used in the latter experiments might reflect an induction of a relative copper deficiency. Copper deprivation elevates plasma cholesterol (see Chapter 10).

In addition to the *in vitro* studies above, another showed that vanadyl sulfate treatment of a rat liver fraction containing microsomes and soluble enzymes inhibited the incorporation of label from $[1-^{14}C]$acetate into total lipids, cholesterol, free fatty acids, lysophosphatidyl-choline, and diphosphatidylglycerol (6). Vanadium did not inhibit incorporation into triglycerides, phosphatidylcholine, phosphatidylserine, or phosphatidylethanolamine, indicating that vanadium selectively blocked biosyntheses of lipids in rat liver *in vitro*. A decrease in the synthesis of certain phospholipids probably explains the finding of a decreased incorporation of labeled phosphate into liver phospholipids of rats injected with vanadyl sulfate (166). This depressed synthesis apparently was not caused by a lack of fatty acids because, in isolated rat hepatocytes, vanadate stimulated fatty acid synthesis (2).

In summary, depending upon the dose, pharmacological, toxicological, or *in vitro* studies show that vanadium depresses, enhances, or has no effect on the biosynthesis or metabolism of various lipids. Deprivation studies also gave a similar array of findings (see Section III). These confusing findings point strongly to a role for vanadium that influences lipid metabolism. However, the findings also highlight the need for further research to clarify that role. Perhaps this research would help explain statistical studies that indicate low vanadium intakes may be associated with human cardiovascular disease (105).

10. Calcium and Calcified Tissue Metabolism

Radiovanadium injected subcutaneously into mice (168), or intraperitoneally into rats (9,175), was found to be concentrated in the areas of rapid mineralization of bones and tooth dentine. The radiovanadium was incorporated into the tooth structure of rats and retained in the molars up to 90 days after injection (175). The addition of vanadium (and strontium) to specially purified diets has also been found to promote mineralization of the bones and teeth and to reduce the number of carious teeth in rats and guinea pigs (97). Aqueous solutions of V_2O_5 (0.1 mg/kg) given intraperitoneally daily to rabbits with holes drilled in their mandibles accelerated reparative regeneration of bone by stimulating ossification (124). The function of vanadium in developing bone and tooth, if any, is unknown.

The role of vanadium in occurrence of dental caries is also uncertain. Geyer (60) found that vanadium administered as V_2O_5 either orally or parenterally to hamsters fed a cariogenic diet gave a high degree of protection against caries. Kruger (97) similarly reported that vanadium, administered intraperitoneally to rats during the period of tooth development, was effective in reducing the incidence of caries. It is difficult to reconcile those findings with those of other experiments in which the administration of vanadium in the drinking water at varying levels was either unsuccessful in decreasing caries incidence (20,117) or actually increased caries incidence (17,74). For example, Bowen (17) fed water containing 2 ppm vanadium to monkeys for a period of 5 years. Monkeys fed the vanadium had an increased incidence of caries compared with controls with no added vanadium in their water.

Perhaps the highly divergent results in studies on vanadium and dental caries can be explained by the findings of Beighton and McDougall (10). They found that water-borne vanadium altered the microbial composition of tooth-fissure plaque of rats. Vanadium depressed levels of *Streptococcus mutans* and elevated levels of actinomycetes in plaque. Thus, the nature of the effect of vanadium on dental caries probably depends upon whether tooth-fissure plaque contains organisms affected by the element, and if it does, whether the compositional change caused by vanadium is beneficial or detrimental in the development of dental caries.

More recent studies also indicate a relationship between vanadium and calcium metabolism. Vanadate infusion in the dog induced cardiovascular changes including increases in mean arterial pressure, total peripheral resistance, pulmonary arterial pressure and cardiac output, and suppressed plasma renin activity (171). Verapamil, a calcium channel blocker, attenuated the effects of vanadate; this suggests that the cardiovascular and humoral alterations were partially mediated by changes in intracellular calcium. An increase in intracellular calcium, caused by decreased calcium transport by both the plasma membrane and endoplasmic reticulum, was suggested as a mechanism whereby vanadate and vanadyl elicit their insulin-mimetic effects (48). Other studies showed that vanadate stimulated calcium-calmodulin-dependent protein phosphorylation (96), and impaired renal function and induced renal hypertrophy in rats (57); these responses to vanadate were prevented by calcium administration. These findings support the suggested relationship between calcium and vanadium—that vanadate is reduced to vanadyl, which displaces calcium from calmodulin or competes with it for binding sites on the calcium-regulatory protein (39).

11. Other Physiological Effects of Vanadium

Numerous physiological changes induced by vanadium in *in vitro* tissue preparations have been described. These changes include the following. Vanadate inhibits renin secretion from rat kidney slices (32) and excites *Limulus* photoreceptors (56). In a number of cardiac muscle preparations—isolated ventricular and atrial muscles of rat, dog, cat, and rabbit—vanadate induces an increase in contractile force which is called the positive inotropic effect; in left atrial muscle from cat and guinea pig it produces a negative inotropic effect (16,63,157). In other words, the response of cardiac muscle is dependent upon the species and the area of the heart. An explanation for this variable response could be that it is dependent on whether the stimulatory (insulin-like) or inhibitory action of vanadate predominates in the various heart cell types (4,189). Intravenous injection of vanadate causes a hypertensive vascular response in dogs (84) and increased excretion of water, sodium, potassium, calcium, phosphate, bicarbonate, and chloride by the kidney in rats (45). Most likely, all the *in vitro,* and many *in vivo* pharmacological, effects of vanadate can be explained by changes in cellular sodium, potassium, and calcium. In whole organisms, whether these cellular changes occur and are of physiological significance at normal, low-level vanadium exposure remains undetermined.

Another possibly pharmacological effect of vanadium that may reflect physiologic action is its apparent anticarcinogenic action. Thompson *et al.* (176) found that the induction of murine mammary carcinogenesis in rats by 1-methyl-1-nitrosourea was blocked by feeding 25 μg of vanadyl per gram diet during the post initiation stages of the neoplastic process. The vanadyl sulfate treatment reduced both tumor incidence and the average number of tumors per rat and prolonged the median cancer-free time without inhibiting overall growth of the animals.

IV. VANADIUM REQUIREMENTS AND SOURCES

Failure to define the conditions that induce reproducible deficiency in animals has prevented the establishment of vanadium essentiality. Moreover, if there is a requirement for vanadium, it probably is very small. However, food contains very little vanadium, the metabolism of which apparently is affected profoundly by other dietary components; thus it is possible that vanadium intake is not always optimal. That possibility demonstrates the need to clarify the issue of whether vanadium has a biological function, and if it does, the conditions that produce vanadium deficiency, and the dietary components and their mechanisms that affect vanadium metabolism.

Recent studies based on fairly reliable analytical methodology show that the vanadium content of most foods is low. Myron *et al.* (119) used atomic absorption to show that beverages, fat and oils, and fresh fruits and vegetables contained the least vanadium, ranging from <1 to 5 ng/g. Whole grains, seafood, meats, and dairy products were generally within a range of 5 to 30 ng/g. Prepared foods ranged from 11 to 93 ng/g, while dill seed and black pepper contained 431 and 987 ng/g, respectively. Byrne and Kosta (21) obtained similar results using neutron activation analyses. They found most fats and oils, fruits, and vegetables contained <1 ng/g. Cereals, liver, and fish tended to have intermediate levels of about 5 to 40 ng/g. Only a few items such as spinach, parsley, mushrooms, and oyster contained relatively high amounts of vanadium. Shellfish apparently are a rich source of vanadium because several types were found to contain >100 ng/g vanadium on a fresh basis (90), or >400 ng/g vanadium on a dry basis (14,21). Byrne and Kosta (21) stated that their analyses indicated that the daily dietary intake of vanadium is in the order of a few tens of micrograms and may vary widely. Myron *et al.* (120) found that nine institutional diets supplied 12.4 to 30.1 μg of vanadium daily, and intake averaged 20 μg. Unreliable analytical procedures probably were behind the apparently erroneous report that the daily vanadium intake for humans is 1 to 4 mg (155).

Information on the vanadium content of animal feeds is limited. Mitchell (110,111) reported vanadium concentrations ranging from <30 to 160 ng/g for red clover, and from <30 to 110 ng/g for rye grass. More than half the pasture samples contained 30–70 ng/g vanadium on the dry basis. A level of 60 ng/g was obtained for oats (grain) and 120 ng/g for oat straw. Berg (12) reported 50 ng/g vanadium for corn, 80 ng/g for soybean meal, and 2700 ng/g for herring fish meal. Vanadium concentrations of 28 to 55 ng/g were found in the seeds of wheat, barley, oats, and peas from plants grown in nutrient solution, and these were substantially increased when vanadate was added to the nutrient solution (188). From the standpoint of toxicity, the vanadium content of phosphate rock used for animal feeding is of concern. According to Ammerman *et al.* (5), phosphorus supplements containing as high as 0.25% vanadium pentoxide (1400 ppm vanadium) have been reported.

V. VANADIUM TOXICITY

Vanadium can be a relatively toxic element to some animal species. Some years ago Franke and Moxon (58) found the relative toxicity of five different elements, fed at the level of 25 μg/g diet to rats, to be in the ascending order of arsenic, molybdenum, tellurium, vanadium, and selenium. Dietary vanadium concentrations of 25 μg/g were toxic to rats, and at the vanadium intake of 50 μg/g diet, the animals exhibited diarrhea and mortality. The toxicity of ingested vanadium (as vanadate) is similar in chicks. In chicks, calcium vanadate fed at the level of 30 μg vanadium per gram of practical ration depressed growth, and at the level of 200 μg/g, caused high mortality (146). It soon became apparent that dietary composition affected vanadium toxicity because Nelson *et al.* (123) reported that chicks tolerated vanadium intakes of 20–35 μg/g diet and that only higher amounts induced growth depression. Subsequently, a total of 13 μg of vanadium as ammonium metavanadate per gram of diet was shown to depress growth in chicks (11). Hathcock *et al.* (72) found depressed growth and high mortality in chicks fed 25 μg of vanadium as either ammonium metavanadate or vanadyl sulfate per gram of diet containing 60% dried skim milk and 32.5% glucose. Berg (12) found that 20 μg vanadium fed as sodium metavanadate per gram diet to be much more toxic, as judged by growth depression, to chicks fed sucrose–soybean meal, sucrose–herring fish meal, or corn–soybean diets than to chicks fed a corn–herring fish meal diet. He further found the toxicity of the sucrose-containing diets could be gradually reduced by stepwise replacement of the sucrose with graded levels of corn, and that the inclusion of 5% cottonseed meal or dehydrated grass or 0.5% ascorbic acid, in the sucrose–herring fish meal diet markedly reduced vanadium toxicity. In contrast, adverse effects of van- adium on interior egg quality, as measured by Haugh units, were exacerbated by feeding chickens ascorbic acid, and unaffected by feeding cottonseed meal (144).

Other substances shown to reduce dietary vanadium toxicity, judged by growth depression, are EDTA (72,80), chromium (65,72), protein (79), ferrous iron (81), chloride (83), and perhaps aluminum hydroxide (191). The mecha- nisms involved in the ameliorating action of these substances on vanadium toxicity are uncertain. Perhaps various substances were partly effective in alter- ing absorption or availability of vanadium by affecting its form in the gastroin- testinal tract. Roshchin *et al.* (148) found that the oral LD_{50} for mice was much higher with V_2O_5 and VCl_3 than with V_2O_3. Mountain *et al.* (116) found dietary vanadium pentoxide was far less toxic to rats and rabbits than sodium vanadate. Parker and Sharma (136) found that the tissue residue of vanadium was always higher in animals fed 50 μg of vanadium as sodium orthovanadate per gram of diet than in those given the same dose as vanadyl sulfate. These findings suggest a difference between forms and oxidation states of vanadium in respect to sol- ubility in biological fluids, absorption from the gut, or metabolism and elimina- tion by the animal.

The toxicity of vanadium not only depends upon diet composition, but also upon age and animal species. Rising *et al.* (144) found that egg mass output and egg numbers were reduced in old hens fed 15–20 μg vanadium per gram diet; these criteria were unaffected in young hens fed 30 μg vanadium per gram diet. In another study, the only sign of toxicity seen in laying chickens fed 300 μg vanadium per gram diet was a severe depression in egg production (66). This level of vanadium would result in high mortality in growing chicks (65,146). Amounts of vanadium up to 500 μg/g diet were tolerated by mature rats if the mineral was gradually increased to that level, but younger animals apparently were more vulnerable (169). Animal species more tolerant to vanadium than the chick and rat apparently include *Coturnix* (quail), sheep, and perhaps human. Hafez and Kratzer (66) found that the addition of up to 300 μg vanadium per gram of ration fed to laying *Coturnix* did not cause any significant change in egg production, egg weight, or egg yolk cholesterol content. Furthermore the same level of vanadium fed to growing male *Coturnix* chicks did not cause any significant growth depression or mortality. Lambs fed vanadium (as ammonium metavanadate) supplements of 0, 10, 100, 200, 400, or 800 μg/g of corn–soybean meal diet showed clinical toxicity signs after 84 days with only the two highest supplements (68).

Dietary supplements up to and including 200 μg/g caused only small initial negative effects on feed intake and on increased vanadium content in organs. Dimond *et al.* (50) gave ammonium vanadyl tartrate orally to six humans for 6 to 10 weeks in amounts ranging from 4.5 to 18 mg vanadium/day with no toxic effects other than some cramps and diarrhea at the larger dose levels. Schroeder *et al.* (155) fed patients 4.5 mg vanadium/day as the oxytartarovanadate for 16 months with no signs of intolerance becoming apparent. However, 4.5 mg vanadium/day represents a dietary level of only 11 ppm if it is assumed that the subjects consumed about 400 g of dry matter daily. That level, as described above, is not particularly toxic to other animals either.

The signs of vanadium toxicity vary among species and with the dose. As described above, some of the more consistent signs of toxicity include depressed growth, elevated tissue vanadium, diarrhea, depressed food intake, and death. Other reported manifestations of high dietary vanadium include the following: In rats, the cystine content of hair is decreased (115), levels of coenzyme A (104) and coenzyme Q (3) are reduced, brain protein synthesis is inhibited (112), monoamine oxidase activity is stimulated (141), and oxidative phosphorylation is uncoupled (73). In laying chickens, there is a reduction in egg albumen quality, egg production, egg weight, and shell quality (55,135,160). Of interest, vanadium apparently expresses its toxicity in laying chickens in several ways because the protective effects of different dietary changes and additives differentially affected the loss of albumen quality, egg production, body weight, and egg weight (135). Finally, high dietary vanadium might make animals more susceptible to bacterial infections (80) and enhance abnormalities caused by deficiency of other nutrients (190).

REFERENCES

1. Ádám-Vizi, V., Váradi, G., and Simon, P. (1981). *J. Neurochem.* **36,** 1616–1620.
2. Agius, L., and Vaartjes, W. J. (1982). *Biochem. J.* **202,** 791–794.
3. Aiyar, A. S., and Sreenivasan, A. (1961). *Proc. Soc. Exp. Biol. Med.* **107,** 914–916.
4. Akera, T., Temma, K., and Takeda, K. (1983). *Fed. Proc., Fed. Am. Soc. Exp. Biol.* **42,** 2984–2988.
5. Ammerman, C. B., Miller, S. M., Fick, K. R., and Hansard, S. L., II. (1977). *J. Anim. Sci.* **44,** 485–508.

6. Anekwe, G. E. (1981). *IRCS Med. Sci.: Libr. Compend.* **9**, 211–212.
7. Anke, M., Groppel, B., Kronemann, H., and Führer, E. (1983). *In* "4th Spurenelement-Symposium 1983" (M. Anke, W. Bauman, H. Bräunlich, and Chr. Brückner, eds.), pp. 135–141. Friedrich-Schiller-Univ., Jena.
8. Basiouny, F. M. (1984). *J. Plant Nutr.* **7**, 1059–1073.
9. Bawden, J. W., Deaton, T. G., and Chavis, M. (1980). *J. Dent. Res.* **59**, 1643–1648.
10. Beighton, D., and McDougall, W. A. (1981). *Arch. Oral Biol.* **26**, 419–425.
11. Berg, L. R. (1963). *Poult. Sci.* **42**, 766–769.
12. Berg, L. R. (1966). *Poult. Sci.* **45**, 1346–1352; Berg, L. R., and Lawrence, W. W. (1971). *Poult. Sci.* **50**, 1399–1404.
13. Blalock, T. L., and Hill, C. H. (1983). *Fed. Proc., Fed. Am. Soc. Exp. Biol.* **42**, 1135.
14. Blotcky, A. J., Falcone. C., Medina, V. A., Rack, E. P., and Hobson, D. W. (1979). *Anal. Chem.* **51**, 178–182.
15. Bogden, J. D., Higashino, H., Lavenhar, M. A., Bauman, J. W., Jr., Kemp, F. W., and Aviv, A. (1982). *J. Nutr.* **112**, 2279–2285.
16. Borchard, U., Fox, A. A. L., Greeff, K., and Schlieper, P. (1979). *Nature (London)* **279**, 339–341.
17. Bowen, W. H. (1972). *J. Ir. Dent. Assoc.* **18**, 83.
18. Briskin, D. P., Thornley, W. R., and Poole, R. J. (1985). *Arch. Biochem. Biophys.* **236**, 228–237.
19. Brown, D. J., and Gordon, J. A. (1984). *J. Biol. Chem.* **259**, 9580–9586.
20. Büttner, W. (1963). *J. Dent. Res.* **42**, 453–460.
21. Byrne, A. R., and Kosta, L. (1978). *Sci. Total Environ.* **10**, 17–30.
22. Cantley, L. C., Jr., Ferguson, J. H., and Kustin, K. (1978). *J. Am. Chem. Soc.* **100**, 5210–5212.
23. Cantley, L. C., Jr., Josephson, L., Warner, R., Yanagisawa, M., Lechene, C., and Guidotti, G. (1977). *J. Biol. Chem.* **252**, 7421–7423.
24. Cantley, L. C., Jr., Resh, M., and Guidotti, G. (1978). *Nature (London)* **272**, 552–554.
25. Carmichael, A., and Vincent, J. S. (1979). *FEBS Lett.* **105**, 349–352.
26. Carpenter, G. (1981). *Biochem. Biophys. Res. Commun.* **102**, 1115–1121.
27. Carreras, J., Climent, F., Bartrons, R., and Pons, G. (1982). *Biochim. Biophys. Acta* **705**, 238–242.
28. Casey, J. D., and Chasteen, N. D. (1980). *J. Inorg. Biochem.* **13**, 111–126, 127–136.
29. Catalán, R. E., Martínez, A. M., and Aragonés, M. D. (1980). *Biochem. Biophys. Res. Commun.* **96**, 672–677.
30. Chasteen, N. D., and Theil, E. C. (1982). *J. Biol. Chem.* **257**, 7672–7677.
31. Chasteen, N. D., White, L. K., and Campbell, R. F. (1977). *Biochemistry* **16**, 363–368.
32. Churchill, P. C., and Churchill, M. C. (1980). *J. Pharmacol. Exp. Ther.* **213**, 144–149.
33. Clausen, T., Andersen, T. L., Stürup-Johansen, M., and Petkova, O. (1981). *Biochim. Biophys. Acta* **646**, 261–267.
34. Climent, F., Bartrons, R., Pons, G., and Carreras, J. (1981). *Biochem. Biophys. Res. Commun.* **101**, 570–576.
35. Cornelis, R., Mees, L., Hoste, J., Ryckebusch, J., Versieck, J., and Barbier, F. (1979). *In* "Nuclear Activation Techniques in the Life Science 1978," pp. 165–177; IAEA-SM-227/25. International Atomic Energy Agency, Vienna.
36. Cornelis, R., Versieck, J., Mees, L., Hoste, J., and Barbier, F. (1981). *Biol. Trace Elem. Res.* **3**, 257–263.
37. Coulombe, R. A., Jr., Briskin, D. P., Keller, R. J., and Sharma, R. P. (1985). *Fed. Proc., Fed. Am. Soc. Exp. Biol.* **44**, 1349.
38. Crane, F. L. (1975). *Biochem. Biophys. Res. Commun.* **63**, 355–361.

39. Cruz-Soto, M., Benabe, J. E., López-Novoa, J. M., and Martínez-Maldonado, M. (1984). *Am. J. Physiol.* **247**, F650–F655.
40. Curran, G. L., Azarnoff, D. L., and Bolinger, R. E. (1959). *J. Clin. Invest.* **38**, 1251–1261.
41. Curran, G. L., and Burch, R. E. (1968). *In* "Trace Substances in Environmental Health–1" (D. D. Hemphill, ed.), pp. 96–104. Univ. of Missouri Press, Columbia.
42. Curzon, M. E. J., and Crocker, D. C. (1978). *Arch. Oral Biol.* **23**, 647–653.
43. Da Luz Duque, A., Dieryck, P., Lambert, M., Svoboda, M., and Christophe, J. (1980). *Arch. Int. Physiol. Biochim.* **88**, B127–B128.
44. Danskin, G. P. (1978). *Can. J. Zool.* **56**, 547–551.
45. Day, H., Middendorf, D., Lukert, B., Heinz, A., and Grantham, J. (1980). *J. Lab. Clin. Med.* **96**, 382–395.
46. Degani, H., Gochin, M., Karlish, S. J. D., and Shechter, Y. (1981). *Biochemistry* **20**, 5795–5799.
47. DeKoch, D. J., West, D. J., Cannon, J. C., and Chasteen, N. D. (1974). *Biochemistry* **13**, 4347–4354.
48. Delfert, D. M., and McDonald, J. M. (1985). *Fed. Proc., Fed. Am. Soc. Exp. Biol.* **44**, 1429.
49. Dick, D. A., Naylor, G. J., and Dick, E. G. (1982). *Pschol. Med.* **12**, 533–537.
50. Dimond, E. G., Caravaca, J., and Benchimol, A. (1963). *Am. J. Clin. Nutr.* **12**, 49–53.
51. Dingley, A. L., Kustin, K., Macara, I. G., and McLeod, G. C. (1981). *Biochim. Biophys. Acta* **649**, 493–502.
52. Dowdy, R. P., and Topping, S. (1981). *Fed. Proc., Fed. Am. Soc. Exp. Biol.* **40**, 886.
53. Edel, J., Pietra, R., Sabbioni, E., Marafante, E., Springer, A., and Ubertalli, L. (1984). *Chemosphere* **13**, 87–93.
54. Erdmann, E., Werdan, K., Krawietz, W., Lebuhn, M., and Christl, S. (1980). *Basic Res. Cardiol.* **75**, 460–465.
55. Eyal, A., and Moran, E. T., Jr. (1984). *Poult. Sci.* **63**, 1378–1385.
56. Fein, A., and Corson, D. W. (1981). *Science* **212**, 555–557.
57. Fernández-Repollet, E., and Martínez-Maldonado, M. (1985). *Fed. Proc., Fed. Am. Soc. Exp. Biol.* **44**, 524.
58. Franke, K. W., and Moxon, A. L. (1937). *J. Pharmacol. Exp. Ther.* **61**, 89–102.
59. Gentleman, S., Reid, T. W., Martensen, T. M., and Chader, G. J. (1985). *Fed. Proc., Fed. Am. Soc. Exp. Biol.* **44**, 1038.
60. Geyer, C. F. (1953). *J. Dent. Res.* **32**, 590–595.
61. Gibson, R. S., and Dewolfe, M. S. (1980). *Nutr. Rep. Int.* **21**, 341–349.
62. Gilbert, K., Kustin, K., and McLeod, G. C. (1977). *J. Cell. Physiol.* **93**, 309–312.
63. Hackbarth, I., Schmitz, W., Scholz, H., Erdmann, E., Krawietz, W., and Philipp, G. (1978). *Nature (London)* **275**, 67.
64. Hafez, Y., and Kratzer, F. H. (1976). *J. Nutr.* **106**, 249–257.
65. Hafez, Y. S. M., and Kratzer, F. H. (1976). *Poult. Sci.* **55**, 918–922.
66. Hafez, Y. S. M., and Kratzer, F. H. (1976). *Poult. Sci.* **55**, 923–926.
67. Hamilton, E. I., Minski, M. J., and Cleary, J. J. (1972/1973). *Sci. Total Environ.* **1**, 341–374.
68. Hansard, S. L., II. Ammerman, C. B., Fick, K. R., and Miller, S. M. (1978). *J. Anim. Sci.* **46**, 1091–1095.
69. Hansard, S. L., II, Ammerman, C. B., and Henry, P. R. (1982). *J. Anim. Sci.* **55**, 350–356.
70. Harris, D. C., and Gelb, M. H. (1980). *Biochim. Biophys. Acta* **623**, 1–9.
71. Harris, W. R., Friedman, S. B., and Silberman, D. (1984). *J. Inorg. Biochem.* **20**, 157–169.
72. Hathcock, J. N., Hill, C. H., and Matrone, G. (1964). *J. Nutr.* **82**, 106–110.
73. Hathcock, J. N., Hill, C. H., and Tove, S. B. (1966). *Can. J. Biochem.* **44**, 983–988.

74. Hein, J. W., and Wisotsky, J. (1955). *J. Dent. Res.* **34,** 756.
75. Henze, M. (1911/1913/1914). *Hoppe-Seyler's Z. Physiol. Chem.* **72,** 494–501; **79,** 215–228; **86,** 340–344.
76. Heyliger, C. E., Tahiliani, A. G., and McNeill, J. H. (1985). *Science* **227,** 1474–1477.
77. Hill, C. H. (1979). *In* "Chromium in Nutrition and Metabolism" (D. Shapcott and J. Hubert, eds.), pp. 229–240. Elsevier, Amsterdam.
78. Hill, C. H. (1979). *J. Nutr.* **109,** 84–90.
79. Hill, C. H. (1979). *J. Nutr.* **109,** 501–507.
80. Hill, C. H. (1980). *J. Nutr.* **110,** 433–436.
81. Hill, C. H. (1981). *Fed. Proc., Fed. Am. Soc. Exp. Biol.* **40,** 715.
82. Hill, C. H. (1985). *Fed. Proc., Fed. Am. Soc. Exp. Biol.* **44,** 751.
83. Hill, C. H. (1985). *Nutr. Res., Suppl. 1,* 555–559.
84. Hom, G. J., Chelly, J. E., and Jandhyala, B. S. (1982). *Proc. Soc. Exp. Biol. Med.* **169,** 401–405.
85. Hopkins, L. L., Jr., and Mohr, H. E. (1971). *In* "Newer Trace Elements in Nutrition" (W. Mertz and W. E. Cornatzer, eds.), pp. 195–213. Dekker, New York.
86. Hopkins, L. L., Jr., and Mohr, H. E. (1974). *Fed. Proc., Fed. Am. Soc. Exp. Biol.* **33,** 1773–1775.
87. Hopkins, L. L., Jr., and Tilton, B. E. (1966). *Am. J. Physiol.* **211,** 169–172.
88. Hori, C., and Oka, T. (1980). *Biochim. Biophys. Acta* **610,** 235–240.
89. Hudgins, P. M., and Bond, G. H. (1979). *Res. Commun. Chem. Pathol. Pharmacol.* **23,** 313–326.
90. Ikebe, K., and Tanaka, R. (1979). *Bull. Environ. Contam. Toxicol.* **21,** 526–532.
91. Jervis, R. E., Evans, G. J., and Hewitt, P. J. (1985). *Nutr. Res., Suppl. 1,* 627–633.
92. Jones, T. R., and Reid, T. W. (1984). *J. Cell. Physiol.* **121,** 199–205.
93. Kasuga, M., Fujita-Yamaguchi, Y., Blithe, D. L., and Kahn, C. R. (1983). *Proc. Natl. Acad. Sci. U.S.A.* **80,** 2137–2141.
94. Kosta, L., Byrne, A. R., and Dermelj, M. (1983). *Sci. Total Environ.* **29,** 261–268.
95. Krawietz, W., Downs, R. W., Jr., Spiegel, A. M., and Aurbach, G. D. (1982). *Biochem. Pharmacol.* **31,** 843–848.
96. Křivánek, J. (1984). *Neurochem. Res.* **9,** 1627–1640.
97. Kruger, B. J. (1958). *Austr. Dent. J.* **3,** 298–302.
98. Kustin, K., and Macara, I. G. (1982). *Comments Inorg. Chem.* **2,** 1–22.
99. Macara, I. G. (1980). *Trends Biochem. Sci.* **5,** 92–94.
100. Macara, I. G., Kustin, K., and Cantley, L. C., Jr. (1980). *Biochim. Biophys. Acta* **629,** 95–106.
101. Macara, I. G., and McLeod, G. C. (1979). *Biochem. J.* **181,** 457–465.
102. Macara, I. G., McLeod, G. C., and Kustin, K. (1979). *Comp. Biochem. Physiol.* **62A,** 821–826.
103. Marumo, F., Tsukamoto, Y., Iwanami, S., Kishimoto, T., and Yamagami, S. (1984). *Nepron* **38,** 267–272.
104. Mascitelli-Coriandoli, E., and Citterio, C. (1959). *Nature (London)* **183,** 1527–1528.
105. Masironi, R. (1969). *Bull. W.H.O.* **40,** 305–312.
106. McKeehan, W. L., McKeehan, K. A., Hammond, S. L., and Ham, R. R. (1977). *In Vetio* **3,** 399–416.
107. Meisch, H.-U., and Bauer, J. (1978). *Arch. Microbiol.* **117,** 49–52.
108. Meisch, H.-U., Becker, L. J. M., and Schwab, D. (1980). *Protoplasma* **103,** 273–280.
109. Meisch, H.-U., Hoffmann, H., and Reinle, W. (1978). *Z. Naturforsch. C. Biosci.* **33C,** 623–628.
110. Mitchell, R. L. (1957). *Research (London)* **10,** 357–362.

111. Mitchell, R. L. (1957). *In* "Trace Analysis, Symposium Trace Analysis, N.Y. Acad. Med. 1955" (J. H. Yoe and H. J. Koch, eds.), pp. 398–412. Wiley, New York.

112. Montero, M. R., Guerri, C., Ribelles, M., and Grisolia, S. (1981). *Physiol. Chem. Phys.* **13**, 281–287.

113. Moo, S. P., and Pillay, K. K. S. (1983). *J. Radioanal. Chem.* **77**, 141–147.

114. Moran, J. K., and Schwarz, K. (1978). *Fed. Proc., Fed. Am. Soc. Exp. Biol.* **37**, 671.

115. Mountain, J. T., Delker, L. L., and Stokinger, H. E. (1953). *Arch. Ind. Hyg. Occup. Med.* **8**, 406–411.

116. Mountain, J. T., Wagner, W. D., and Stokinger, H. E. (1959). *Fed. Proc., Fed. Am. Soc. Exp. Biol.* **18**, 425.

117. Muhler, J. C. (1957). *J. Dent. Res.* **36**, 787–794.

118. Myers, V. C., and Beard, H. H. (1931). *J. Biol. Chem.* **94**, 89–110.

119. Myron, D. R., Givand, S. H., and Nielsen, F. H. (1977). *Agric. Food Chem.* **25**, 297–300.

120. Myron, D. R., Zimmerman, T. J., Shuler, T. R., Klevay, L. M., Lee, D. E., and Nielsen, F. H. (1978). *Am. J. Clin. Nutr.* **31**, 527–531.

121. Naylor, G. G., and Smith, A. H. W. (1982). *Lancet* **I**, 395–396.

122. Naylor, G. J. (1980). *Neuropharmacology* **19**, 1233–1234.

123. Nelson, T. S., Gillis, M. B., and Peeler, H. T. (1962). *Poult. Sci.* **41**, 519–522.

124. Nemsadze, O. D. (1977). *Stomatologiya (Moscow)* **56**, 1–5.

125. Nielsen, F. H. (1980). *In* "Advances in Nutritional Research–3" (H. H. Draper, ed.), pp. 157–172. Plenum, New York.

126. Nielsen, F. H. (1980). *In* "Inorganic Chemistry in Biology and Medicine, ACS Symp. Series 140" (A. E. Martell, ed.), pp. 23–42. Am. Chem. Soc., Washington, D.C.

127. Nielsen, F. H. (1984). *Proc. N.D. Acad. Sci.* **38**, 57.

128. Nielsen, F. H. (1985). *J. Nutr.* **115**, 1239–1247.

129. Nielsen, F. H., and Ollerich, D. A. (1973). *Fed. Proc., Fed. Am. Soc. Exp. Biol.* **32**, 929.

130. Nielsen, F. H., Urich, K., and Uthus, E. O. (1984). *Biol. Trace Elem. Res.* **6**, 117–132.

131. Nishigori, H., Alker, J., and Toft, D. (1980). *Arch. Biochem. Biophys.* **203**, 600–604.

132. Novakova, S., Nikolchev, G., Angelieva, R., Dinoeva, S., and Mautner, G. (1981). *Gig. Sanit.* 58–59.

133. Oberg, S. G., Parker, R. D. R., and Sharma, R. P. (1978). *Toxicology* **11**, 315–323.

134. Ordzhonikidze, E. K., Roschin, A. V., Shalganova, I. V., Bogomazov, M. Ya., and Kazimov, M. A. (1977). *Gig. Tr. Prof. Zabol.* **6**, 29–34. (In Russian.)

135. Ousterhaut, L. E., and Berg, L. R. (1981). *Poult. Sci.* **60**, 1152–1159.

136. Parker, R. D. R., and Sharma, R. P. (1978). *J. Environ. Pathol. Toxicol.* **2**, 235–245.

137. Parker, R. D. R., Sharma, R. P., and Oberg, S. G. (1980). *Arch. Environ. Toxicol.* **9**, 393–403.

138. Peabody, R. A., Wallach, S., Verch, R. L., and Lifschitz, M. L. (1977). *In* "Trace Substances in Environmental Health–11" (D. D. Hemphill, ed.), pp. 297–304. Univ. of Missouri Press, Columbia.

139. Peabody, R. A., Wallach, S., Verch, R. L., and Lifschitz, M. (1980). *Proc. Soc. Exp. Biol. Med.* **165**, 349–353.

140. Pépin, G., Bouley, G., and Boudene, C. (1977). *C.R. Hebd. Seances Acad. Sci., Ser. D* **285**, 451–454.

141. Perry, H. M., Jr., Tietlebaum, S., and Schwartz, P. L. (1955). *Fed. Proc., Fed. Am. Soc. Exp. Biol.* **14**, 113–114.

142. Phillips, T. D., Nechay, B. R., and Heidelbaugh, N. D. (1983). *Fed. Proc., Fed. Am. Soc. Exp. Biol.* **42**, 2969–2973.

143. Rifkin, R. J. (1965). *Proc. Soc. Exp. Biol. Med.* **120**, 802–804.

144. Rising, R., Maiorino, P. M., and Reid, B. L. (1985). *Fed. Proc., Fed. Am. Soc. Exp. Biol.* **44**, 761.
145. Rizk, S. L., and Sky-Peck, H. H. (1984). *Cancer Res.* **44**, 5390–5394.
146. Romoser, G. L., Dudley, W. A., Machlin, L. J., and Loveless, L. (1961). *Poult. Sci.* **40**, 1171–1173.
147. Roshchin, A. V., Ordzhonikidze, E. K., and Shalganova, I. V. (1980). *J. Hyg. Epidemiol. Microbiol. Immunol.* **24**, 377–383.
148. Roshchin, I. V., Il'nitskaya, A. V., Lutsenko, L. A., and Zhidkova, L. V. (1965). *Fed. Proc., Fed. Am. Soc. Exp. Biol.* **24**, T611–T613.
149. Ryan, D. E., Holzbecker, J., and Stuart, D. C. (1978). *Clin. Chem.* **24**, 1996.
150. Sabbioni, E., and Marafante, E. (1978). *Bioinorg. Chem.* **9**, 389–407.
151. Sabbioni, E., and Marafante, E. (1981). *J. Toxicol. Environ. Health* **8**, 419–429.
152. Sabbioni, E., and Marafante, E. (1981). *In* "Trace Element Metabolism in Man and Animals (TEMA-4)" (J. McC. Howell, J. M. Gawthorne, and C. L. White, eds.), pp. 629–631. Austr. Acad. Sci., Canberra.
153. Sabbioni, E., Marafante, E., Amantini, L., Ubertalli, L., and Birattari, C. (1978). *Bioinorg. Chem.* **8**, 503–515.
154. Sakurai, H., Goda, T., and Shimomura, S. (1982). *Biochem. Biophys. Res. Commun.* **107**, 1349–1354.
155. Schroeder, H. A., Balassa, J. J., and Tipton, I. H. (1963). *J. Chronic Dis.* **16**, 1047–1071.
156. Schwabe, U., Puchstein, C., Hannemann, H., and Söchtig, E. (1979). *Nature (London)* **277**, 143–145.
157. Schwartz, A., Adams, R. J., Grupp, I., Grupp, G., Holroyde, M. J., Millard, R. W., Solaro, R. J., and Wallick, E. T. (1980). *Basic Res. Cardiol.* **75**, 444–451.
158. Schwarz, K., and Milne, D. B. (1971). *Science* **174**, 426–428.
159. Seargeant, L. E., and Stinson, R. A. (1979). *Biochem. J.* **181**, 247–250.
160. Sell, J. L., Arthur, J. A., and Williams, I. L. (1982). *Poult. Sci.* **61**, 2112–2116.
161. Shand, C. A., Aggett, P. J., and Ure, A. M. (1985). *In* "Trace Elements in Man and Animals (TEMA-5)" (C. F. Mills, I. Bremner, and J. K. Chesters, eds.), pp. 642–645. Commonwealth Agric. Bureaux, Farnham Royal, U.K.
162. Shechter, Y., and Karlish, S. J. D. (1980). *Nature (London)* **284**, 556–558.
163. Shuler, T. R., and Nielsen, F. H. (1983). *Proc. N.D. Acad. Sci.* **37**, 88.
164. Simonoff, M., Llabador, Y., Peers, A. M., and Simonoff, G. N. (1984). *Clin. Chem.* **30**, 1700–1703.
165. Singh, J., Nordlie, R. C., and Jorgenson, R. A. (1981). *Biochim. Biophys. Acta* **678**, 477–482.
166. Snyder, F., and Cornatzer, W. E. (1958). *Nature (London)* **182**, 462.
167. Somerville, J., and Davies, B. (1962). *Am. Heart J.* **64**, 54–56.
168. Söremark, R., and Üllberg, S. (1962). *In* "Use of Radioisotopes in Animal Biology and the Medical Sciences–2, Proc. Conf., Mexico City, 1961" (M. Fried, ed.), pp. 103–114. Academic Press, New York.
169. Strasia, C. A. (1971). Ph.D. thesis, University Microfilms, Ann Arbor, Michigan.
170. Stroop, S. D., Helinek, G., and Greene, H. L. (1982). *Clin. Chem.* **28**, 79–82.
171. Sundet, W. D., Wang, B. C., Hakumäki, M. O. K., and Goetz, K. L. (1984). *Proc. Soc. Exp. Biol. Med.* **175**, 185–190.
172. Svoboda, P., Teisinger, J., Pilar, J., and Vyskocil, F. (1984). *Biochem. Pharmacol.* **33**, 2485–2491.
173. Swarup, G., Cohen, S., and Garbers, D. L. (1982). *Biochem. Biophys. Res. Commun.* **107**, 1104–1109.

174. Tamura, S., Brown, T. A., Whipple, J. H., Fujita-Yamaguchi, Y., Dubler, R. E., Cheng, K., and Larner, J. (1984). *J. Biol. Chem.* **259**, 6650–6658.
175. Thomassen, P. R., and Leicester, H. M. (1964). *J. Dent. Res.* **43**, 346–352.
176. Thompson, H. J., Chasteen, N. D., and Meeker, L. D. (1984). *Carcinogenesis* **5**, 849–851.
177. Thürauf, J., Schaller, K. H., Syga, G., and Weltle, D. (1978). *Wiss. Umwelt* 84–88.
178. Thürauf, J., Syga, G., and Schaller, K.-H. (1978). *In* "Biological Monitoring" pp. 165–171. Gentner, Stuttgart, FRG.
179. Tolman, E. L., Barris, E., Burns, M., Pansini, A., and Partridge, R. (1979). *Life Sci.* **25**, 1159–1164.
180. Török, T. L., Rubányi, G., Vizi, E. S., and Magyar, K. (1982). *Eur. J. Pharmacol.* **84**, 93–97.
181. VanEtten, R. L., Waymack, P. P., and Rehkop, D. M. (1974). *J. Am. Chem. Soc.* **96**, 6782–6785.
182. Vanoeteren, C., Cornelis, R., Versieck, J., Hoste, J., and De Roose, J. (1982). *Radioanal. Chem.* **70**, 219–238.
183. Vijaya, S., and Ramasarma, T. (1984). *J. Inorg. Biochem.* **20**, 247–254.
184. Vilter, H. (1984). *Phytochemistry* **23**, 1387–1390.
185. Vyskocil, F., Teisinger, J., and Dlouhá, H. (1980/1981). *Nature (London)* **286**, 516–517; **294**, 288.
186. Ward, N. I., Bryce-Smith, D., Minski, M., and Matthews, W. B. (1985). *Biol. Trace Elem. Res.* **7**, 153–159.
187. Webb, D. A. (1956). *Pubbl. Staz. Zool. Napoli* **28**, 273–288.
188. Welch, R. M., and Cary, E. E. (1975). *Agric. Food Chem.* **23**, 479–482.
189. Werdan, K., Bauriedel, G., Fischer, B., Krawietz, W., Erdmann, E., Schmitz, W., and Scholz, H. (1982). *Biochim. Biophys. Acta* **687**, 79–93.
190. Whanger, P. D., and Weswig, P. H. (1978). *Nutr. Rep. Int.* **18**, 421–428.
191. Wiegmann, T. B., Day, H. D., and Patak, R. V. (1982). *J. Toxicol. Environ. Health* **10**, 233–245.
192. Wilhelm, C., and Wild, A. (1984). *J. Plant Physiol.* **115**, 115–124.
193. Williams, D. L. (1973). Ph.D. thesis, University Microfilms, Ann Arbor, Michigan.
194. Yukawa, M., Suzuki-Yasumoto, M., Amano, K., and Terai, M. (1980). *Arch. Environ. Health* **35**, 36–44.
195. Zemková, H., Teisinger, J., and Vyskočil, F. (1982). *Biochim. Biophys. Acta* **720**, 405–410.

10

Copper

GEORGE K. DAVIS[1]

University of Florida
Gainesville, Florida

WALTER MERTZ

U.S. Department of Agriculture
Agricultural Research Service
Beltsville Human Nutrition Research Center
Beltsville, Maryland

I. COPPER IN ANIMAL TISSUES AND FLUIDS

A. Early History

Archeological records suggest that copper (Cu) use by humans antedated history by thousands of years. Historical records show that copper and copper compounds had been used medicinally at least as early as 400 BC (299). Lehman (280) reported that many compounds of copper were tried on many diseases, with little success, during the nineteenth century. Prasad (363) has noted that the presence of copper in plants and animals was recognized well over 150 years ago. Publications detailing the biological aspects of copper in nutrition, biochemistry, enzymology, toxicology, and environmental sciences have appeared in the literature from countries around the world. Extensive reviews covering many different aspects of copper in biological systems have been published (299,336,339,347,348,363,398,413,424).

---

[1]Present address: 2903 S. W. Second Court, Gainesville, Florida 32601.

Abderhalden (1) had recognized that the anemia produced in rats on a milk diet could not be prevented or cured by the addition of inorganic iron to the diet. However, it was not until the work of McHargue (308,309), who reported the association of copper with substances containing the fat-soluble vitamins, and the classical work of Hart *et al.* (189), following on the observations of Abderhalden, that it was widely recognized that rats on a milk diet required copper in addition to iron to cure the anemia which developed. This early work which firmly established the need for copper in the formation of hemoglobin, in addition to the iron requirement, has been reviewed by Elvehjem (123), Schultze (402), and Mason (299). As investigations into the role of copper in animal metabolism have continued, it has become evident that copper functions as a critical element in other metabolic pathways that are of great importance in the practical husbandry of animals that may be subjects to diet deficient or toxic with respect to copper. Soon after the work of McHargue and the Wisconsin workers, naturally occurring copper deficiency was noted in the cattle of Florida (333) and in the sheep and cattle of the Netherlands (416,417). Recognition of the importance of copper in the metabolism of plants and animals led to the identification of copper as a critical element responsible for problems that went by many local names in many parts of the world. Owen (348) has compiled a list of the different names which have been given to copper deficiency in animals and plants. Some examples include: bush sickness, enzootic ataxia, gray spec disease, pining, salt sick, swayback, peat scours, and falling disease.

Copper deficiency anemia occurs in all species of animals. Bone disorders occur in many species such as rabbits, chicks, pigs, dogs, and young horses (21,22). The nerve disorder of lambs, included in the term enzootic neonatal ataxia which also occurs in goats, pigs, and guinea pigs, has been identified in many locations. The achromotrichia which is characteristic of copper deficiency in rats, rabbits, guinea pigs, cats, dogs, cattle, and sheep does not appear in swine. (264,265). The failure of keratinization which has been seen in hair, fur, and wool of copper-deficient rats, rabbits, guinea pigs, dogs, cattle, and sheep, in mutant mice, and in Menkes' disease of humans has been widely recognized but was first pointed out by Bennetts (30) as it occurred in the wool of sheep.

Copper deficiency has been identified as the cause of reproductive failure of rats, guinea pigs, poultry, cattle, and sheep.

Cardiovascular disorders associated with copper deficiency have been noted as the cause of sudden death in cattle in Western Australia and in Florida (31, 33,34,104). Cardiovascular disorders have also been noted in chicks, in other poultry, in swine, in guinea pigs, and in rats. Failure of elastin and collagen biosynthesis resulting from a reduction in lysyl oxidase activity and reduced strength of cardiac and arterial tissues caused by copper deficiency has resulted in ruptures as the immediate cause of death (204,299,338). Mason (299) in his extensive review of copper metabolism and the requirements of humans has detailed the recognition of the role of copper-containing enzymes in human and animal metabolism.

Copper as a toxic agent has a long history which has been reviewed by Mason, Owen, and Bremner (51,299,348) and in the publications of the National Research Council (329,330). It was the report by Ferguson *et al.* (136) on the action of molybdenum that increased the attention given to the importance of trace element interaction in copper metabolism. In the reports of Dick (110,111), the role of molybdenum and sulfate was emphasized as a control of copper toxicity and as having a close relationship to copper metabolism in animals. Many subsequent studies have demonstrated the need for a balance between copper, molybdenum, sulfate, zinc, and iron if absolute or functional deficiency or toxicity is to be avoided.

B. Copper in Foods and Feeds

The copper content of food and of feeds is subject to many different influences. As a result, the copper content of a given material will reflect its origin, the conditions under which it was produced, handled, and prepared for use. Consequently, analytical values published for a specific food or feed must be used with caution when compared with similar products of slightly different backgrounds. Plant materials show variations in copper content that may be influenced by the copper content of the soils on which they are produced as well as the type and pH of the soil, by the copper compounds present in the soil, and by other metallic and organic residues. In addition to the species and variety of the plant, the quantity and availability of the copper in the soil and environment may influence the copper content of the edible portions.

Seasonal variations and weather can also influence the amount of copper that may be taken up by the plant, introducing still an additional variation in copper content of the human or animal food source. In irrigated areas, the copper content of the water source may be a factor in determining the copper content of the growing plant. Since copper compounds are commonly used in fungicides and fertilizers, these, too, can make a difference in the copper content of the ultimate food or feed product. Foods that are processed from harvesting to the final preparation for consumption may have the copper content increased or decreased by the processes.

Additional factors may influence the copper content of animal tissues that are used for food. In addition to the copper consumed by the animals in diet and water, species variation, age, and tissue accumulation are factors in the ultimate content of copper of the food presented for consumption. Animals raised in areas subject to copper contamination may have increased copper in their tissues. Some special environmental situations can also have a marked influence on the copper content of animal products. Seafood products will reflect the geographic variation in copper content, probably due to the amount of copper in the water in which they have been located. This is especially true of such products as oysters and clams.

Recognizing the possibility of extreme variations in copper content of foods and feeds, it is interesting to note that there have been a number of reports published indicating that human consumption of copper in many countries may be well below the "range of estimated safe and adequate intakes" of 2–3 mg of copper per day for adult humans (56,240,261,263,290,399,487).

Animals may be subject to high copper intakes, whether given to promote growth, as with swine, or restricted to feeds with relatively high copper content, such as sheep grazing on forages containing high levels of copper. Animals under such conditions can exhibit especially high copper levels in organ meats such as liver and kidney.

Analysis of the copper content of foods has been extensive, and reference can be made to tables prepared by Pennington and Calloway (356), by Koivistoinen (266), Wissenschaftliche Tabellen Geigy, Ciba Geigy (489), National Research Council (329,332), Fonnesbeck and Lloyd (148), and Latin American Tables of Feed Composition (276).

To give some suggestions of the ranges within which foods and feeds can vary, Tables I and II are presented with the observation that in any given situation the copper values of a specific product may deviate greatly from the values presented in different tables, reflecting the influence of the factors such as those referred to above.

C.   Copper in Tissues

The copper content of the healthy adult human body is on the order of 50–120 mg (413). Newborn and very young animals are normally richer in copper per unit of body weight than adults of the same species (72,73,98,152,151, 163,192,407,483). The newborn levels, with some species variation, are largely maintained throughout the suckling period, followed by a steady fall during growth to the time when adult values are reached. Owen (347) has examined the relationship of the copper concentration of various tissues and notes that tissues other than eye, skin, tongue, bone, and liver are found to be comparable, averaging about 2 μg/g in the fat-free tissue. The distribution of total-body copper varies with species, age, and copper status of the animal (347,348,364). Cartwright and Wintrobe (72,73) studied the distribution of copper in the tissues of five normal humans. They found a total of 23 mg of copper in the liver, heart, kidneys, spleen, and brain. They reported 8 mg in the liver and 8 mg in the brain. It seems from later work that the proportion of the total copper in the brain may be somewhat smaller. Miyata et al. (322) have given the distribution of copper in the

Table I

Some Food Copper Values[a]

Food	Copper content (mg/kg)
Milk	
Human colostrum	0.57 (0.24–0.76)
Human	0.2–0.76
Cow	0.1–0.88
Nonfat dry	0.7
Liver	
Bovine	157
Lamb	56
Kidney	
Bovine	2.1–4.3
Muscle meat	
Bovine	0.1–1.8
Pork	0.1–9.1
Cereals	
Corn (maize) products	0.6–16.6
Wheat products	3.3–36.0
Rice products	0.6–3.1
Wheat bread	2.9
Whole-wheat bread	3.4
Vegetables	
Potatoes	0.48–16.0
Potato chips	2.2–3.6
Sweet potato	0.15
Carrot	0.37–0.62
Broccoli	0.68–0.87
Peas	1.9–2.4
Lettuce	0.1–2.9
Tomatoes	0.1–3.4
Sweet corn	0.02–0.15
Cabbage	0.1–1.7
Seafoods	
Oysters	0.3–16.0
Tuna	0.1–1.2
Salmon	0.5–0.8
Shrimp	2.0–2.9
Rainbow trout	0.1–3.3
Flounder	0.1–2.5
Catfish	1.4–2.5
Fruits	
Apples	0.1–2.3
Bananas	0.7–3.0
Grapes	0.74–1.5
Peaches	1.1–1.4

(*continued*)

Table 1 (*Continued*)

Food	Copper content (mg/kg)
Pineapple	0.86–0.96
Prunes	3.7–5.0
Raisins	2.7–4.1
Oranges	0.8–0.9
Grapefruit	0.3–0.6
Ice cream	
Chocolate	0.3–3.4
Vanilla	0.1–0.9
Strawberry	0.1–1.4
Nuts	
Peanuts	2.7–9.6
Pecans	9.7–13.6
Walnuts	2.0–13.9
Sunflower seeds	14.3–19.0

[a] Values adapted from Pennington and Calloway (356); Koivistoinen (266); Alvarez (11); and Ciba-Geigy (489).

Table II
Copper in Some Animal Feeds[a]

Feed	Copper content (mg/kg, dry matter basis)
Alfalfa (*Medicago sativa*), dehydrated	3.2–9.0
White clover (*Trifolium repens*) hay	12.7–18.4
Soybean (*Glycine max*) hay	8.0–11.5
Wheat (*Triticum aestivum*) straw	3.0–3.7
Oat (*Avena sativa*) straw	7.2–11.0
Corn (*Zea mays*) stover	1.7–5.0
Cottonseed meal (*Gossypium* spp.), 41% protein	20.2–22.0
Soybean meal	9.0–22.3
Bermuda grass (*Cynodon dactylon*)	4.9–9.2
Kentucky bluegrass (*Poa pratensis*)	1.6–12.7
Lespedeza (*Lespedeza striata*)	6.0–10.2
Orchard grass (*Dactylis glomerata*)	2.3–9.8
Ryegrass (*Lolium* spp.)	5.6–15.0
Wheat grain	6.3–13.9
Oat grain	6.3–11.0
Corn grain	3.3–22.3

[a] Adapted from Fonnesbeck and Lloyd (148); Koivistoinen (266); National Research Council *Copper* (329); Latin American Tables of Feed Composition (276).

various parts of the brain, with values ranging from 3.3 to 38.8 μg/g of dry matter. The data reported by Cunningham (95) for a range of species also indicate that a higher proportion of the total-body copper exists in the liver than in the brain.

In ruminants which have a high capacity for hepatic storage, the proportion in the liver can be very high. Dick (112) found the total-body copper of two adult sheep with very high liver copper concentrations to be distributed as follows: liver, 72–79%; muscles, 8–12%; skin and wool, 9%; and skeleton, 2%. It is further apparent from study of the neonate and adult bovines that by far the largest portion of the total-body copper occurs in the liver of that species (40,342,483). Tissues containing relatively high concentrations of copper would include the liver, brain, heart, and hair (55,61,469,490,491). Tissues with intermediate copper concentrations would include pancreas, skin, muscles, spleen, and bones. Tissues low in copper concentrations include pituitary, thyroid, thymus, prostate gland, ovary, and testes (21,55,152,303,404). The liver, blood, spleen, lungs, brain, and bones are particularly responsive to variations in dietary copper intakes, whereas endocrine glands, muscles, and heart are much less affected (199,284,388). Some typical copper concentrations in adult human tissues can be cited: (all in μg/g wet weight) liver, 14.7 ± 0.9; brain, 5.6 ± 0.2; lung, 2.2 ± 0.2; kidneys, 2.1 ± 0.4; ovary, 1.2 ± 0.3; testis, 0.8 ± 0.2; lymph nodes, 0.8 ± 0.06; muscle, 0.7 ± 0.02 (199,322,329).

To some extent the tissues of newborn animals reflect the copper status of the dams. Some copper values for newborn calves are (in μg/g dry matter): liver, 490; heart, 14.9; kidney, 13.8; brain, 9.4; lung, 5.8; muscle, 4.6; bone, 1.2 (39,163). Hennig *et al.* (199) found that 18 μg/g of copper in the brain was normal for cows. Brain copper levels <9 μg/g were characteristic of cows and sheep showing copper deficiency symptoms with decreased copper concentrations in liver, kidneys, rib, blood serum, and hair. Factors impinging on the copper status of liver, blood, and hair copper concentrations are given further consideration later in this chapter. Copper concentrations in the pigmented portions of the eye are quite high (42,43,87,303). Quite large differences exist between species, but the eye tissue can be placed in a similar descending order of copper concentrations as follows: iris, choroid, vitreous humor, aqueous humor, retina (minus pigmented epithelium), optic nerve, cornea, sclera, and lens. Levels as high as 105 μg/g for the iris and 88 μg/g (dry basis) for the choroid of the eyes of freshwater trout and 50 to 13.5 μg/g for these tissues in sheep's eyes have been reported. Copper is associated particularly with the melanins and is largely bound to protein. The role of copper in these sites has not yet been explained.

A range of from 0.07 to 208 μg/g copper was obtained by Losee (287) in a study of dental enamel, with a mean of 6.8 ± 4.0 μg/g. These values for human enamel suggest the wide range that has been found. The inner layers of dental enamel had a copper concentration of 11.3 μg/g and the outer layers 9.5 μg/g

(179,287). When animals are subjected to high dietary intakes of copper—as, for example, when swine are given high intakes to promote growth, or sheep are grazed on forages containing high levels of copper—the copper content of organ meats such as liver and kidney can be excessive.

D. Copper in the Liver

Many factors influence copper concentration in the liver. These include species, age of the animal, chemical composition of the diet, and various disease conditions. Sex does not appear to influence the liver copper concentration, although the Australian salmon (*Arripis trutta*) female carries a higher level than the male (25). Individual variation is high in all species. Alvarez has described the U.S. Bureau of Standards reference sample of bovine liver which contains 157 μg/g copper on a dry matter basis (11).

1. Effect of Species

Although diets high in copper can cause marked increases in the liver copper concentration, most nonruminant species contain between 10 and 50 μg/g of copper on a dry matter basis. A high proportion of these contain between 15 and 30 μg/g (25,287,373).

Species as unrelated and with characteristic environments such as humans, rats, rabbits, cats, dogs, foxes, pigs, kangaroos, whales, snakes, crocodiles, domestic fowls, turkeys, sharks, and sea herring regularly exhibit such values. Sheep and cattle characteristically have higher values as do ducks, frogs, and certain fish (42,43). The normal range for these species is between 100 and 400 μg/g on a dry matter basis. The differences are not all related to dietary intake and may be related to differences in inherited patterns of copper excretion. Domestic fowls, turkeys, and ducks consume similar diets but the domestic fowls carry much lower liver copper concentrations than do ducks. These differences are maintained even when dietary intakes are increased by two and five times (26).

Sheep and cattle appear to have a superior capacity to bind copper in the liver as well as a lower capacity for excretion (79). Blood copper values do not rise in these species in the same way as they do in rats on increased copper intake, although very high intakes can override the capacity for copper fixation in the liver by ruminants, with consequent increases in blood values (79,319). In a study by Charmley and Symonds, pigs excreted 80–90% of infused copper as compared to only 4% excreted by steers (79).

2. Effect of Age

The newborn of most species, including humans, have liver copper concentrations which are higher than the values found in adults. In sheep, newborn liver copper values are lower than in adults and copper concentration continues to increase throughout life. The liver copper concentrations of newborn calves are comparable to those found in adult cattle. The extent of intrauterine copper storage and the time of maximal liver copper concentration varies with the species. In the rat, guinea pig, rabbit, dog, and human the peak occurs at or very shortly after birth (108,407). In the pig the peak occurs slightly earlier in embryonic life, while in the bovine liver copper concentrations do not rise greatly during gestation. The decline in rat liver concentration as the animal matures is accompanied by changes in the copper distribution among the subcellular fractions. At birth >80% of the total is present in the nuclear and mitochondrial fractions, while in the adult rat the supernatant contains about one-half of the total copper content (126,129,173,236,275,334,393).

In adult animals a major portion of the liver copper is in the cytosol, where the copper is bound to the enzyme superoxide dismutase (SOD) and a low molecular weight protein similar to metallothionein (126). A protein subfraction containing more than >4% of copper (neonatal hepatic mitochondrial cuprein) has been isolated from the mitochondrial fraction of newborn bovine liver (360) and a similar substance, though with a slightly lower copper content, from a normal newborn human liver (361). Significant amounts of this compound are not found in adult human or bovine liver except in patients with Wilson's disease (hepatolenticular degeneration) (408). A copper thionein with a copper-binding constant four times greater than that of normal subjects has been demonstrated in the livers of such subjects (129). Neonatal hepatic mitochondrial cuprein has a very high cysteine content (129) and an amino acid composition similar to that of metallothionein. This led Porter to suggest a relationship between the two compounds (359). Rupp and Weser have produced evidence that the copper thionein is possibly the low molecular weight form of neonatal hepatic mitochondrialcuprein (389).

The forms and distribution of the copper in the livers of both rats and ruminants have been further studied by Bremner with animals of varying copper and zinc status. The presence of several low molecular weight copper- and zinc-binding fractions was demonstrated in rat and ruminant liver cytosol including the metallothioneinlike protein. These varied in the amounts and metal contents with varying copper and zinc status of the animal. They are probably involved in the cellular metal detoxification mechanisms (50).

3. Effect of Diet

Liver copper concentrations have been used as an indicator of the copper status of animals; they tend to indicate a deficiency when very low and toxicity when very high. Rats and pigs suffering from milk anemia have subnormal liver copper levels (404,488). Copper-deficient rats have low liver copper (116,319). Chickens and dogs also show low liver copper when subjected to copper-deficient diets

(21,22,387,388). Low liver copper in sheep and cattle is characteristic of these animals when grazing copper-deficient pastures (31,329,481,490). The level of the copper in the diet which will maintain a normal plasma copper level will vary considerably depending on the amount of molybdenum and sulfate, zinc, and iron that are present in the diet. While Claypool *et al.* (83) estimated that 40 μg/g (dry weight) would be adequate to maintain normal copper in the plasma in cattle, such an estimate should not be relied on unless dietary levels of molybdenum and sulfate are also considered.

Bennetts and Beck (31) found the liver copper of five ataxic lambs from copper-deficient ewes to range from 4 to 8 μg/g on a dry weight basis. This compares with values of 120–350 μg/g for normal lambs from ewes receiving adequate copper. Supplementary copper administered during pregnancy raises fetal copper to normal, when the mother is copper deficient, but has little effect on liver copper storage of the newborn when the mother is already receiving adequate copper (163). Ruminants and nonruminant animals differ in their response to high levels of dietary copper intakes. Owen (348) has reviewed the literature on both copper deficiency and toxicity as it occurs in many different species. Dick (112,113) studied liver copper storage in sheep ingesting graded increments of copper from 3.6 to 33.6 mg/day for 177 days. The liver copper levels increased steadily from 562 μg/g (dry matter) at the lowest intake to 2340 μg/g at the highest copper intake. The proportion of the copper stored in the liver was uniform at intakes up to 18.6 mg/day, and the increase in liver storage was linear. Copper supplementation of the normal diet of rats has no comparable effect on liver storage until high intakes are reached. At this threshold, which has been reported as 1 mg/day or 200 μg/g in the diet, liver copper levels increase rapidly, apparently due to overloading of the excretory mechanism, but do not reach those found in sheep by Dick (112,113). Even at high levels rats appear to adjust and the liver levels tend to decrease with continued consumption of high levels of copper in the diet (44,51,194,319).

In studies of copper loading in the rat, nuclei and mitochondria were found to hold most of the excess copper, with the microsomes and cytoplasm accumulating much less (173,275). Under different conditions a linear increase in the amount of copper in each intracellular fraction of the total-liver copper was observed by Milne and Weswig (319). The relative amount in the mitochondrial soluble fraction remained essentially constant, while the relative amount in the microsomes and in the debris increased from the depleted to the copper-supplemented groups. Liver copper levels are affected by other dietary factors, which influence copper retention in the body through their effect on copper absorption, excretion, or both. The storage of copper in the livers of sheep and cattle can be reduced significantly by an increase in dietary molybdenum and sulfate (112–114,441,490,491).

Copper retention in the liver is also influenced by the level of zinc, cadmium, calcium carbonate, and iron in the diet (226,301). A marked zinc–copper antagonism is evident both when copper is limiting and in copper toxicosis (443). A significant inverse correlation between hepatic iron and copper concentrations has been demonstrated in rats (427). Rats consuming an iron-deficient diet accumulated a high level of copper in the liver in 7–8 weeks. Rats fed a copper-deficient diet accumulated an excessive amount of iron. Ascorbic acid has been shown to have an adverse effect on copper deficiency, increasing the severity of this condition and reducing copper absorption (137,146,147,192,249,265,273, 320,342,363,443,492). In addition to trace element interactions with copper, there is evidence that protein levels in the diet as well as the level of sulfur amino acids may influence the liver copper concentration in animals (411,412). In humans there is the possibility that various drugs may influence the copper absorption and utilization (478).

4. Effect of Disease

Abnormally high liver copper levels are characteristic of a number of diseases in humans. These include primary biliary cirrhosis, Mediterranean anemia, hemochromatosis, cirrhosis and yellow atrophy of the liver, tuberculosis, carcinoma, severe chronic diseases accompanied by anemia, and Wilson's disease (71,354). In rats, depletion of copper has resulted from acute and chronic infections due to increased hepatic synthesis and secretion of ceruloplasmin by the liver (28). An effect has also been noted on ceruloplasmin activity in chickens when injected with the endotoxin of *Escherichia coli* (65).

In humans, infections do not appear to have the same effect on liver synthesis and secretion of ceruloplasmin that occur in rats (418).

E. Copper in Blood

Extensive reviews of copper in the blood of humans and animals may be found in Owen (347), in Mason (299), and in Sigel (413). In plasma (or serum) most of the copper is bound to ceruloplasmin. This has been estimated variously as from 90 to 93% (72,73,198), but Smith noted that in aborigine children only 85% of the copper was present in ceruloplasmin (422). The remaining copper in the serum is less firmly bound in large part to albumin and to a still lesser extent to amino acids (334). In normal mammalian blood approximately one-half of the copper is present in the erythrocytes. Much smaller amounts are present in the white blood cells and platelets. In the erythrocytes the copper is present both as a labile pool much like that in the plasma and in a more firmly bound form, superoxide dismutase (SOD). The labile pool is copper complexed with amino acids representing about 40% of the copper in the erythrocytes. The 60% of the copper bound in the enzyme SOD provides a good measure of copper status in humans (29), in cattle (300), in mice (366), and in rats (353). A very small part of the copper is bound to a pink copper-binding protein (379).

In seeking a better way to measure the copper status in humans and in animals, the determination of SOD in erythrocytes has proved to be somewhat better than the determination of total-blood copper (300). Numerous investigators have examined the differences in the distribution of copper in blood in individuals of the same species, in different species, and those occurring under different conditions such as pregnancy. These have been discussed in reviews by Owen (347), Mason (299), Scheinberg (398), Prasad (363), and Smith (422).

Williams (485) has pointed out that there are differences between white and black men in the serum concentration of copper (96.5 ± 11.5 versus 118.1 ± 32.7 µg/dl, respectively, $p < 0.01$), but not between white and black women (118.5 versus 118.4 µg/dl). Hatano et al. (192), examining the copper levels in erythrocytes and plasma of healthy Japanese children and adults, did not note the differences that Smith reported as existing between Caucasian and aboriginal children (422). With such a high proportion of the plasma copper existing as ceruloplasmin, it is not surprising that highly significant correlations have been demonstrated between ceruloplasmin levels in serum and whole-blood copper (71,304,453). Chickens and turkeys have a much smaller proportion of the normally low plasma copper present as ceruloplasmin (430,480).

Ceruloplasmin is a true oxidase (ferroxidase) involved in iron utilization and in promoting iron saturation of transferrin in the plasma (345). It does not play a significant role in copper transport, because the amount of ceruloplasmin exchanged daily is small compared with the amount of copper absorbed from the intestinal tract. The albumin-bound "direct-reacting" copper of the plasma constitutes true transport copper (176). In addition to ceruloplasmin and albumin-bound copper, a small proportion of the copper exists in combination with amino acids (334) and as copper enzymes, the concentration of which varies with the copper status of the animal. A high molybdenum intake results in a portion of the plasma copper existing in the form of a copper–molybdenum complex (249). In mammalian blood the white blood cells contain about one-fourth the concentration of copper that exists in erythrocytes, but the ratio of erythrocytes to leukocytes and platelets is such that the copper content of leukocytes and platelets is generally considered to be insignificant as a proportion of the total-blood copper (71). The marked effect of molybdenum and sulfate on the distribution of copper, especially in ruminants (227) is referred to in the chapter on molybdenum (see Mills and Davis, Chapter 13 in this volume).

1. Normal Copper Levels

The normal range of copper in the blood of healthy animals can be given as from 0.5 to 1.5 µg/ml, with a high proportion of the values lying between 0.8

and 1.2 μg/ml (26). A normal range of approximately one-half the values for mammalian blood has been found for poultry, fish, frogs, and marsupials (26).

Within any given species, more individual variation will occur than appears in diurnal or day-to-day variations. Cartwright (71) found this to be the case in humans, and this has also been demonstrated in sheep, where breed differences can also be very significant (193,483,492). Finnish Landrace sheep have markedly lower plasma copper concentrations than merino sheep. These differences become important as they appear to influence the ability of different breeds to utilize pastures that may be high or low in copper content (122,160,453,483, 490,491). An association between hemoglobin type and copper concentration has also been demonstrated (483). Corpuscular copper levels as distinct from plasma copper levels are less affected by hemoglobin type differences. Plasma copper does not increase after meals or decrease during fasting, and there appears to be no cyclic pattern in humans (71). In most species whole-blood copper levels are similar in males and females, but plasma copper is slightly higher in white human females than in white human males (485).

Serum copper levels are greatly elevated in women taking oral contraceptives and during pregnancy. The effect of oral contraceptives on serum copper has been studied on a number of occasions. Halstead *et al.* (182) found that copper values of 1.18 ± 0.2 μg/ml in controls compared with values of 3.0 ± 0.7 μg/ml in women who were taking oral contraceptives. In another study, Hambidge and Droegemuller (184) noted that women using contraceptives had plasma copper of 221 ± 13.8 μg/dl compared with 107.4 ± 5.1 for controls. This is apparently an estrogen effect, since the administration of estradiol in humans (234) significantly increases plasma copper levels. The use of copper intrauterine devices is not associated with an increase in copper concentration in the blood in women, although the copper levels in endometrium and cervical mucus increased (200,448), and decreases in endometrial protein and RNA content have been reported (200). An estrogen effect is further apparent, since the plasma copper level is increased at sexual maturity and the onset of lay in pullets (207). Plasma copper in 19-week-old mature pullets ranged from 11 to 15 μg/dl but rose to 25–31 μg/dl during the following 6 weeks as the birds began laying. There was no further increase after 25 weeks of age.

2. Effect of Pregnancy and Neonatal Growth

King and Wright (240) found that pregnant women retained more copper either through increased absorption or decreased excretion than did nonpregnant women. With this increased retention serum copper tends to double during pregnancy (132,182,184,335,469). In a study of pregnant women, Nielsen observed that the copper in serum increased from the nonpregnant status of 1.2 μg/ml to 2.7 μg/ml at 3 months of pregnancy. Vir *et al.* (469) found copper levels of 1.71–1.79 μg/ml during pregnancy, as compared with 0.98 μg/ml in nonpregnant women. In uncomplicated pregnancies, Hambidge and Droegemuller (184) noted plasma copper values of 162 ± 6.1 μg/dl at 16 weeks and 192.1 ± 5.4 μg/ml at 38 weeks of gestation. The levels returned to nonpregnant values within a few weeks of parturition. Kiilholma et al. (239) noted that during pregnancy serum copper concentrations increased from 0.91 to 1.48, 1.91, and 2.20 μg/ml during the first, second, and third trimesters, respectively. Five weeks postpartum the serum copper was 1.09 μg/ml.

Henkin et al. (198) found that infant serum copper and ceruloplasmin levels rise rapidly during the first weeks of life, fall below adult levels at 2 months of age, rise to the adult range at 3 months of age, and increase above adult range by 8 months and continue at that level throughout the remainder of infancy. These changes were related to nondiffusible, that is, ceruloplasmin copper. In preterm infants given parenteral nutrition at 20 μg/kg/day copper, there was an increase in serum copper concentration starting the first week. Concentrations were significantly higher at days 14 and 21 than the corresponding values in infants given nutrition enterally. In part the rise in copper was due to ceruloplasmin in plasma from blood transfusions. The serum copper and ceruloplasmin levels of the full-term human infant are about one-third those of the adult range (132,198,397).

A different pattern of blood copper distribution from that described for women and babies is apparent in ewes and lambs. Gooneratne and Christensen (163) found that fetal liver copper in bovines increased from 201.9 ± 87.7 at 30–59 days to 390.6 ± 99.8 mg/kg (dry matter) at term. The maternal liver copper decreased from 50.7 ± 68.9 to 16.2 ± 19.3 during this period. Studies with grazing ewes showed a similar decline in pregnancy in the levels of whole-blood and plasma copper and in ceruloplasmin levels (64), while Howell et al. (218) found that the blood copper and ceruloplasmin levels rose in the ewe at parturition, reaching the highest levels 1 week after lambing. In lambs blood copper and ceruloplasmin levels were low at birth but were within the normal adult range after 1 week of age. Wiener et al. (483) noted that the breed of dam and breed of lamb both were involved in the copper status of the lamb when born. Breed differences in sheep have been particularly marked with regard to liver storage and copper utilization. This difference in breed was also shown by Parry et al. (351), who found that during pregnancy, Dorset, Clun, and Finn breeds differed markedly with regard to copper homeostasis. Blood copper and ceruloplasmin levels are low at birth in lambs and quickly elevate to the normal adult range.

In the bovine, blood and plasma copper levels are lower and erythrocyte levels are higher in newborn calves than in their mothers (163). Ceruloplasmin is absent from the serum of baby pigs at birth but is synthesized during the first 3 days of life, and the piglet shows no significant difference from that synthesized by the adult (78).

3. Effect of Diet

Subnormal levels of dietary copper are reflected in subnormal blood copper concentrations in all species that have been studied. In sheep and cattle copper values consistently below 0.6 μg/ml in whole blood or plasma are indicative of copper deficiency. Levels as low as 0.1 μg/ml have been reported in these species grazing copper-deficient pastures (31). Copper levels of 0.2 μg/ml in the blood of pigs have been reported, but erythrocyte copper is reduced to a lesser extent than in serum. (274).

In copper-deficient rats a decline in erythrocyte copper and ceruloplasmin, and "direct-reacting" copper of the plasma has been observed (44), but no such decline was seen in this species in a study by Dreosti and Quicke (116), even in severe copper deficiency. Ceruloplasmin estimations possess advantages over whole-blood or plasma copper determinations for detecting copper deficiency because of the relative stability of the enzyme, the technical convenience of the assay, the smaller serum sample required, and the avoidance of copper contamination problems (453). The erythrocyte SOD may be an even superior reflection of copper status (29).

Elements such as zinc, cadmium, and iron that depress copper absorption can reduce plasma copper concentrations when ingested at high dietary levels. The effect of molybdenum and sulfate depends on the status of the animal with respect to these nutrients and copper. Molybdenum can reduce blood copper levels, and high and prolonged intakes of molybdenum and sulfate can cause striking changes in blood levels of copper and copper distribution in the various blood components in sheep (38,96,122,171,212,227,300,342). Sheep maintained on a daily intake of 120 mg of molybdenum and 7.4 g of sulfate for 29 months maintained plasma total copper concentrations at twice the experimental level, and plasma direct-reacting copper was 10 times this level. Plasma ceruloplasmin and the ultrafilterable copper levels were not significantly decreased, but the copper concentration in the red cells was reduced to one-tenth the normal level. All the copper in the plasma of sheep on the high molybdenum plus sulfate intakes could be accounted for in terms of direct-reacting ceruloplasmin and ultrafilterable copper. It is apparent that there was a marked redistribution of the copper from erythrocytes to direct-reacting copper in plasma. Milne and Weswig (319) observed no increases in plasma copper in rats when the copper content of the diet was raised from 10 μg/g copper to 50 μg/g, but when copper content was raised to 100 μg/g in the diet, the plasma copper concentration increased from 1.13 to 2.34 μg/ml.

Dietary copper concentrations of ≥100 μg/g were similarly necessary in the diet of pigs to produce significantly elevated levels of plasma copper. At highly toxic intakes of 750 μg/g a severe hypercupremia occurs in pigs which can largely be prevented by the concurrent administration of 500 μg/g of zinc and additional iron (340,443). Hypercupremia occurs in other species as a consequence of extremely high dietary intakes of copper (120), and extremely high levels can occur in sheep shortly before the hemolytic crisis. In rats copper levels tend to decrease with continued high levels of copper as the animals appear to adapt to the high levels. In studies of copper loading in the rat, the nuclei and mitochondria were found to hold most of the excess copper, with the microsomes and cytoplasm accumulating much less. The amounts in the mitochondria and the soluble fraction remain relatively constant, while the amount in the microsomes increases (173,275).

4. Effect of Disease

The abnormal wool seen in sheep that are deficient in copper suggested a possible role of copper in the X-chromosome-linked fatal disorder of male infants, called Menkes' disease. The kinky, depigmented hair parallels the abnormal wool seen in sheep (100,101,103). Supplemental copper given in Menkes' disease has little effect other than an increase of copper and ceruloplasmin in the serum. A low serum copper and low ceruloplasmin content of the serum are characteristic of the Menkes' kinky-hair syndrome.

Acute and chronic infections in humans due to viral or bacterial organisms, in leukemia, Hodgkin's disease, various anemias, "collagen disorders," hemochromatosis, lymphomas, and several other malignancies, as well as myocardial infarctions, usually result in an increase in serum copper and ceruloplasmin (326,354,355,389,410,488). Hypocupremia is associated with nephrosis and Wilson's disease (74). Hypocupremia also occurs in kwashiorkor and cystic fibrosis associated with low protein intakes (278). Normally up to 95% of the copper in human blood serum is bound to ceruloplasmin, although patients with Wilson's disease may have little or no ceruloplasmin and are therefore an exception because their serum copper can be quite high (395,396). In Wilson's disease serum copper has been shown to be as low as 0.61 ± 0.21 μg/ml as compared with normal values of 1.14 ± 0.17 μg/ml. Almost all Wilson's disease patients have <23 mg of ceruloplasmin per deciliter of serum, which is the lower limit of normal values (75). The diagnostic value of ceruloplasmin activity is poor for Wilson's disease, since some individuals without Wilson's disease also exhibit abnormally low levels of ceruloplasmin. Ceruloplasmin is reduced in all normal neonates, in Wilson's disease, and in Menkes' syndrome (103,395,396). In addition to the low copper and ceruloplasmin in the serum, degenerative changes occur in the brain, arteries, hair, and bone (101,103). This is related to the defect of copper transport from the intestine mucosal cells to the blood. A similar condition is also found in X-linked mottled mutant or brindled mice (221,366,368,371).

Patients with rheumatoid arthritis, rheumatic fever, lupus erythematosus, lymphoma, carcinoma, various liver diseases, and many different infections have all been reported to exhibit increased concentrations of serum copper and ceruloplasmin (329). The increases are associated with a redistribution of copper from the liver to the blood, believed to be initiated by a leukocytic endogenous mediator which stimulates the liver to synthesize additional quantities of ceruloplasmin (28). Serum copper is also significantly elevated in pellagrins. Krishnamachari (271) found copper values of 166.6 ± 7.95 μg/dl in pellagrins as compared with 103.9 ± 5.95 μg/dl for normal controls. Following treatment of the subjects the activity of ceruloplasmin was unaltered in the disease.

Elevated ceruloplasmin values have been observed in sheep by McCosker (304,305) in several disease conditions. Chicks infected with *Salmonella gallinarum* may show a 6-fold increase in ceruloplasmin activity (430). A similar increase was seen with other stressors including adrenocorticotrophin (ACTH) and hydrocortisone. A two to five times increase in ceruloplasmin levels has been obtained in chicks by intravenous injection of endotoxin preparations from three strains of *E. coli* into pathogen-free fowls (65). Starcher and Hill (430) maintain that any stress or any condition resulting in elevated corticosteroid levels could increase ceruloplasmin concentrations in chicks. On the other hand, Henkin (197) demonstrated an inverse relationship between plasma cortisol and serum copper concentrations in human patients and cats, and a direct relation between cortisol and urinary excretion of copper (197).

F. Copper in Milk

The copper content of human milk decreases as lactation progresses. Dorea *et al.* (115), in a study with Brazilian women, noted that at 15 days postpartum, human milk contained 0.75 μg/ml copper while at 90 days postpartum the value was 0.57 μg/ml. The ash of the milk contained 0.4–0.66 mg/g. Kirsten *et al.* (250), in a study of the breast milk of Cape Town mothers, found foremilk copper levels at 3 days 57 ± 74.8 μg/dl, with a decline to 28 ± 29.7 μg/dl at 36 weeks. At the end of the first week of lactation, Cavell and Widdowson (77) reported a mean of 0.62 with a range of 0.51–0.77 μg/ml for the milk of 10 women. Additional values are given in the table on copper in food (Table I). Dang *et al.* (99), in a study of the colostrum of 10 tribal women and 19 urban women, noted that the tribal women had colostrum copper levels of 0.27 μg/g, while the urban women had concentrations of 0.91 μg/g. Apparently this reflected the lack of copper in the diet of the tribal women.

In all species, colostrum is substantially higher in copper than milk (356), and in all species there appears to be a decline in copper level throughout lactation. The copper content of milk differs with species, stage of lactation, and copper nutriture of the animal. Rat milk is exceptionally righ in copper. Values range from 2.8 to 3.3 μg/ml in copper-sufficient animals and 1.7 to 2.8 μg/ml in copper-deficient rats (116). The milk copper level of normal ewes declined from 0.6 to 0.2 μg/ml in early lactation to 0.04 to 0.16 μg/ml several months later. The copper level in mare's milk fell from a mean of 0.36 μg/ml in the first week of lactation to 0.17 μg/ml, several weeks later (148). It appears that human milk is appreciably richer in copper than cow's milk (356,358). Murthy and Rhea (328) gave a range of 0.04–0.15 μg/ml and a national mean of 0.089 μg/ml for market cow's milk for U.S. cities, as compared to a mean of 0.24 ± 0.08 μg/ml for 22 samples of human milk. Additional values may be found in Fonnesbeck and Lloyd (148), Pennington and Calloway (356), Koivistoinen (266), and Wissenschaftliche Tabellen Geigy (489).

Copper levels in milk from ewes and cows grazing deficient pastures were as low as 0.01–0.02 μg/ml as reported by Beck (24). Adding copper to diets already adequate in copper has little effect on the copper content of milk of cows or goats (124) or women (327). However, Dunkley et al. (117) obtained a substantial elevation of milk copper for at least 4 weeks following subcutaneous injection of cows with 300 mg of copper as copper glycinate. This was achieved without any increase in the incidence of spontaneous oxidized flavor in the milk, probably because of the small amount of copper associated with the milk fat in early lactation. After 2–4 weeks of lactation only about 15% of the copper in cow's milk is associated with the fat, whereas after 15 weeks the proportion associated with fat rises to 35% (244).

G.   Copper in Hair and Wool

The ready availability has made hair a tempting material for analysis in the hope that it would provide a measure of the copper status of individuals or groups, or of animal species. However, the extreme individual variation, differences in sample preparation, and exogenous contamination have limited the usefulness of hair copper determinations. Gibson and Gibson (157), for example, after the analysis of over 800 samples of hair and serum, were unable to find any correlation at any time. Although the usefulness of hair analysis is very limited, there have been many investigations of copper in hair as related to individual copper status. Vir et al. (469) could find no significant difference in hair during pregnancy, even though serum copper rose from 0.98 to 1.79 μg/ml. The hair contained 14.3 μg/g of copper. Bradfield et al. (46) examined hypochromotrichia in hair which had copper values from 27 ± 14 to 32 ± 17 μg/g but the differences were not significant. Hair color does not appear to influence the copper concentration (115). Vuori et al. (470), in examining Finnish children from rural and urban locations, noted that the hair copper ranged from 9.7 μg/g in 3-year-old urban girls to 29.8 μg/g in an 18-year-old rural girl. Matsuda et al. (301) examined the effect of copper in formulas on the concentrations of copper in hair and could find no significant difference in copper content of hair related to differences in the copper content of the formulas. Gordon (167) looked at the influence of sex and age as related to the copper content of hair, but the differences were not significant. Hambidge, in 1973 (183), found that the proximal sections ranging from 1 to 5 cm of hair of 27 healthy human subjects averaged 11.8 μg/g while the most distal sections averaged 20.7 μg/g. The highest mean concentration of that part of the hair shaft that was exposed to the external environment for the longer duration suggests that exogenous copper contributes to the hair content of this element. Hambidge maintains that "the interpretation of analytical data on hair copper requires great caution" and that "analyses should be limited to recent grown hair within 1 to 2 cm of the scalp." From the results of a study of human hair as a biopsy material for the assessment of copper status, Klevay (252) concluded that only the age- and sex-matched individuals or groups should be compared. He observed a fall in the copper content of the hair during the first decade of life, while in older children and in adults, female hair contained more copper than male hair.

Reinhold *et al.* (380) found that the copper content of human hair does not rise significantly with age as occurs in the hair of rats. Earlier claims that the copper concentration of pigmented hair is higher than in unpigmented hair have not been confirmed by later studies (115). It seems also that kwashiorkor is not necessarily accompanied by subnormal hair copper levels (278).

The copper concentration of wool appears to be extremely variable. Wooliams *et al.* (490,491) examined the copper content of wool as related to breed and to the liver and plasma concentrations of copper. Burns *et al.* (61) reported a mean level of 25 μg/g copper from 50 washed fleeces with a coefficient of variation of 100%. Cunningham and Hogan (97) obtained levels ranging from 8.3 to 13.3 μg/g, and Healy *et al.* (195) found 42–147 μg/g in small groups of New Zealand sheep. By contrast, Bingley (38) found the washed wool of two housed experimental sheep on a good diet to contain only about 2–4 μg/g of copper on a dry basis. These levels were reduced to less than one-third when the animals were liberally dosed with molybdate and sulfate. It seems that the level of copper in wool is highly dependent on the nature of the diet. Contamination is probably a further factor contributing to the reported variability.

In a study of cattle hair, an average copper value of 9 μg/g (97) was found. Values <8 μg/g may be indicative of copper deficiency in cattle, but Cunningham and Hogan found no relationship between the copper content of the hair and the levels of copper in the diet or the liver. A reduced copper content of the hair was apparent with increased dietary molybdenum. In a study with rats there was no significant decline of hair copper in copper deficiency (116).

H. Copper in the Avian Egg

The application of modern analytical methods has resulted in much lower values for copper concentrations in hen's eggs than previously accepted. Varo *et al.*

(467) reported a mean copper level of 0.7 $\mu g/g$ fresh weight, with a range of 0.68–0.73. This is very close to an average of 0.62 $\mu g/g$ found in a U.S. study (90). Pennington and Calloway's literature survey reported a mean of 1.0 ± 0.7 $\mu g/g$ edible portion, with a range of 0.2–2.3 $\mu g/g$ (356).

The extent to which these concentrations are affected by dietary copper intakes has apparently not been studied, although Panic et al. (350) found that the copper-64 (^{64}Cu) content of egg yolk was positively associated with plasma ^{64}Cu level in hens following ^{64}Cu intramuscular injections. This work also reported considerable deposition of the copper in the egg white. The copper concentration in the yolk was about two-thirds and the white about one-third of the total copper concentration in the whole egg. The copper in the egg is transferred readily to the developing embryo; this has been examined by McFarlane and Milne (306) and by Dewar et al. (109).

II. COPPER METABOLISM

A. Absorption

Depending on the species studied, copper can be absorbed in all segments of the gastrointestinal tract. Although sites in the small intestine appear to play the major role of copper absorption, a substantial absorptive activity has been demonstrated for the human stomach (63) and in the sheep large intestines (168). Studies in human subjects and animal species have shown that the intestinal absorption of copper is regulated by the nutritional status of the individual, but also that it is influenced by the chemical form in which the element is present and by a substantial number of interactions with other dietary factors that affect bioavailability. Although little is known about the influence of age on copper absorption in different animal species, it has been shown that young suckling lambs absorb copper at a much higher rate than mature sheep (439). Thus, it is not surprising that estimates of the fractional copper absorption in humans vary from a low of 25 to a high of 70% (72,241,437,465). Because the intestines excrete copper as well as absorb it, there is a large difference between true and apparent absorption, and isotope methods are required to measure the former. Cousins (92) has reviewed some of the interacting dietary factors that affect the absorbability of copper. He pointed out the specific effects of complexing and chelating agents, for example, the fact that L-amino acid complexes are absorbed to a greater degree than complexes of D-amino acids (246). Of great practical importance in animals is the copper–thiomolybdate interaction, whereas the interaction of ascorbic acid with copper may create practical problems in humans (80,461).

Both compounds significantly depress the biological availability of copper; it is, however, not known whether this interaction occurs exclusively at the level of intestinal absorption.

Although there is as yet no complete understanding of the biochemical mechanisms involved, there is good evidence that the intestinal absorption of copper is regulated by the need of the organism, and that metallothionein in the epithelial cells of the intestine may play a key role in that regulation. The absorption of copper in several experimental species is higher in the presence of a copper deficiency than it is in adequate nutritional status (94,405,460).

Copper appears to be absorbed by two mechanisms, one saturable and the other unsaturable, suggesting active transport for the former and simple diffusion for the latter (54). As is also true for other transport systems, low concentrations of dietary copper are predominantly transported via the saturable, active pathway, whereas the diffusion process comes into play at higher concentrations. Cousins (92) has reviewed the potential role of metallothionein as a regulator of copper absorption and as the site of intestinal interaction between copper and zinc. He postulated that excesses of intracellular copper or zinc induce the synthesis of thionein. The latter would bind the intracellular ions and remove them from further transport into the portal circulation until the epithelial cell is eventually sloughed off and the metals become part of the intestinal contents again. It is known that the affinity of metallothionein for copper is much greater than for zinc; therefore, the expected regulation of transport by metallothionein would be much greater for copper than for zinc. On the other hand, zinc is a more potent inducer of metallothionein synthesis than is copper; thus it appears that in the intestinal interaction zinc plays the more active and copper the more passive role. Although this proposed mechanism is consistent with several observations of the effects of large excesses of the interacting metals, its applicability for physiological situations of more balanced intake remains to be defined.

Other endogenous influences on copper absorption are as yet unidentified substances present in bile, pancreatic juice, and other intestinal and gastric secretions, which have been shown to depress copper absorption in a number of experimental systems (161,228). Administration of estradiol to female rats was found to decrease the absorption of copper, in spite of increased plasma ceruloplasmin activity (85). On the other hand, copper absorption was found to be increased in tumor-bearing rats (85) and in streptozotocin-diabetic rats, suggesting that insulin or glucagon or other factors playing a role in diabetes may affect copper metabolism, but the mechanisms mediating these influences are still unknown.

In addition to these endogenous factors regulating copper absorption, there are powerful external influences that affect the biological availability of copper. Underwood (458) described the importance of these external influences in the statement: ''Copper absorption and retention is so strongly influenced by a number of other mineral elements and dietary components that a series of minimum copper requirements exists, depending on the extent to which these influencing factors are present or absent from the diet, and on their criteria of sufficiency in blood.'' The nature of the copper species in animal feeds and human diets is unknown, but the chemical form in which copper is ingested undoubtedly influences the availability for absorption strongly. Differences in biological copper availability between fresh and dried plant feeds have been demonstrated independently by at least two groups (190,314–316). Mills and collaborators were able to show that neutral or anionic copper complexes extracted from fresh herbage were better utilized by copper-deficient rats than equivalent amounts of copper sulfate. Kirchgessner and collaborators demonstrated significant and biologically meaningful differences of copper absorption in *in vitro* and *in vivo* systems, depending on the ligands present in the complexes. It is not known whether organic ligands, such as amino acids, function only by protecting the metal against precipitation in the milieu of the intestines and give off the copper at the brush border of the epithelial cells, or whether certain complexes are absorbed and transported intact (246,247).

Of the organic dietary ingredients interacting with copper, ascorbic acid is outstanding by its depression of copper bioavailability. This has been demonstrated in chicks, rabbits, rats, and humans (68,146,202,219,461). This effect may be mediated through the reduction by ascorbic acid to cuprous ion, or through the formation of a very stable complex, or both. The popularity of supplementing dietary intakes with high amounts of ascorbic acid may thus have a negative impact on the copper status of population groups in several industrialized countries. Although copper forms stable complexes with phytates, the depressing effect of these substances is greater on zinc than it is on copper; it may be negligible under practical conditions in most Western countries (324). Many amino acids as well as other organic acids occurring in foods and feeds stimulate copper absorption. This is consistent with the finding by several investigators that dietary protein generally enhances copper availability (125,172,302). The role of fiber in human dietaries in copper absorption appears to be minor under most practical conditions; amounts of neutral detergent fiber up to 27 g/day did not significantly affect copper balance (237). Dietary fructose strongly depresses the biological availability of marginal amounts of copper in the diets of experimental animals, but reduced absorption is only partly responsible for the observed effects in the intact animals (143).

Of the interactions between copper and inorganic constituents, that with thiomolybdate is of outstanding practical importance in ruminant species in which a high environmental concentration of molybdenum and sulfur can induce copper deficiency in spite of copper intakes that would otherwise be adequate. Copper–molybdenum interactions may also have some importance in the human diet. Deosthale and Gopalan have demonstrated that high but not excessive dietary intakes of molybdenum (0.54 mg/day) are associated with increased urinary loss of copper (107). A series of reports from the Soviet Union have described a goutlike syndrome together with elevated concentrations of blood xanthine oxidase and uric acid in the population of an area with high environmental exposure to molybdenum. The syndrome was not found in another area in which equally high molybdenum concentrations were associated with elevated copper intake (268). Krishnamachari *et al.* (272) have postulated that a relative copper deficiency induced by excessive dietary molybdenum, together with low calcium and high fluorine intakes, may be responsible for a typical syndrome of genu valgum endemic in a region in India. The other known interactions that affect copper absorption, those with calcium, iron, and zinc, may also be of practical importance in human and animal nutrition. The interaction between copper and calcium appears to be complex, possibly also involving the calcium–zinc antagonism. This might explain apparently contradictory results of animal experiments. For example, dietary calcium depressed the absorption of copper in mice (450), but enhanced copper toxicity in pigs (443) and give a degree of protection against meat anemia in mice. These effects may be mediated under some conditions by a direct interaction between calcium and zinc, reducing the antagonistic action of the latter element on copper status.

A strong antagonism between copper and zinc has been clearly established in numerous animal experiments (128,201,203,405,462). That antagonism appears to play a role in human subjects as well. Brewer and associates detected signs of copper deficiency in patients who received pharmacological doses of zinc as a treatment for sickle cell anemia, who promptly responded to copper supplementation (53). Supplementation of human subjects with 150 mg of zinc per day for 5 weeks resulted in a significant decline of the high-density lipoprotein (HDL) cholesterol fraction (210), possibly mediated through a zinc–copper antagonism. That antagonism is also the feature of Klevay's hypothesis linking marginal copper status with an increased risk for cardiovascular diseases, to be discussed later (255). The interaction of copper with iron rests predominantly in the copper requirement for proper iron utilization and results in complex additional interactions with other dietary factors. For example, the beneficial effect of ascorbic acid on iron status, mediated through the stimulation of nonheme iron absorption, is negated by the reduced absorption of copper, with the resulting impairment of the utilization of the absorbed iron.

Antagonistic interactions between copper and excessive amounts of cadmium, silver, and mercury have been described (205,233,459) but have more scientific than practical importance. The reduction of glutathione peroxidase activity in copper-deficient but selenium-adequate rats suggests an interaction between these two elements, but the mechanism of such interaction is unknown (141,230).

Of the known interactions that affect copper absorption, those with calcium, zinc, and iron may be of practical importance in human and animal nutrition. The depression of copper absorption by high dietary calcium levels in mice may be due to changes of pH in the intestines; on the other hand, high-calcium diets have been shown to enhance copper toxicity in pigs (443).

The molecular mechanisms which account for the transport of absorbed copper within the intestinal cell and their regulation are largely unknown. Metallothionein, which plays an important role in the intermediary metabolism of copper, may also have a function in the intestines (92). The fact that metallothionein synthesis is readily induced by high levels of dietary zinc (339), and the greater affinity of that protein for copper than for zinc, could explain the fact that excessive levels of dietary zinc depress copper absorption. The newly synthesized metallothionein would preferentially bind copper and make it unavailable for further transport into the circulation. However, the same authors pointed out that these mechanisms were not effective at lower, more nearly physiological concentrations of copper and zinc in the diet. Evans and Hahn (127) demonstrated a number of copper-binding substances in rat intestine, in addition to a high molecular weight compound, presumably metallothionein, but the function of these compounds is yet unknown. The results of investigations into Menkes' kinky-hair disease in children (103), and a mouse mutant with defective copper metabolism (221), indicate the existence of two different mechanisms, one involving copper transport from the intestinal lumen into the mucosal cells and one from the mucosal cells to the plasma. In these two conditions copper readily enters the mucosal cells where it builds up, but enters the bloodstream in inadequate quantities due to an undefined genetic defect or defects in this phase of copper absorption.

B. Intermediary Metabolism

Although the great majority of copper in the plasma is bound to ceruloplasmin, there is much evidence that newly absorbed copper is transported from the intestines loosely bound to albumin or certain amino acids. In this form the element is readily available to the liver and other tissues (476), in contrast to the much more tightly regulated distribution of ceruloplasmin-bound copper. This difference may explain the tissue damage by copper in Wilson's disease, in which the latter transport protein is lacking.

The liver is the central organ of copper metabolism; its concentrations reflect intake and the copper status of the organism (52). The uptake of copper into the hepatocytes has been studied in *in vitro* systems (392,476). The uptake is rapid, follows first-order kinetics, is saturable, and is temperature dependent. On the other hand, transport is not inhibited by the metabolic poison 2,4-dinitrophenol, whereas Weiner and Cousins observed inhibition with other inhibitors, cyanide and *n*-ethylmaleimide. Saltman *et al.* (392) did not observe competition with copper uptake by other metal ions; on the other hand, the presence of amino acids stimulated uptake (186). These observations suggest a specific, facilitated diffusion mechanism for copper, possibly through the formation of certain copper–amino acid complexes.

Once within the cell, more than half of the accumulated copper is found in the supernatant fraction and one-fourth within the nuclei (419). Upon gel filtration the supernatant copper was present in two major fractions. It first appeared in a fraction of approximately 10,000 daltons, believed to represent metallothionein, from where it was slowly transferred to a fraction of approximately 35,000 daltons, possibly representing SOD and other copper enzymes (347,463).

Ceruloplasmin is the carrier for the tissue-specific export of copper from the liver to the target organs. It is, together with metallothionein, the dominant component of intermediary copper metabolism, mediator of the many physiological and pathological influences on copper metabolism. Research has led to important hypotheses relating to the important role of that substance in the defense of the organism against metabolic, neoplastic, infectious, inflammatory, and other forms of stress (92,150).

C. Excretion

In all species studied, a high proportion of ingested copper appears in the feces. Most of this is unabsorbed copper, but active excretion also occurs via the bile. Cartwright and Wintrobe (72,73) have estimated that of an assumed human dietary intake of 2–5 mg daily, 0.6–1.6 mg (32%) is absorbed, and 0.5–1.3 mg is excreted in the bile. These estimates may need reevaluation in the light of subsequent findings of a much lower average copper intake (1 mg) in the U.S. population and of the impact of chemical form and dietary interactions on the absorbability of copper. The biliary system is also the major pathway of copper excretion in pigs and dogs (41,291), mice (159), and poultry (26). Intravenous injection of copper, resulting in elevated blood and tissue copper levels, increases copper excretion in the bile and hence the feces but does not normally raise copper output (291). In patients with liver cirrhosis accompanied by biliary obstruction, urinary copper excretion is increased, but this does not occur in Laennec's cirrhosis without significant biliary obstruction (23). In Wilson's disease bile copper concentrations are not greatly raised, despite the presence of a markedly elevated hepatic copper pool (63).

Copper concentrations in human urine as reported in the literature vary considerably. While analytical uncertainties may account for some of the variation, modern, well-controlled studies conducted with careful observation of quality control and avoidance of contamination have reported "normal" ranges of 3.8–66 μg/liter (181,293). A daily copper excretion of 5–50 μg in the urine of adults is considered normal (475). Urinary excretion also is influenced by adrenocorticosteroid hormones. Patients with untreated adrenocortical insufficiency exhibited significantly higher serum copper but significantly lower urinary copper output than controls (197), and normal volunteers given ACTH decreased their serum concentration and increased urinary copper excretion. High serum and biliary copper and low urinary copper excretion was observed in adrenalectomized and hypophysectomized cats, and the situation was reversed with hormone therapy. The direct relationship between plasma cortisone and urinary copper excretion is apparent from these studies.

Negligible amounts of copper are lost in the sweat (185) and comparatively small amounts in the normal menstrual flow. An average of 0.5 mg copper per period has been estimated (282), which indicates that a woman's copper status, unlike her iron status, is not compromised by menstruation.

III. COPPER DEFICIENCY AND FUNCTIONS

In order to understand the many facets of copper deficiency in different species, it is necessary to recall the widely accepted definition of deficiency as a consistent and reproducible impairment of a biological function from normal to subnormal (312). A deficiency may be preceded or accompanied by the "depletion phase," during which copper concentrations in body fluids and tissues decline, but these levels are not valid as the sole criteria of deficiency. Copper in the circulation is subject to so many nonnutritional influences, hormonal or infectious, that it is not useful as a diagnostic criterion. There is an increasing knowledge of interactions between copper and other dietary factors, for example, with dietary molybdenum and sulfur or with dietary carbohydrate, and such interactions determine the fate of the organism much more strongly than does the concentration of copper in the circulation.

As is true for many other trace elements, the great diversity of signs of copper deficiency in different species cannot be totally explained on the basis of our knowledge of the element's biochemical functions. Of the many known copper proteins, four enzyme systems may play key roles in the development of deficiency: (1) the ferroxidase activity of ceruloplasmin explains in part the disturbances of hematopoiesis in copper deficiency; (2) the monoamine oxidase enzymes may account for the role of copper in pigmentation and control of neurotransmitters and neuropeptides; (3) lysyl oxidase definitely is essential for maintaining the integrity of connective tissue, a function that explains the disturbances in lungs, bones, and the cardiovascular system; and (4) the copper enzymes cytochrome c oxidase and SOD play a central role in the terminal steps of oxidative metabolism and the defense against the superoxide radical, respectively. These functions have been postulated to account for the disturbances of the nervous system as seen in neonatal ataxia in several animal species. In addition, it has been demonstrated that the copper status of animals (377) and humans (37) strongly affects the levels of neuropeptides, either through an effect on biosynthesis or on release. Bhathena and collaborators have shown that a mild, marginal copper deficiency in human subjects is associated with markedly reduced levels of plasma enkephalins, progressive increases in the pancreatic concentrations of leucine enkephalin-containing peptides, together with a decrease in free leucine and methionine enkephalins. The reader is referred to a review of that subject by Bhathena and Recant (36). Although this knowledge can explain the development of individual signs of copper deficiency, the understanding of the sequence of events in different species must await the results of future research.

A. Anemia and Iron Metabolism

Anemia is a common expression of severe, prolonged copper deficiency in most species studied. A notable exception is the brindled mouse, the genetic analog of Menkes' disease in humans, which does not exhibit anemia despite a grossly disturbed copper metabolism (367). In 1952, Gubler *et al.* (176) produced evidence of impaired absorption, mobilization, and utilization of iron in copper-deficient pigs. Subsequently it was shown (345) that ceruloplasmin catalyzes the oxidation of iron from the divalent to the trivalent state, required before the element can be coordinated to the transport protein, transferrin. Because of that enzyme activity ceruloplasmin is also termed ferroxidase. Freiden (149) has pointed out that the

> iron absorption by the mucosal cells appears to be unaffected in copper deficiency but release of iron into plasma is impaired. The release of iron from the liver parenchmal cell is also reduced in copper deficiency. The most significant pathway, quantitatively, is the mobilization of iron from the R-E system into the plasma. This is lowered in copper deficiency and can be restored rapidly by injection of CP, or more slowly, by copper administration.

This attractive hypothesis, however, does not explain all observations. As is true in the brindled mouse, some cases with Wilson's or Menkes' disease with low plasma ceruloplasmin levels were found not to be anemic. Increased breakdown of heme was postulated as a possible source of the increased deposition of iron in the liver (486). Cohen *et al.* reduced the serum copper and ceruloplasmin levels in rats by feeding chelating agents but were unable to detect anemia in these animals (86). Weisenberg *et al.*, were able to reverse the anemia and growth retardation of copper-deficient rats by increasing dietary iron alone, which did not significantly affect the levels of ceruloplasmin (477). The results of Fields and collaborators are also difficult to reconcile with the ferroxidase theory. These investigators did not observe anemia in two copper deficiency systems: copper-deficient diets reduced ceruloplasmin to undetectable levels in rats regardless of the source of dietary carbohydrate, but anemia was seen only in the animals fed sucrose or fructose, not in those fed starch. Both male and female rats fed copper-deficient, fructose-containing diets had nondetectable ceruloplasmin levels, but only the male rats became anemic whereas the females remained normal (138–145). These observations do not speak against an important role of ceruloplasmin in iron metabolism, but they suggest that additional mechanisms are operational. The exact nature of these mechanisms is as yet unknown, although it is unlikely that there is a copper requirement for the enzymes of the biosynthetic heme pathway (279a). Copper appears to be required for maintaining the integrity of red blood cells in the circulation; it has been shown that the survival of red blood cells is shortened in the copper-deficient pig (56).

B. Bone Disorders

Copper is essential for proper bone metabolism, and skeletal abnormalities occur in copper-deficient rabbits (220), chicks (67,154,387,388), pigs (274,286), dogs (21,22,481), foals (30), and children (170). On the other hand, rats and several ruminant species appear to be relatively resistant (219,317). Suttle et al. (444) were unable to induce osteoporosis or depressed osteoblastic activity by depleting lambs for 8 months starting at the age of 5 months, even though these animals showed other signs of copper deficiency. The same group of workers did detect osteoporosis with reduction or cessation of osteoblastic activity only under more severe and prolonged conditions of intrauterine copper deficiency in lambs that were borne of ewes that had been given copper-deficient diets for 2 years. It appears that osteoblastic activity is depressed in copper deficiency in all species studied, but its effects on the various parameters of bone formation vary widely among species. For example, poor mineralization of bones has been reported in mice fed a meat diet (177,178) and a human infant with nutritional copper deficiency (447), whereas the bone contents of the major minerals in copper-deficient pigs (21,22) and chicks (387,388) were found to be normal. Supplementation of a milk-based diet with 8 and 50 $\mu g/g$ of copper delayed and attenuated the appearance of scoliosis in a line of chickens susceptible to that condition (344). The same authors also reported spinal deformities in 20–30% of normal chicks fed a low-copper (0.5 $\mu g/g$) diet. In young dogs rendered severely copper deficient, a gross bone disorder develops with fractures and severe deformities (21,22,466). Copper deficiency in children recovering from protein–energy malnutrition resulted in widened epiphyseal cartilages (170) in addition to anemia and neutropenia, but not gross deformities.

The primary biochemical lesion in the bones of copper-deficient animals is probably a reduction in the activity of the copper enzyme amine or lysyl oxidase, leading to diminished stability and strength of bone collagen as a result of impaired cross-linkage of their polypeptide chains, in a manner analogous to the effect on aortic elastin described in Section III,H. A marked reduction in amine oxidase activity occurs in the bones of copper-deficient chicks, and collagen extracted from such bones is more easily solubilized than collagen from control bones (387,388), similar to the reduction of lysyl oxidase and of elastin in the lungs of that species (188,431). Since the solubility of collagen, like elastin, is inversely related to the degree of cross-linking (62), there seems little doubt that the structural integrity of the collagen is impaired through a failure of establishment of the required cross-linkages. Later work has shown differences in the concentrations of individual glycosaminoglycan species, in the presence of normal total glycosaminoglycans in copper-deficient lungs of chicks (375). As the metabolism of those latter substances is strongly dependent on dietary silicon (see Carlisle, Chapter 7 in Vol. 2), that observation is a reminder of the many minerals and trace elements that are required for normal bone formation. Better knowledge of these interactions and of the requirements for the interacting factors by humans and animals may be able in the future to explain the differences in the expression of copper deficiency in the bone of different species.

C. Diarrhea (Scouring) of Cattle

Diarrhea is not a common manifestation of copper deficiency in most species and does not always occur in cattle in copper-deficient areas (104). Intermittent scouring in cattle occurs in the severely copper-deficient areas in Australia, where molybdenum pasture levels are not high (34), and has been reported in England (5). Scouring was such a prominent feature of a copper-responsive disease occurring in parts of Holland that it was designated ''scouring disease'' (416,417). Molybdenum has been incriminated as the primary causal factor in ''peat scours'' in New Zealand (96), but it seems unlikely that a high dietary molybdenum–copper ratio can account for the occurrence of scouring in cattle in other copper-low, molybdenum-normal areas. Of interest in this connection are the findings of Fell and co-workers (133) with young Friesian steers showing clinical signs of copper deficiency. A marked depletion of cytochrome oxidase in the epithelium of the duodenum, jejunum, and ileum with partial villus atrophy in the duodenum and jejunum was found. Enterocytes from these parts of the small intestine showed mitochondrial abnormalities ranging from slight swelling to marked localized dilation. While no clear relationship between the histochemical and ultrastructural changes and diarrhea was apparent, these observations represent the first indications of pathological changes, which could be of significance to the incidence of diarrhea in copper-deficient cattle.

D. Neonatal Ataxia

A nervous disorder of lambs characterized by uncoordination of movement has long been recognized in various parts of the world and given local names such as ''swayback'' and lamkruis.'' All these conditions are pathologically identical, and the term enzootic neonatal ataxia can properly be applied to them. In 1937 Bennetts and Chapman (32) showed that the ataxia of lambs occurring in Western Australia was associated with subnormal levels of copper in the pastures and in the blood and tissues of both ewes and affected lambs, and could be prevented by copper supplementation of the ewe during pregnancy (31). Subsequent investigations in several countries confirmed the efficacy of copper supplements to the ewe in preventing the ataxia (8). In some areas the incidence of ataxia could not always be explained in terms of a simple dietary copper deficiency (223,409). Thus in the swayback areas of England the copper content of the pastures can be in the normal range, and in Scotland no correlation was found between the copper status of the animal and the severity of the ataxia (19), although many of the reported values for copper in the pastures, blood, and tissues of affected lambs and their mothers are close to the copper deficiency levels observed in Australia. The possibility that swayback is a copper deficiency induced by molybdenum received little support from the study of Allcroft and Lewis (4), but later Alloway (9), in an investigation of copper and molybdenum in swayback pastures, has produced evidence which suggests that in some cases a molybdenum-induced hypocuprosis may be a contributory factor in the incidence of the disease. Under experimental conditions neonatal ataxia has been produced by maintaining ewes of low-copper status on high molybdate and sulfate diets (135,442), and by feeding stall-fed ewes a semipurified copper-deficient diet, adequate and "normal" in other respects (283).

Above-normal lead intakes may sometimes be involved in the incidence of swayback, as suspected earlier (223), especially since 1972 studies with rats have demonstrated a reciprocal antagonism between lead and copper at the absorptive level (251).

Neonatal ataxia in copper-deficient goats (349,401) and lesions in the central nervous system (CNS) in copper-deficient newborn guinea pigs (131) have been observed. Ataxia was previously observed in young pigs, associated with low liver copper levels and "demyelination of all areas of the spinal cord except the dorsal areas" (235,484), and a subclinical myelopathy resembling that of swayback in lambs has been observed in copper-deficient miniature swine (66). Demyelination of the spinal cord, extending into the medulla and cerebellum in young pigs, with ataxia and subnormal liver copper concentrations have been described (66,307), and Carlton and Kelly (69) reported neural lesions in the offspring of female rats fed a copper-deficient diet, accompanied by behavioral disturbances in some animals.

Two types of ataxia occur in lambs under field conditions: the neonatal form in which the lambs are affected when born, and a delayed type in which signs may not appear for some weeks. Uncoordinated movements of the hindlimbs, a staggering gait, and swaying of the hindquarters are evident as the disease develops. Some of the lambs affected at birth are unable to rise and soon die.

Neonatal ataxia (swayback) in lambs was initially characterized as a demyelinating encephalopathy, with cavitation of the white matter of the cerebral hemispheres leading to collapse of these structures (223). Barlow *et al.* (19) found no cerebral lesions in half the ataxic lambs examined, whereas cell necrosis and nerve fiber degeneration occurred in the brain stem and spinal cord of all the affected animals. The pathological criteria were given as ''cavitation or gelatinous lesions of the cerebral white matter and/or a characteristic picture of chromatolysis, neurone necrosis and myelin degeneration in the brain stem and spinal cord.'' The concept that the significant changes of swayback are those of necrosis and degeneration of neurons accompanied by changes in spinal cord nerve fibers has been supported by Howell (211). Later investigations always revealed chromatolysis and degenerative changes in the large nerve cells of the brain stem and spinal cord of swayback sheep (16,17,18). Mills and co-workers (135,318) found the ataxia to be associated with degeneration of the nuclei of the large motor neurons of the red nucleus in the brain stem. Howell *et al.* (217) also observed changes in the neurons of the brain and spinal cord of ataxic lambs, but products of degenerating myelin were not detected. It was suggested that the lesion in the white matter of the spinal cord may be one of myelin aplasia.

Myelin aplasia, rather than myelin degeneration, is compatible with the biochemical evidence from ataxic lambs. The primary lesion in this condition is now generally accepted as a low-copper content in the brain leading to a deficiency of cytochrome oxidase, the copper-containing terminal respiratory enzyme, in the motor neurons (134). Howell and Davison (212) first demonstrated a significant lowering of copper content and cytochrome oxidase in the brain of swayback lambs. Mills and Williams (318) also found a lower cytochrome oxidase activity in the brain stem of swayback than normal lambs and provided evidence that brain copper levels are of greater significance to the integrity and function of the CNS than are liver copper levels. None of the 18 ataxic lambs they examined had a brain copper content as high as 3 μg/g dry weight, whereas of the 37 animals in the normal group only two had <3 μg/g. The greatest reduction in cytochrome oxidase activity occurs in those groups of nerve cells—that is, the large motor neurons of the red nucleus and ventral horns of the gray matter in the spinal cord—which show the morphological lesions of the disease (16,134).

In studies of copper deficiency in rats, Gallagher *et al.* (155) demonstrated depressed cytochrome oxidase activity in the brain, as well as in the heart and liver, together with a depression in phospholipid synthesis. It was concluded that the loss of cytochrome oxidase activity results from a failure of synthesis of its prosthetic group, heme, and that this must be regarded as a basic function of copper. Positive linear and/or quadratic responses in brain, liver, and tibia ossifi activities were later demonstrated by Evans and Brar (130) when graded increments of dietary copper were given to rats receiving a copper-low diet.

Gallagher and Reeve (156) have reported a direct causal relationship between the loss of cytochrome oxidase and impaired phospholipid synthesis. Liver mito-chondria were found to be unable to synthesize phospholipids significantly under anaerobic conditions with or without added electron transfer systems unless adenosine triphosphate (ATP) was added. It appears that the loss of cytochrome oxidase in copper deficiency leads to depressed phospholipid synthesis by liver mitochondria, by interfering with the provision of endogenous ATP to maintain an optimal rate of phospholipid synthesis. On this basis the particular sensitivity of the lamb to neonatal ataxia could be explained in the following terms: at the critical period in late gestation when myelin is being laid down most rapidly in the fetal lamb, copper deficiency causes a depression in cytochrome oxidase activity which leads to inhibition of aerobic metabolism and phospholipid syn-thesis. This leads to inhibition of myelin synthesis, since myelin is composed largely of phospholipids.

E. Pigmentation of Hair and Wool

Achromotrichia is one of the manifestations of copper deficiency in rats, rabbits, guinea pigs, cats, dogs, cattle, and sheep, but has not been observed in the pig. Lack of pigmentation in the fur of the rabbit (423) and in the wool of sheep is a more sensitive index of copper deficiency than anemia. The pigmenta-tion process is so susceptible to changes in the copper status of the sheep that once copper-deficient condition has been established, alternating bands of pig-mented and unpigmented wool fibers can be produced accordingly as copper is added to or withheld from the diet. Even on fairly high copper intakes it is possible to block the functioning of copper in the pigmentation process within 2 days by raising the molybdate and sulfate intakes sufficiently. Dick (112) sug-gests that this effect may take place within hours of giving the first dose of these compounds. A breakdown in the conversion of tryosine to melanin is the proba-ble explanation of this failure of pigmentation, since this conversion is catalyzed by copper-containing polyphenyloxidases (376).

F. Impaired Keratinization

Changes in the growth and appearance of hair, fur, and wool have been noted in copper-deficient rats, rabbits, guinea pigs, dogs, cattle, and sheep, in mutant mice, and in Menkes' kinky-hair syndrome in children arising from genetic defect in copper metabolism.

A reduction in the quantity and quality of wool produced by sheep in copper-deficient areas was noted early in the Australian investigations (30,298). The lowered wool weights are probably an expression of an inadequate supply of substrate to the wool follicles, consequent on a reduced feed intake by the copper-deficient animals. The deterioration in the process of keratinization, signified by the failure to impart crimp to the fibers, is a specific effect of copper deficiency.

As the sheep's reserves of copper are depleted and its blood copper level falls, the crimp in the wool becomes less distinct in the newly grown staple, until the fibers emerge as almost straight hairlike growths, to which the descriptive terms "stringy" or "steely" wool have been given. The tensile strength of steely wool is reduced, the elastic properties are abnormal, and it tends to set permanently when stretched (294). A spectacular restoration of the crimp and physical properties of the wool occurs when copper supplements are given. These abnormalities are more obvious in merino wool, which is normally heavily crimped, than in British-breed wool. However, Lee (279) has observed fleece abnormalities in four British breeds comparable with those encountered in copper-deficient merinos. Furthermore, loss of wool crimp has been reported in crossbred ewes as an early indication of copper deficiency (445).

The characteristic physical properties of wool, including crimp, are dependent on the presence of disulfide groups that provide the cross-linkages or bonding of keratin and on the alignment or orientation of the long-chain keratin fibrillae in the fiber. Both of these are adversely affected in copper deficiency. Straight steely wool has more sulfhydryl groups and fewer disulfide groups than normal (294). An essentially similar pattern has been demonstrated in the hair of patients with Menkes' kinky-hair syndrome (100). It seems therefore that copper is required for the formation or incorporation of disulfide groups in keratin synthesis. Furthermore, wool from copper-deficient sheep contains more N-terminal glycine and alanine, and sometimes more N-terminal serine and glutamic acid, than normal wool, indicating that a lack of copper can interfere with the arrangement of the polypeptide chain in keratin synthesis (60).

The phenotypic similarities, especially of hair growth, between mice homozygous for the autosomal recessive mutant gene crinkled (cr) and copper-deficient animals led Hurley and Bell (222) to examine the effect of copper supplementation during pregnancy and lactation on the expression of the gene in homozygous mutants. The feeding of a high-copper diet (500 μg/g) during this time was found to double the survival of the mutant mice to 30 days of age, prevent their characteristic lag in pigmentation, and produce near-normal skin and hair development. The interaction of a gene and a trace metal in development is strikingly illustrated by this finding.

G. Infertility

In female rats and guinea pigs, copper deficiency results in reproductive failure due to fetal death and resorption (119,180,213,214). The estrous cycles remain unaffected and conception appears to be uninhibited. Normal fetal development in copper-deficient rats has been shown to cease on the thirteenth day of pregnancy when the fetal tissues were disintegrating (213). Necrosis of the placenta became apparent on the fifteenth day of gestation.

Hens fed a severely copper-deficient diet (0.7–0.9 μg/g) for 20 weeks displayed reduced egg production and subnormal levels of copper in the plasma, liver, and eggs, while hatchability dropped rapidly and approached zero in 14 weeks (394). The embryos from these hens exhibited anemia, retarded development, a high incidence of hemorrhage after 72–96 hr of incubation, and a reduction in monoamine oxidase activity. The anemia, hemorrhages, and mortality are probably due to defects in red cell and connective tissue formation during early embryonic development.

Low fertility in cattle grazing copper-deficient pastures, associated with delayed or depressed estrus, has been observed in several areas (5,33–35,464), and infertility, associated in some cases with aborted small dead fetuses, has been demonstrated in experimental copper deficiency in ewes (214).

H. Cardiovascular Disorders

The first evidence of cardiac lesions in copper deficiency emerged from studies of a disease of cattle occurring in Western Australia known locally as ''falling disease'' (33–35). The essential lesion of this disease is a degeneration of the myocardium with replacement fibrosis. The sudden deaths are believed to be due to acute heart failure, usually after mild exercise or excitement. The disease is completely preventable by copper therapy of the animals or by treatment of the pastures with copper compounds to raise their inherently very low copper levels (1–3 μg/g) to normal. Falling disease has never been observed in sheep or horses grazing the same untreated areas and occurs only rarely in cattle in copper-deficient areas elsewhere (104).

Sudden cardiac failure associated with cardiac hypertrophy has been reported in copper-deficient pigs (175,343) and rats (2,138,162). In the latter species a marked reduction of cytochrome oxidase activity in heart and brain and hypertrophy of heart and spleen has been described (2). In 1961 O'Dell and co-workers (338) demonstrated a derangement in the elastic tissue of the aortas of copper-deficient chicks, and found the mortality in these animals to be caused by a rupture of the major blood vessels. These findings were confirmed (361,414) and extended to other species, pigs (70,91) and guinea pigs (131,343).

Studies in several laboratories have combined to elucidate the role of copper in elastin and collagen biosynthesis. The most significant findings may be given as follows. The elastin content of the aorta of copper-deficient pigs and chicks (431) is decreased. The elastin from such animals contains an elevated content of lysine and less desmosine and isodesmosine than that of normal animals (313,337). Desmosine, a tetracarboxylic tetraamino acid, and its isomer isodesmosine are the key cross-linkage groups in elastin (352), and at least two and possibly four lysine residues condense to form these substances (449). For this reaction to take place, the ε-amino group of the lysine residues needs to be removed and the carbon oxidized to an aldehyde—a reaction catalyzed by amino or lysyl oxidases, which are copper-containing enzymes (57,206). Harris (187) has shown that copper deficiency in chicks severely depresses lysyl oxidase activity in aortic tissue, and that copper is a key regulator of lysyl oxidase activity in the aorta and may be a major determinant of the steady-state levels of the enzyme in that tissue.

The role of copper in the formation of aortic elastin can best be stated in the words of Hill *et al.* (204) as follows:

> The primary biochemical lesion is a reduction in amino-oxidase activity of the aorta. This reduction in enzymatic activity results, in turn, in a reduced capacity for oxidatively deaminating the epsilon-amino group of the lysine residues in elastin. The reduction in oxidative deamination results, in turn, in less lysine being converted to desmosine. The reduction in desmosine, which is the cross-linkage group of elastin, results in fewer cross-linkages, in this protein, which in turn results in less elasticity of the aorta.

While the findings relating to lysyl oxidase offer a logical explanation and mechanism for the weakening of cardiac and aortic structures, results of several studies suggest additional, complementary disturbances. Electrocardiographic changes, accompanied by decreased levels of ATP and phosphocreatine, together with disruption of mitochondrial structures preceded gross pathological lesions in the heart tissue of copper-deficient rats (267). Prohaska and Heller (369) reported substantially reduced norepinephrine levels in the left ventricle of copper-deficient rats, possibly explaining decreased coronary resistance and reduced systolic pressure (369). Decreased systemic blood pressure in copper-deficient rats has been independently reported by two groups of investigators (140,310,493). Conversely, Liu and Medeiros reported that excessive dietary copper intake is associated with increases of systolic blood pressure in the rat (285). As has been emphasized repeatedly in this chapter, the type of dietary carbohydrate strongly affects the consequences of copper-deficient diets for the cardiovascular system of rats and pigs (381,432). The changes in the cardiac conduction mechanism reported in some animal experiments are consistent with a report of tachycardia and extrasystolic beats in human volunteers fed a low-copper diet. This will be discussed later in this chapter (382).

I. Copper Deficiency and the Immune System

There is a long history of empirical observations, consistent with recent experimental results suggesting a strong antiinflammatory action of copper, possibly related to its moderating effects in autoimmune diseases, such as rheumatoid arthritis. As it is not clear whether such effects are dependent on the nutritional copper status or are strictly pharmacological, they will not be further discussed in this chapter. The reader instead is referred to extensive reviews of that field (425,426).

Experimental results have established that the nutritional copper status strongly affects basic processes of the immune system in at least three animal species. Neutrophil cells from copper-deficient Friesian steers showed an impaired ability to kill *Candida albicans* organisms (45). Prohaska and associates demonstrated that the reduced immune defense in mice and rats is associated with suppressed splenic T cell function in copper deficiency, through a direct effect of the deficiency on the biochemical function and morphology of lymphoid cells (288,370,371). The same group also demonstrated a significant reduction of survival rate to 15% of the controls in male mice injected with leukemia cells (289). For a detailed discussion of the role of copper in the immune system, the reader is referred to Cousins' review (92).

J.   Lipid Metabolism and Oxidative Stress

Although many studies have demonstrated strong effects of copper deficiency in experimental animals on the metabolism of lipids, cholesterol, and on peroxidation, there are also conflicting reports. Those apparent discrepancies can possibly be reconciled by taking into consideration the many hormonal and dietary factors that enhance or protect against the effects of low-copper diets, for example, the nature of dietary carbohydrate. A participation of copper in the desaturation reactions of fatty acids has been postulated on the basis of decreased ratios of monounsaturated to saturated fatty acids in plasma and various tissues (94,471). Yet, other studies reported increased ratios of unsaturated to saturated fatty acids in the liver of copper-deficient rats (415), or no change in the unsaturation index of fatty acids in the liver mitochondria of severely copper-deficient rats and no changes in the swelling rate *in vitro* of the mitochondria (277). On the other hand, Balevska *et al.* (13) observed a decrease in the relative levels of linoleic acid in the membrane phospholipids of copper-deficient mitochondria.

Although the diverse reported changes of fatty acid composition of organs and organelles may contribute to the increased peroxidative status of copper-deficient animals, it is predominantly the role of copper in the SOD enzyme system and a yet unexplained interaction with the selenium enzyme glutathione peroxidase that accounts for these effects. Twofold increases of lipid peroxidation in liver mitochondria and microsomes of copper-deficient rats were associated with decreases in the activity of SOD, catalase, and glutathione peroxidase (12). These authors reported similar findings in rat erythrocyte membranes (390). Fields *et al.* confirmed these results in their experiments, again emphasizing the deficiency-promoting effect of dietary fructose (141). Jenkinson *et al.* investigating the health consequences of the increased susceptibility of copper-deficient rats to peroxidative damage reported significantly increased mortality in copper-deficient rats exposed to an 85% oxygen atmosphere (231).

Previous conflicting reports regarding the effect of copper deficiency on the selenium enzyme glutathione peroxidase can be reconciled on the basis of the observation of Fields *et al.* of the importance of dietary fructose on the expression of copper deficiency (141). In general, inhibition of glutathione peroxidase activity was observed by various authors only when there were additional signs of copper deficiency (12,230,372).

At present there is no evidence for an interaction between copper and selenium at the intestinal level; the investigation of a potential reduction of glutathione peroxidase by the products of SOD deserves attention.

Although the role of copper in cholesterol metabolism has been controversial in the past, it is now evident that this role is strongly determined by dietary interactions. The controversies in the past may be resolved by attention to the effects of dietary carbohydrates (139), proteins, and zinc (433). These results indicate that true copper deficiency, induced by whatever means, results in an elevation of serum cholesterol. The mechanism of these changes is unknown at this time; it has been hypothesized that the hypercholesterolemia of copper-deficient rats may be due to an impairment of the cholesterol degradation process (281). The importance of these findings for human nutrition is emphasized by the observation that excessive zinc supplementation with approximately 10 times the Recommended Dietary Allowances results in significant changes in the composition of plasma cholesterol, with a significant reduction of the HDL fraction (210).

K. Copper Deficiency and Glucose Metabolism

Several studies have demonstrated an impairment of glucose metabolism in copper-deficient rats and humans (84,191,257,258). Fields *et al.* (142) demonstrated that the effect of copper deficiency on glucose metabolism is strongly dependent on the source of dietary carbohydrate with copper. They also demonstrated that the reduced glucose tolerance was due to a reduced lipogenesis and glucose oxidation rate in copper-deficient animals. *In vitro* studies demonstrated that both parameters could be increased by copper supplementation *in vitro*. The hypothesis that copper affects the insulin binding and, secondarily, glucose transport in isolated rat adiposides *in vitro* remains to be confirmed (145).

IV. COPPER REQUIREMENT

The manifestation of the copper status in physiological or biochemical functions is so strongly influenced by other mineral elements and dietary components that a series of minimum copper requirement exist, depending on the extent to which these influencing factors are present or absent from the diet, and on the criteria of adequacy employed. Some interactions may work at the intestinal, absorptive level; others may be of a much more general nature. For example, the presence of fructose or sucrose in the diet of rats and pigs affects the absorption, tissue levels, and excretion of copper in the animals to a much lesser degree than would be expected in view of the all-or-none effect on survival. In addition, sex in at least one species, the rat, determines death or survival on a marginally copper-deficient diet: males are much more susceptible to death from heart disease than are females, a situation reminiscent of a similar difference in human cardiovascular disease (144,436). The biochemical basis for these interactions as well as for the species differences in copper requirement, to be discussed subsequently, are subject to intensive study but are as yet largely unknown.

A. Rats and Guinea Pigs

Copper deficiency develops rapidly in young rats and more slowly in adult rats on diets containing ≤ 1 μg/g copper. Mills and Murray (317), employing a diet containing only 0.3–0.4 μg/g copper, established the following tentative minimum requirements for rats of about 70 g live weight fed 10 g diet daily: 1 μg/g for hemoglobin production, 3 μg/g for growth, and 10 μg/g for melanin production in hair. The minimum requirements for reproduction and lactation were not investigated, but at a level of 50 μg/g copper these requirements were stated to be fully met through at least one generation. Spoerl and Kirchgessner (428) have shown that the copper requirements for pregnancy and lactation in rats are fully met on a starch–casein diet containing 8 μg/g copper.

Everson and co-workers (131) found that young female guinea pigs fed a diet providing 0.5–0.7 μg/g copper grew well at first, and their reproduction was equal to that of controls receiving a 6 μg/g copper diet. Eventually they displayed a mild anemia, their hair coats became wiry and depigmented, and the growth of their offspring began to slow down at about the twelfth day. Surviving animals maintained on the copper-deficient diet were markedly stunted by 50 or 60 days of age.

B. Pigs

No differences due to treatment were observed in baby pigs fed diets containing 6, 16, or 106 μg/g copper (456). It was therefore assumed that 6 μg/g copper is adequate for growth in such animals. It was later concluded that 4 μg/g of dietary copper is adequate for growing pigs up to 90 kg live weight (3). All ordinary swine rations contain appreciably higher levels of copper than the 4 or 6 μg/g just quoted. Cereal grains contain 4–8 μg/g copper and the leguminous seeds and oilseed meals provided as protein supplements generally contain 15–30 μg/g.

In 1955 Barber and associates (14) found that the addition of copper sulfate to a normal ration at the rate of 250 μg/g copper resulted in increased rate of weight gains in growing pigs. In 1967 Braude (47) assessed the results of trials carried out in Great Britain and concluded that daily live-weight gain can be accelerated by about 8% and the efficiency of food use by about 5.5% by the inclusion of 250 μg/g copper in pig rations. In a large trial carried out in 21 centers such supplementation was found to increase the mean daily live weight by 9.7% and the efficiency of feed use by 7.9% (49). The results of a further trial in which copper sulfate supplements at 150, 200, and 250 mg copper per kilogram diet were compared revealed significant improvements from all three levels, but the best performance and the highest increase in profitability came from the 250-μg/g treatment (48). The response in weight gain is related to the amount of copper in the gut, and copper sulfate is more effective than copper sulfide or oxide. However, it is the copper and not the sulfate radical that is effective (15,41). Copper apparently does not stimulate growth by suppressing or increasing bacterial multiplication in the gastrointestinal tract, because differences between copper-supplemented and unsupplemented pigs in the bacterial flora of the feces (153), or throughout the alimentary canal (420), have not been observed. Kirchgessner and co-workers (245,248) have demonstrated an improvement in the digestibility of proteins in young pigs after dosing with copper and have also shown that appropriate concentrations of cupric ions activate pepsin and raise peptic hydrolysis. These important findings could provide at least a partial explanation for the growth-stimulating effect of the added dietary copper.

The extremely high copper requirements of pigs for maximum growth rate and efficiency of feed use apparent from the trials just described are not valid under all conditions. In several trials carried out elsewhere copper supplementation of normal rations at the rate of 250 μg/g copper has been shown to (1) reduce live-weight gains and efficiency of feed use (20), (2) lead to anemia and copper toxicosis in some animals (383,473), (3) cause mortality and result in skin lesions similar to those of parakeratosis and rectifiable by zinc supplements (341), and (4) induce anemia rectifiable by increasing the iron content of the ration (59,158).

It is apparent that the discrepancies in the reported responses to high copper levels relate primarily to differences in the zinc and iron levels of the basal rations, and that these high copper levels can only be satisfactorily and safely exploited if the zinc and iron contents of the pigs' diets are higher than those found adequate at lower copper intakes. This results from a competitive antagonism between copper and iron and copper and zinc at the absorptive level as discussed in the chapter on iron (Morris, Chapter 4, this volume) and in the chapter on zinc (Hambidge *et al.*, Chapter 1, Vol. 2). Differences in the calcium content of the rations constitute a further factor of potential nutritional importance. High calcium intakes reduce zinc availability, thus enhancing the possibility of copper toxicity from copper-supplemented pig rations and introducing a three-way interaction between calcium, zinc, and copper (443). Heavy copper supplementation of swine rations is clearly a matter of considerable complexity. It may also pose environmental problems, because dietary copper is concentrated in animal wastes, resulting in substantial increases in soil and water copper concentrations when the wastes are discharged into bodies of water or onto land (105). Chronic copper poisoning in sheep grazing herbage dressed with the liquid manure of pigs fed copper-supplemented rations has been reported from Holland (457). The copper in slurry from a piggery where such high-copper rations were fed is highly available and potentially toxic to sheep (365).

C. Poultry

Definitive data on the minimum copper requirements of chicks for growth or hens for egg production have not been reported, nor is there any evidence that poultry have high copper requirements, relative to mammals, as they have for manganese. Diets containing 4–5 $\mu g/g$ copper can be considered adequate, so long as these diets do not contain excessive amounts of elements that are metabolic antagonists of copper, such as iron, zinc, cadmium, and molybdenum. All poultry rations composed of normal feeds are likely to contain ≥ 5 $\mu g/g$ copper.

The results of experiments in which poultry are fed copper supplements at rates equivalent to those used to stimulate the growth of pigs (100–250 $\mu g/g$ copper) are equivocal. Some studies have shown no significant improvement in growth rate, while others have demonstrated a positive response between 75 and 225 $\mu g/g$ copper, with an inhibition above 300 $\mu g/g$ (421). Later King observed a slight but nonsignificant increase in growth rate from 100 $\mu g/g$ copper as copper sulfate in broiler chicks over a 9-week period and a significant increase in the growth rate of ducklings 8–63 days of age from the same level of copper supplementation (243). A feature of this work is the finding that the ceca of the copper-treated birds are significantly smaller and thinner, both as a proportion of body weight and as the weight of unit length (243), and that the intestinal wall is thinned (242). It was suggested that this contributes to the growth-promoting effect by facilitating the uptake of nutrients or by exposing more glandular tissue, thus allowing a more rapid contact of the ingesta by digestive ferments. The extent to which variations in dietary intakes of the interacting elements zinc and iron contribute to the varying results reported with chicks is not clear, but the source of the protein can be important. Jenkins and co-workers (229) obtained significant increases in chick growth when 250 $\mu g/g$ copper as copper sulfate was added to a diet composed mainly of wheat and fish meal with added tallow, but observed either no growth response or a depression when this amount of copper was added to a maize–soybean diet.

D. Sheep and Cattle

Under appropriate conditions crossbred sheep can be maintained in copper balance at intakes as low as 1 mg/day, which is equivalent to about 1 μg/g copper in the dry diet (113). In most environments conditions exist that impose a substantially higher copper requirement on sheep (and cattle) than 1 μg/g. Thus Marston and associates (295–297) found that a copper supplement of 5 mg/day, bringing the total dietary copper intake to about 8 μg/g, was insufficient to ensure normal blood copper levels and wool keratinization in all merino sheep grazing pastures grown on the calcareous soils of South Australia. Under these conditions, where there is a high consumption of calcium carbonate from the environment and moderate intakes of molybdate and sulfate from the herbage, the minimum copper requirements of wool sheep are close to 10 mg/day, or about 10 μg/g of the dry diet. In Western Australia pastures containing 4–6 μg/g copper (dry basis) and with molybdenum concentrations generally <1.5 μg/g provide sufficient copper for the full requirements of cattle and British-breed and crossbred sheep (27). To avoid defective keratinization of the wool, merino sheep require a minimum of 6 μg/g copper in the dry pastures. However, even the relatively low levels of molybdenum in these pastures may be having some effect on the copper requirements, as Dick (113) has shown that molybdenum intakes as low as 0.5 mg/day can adversely affect copper retention in the sheep, provided the sulfate intake is higher. Suttle (440) later also demonstrated effects on copper metabolism from molybdenum intakes much smaller than those commonly conceived as significant. These findings and the importance of the copper–molybdenum dietary ratio are considered in the chapter on molybdenum (Mills and Davis, Chapter 13, this volume).

The critical significance of the relative intakes of copper and molybdenum is further apparent from studies by Thornton and co-workers (451) in England. They found evidence of copper deficiency, responsive to copper therapy, in cattle grazing pastures "normal" in copper content (7–14 μg/g) and with molybdenum contents within the 3–20 μg/g range. Many of the animals showed no marked clinical signs of hypocuprosis, although unthriftiness and poor fertility were stated to be common in the affected area. All these investigations emphasize the difficulty of assigning precise minimum copper requirements and the impossibility of basing adequacy of copper intakes alone.

The most widely practiced means of providing adequate copper to grazing stock in many areas is to apply copper-containing fertilizers to the pastures. Australian experience indicates that 5–7 kg/hectare of copper sulfate, or its copper equivalent in copper ores, is sufficient to raise the copper in the herbage to adequate levels for several years. In some areas significant increases in total herbage production are coincidentally achieved. Under range conditions where fertilizers are not normally applied, or on calcareous soils where copper absorption by plants is poor, salt licks containing 0.5–1.0% copper sulfate may be supplied. Such licks are usually consumed by sheep and cattle in copper-deficient areas in sufficient quantities and with sufficient frequency to maintain adequate copper intakes.

Intramuscular or subcutaneous injections of copper glycinate or copper calcium ethylenediaminetetraacetate (CuCa edetate) at intervals of 3–6 months, in doses of 30–40 mg copper for sheep or 120–240 mg for cattle, have been found satisfactory (325,438). A single injection of 100 mg of CuCa edetate in copper-deficient cattle was found adequate to promote and sustain satisfactory blood copper levels for 6 months (451), and a single injection in midpregnancy of 40–50 mg copper, either as copper glycinate, methionate, or CuCa edetate, was similarly found effective in preventing swayback in the lambs and in maintaining satisfactory blood and liver copper concentrations in the ewes and the lambs (196). Some workers have reported deaths of ewes given subcutaneous CuCa edetate even when the injections are at the recommended levels (6,225).

V.   COPPER IN HUMAN HEALTH AND NUTRITION

A.   Dietary Deficiencies

The statement by Cartwright and Wintrobe (73) that clinical copper deficiency is rare in human subjects, is still valid today. Severe copper deficiency caused by inadequate copper intake under other than experimental and clinical conditions is known to occur only in infants. Dietary deficiency was first diagnosed in malnourished children in Peru during rehabilitation with milk powder low in copper. As they recovered from malnutrition they developed anemia, neutropenia, and disturbances of bone formation, together with decreased plasma levels of ceruloplasmin and copper. All these signs responded to copper supplementation (88,89). Several reports have described signs of copper deficiency in premature infants and in infants and adults maintained on total parenteral alimentation for various medical reasons (10,118,174).

Other forms of hypocupremia, resulting in mild copper deficiency, are unrelated to the dietary copper intake. One syndrome can result from excessive protein loss due to nephrosis or tropical sprue or from inadequate protein intake, as in kwashiorkor. Another, much more severe nonnutritional copper deficiency is the Menkes' steely-hair syndrome, a hereditary disease in which intestinal copper absorption, but also subsequent utilization, is inadequate (311). The underlying cause is as yet unknown. The syndrome resembles in some ways the dietary copper deficiency observed in Peruvian children, with low copper levels in the organism and bone changes; in addition, there are signs of mental disease and of specific structural changes in hair. The disease, in contrast to dietary deficiency, does not respond to copper supplementation and is considered incurable at the present state of our knowledge. Copper deficiency in adults has been reported in patients on total parenteral alimentation (468) and, interestingly, in sickle cell anemia patients who were given high-zinc supplements (150 mg/day) for almost 2 years (363,364). The zinc–copper antagonism will be discussed later in this chapter.

Klevay et al. reported three cases of experimentally induced copper deficiency in healthy adult men. The first, a 29-year-old, was fed a diet containing 0.8 mg copper for 105 days, during which plasma copper declined to 48 μg/dl. The SOD in erythrocytes declined to approximately 20% of its original value, but there was no indication of anemia or neutropenia. During the depletion period, the serum cholesterol increased from 202 to 234 mg/dl. The experiment was terminated when 24-hr ECGs revealed six ventricular and some superventricular premature beats, and resupplementation with copper was begun. After 2 weeks all abnormalities had disappeared. The authors concluded that the copper requirement of that patient exceeded 0.8 mg/day (259). The other two subjects, healthy young males, were kept on a low-copper diet furnishing 0.78 mg and 2500 kcal/day, for 30 and 100 days, respectively. Although the erythrocyte SOD appeared to be lower and plasma cholesterol higher during depletion than during the control period, there was no significant correlation between these parameters and the time of depletion. The hematological parameters were not affected by copper status. Plots of serum glucose concentrations following an intravenous glucose injection showed significantly higher glucose concentrations during the copper depletion phase than during either pretest or repletion phase, suggesting impaired glucose utilization. These differences, however, were not reflected in the glucose removal rates, calculated as the slope of the regression of serum glucose concentrations versus time, following the injection.

Reiser et al. (382) fed 23 male volunteers a diet containing 1 mg of copper and 2850 kcal/day (copper density 0.35 mg/1000 kcal) for a period of 11 weeks. This study had been designed to repeat in human subjects the observations consistently made in rats that fructose-containing diets markedly aggravate the consequences of marginal copper deficiency, in contrast to the protective effect of starch. The study, originally planned for 14 weeks, had to be terminated prematurely because of the occurrence of cardiovascular disturbances in 4 of the 23 subjects (one coronary infarct, two incidences of severe tachycardia, and one occurrence of extrasystolic beats) at the end of 11 weeks of depletion, although neither serum ceruloplasmin activity nor copper concentration had significantly declined. In contrast, the erythrocyte SOD levels were significantly depressed by the deficiency treatment and also by the substitution of starch in the diet by fructose. The SOD activities of the fructose-fed subjects were significantly lower than those of the starch-fed subjects and increased again significantly following repletion with copper.

Because none of the experimental studies reported have produced the classical signs of copper deficiency of anemia, leukopenia, and bone changes, the question arises whether the observed signs of heart-related dysfunctions should be considered more sensitive indicators of copper deficiency. The fact that such changes have been observed independently by two investigators would reinforce this assumption, as does the similarity of the deficiency signs described for humans with those of experimental animals.

B. Requirements and Intakes

An expert committee of the World Health Organization (479) has recommended a daily copper intake of 80 μg/kg body weight for infants. The same intake is recommended by the Committee on Recommended Dietary Allowances (331). This figure is substantially higher than the intake of 250 μg during the first month of life in breast-fed infants as determined by Casey *et al.* (76) or the 180 μg that can be calculated on the basis of the data of Picciano and Guthrie (357). It must be realized, however, that the official recommendations cited above address not only the need of breast-fed infants but also those raised on formula diets. The latter formulations presumably contain the copper in a less bioavailable form than that present in human milk. The recommendations for children and adults are based on balance studies and on those intakes that equal the daily copper loss in the subjects studied. It must be emphasized, however, that most balance studies do not take into account the obligatory loss of copper via the skin and its appendices. Thus, the balance method probably underestimates the daily requirement. Taking into account the results of such balance studies, especially those of Klevay *et al.* (263) in which 1.3 mg/day copper were needed to compensate for obligatory copper losses, the Food and Nutrition Board established an estimated Range of Safe and Adequate Intakes for copper of 2–3 mg (331), which is similar to the recommendation of a WHO expert committee of 30 μg/kg/day for an adult male (479).

While many of the older analytical studies suggested that most Western diets furnish amounts within that range, practically all later studies have clearly demonstrated that this is not the case. Klevay *et al.* (262) in their survey of 1979 found many daily diets in the United States containing considerably less than the 2 mg/day recommended, and in a careful analytical study of self-selected duplicate diets the mean intake was 1 mg/day (208). The copper intake of self-selected diets in the human study mentioned above was also close to 1 mg/day. Although there is good evidence that such figures are characteristic for typical diets consumed in the United States, they are probably not representative of diets in other countries, where organ meats are accepted as normal parts of the daily diet, where there is a substantial consumption of seafoods, or where foods of vegetable origin, such as various nuts rich in copper, are easily available (see Table I).

Very little is known about the bioavailability to human subjects of copper in different foods, and dietary interactions that affect copper are not well quantified. Although several interactions have been shown to affect copper status of experimental animals, the results of human studies are much less convincing, except for the interaction between excessive amounts of ascorbic acid and dietary copper. Finley and Cerklewski (146) demonstrated a significant decline of serum copper and of serum ceruloplasmin during 6 months of supplementation of 13 young men with 1500 mg of ascorbic acid daily. Cessation of the vitamin C supplementation resulted in a return of these depressed values toward normal. Adverse health effects, however, were not noted during the experiment. The effects of dietary fiber and phytate in amounts found in typical Western diets on copper status of human subjects are probably minimal (237,324,455). Whether the zinc–copper antagonism, demonstrated in animals, has practical importance in human dietaries is not known. The supplementation with 150 mg of zinc daily for 5 weeks has resulted in a significant depression of HDL cholesterol in human subjects. It has been postulated that this effect might be mediated through a copper–zinc interaction. There is as yet no direct evidence for the importance in humans of the copper–carbohydrate interaction, amply demonstrated in experimental animals.

The importance of drinking water, particularly soft water, as a source of copper to humans was stressed by Schroeder *et al.* (400). These workers observed a progressive increase in the copper concentration of waters from brooks, reservoirs, hospitals, and private homes. They maintained that some soft waters, with their capacity to corrode copper pipes, could raise intakes by as much as 1.4 mg/day, whereas hard waters would reduce this increment to ≤ 0.05 mg. Robinson *et al.* (384), working in New Zealand, have also shown that soft water can contribute appreciable quantities of copper. They calculated that if the beverages consumed by one individual had been made up with soft water from the cold tap, 0.4 mg/day copper would have been obtained from this source alone and no less than 0.8 mg/day if obtained from the hot tap. It is obvious that the drinking water cannot be neglected as a source of dietary copper, even though it is difficult to quantify. The possible occurrence of chronic copper poisoning in infants as a result of contamination of water supplies by copper pipes has been the subject of two reports (391,472).

Adverse effects of copper in drinking water have also been suggested in an epidemiological study bv Punsar *et al.* (374), who correlated morbidity and mortality from cardiovascular disease in Finland with the analyzed content of many trace elements in the drinking water of the respective households. Copper was positively correlated with cardiovascular diseases; the only other significant correlation, a negative one, was with chromium.

C. The Zinc–Copper Dietary Ratio as a Risk Factor in Coronary Heart Disease

Klevay (255) has hypothesized that an imbalance in zinc and copper metabolism contributes to the risk of coronary heart disease (CHD). This hypothesis was initially based on the results of experiments with rats consuming a cholesterol-free diet in which the intakes of zinc and copper were varied by varying the ratio of salts of these elements in the drinking water. Water with a zinc–copper ratio of 40 consistently and significantly produced higher concentrations of cholesterol in the plasma than did water with a ratio of 5. Subsequent investigations by this worker has given added substance to this hypothesis and has related variations in the zinc–copper ratio to other risk factors in CHD. For example, a relationship between the amount of fat and the ratio of zinc to copper of foods ($x^2 = 13.5$, $p < 0.001$) was demonstrated (253). Some U.S. meals and diets of considerable variability were shown to have zinc–copper ratios in excess of those which produce hypercholesterolemia in rats (256), and the mortality rate for CHD and the ratio of zinc to copper in milk of 47 cities in the United States were correlated ($r = 0.354$, $p < 0.02$) (254). Attention was also drawn to the higher zinc–copper ratio in cow's milk than in breast milk. It was suggested that one of the benefits that might be gained from a return to breast-feeding is a reduction in CHD. However, Walravens and Hambidge (474) found no effect on the plasma cholesterol levels of infants fed on a milk formula in which the zinc–copper ratio was raised from 5 : 1 to 17 : 1. In this connection it is pertinent to mention the work of Osborn (346), who studied the coronary arteries of 109 people up to the age of 20 and concluded that the arteries of those who had not been breast-fed had the greatest reductions in luminal size. Such an association by chance was calculated to be unlikely ($p < 0.005$) (254).

Although many of the epidemiological data support Klevay's hypothesis of high zinc–copper ratios as risk factors for CHD, the same author has subsequently placed more emphasis on copper deficiency per se and has emphasized the similarities between several experimental animal systems and human subjects with CHD (261). He stressed the importance of glucose intolerance, hypercholesterolemia, abnormal ECGs, hyperuricemia, and hypertension, all induced by copper deficiency, as predictive of risk of cardiovascular disease in humans. Later work has resulted in better knowledge of the interactions of copper with other inorganic and organic constituents of the diet. Such interactions are now known to determine the fate of animals fed low-copper diets. They may also explain some of the disagreements among researchers in the past concerning the expression of copper deficiency in various experimental systems.

VI. COPPER TOXICITY

Copper toxicity in animals has been recognized for a great many years and, as with other metals, copper toxicity can occur in all animal species. In humans, acute copper toxicity is usually associated with accidental consumption of copper sulfate, usually by children, or as used as a means of suicide as has been reported in India (82). In such poisoning there is a metallic taste in the mouth associated with epigastric pain, nausea, vomiting, and diarrhea (299). In severe cases vascular collapse and death may occur.

Chronic copper poisoning can occur when the mechanism for sequestering copper in the liver is exceeded. When this occurs there is a hemolytic crisis much like that occurring in sheep. The copper toxicity of Wilson's disease develops with copper accumulation in the renal tubules, cornea, brain, and other organs, resulting in damage to these structures (396,398). The application of copper salts to denuded surfaces can result in copper toxicosis (209,292).

In some breeds of sheep, continued ingestion of copper at levels as low as 12 $\mu g/g$ of the diet can result in copper toxicosis (329). Sheep are relatively more susceptible to copper toxicosis than other domestic species, and it has been noted that copper chloride is two to four times more toxic than copper sulfate. Sheep have been poisoned by 20–100 mg of copper per kilogram when given as a single dose. Sheep consuming excessive amounts of copper over a period of weeks to months with low molybdenum intake usually do not show toxic symptoms until the liver concentration of copper exceeds 150 $\mu g/g$ wet weight. When this level of copper in the liver has been reached, the animal becomes listless, weak, tremulous, and anorexic. Hemoglobinemia, hemoglobinuria, and icterus develop and mortality is often $>75\%$.

Animal species differ greatly in their tolerance of high levels of copper in the diet. Rats appear to be particularly tolerant, according to Boyden et al. (44), who reported normal growth and health in rats maintained on diets containing 500 $\mu g/g$ of copper, about 100 times the normal level, despite a 14-fold increase in liver copper content. The actual level of copper in the liver never attained the levels seen in chronic copper poisoning in sheep.

Pigs and poultry also appear to be comparatively tolerant of high dietary copper. Diets containing 500 $\mu g/g$ of copper will slow growth and egg production and 1200 $\mu g/g$ can prove fatal for laying fowl (232,434). Christmas and Harms (81) found that 900 $\mu g/g$ of copper in the diet of turkeys was slightly toxic, but that 2400 $\mu g/g$ proved fatal.

Zinc, iron, cadmium, and molybdenum in diets or supplements all interact with copper metabolically. For this reason it is impossible to give a maximum or minimum tolerable dietary copper level based on copper content alone. The level of protein and the level of the interacting metal ion and sulfate all can influence the absorption and utilization of copper. Consequently, any statement about copper levels in the diet as related to toxicity has to be made in the context of the presence of these other substances in the diet.

A. Copper Toxicity in Pigs

Reports that increased growth rates of swine could be obtained with diets containing up to 250 μg/g of added copper, prompted a number of investigations of the potential toxicity of these high levels of copper. It was found that copper toxicity would result if suitable amounts of zinc and iron were not added at the same time as the supplemental copper (7,59,383,443). Suttle and Mills demonstrated that, even with dietary levels of 250, 425, and 750 μg/g of copper, all of the signs of copper toxicity which occurred on these levels could be eliminated by providing simultaneously an additional 150 μg/g of zinc and 150 μg/g of iron to the diets containing 250 and 425 μg/g of copper, but that an additional 500 μg/g of zinc and 750 μg/g of iron were necessary to eliminate jaundice and to produce normal serum copper and aspartate aminotransferase values when 750 μg/g of copper were present in the diet. Iron was necessary to protect against the development of anemia. The role of zinc and iron in preventing copper toxicosis when high levels of copper are fed would appear to be explained by the competition between the metal ions for protein-binding sites in the absorption process (50). Additional protein may also have a beneficial effect (473).

B. Copper Toxicity in Ruminants

Sheep appear to be the most susceptible of the domestic animals to copper toxicosis. Chronic copper poisoning occurred in grazing sheep in Australia where the copper content of the soil was high, when grazing on pastures such as subterraneum clover on acid soils (*Trifolium subterraneum*), when sheep grazed in orchards and vineyards sprayed with copper compounds and when pastures had been sprayed with copper sulfate as a molluscicide (58,169,329). Dry feed has also resulted in copper toxicity (270) or when given in a mineral salt mixture containing a recommended level of copper, free choice (269).

Copper poisoning has occurred in sheep when the copper content of the soil and pastures was abnormally high or when, in the presence of normal copper levels the molybdenum was very low. Copper toxicity may occur (58,329) in sheep with liver damage caused by the plant *Heliotropium europaeum*. When sheep have received diets that were actually formulated for pigs with added copper to stimulate growth, copper toxicosis has developed (446).

Copper toxicosis occurs in sheep when the animals consume copper in excess of their requirement over a period of time that varies with the level of copper that is consumed. It may occur under intensive rearing on a cereal-based diet with added copper (452). It can occur when sheep graze plants that may contain as much as 50–60 μg/g of copper on a dry basis, as can happen when the plants are grown on acid-type soils. There is a breed difference in response with some breeds more susceptible to copper toxicosis than others (490,491). Merinos appear to be less susceptible than other breeds. When sheep are grazed on soils where subterraneum clover is grown, and the clover may contain as much as 10–15 μg/g of copper and very little molybdenum, sheep may develop chronic copper toxicity. Overcoming this problem by providing molybdenum and sulfate in salt licks or by dosing the animals with ammonium molybdate and sodium sulfate has been successful (110, 385,386). Liver copper levels were reduced and mortality was prevented.

Bull *et al.* (58) have reported on the enzootic and hemolytic jaudice that occurs when sheep are raised on the plant *H. europaeum,* which causes damage to the liver parenchyma due to the presence of the alkoloids heliotrine and lasiocarpine. The parenchyma cells and nuclei increase in size, have a shortened life and are not replaced by new liver cells. As the liver becomes atrophic, copper retention is extremely high, often over 1000 μg/g on a dry, fat-free basis, and death results from copper poisoning. Howell and Kumaratilake (215) have discussed the morphological distribution of copper-loaded hepatic parenchyma cells in toxicosis of sheep, and Gooneratne and Howell (165) have examined the pathology of hepatic and renal cells in chronically poisoned sheep.

There are two phases in the process of copper toxicosis. During the first phase there is an accumulation of copper in the tissues with a significant rise in serum transaminase and lactic dehydrogenase. This occurs several weeks before the hemolytic crisis develops (386,454). The action of thiomolybdate in reducing the copper in liver and therefore preventing the mortality from copper toxicosis has led to attempts to ascertain changes in blood that occur prior to the crisis. Serum glutamic oxaloacetic transaminase (SGOT) determination has been proposed as a means of early detection. The hemolytic crisis is accompanied by a marked rise in copper levels, by falls in hemoglobin and glutathione concentrations, while the methemoglobin rises (453). There is also a sudden dramatic rise in creatine phosphokinase (CPK) levels in serum, suggesting that the changes occur in the muscle cell membranes as well as in the liver and kidney and brain. Ishmael *et al.* (224) noted that there was a swelling and necrosis of the hepatic parenchyma cells and Kupffer cells about 6 weeks before the hemolytic crisis. These cells are rich in acid phosphatase and contain PAS-positive diastase-resistant material as well as copper. In addition, the liver parenchyma exhibits a reduced activity of ATPase, nonspecific esterases, succinic tetrazolium reductase, and glutamic dehydrogenase, all pointing to a functional disturbance of the liver which occurs some time before the hemolytic crisis. Hemolytic crises are associated with focal necrosis of liver tissue, neutrophilia, and high blood urea levels. Kidney and liver copper levels are high and copper levels of the spinal cord are slightly elevated. It has been suggested that the release of copper from the liver which occurs with damage to the liver cells may be the factor which precipitates the hemolytic crisis. Experimental copper toxicosis has demonstrated that the copper in liver and kidneys rises significantly prior to hemolysis (166). In animals that develop hemolysis, degeneration, necrosis, and loss of enzyme activity from the cells of the proximal convoluted tubules become apparent with the functional impairment. The brain and spinal cord also appear to suffer changes with chronic copper poisoning (216,323). There is a change in the ultrastructure of spongiform white matter, but there is no significant involvement of neurons, glia, or blood vessels and no apparent increase in the extracellular space. Howell *et al.* (216) noted that there were astrocytic changes in hemolytic and posthemolytic animals with an increase in the volume of astrocytic nuclei. The authors suggest that the changes that occur in the brain of animals with chronic copper poisoning may be due to the effect of altered metabolic processes of glial transport mechanisms.

It has been observed that young calves will develop copper toxicity on relatively low copper intakes when receiving milk-based diets (406). In cattle the dose required to cause copper toxicity may be 3–12 times that required to prevent deficiency (378). Continued supplementation of diets with high copper levels can result in liver copper concentrations as high as 2000 $\mu g/g$ on a dry matter basis (435). Such levels in cattle are followed by an increase in renal copper with perirenal edema and tubular necrosis. Hemorrhage and hemoglobinuria precede death. In chronic poisoning of cattle the onset is gradual until the onset of sudden crises of hemoglobinuria, hemoglobinemia, and icterus (238). Vitamin A supplements enhance the severity of the toxicosis. Both oral and parenteral routes of application are effective to elevate hepatic copper levels (321). Hair copper levels have been reported to increase from 7 to 22 $\mu g/g$, together with darkening of the hair (106).

ACKNOWLEDGMENTS

When it became evident that repeated delays in the completion of the copper chapter would jeopardize even the once postponed publication date of Vol. 1, Dr. G. K. Davis, on short notice, kindly consented to revise the sections dealing with copper in foods, feeds, and tissues and with toxicology, while the editor updated the remaining sections, with parts of Underwood's text left intact. The substantial help from Drs. M. Fields and M. Failla of the Beltsville Human Nutrition Research Center, and Dr. D. E. Ullrey of Michigan State University is gratefully acknowledged.

REFERENCES

1. Abderhalden, E. (1900). *Z. Biol.* **39,** 482.
2. Abraham, P. A., and Evans, J. L. (1971). *Trace Subst. Environ. Health Proc. Univ. Mo. Annu. Conf., 5th* p. 335.

3. Agricultural Research Council (Great Britain) (1967). *Nutr. Requir. Farm. Livestock* No. 3.
4. Allcroft, R., and Lewis, G. (1957). *J. Sci. Food Agric.* **8**, (Suppl.), S96.
5. Allcroft, R., and Parker, W. H. (1949). *Br. J. Nutr.* **3**, 205.
6. Alcroft, R., Buntain, D., and Rowlands, W. T. (1965). *Vet. Rec.* **77**, 634.
7. Allcroft, R., Burns, K. N., and Lewis, G. (1961). *Vet. Rec.* **73**, 714.
8. Allcroft, R., Clegg, F. G., and Uvarov, O. (1959). *Vet. Rec.* **71**, 884.
9. Alloway, B. J. (1973). *J. Agric. Sci.* **80**, 521.
10. Al-Rashid, R. A., and Spangler, J. (1971). *N. Engl. J. Med.* **285**, 841.
11. Alvarez, R. (1985). *In* "Trace Elements in Man and Animals" (C. F. Mills, I. Bremner, and J. K. Chesters, eds.), p. 665. Commonwealth Agric. Bureaux, Farnham Royal, U.K.
12. Balevska, P. S., Russanov, E. M., and Kassabova, T. A. (1981). *Int. J. Biochem.* **13**, 489.
13. Balevska, P. S., Russanov, E. M., and Stanchev, P. I. (1985). *Biol. Trace Elem. Res.* **8**, 211.
14. Barber, R. S., Braude, R., Mitchell, K. G., and Cassidy, J. (1955). *Chem. Ind. (London)* **21**, 601; Barber, R. S., Braude, R., and Mitchell, K. G. (1955). *Br. J. Nutr.* **9**, 378.
15. Barber, R. S., Braude, R., Mitchell, K. G., and Rook, J. F. (1957). *Br. J. Nutr.* **11**, 70.
16. Barlow, R. M. (1963). *J. Comp. Pathol.* **73**, 51, 61.
17. Barlow, R. M., and Cancilla, P. (1960). *Acta Neuropathol.* **6**, 175.
18. Barlow, R. M., Field, A. C., and Ganson, N. C. (1964). *J. Comp. Pathol. Ther.* **74**, 530.
19. Barlow, R. M., Purves, D., Butler, E. J., and McIntyre, I. J. (1960). *J. Comp. Pathol. Ther.* **70**, 396, 411.
20. Bass, B., McCall, J. T., Wallace, H. D., Combs, G., Palmer, A. Z., and Carpenter, J. E. (1956). *J. Anim. Sci.* **15**, 1230.
21. Baxter, J. H., and Van Wyk, J. J. (1953). *Bull. John Hopkins Hosp.* **93**, 1.
22. Baxter, J. H., Van Wyk, J. J., and Follis, R. H., Jr. (1953). *Bull. Johns Hopkins Hosp.* **93**, 25.
23. Bearn, A. G., and Kunkel, H. (1954). *J. Clin. Invest.* **33**, 400.
24. Beck, A. B. (1941). *Aust. J. Exp. Biol. Med. Sci.* **19**, 145.
25. Beck, A. B. (1956). *Aust. J. Zool.* **4**, 1.
26. Beck, A. B. (1961). *Aust. J. Agric. Res.* **12**, 743.
27. Beck, A. B., and Harley, R. (1951). *West. Aust. Dep. Agric. Leafl.* No. 678.
28. Beisel, W. R., Pekarek, R. S., and Wannemacher, R. W., Jr. (1974). *In* "Trace Element Metabolism in Animals" (W. G. Hoekstra, J. W. Suttie, H. E. Ganther, and W. Mertz, eds.), p. 217. Univ. Park Press, Baltimore.
29. Bennett, F. I., Golden, M. H. N., Golden, B. E., and Ramdath, D. D. (1985). *In* "Trace Elements in Man and Animals" (C. F. Mills, I. Bremner, and J. K. Chesters, eds.), p. 578. Commonwealth Agric. Bureaux, Farnham Royal, U.K.
30. Bennetts, H. W. (1932). *Aust. Vet. J.* **8**, 137, 183.
31. Bennetts, H. W., and Beck, A. B. (1942). *Aust. C.S.I.R.O. Bull.* **147**, 1.
32. Bennetts, H. W., and Chapman, F. E. (1937). *Aust. Vet. J.* **13**, 138.
33. Bennetts, H. W., Beck, A. B., and Harley, R. (1948). *Aust. Vet. J.* **24**, 237.
34. Bennetts, H. W., and Hall, H. T. B. (1939). *Aust. Vet. J.* **15**, 52.
35. Bennetts, H. W., Harley, R., and Evans S. T. (1942). *Aust. Vet. J.* **18**, 50.
36. Bhathena, S. J., and Recant, L. (1987). *In* "Biology of Copper Complexes" (J. R. J. Sorenson, ed.). Humana Press, Clifton, N.J., in press.
37. Bhathena, S. J., Recant, L., Voyles, N. R., Timmers, K. I., Reiser, S., Smith, J. C., Jr., and Powell, A. S. (1986). *Am. J. Clin. Nutr.* **43**, 42.
38. Bingley, J. B. (1974). *Aust. J. Agric. Res.* **25**, 467.
39. Bingley, J. B., and Dufty, J. H. (1972). *Res. Vet. Sci.* **13**, 8.
40. Bohman, V. R., Drake, E. L., and Beherns, W. C. (1984). *J. Dairy Sci.* **67**, 1468.
41. Bowland, J. P., Braude, R., Chamberlain, A. C., Glascock, R. F., and Mitchell, K. G. (1961). *Br. J. Nutr.* **15**, 59.

42. Bowness, J. M., and Morton, R. A. (1952). *Biochem. J.* **51**, 530.
43. Bowness, J. M., Morton, R. A., Shakir, M. H., and Stubbs, A. L. (1952). *Biochem. J.* **51**, 521.
44. Boyden, R., Potter, V. R., and Elvehjem, C. A. (1938). *J. Nutr.* **15**, 397.
45. Boyne, R., and Arthur, J. R. (1981). *J. Comp. Pathol.* **91**, 271.
46. Bradfield, R. B., Hoo, T. S., and Baertl, J. M. (1980). *Am. J. Clin. Nutr.* **33**, 1315.
47. Braude, R. (1967). *World Rev. Anim. Prod.* **3**, 69.
48. Braude, R., and Ryder, K. (1973). *J. Agric. Sci.* **80**, 489.
49. Braude, R., Townsend, M. J., Harrington, G., and Rowell, J. G. (1962). *J. Agric. Sci.* **58**, 251.
50. Bremner, I. (1974). *Q. Rev. Biophys.* **7**, 75.
51. Bremner, I. (1979). *In* "Copper in the Environment, Part 2" (J. O. Nriagu, ed.), p. 285. Wiley, New York.
52. Bremner, I. (1980). *In* "Biological Roles of Copper," p. 23. Elsevier, New York.
53. Brewer, G. J., Shoomaker, E. B., Leichtman, E. A., Kruckeberg, W. C., Brewer, L. F., and Meyers, N. (1979). *In* "Zinc Metabolism: Current Aspects in Health and Disease" (G. J. Brewer and A. A. Prasad, eds.), p. 241. Liss, New York.
54. Bronner, F., and Yost, J. H. (1985). *Am. J. Physiol.* **249**, G108.
55. Brooks, C. B., and Cummings, L. P. (1984). *J. Nutr. Elderly* **3**, 27.
56. Brown, E. D., Howard, M. P., and Smith, J. C., Jr. (1977). *Fed. Proc., Fed. Am. Soc. Exp. Biol.* **36**, 1122.
57. Buffoni, F., and Blaschko, H. (1964). *Proc. R. Soc. London Ser. B* **161**, 153.
58. Bull, L. B., Culvenor, C. C. J., and Dick, A. T. (1968). *Front Biol.* **9** ("The Pyrrolizidine Alkaloids. Their Chemistry and other Biological Properties."). Wiley, New York.
59. Bunch, R. J., Speer, V. C., Hays, V. W., and McCall, J. T. (1963). *J. Anim. Sci.* **22**, 56.
60. Burley, R. W., and de Koch, W. T. (1957). *Arch. Biochem. Biophys.* **68**, 21.
61. Burns, R. H., Johnston, A., Hamilton, J. W., McColloch, R. J., Duncan, W. E., and Fisk, H. G. (1964). *J. Anim. Sci.* **23**, 5.
62. Burnstein, P., Kang, A. H., and Piez, K. A. (1966). *Proc. Natl. Acad. Sci. U.S.A.* **55**, 417.
63. Bush, J. A., Mahoney, J. P., Markowitz, H., Gubler, C. J., Cartwright, G. E., and Wintrobe, M. M. (1955). *J. Clin. Invest.* **34**, 1766.
64. Butler, E. J. (1963). *Comp. Biochem. Physiol.* **9**, 1.
65. Butler, E. J., Curtis, M. J., Harry, E. G., and Deb, J. R. (1972). *J. Comp. Pathol.* **82**, 299.
66. Cancilla, P. A., and Barlow, R. M. (1970). *J. Comp. Pathol.* **80**, 315.
67. Carlton, W. W., and Henderson, W. (1964). *Avian Dis.* **8**, 48.
68. Carlton, W. W., and Henderson, W. (1965). *J. Nutr.* **85**, 67.
69. Carlton, W. W., and Kelly, W. A. (1969). *J. Nutr.* **97**, 42.
70. Carnes, W. H., Shields, G. S., Cartwright, G. E., and Wintrobe, M. M. (1961). *Fed. Proc., Fed. Am. Soc. Exp. Biol.* **20**, 118.
71. Cartwright, G. E. (1950). *In* "A Symposium on Copper Metabolism" (W. D. McElroy and B. Glass, eds.), p. 274. Johns Hopkins Press, Baltimore.
72. Cartwright, G. E., and Wintrobe, M. M. (1964). *Am. J. Clin. Nutr.* **14**, 224.
73. Cartwright, G. E., and Wintrobe, M. M. (1964). *Am. J. Clin. Nutr.* **15**, 94.
74. Cartwright, G. E., Gubler, C. J., and Wintrobe, M. M. (1954). *J. Clin. Invest.* **33**, 685.
75. Cartwright, G. E., Markowitz, H., Shields, G. S., and Wintrobe, M. M. (1960). *Am. J. Med.* **28**, 555.
76. Casey, C. E., Hambidge, K. M., and Neville, M. C. (1985). *Am. J. Clin. Nutr.* **41**, 1193.
77. Cavell, P. A., and Widdowson, E. M. (1964). *Arch. Dis. Child.* **39**, 496.
78. Chang, I. C., Lee, T. P., and Matrone, G. (1975). *J. Nutr.* **105**, 624.
79. Charmley, L. L., and Symonds, H. W. (1985). *In* "Trace Elements in Man and Animals" (C.

F. Mills, I. Bremner, and J. K. Chesters, eds.), p. 339. Commonwealth Agric. Bureaux, Farnham Royal, U.K.

80. Chesters, J. K., Mills, C. F., and Price, J. (1985). *In* "Trace Elements in Man and Animals" (C. F. Mills, I. Bremner, and J. K. Chesters, eds.), p. 351. Commonwealth Agric. Bureaux, Farnham Royal, U.K.

81. Christmas, R. B., and Harms, R. H. (1979). *Poult. Sci.* **58**, 382.

82. Chuttani, H. K., Gupta, P. S., Gulati, S., and Gupta, D. N. (1965). *Am. J. Med.* **39**, 849.

83. Claypool, D. W., Adams, F. W., Pendell, H. W., Hartman, N. A., Jr., and Bone, J. F. (1975). *J. Anim. Sci.* **41**, 911.

84. Cohen, A. M., Teitelbaum, A., Miller, E., Ben-Tor, V., Hirt, R., and Fields, M. (1982). *Isr. J. Med. Sci.* **18**, 840.

85. Cohen, D. I., Illowsky, B., and Linder, M. C. (1979). *Am. J. Physiol.* **236**, E309.

86. Cohen, N. L., Keen, C. L., Lönnerdahl, B., and Hurley, L. S. (1983). *Biochem. Biophys. Res. Commun.* **113**, 127.

87. Cook, C. S., and McGahan, M. C. (1986). *Curr. Eye Res.* **5**, 69.

88. Cordano, A. (1978). *In* "Zinc and Copper in Clinical Medicine" (K. M. Hambidge and B. F. Nichols, Jr., eds.), p. 119. Spectrum, New York.

89. Cordano, A., Baertl, J. M., and Graham, G. G. (1964). *Pediatrics* **34**, 324.

90. Cotteril, O. J., Marion, W. W., and Naber, E. C. (1977). *Poult. Sci.* **56**, 1927.

91. Coulson, W. F., and Carnes, W. H. (1963). *Am. J. Pathol.* **43**, 945.

92. Cousins, R. J. (1985). *Physiol. Rev.* **65**, 238.

93. Crampton, R. F., Matthews, D. M., and Poisner, R. (1965). *J. Physiol. (London)* **178**, 111.

94. Cunnane, S. C., Horrobin, D. F., and Manku, M. S. (1985). *Ann. Nutr. Metab.* **29**, 103.

95. Cunningham, I. J. (1931). *Biochem. J.* **25**, 1267.

96. Cunningham, I. J. (1950). *In* "A Symposium on Copper Metabolism" (W. D. McElroy and B. Glass, eds.), p. 246. Johns Hopkins Press, Baltimore.

97. Cunningham, I. J., and Hogan, K. G. (1958). *N.Z.J. Agric. Res.* **1**, 841.

98. Cymbaluk, N. F., Bristol, F. M., and Christensen, D. A. (1986). *Am. J. Vet. Res.* **47**, 192.

99. Dang, H. S., Jaiswal, D. D., Wadhwani, C. N., Somersunderan, S., and Dacosta, H. (1985). *Sci. Total Environ.* **44**, 177.

100. Danks, D. M., Campbell, P. E., Mayne, V., and Cartwright, E. (1972). *Pediatrics* **50**, 188.

101. Danks, D. M., Campbell, P. E., Stevens, B. J., Mayne, V., and Cartwright, E. (1972). *Pediatrics* **50**, 188.

102. Danks, D. M., Campbell, P. E., Walker-Smith, J., Stevens, B. J., Gillespie, J. M., Blomfield, J., and Turner, B. (1972). *Lancet* **1**, 1100.

103. Danks, D. M., Cartwright, E., Stevens, B. J., and Townley, R. R. W. (1973). *Science* **179**, 1140.

104. Davis, G. K. (1950). *In* "A Symposium on Copper Metabolism" (W. D. McElroy and B. Glass, eds.), p. 216. Johns Hopkins Univ. Press, Baltimore.

105. Davis, G. K. (1974). *Fed. Proc., Fed. Am. Soc. Exp. Biol.* **33**, 1194.

106. Deland, M. P. B., Cunningham, P., Milne, M. L., and Dewey, D. W. (1979). *Aust. Vet. J.* **55**, 493.

107. Deosthale, Y. G., and Gopalan. C. (1974). *Br. J. Nutr.* **31**, 351.

108. Devaraj, M., Patel, A. V., and Janakiraman, K. (1985). *Indian J. Anim. Sci.* **55**, 228.

109. Dewar, W. A., Teague, P. W., and Downie, J. N. (1974). *Br. Poult. Sci.* **15**, 119.

110. Dick, A. T., and Bull, L. B. (1945). *Aust. Vet. J.* **21**, 70.

111. Dick, A. T. (1953). *Aust. Vet. J.* **29**, 233.

112. Dick, A. T. (1954). Doctoral thesis, University of Melbourne, Australia.

113. Dick, A. T. (1954). *Aust. J. Agric. Res.* **5**, 511.

114. Dick, A. T. (1956). *Soil Sci.* **81**, 229.

115. Dorea, J. G., and Essado-Pereira, S. (1983). *J. Indian Chem. Soc.* **60,** 2375.
116. Dreosti, I. E., and Quicke, G. V. (1966). *S. Afr. J. Agric. Sci.* **9,** 365.
117. Dunkley, W. J., Ronning, M., and Voth, J. (1963). *J. Dairy Sci.* **46,** 1059.
118. Dunlap, W. M., James, G. W., III, and Hume, D. M. (1974). *Ann. Intern. Med.* **80,** 470.
119. Dutt, B., and Mills, C. F. (1960). *J. Comp. Pathol. Ther.* **70,** 120.
120. Eden, A. (1940). *J. Comp. Pathol. Ther.* **53,** 90.
121. Eden, A., and Green, H. (1939). *J. Comp. Pathol. Ther.* **52,** 301.
122. Eden, A., Hunter, A. A., and Green, H. H. (1945). *J. Comp. Pathol.* **55,** 29.
123. Elvehjem, C. A. (1935). *Physiol. Rev.* **15,** 471.
124. Elvehjem, C. A., Steenbock, H., and Hart, E. B. (1929). *J. Biol. Chem.* **83,** 27.
125. Engel, R. W., Price, N. O., and Miller, R. F. (1967). *J. Nutr.* **92,** 197.
126. Evans, G. W. (1971). *Nutr. Rev.* **29,** 195.
127. Evans, G. W., and Hahn, C. J. (1974). *In* "Protein–Metal Interactions" (M. Friedman, ed.). Plenum, New York.
128. Evans, G. W., Grace, C. I., and Votava, H. J. (1975). *Am. J. Physiol.* **228,** 501.
129. Evans, J. L., and Abraham, B. A. (1973). *J. Nutr.* **103,** 196.
130. Evans, J. L., and Brar, B. S. (1974). *J. Nutr.* **104,** 1285.
131. Everson, G. J., Tsai, M. C., and Wang, T. (1967). *J. Nutr.* **93,** 533.
132. Fay, J., Cartwright, G. E., and Wintrobe, M. M. (1949). *J. Clin. Invest.* **28,** 487.
133. Fell, B.F., Dinsdale, B.,and Mills, C.F.(1975) *Res. Vet. Sci.* **18,** 274.
134. Fell, B. F., Mills, C. F., and Boyne, R. (1965). *Res. Vet. Sci.* **6,** 10.
135. Fell, B. F., Williams, R. B., and Mills, C. F. (1961). *Proc. Nutr. Soc.* **20,** xxvii.
136. Ferguson, W. S., Lewis, A. H., and Watson, S. J. (1938). *Nature (London)* **141,** 553.
137. Festa, M. D., Anderson, H. L., Dowdy, R. P., and Ellersieck, M. R. (1985). *Am. J. Clin. Nutr.* **41,** 285.
138. Fields, M., Ferretti, R. J., Smith, J. C., and Reiser, S. (1983). *J. Nutr.* **113,** 1335.
139. Fields, M., Ferretti, R. J., Reiser, S., and Smith, J. C. (1984). *Proc. Soc. Exp. Biol. Med.* **175,** 530.
140. Fields, M., Ferretti, R. J., Smith, J. C., and Reiser, S. (1983). *Life Sci.* **34,** 763.
141. Fields, M., Ferretti, R. J., Smith, J. C., and Reiser, S. (1984). *Biol. Trace Elem. Res.* **6,** 379.
142. Fields, M., Ferretti, R. J., Smith, J. C., and Reiser, S. (1984). *J. Nutr.* **114,** 393.
143. Fields, M., Holbrook, J., Scholfield, D., Smith, J. C., Jr., Reiser, S., and Los Alamos Medical Research Group (1986). *J. Nutr.* **116,** 625.
144. Fields, M., Lewis, C., Scholfield, D. J., Powell, A. S., Rose, A. J., Reiser, S., and Smith, J. C. (1986). *Proc. Soc. Exp. Biol. Med.* **183,** 145–149.
145. Fields, M., Reiser, S., and Smith, J. C. (1983). *Nutr. Rep. Int.* **29,** 163.
146. Finley, E. B., and Cerklewski, F. L. (1983). *Am. J. Clin. Nutr.* **37,** 553.
147. Fischer, P. W. F., Giroux, A., and L'Abbé, M. (1984). *Am. J. Clin. Nutr.* **40,** 743.
148. Fonnesbeck, P. V., and Lloyd, H. (1982). "International Feedstuffs Institute Table of Feed Composition." Utah State University Logan, Utah.
149. Frieden, E. (1971). *Adv. Chem. Ser.* **100,** 292.
150. Frieden, E. (1980). *In* "Biological Roles of Copper," p. 93. Elsevier, New York.
151. Friel, J. K., Gibson, R. S., Peliowski, A., and Watts, J. (1984). *J. Pediatr.* **104,** 863.
152. Friendship, R. M., Wilson, M. R., and Gibson, R. S. (1985). *Can. J. Comp. Med.* **49,** 308.
153. Fuller, R., Newland, L. G. M., Briggs, C. A. E., Braude, R., and Mitchell, K. G. (1960). *J. Appl. Bacteriol.* **23,** 195.
154. Gallagher, C. H. (1957). *Aust. Vet. J.* **33,** 311.
155. Gallagher, C. H., Judah, J. D., and Rees, K. R. (1956). *Proc. R. Soc. London Ser. B* **145,** 134, 195.
156. Gallagher, C. H., and Reeve, V. E. (1971). *Aust. J. Exp. Biol. Med. Sci.* **49,** 21.

157. Gibson, R. S., and Gibson, I. L. (1984). *Sci. Total Environ.* **39**, 93.
158. Gipp, W. F., Pond, W. G., Kallfelz, F. A., Tasker, J. B., Van Campen, D. R., Krook, L., and Visek, W. J. (1974). *J. Nutr.* **104**, 532.
159. Gitlin, D., Hughes, W. L., and Janeway, C. A. (1960). *Nature (London)* **188**, 150.
160. Givens, D. D., Hopkins, J. R., Brown, M. E., and Walsh, W. A. (1981). *J. Agric. Sci.* **97**, 497.
161. Gollan, J. L. (1975). *Clin. Sci. Mol. Med.* **49**, 237.
162. Goodman, J. R., Warshow, J. B., and Dallman, R. P. (1970). *Pediatr. Res.* **4**, 244.
163. Gooneratne, R., and Christensen, D. (1985). *In* "Trace Elements in Man and Animals" (C. F. MIlls, I. Bremner, and J. K. Chesters, eds.), p. 334. Commonwealth Agric. Bureaux, Farnham Royal, U.K.
164. Gooneratne, R., Christensen, D., Chaplin, R., and Trent, A. (1985). *In* "Trace Elements in Man and Animals" (C. F. Mills, I. Bremner, and J. K. Chesters, eds.), p. 342. Commonwealth Agric. Bureaux, Farnham Royal, U.K.
165. Gooneratne, S. R., and Howell, J. McC. (1985). *In* "Trace Elements in Man and Animals" (C. F. Mills, I. Bremner, and J. K. Chesters, eds.), p. 187. Commonwealth Agric. Bureaux, Farnham Royal, U.K.
166. Gopinath, C., Hall, G. A., and Howell, J. McC. (1974). *Res. Vet. Sci.* **16**, 57.
167. Gordon, G. F. (1985). *Sci. Total Environ.* **32**, 133.
168. Grace, N. D. (1975). *Br. J. Nutr.* **34**, 73.
169. Gracey, J. F., and Todd, J. R. (1960). *Br. Vet. J.* **116**, 405.
170. Graham, G. G., and Cordano, A. (1976). *In* "Trace Elements in Human Health and Disease" (A. S. Prasad, ed.), Vol. I, p. 363. Academic Press, New York.
171. Gray, L. F., and Daniel, L. J. (1964). *J. Nutr.* **84**, 31.
172. Greger, J. L., and Snedeker, S. M. (1980). *J. Nutr.* **110**, 2243.
173. Gregoriadis, G., and Sourkes, T. (1967). *Can. J. Biochem.* **45**, 1841.
174. Griscom, N. T., Craig, J. N., and Neuhauser, E. B. D. (1971). *Pediatrics* **48**, 883.
175. Gubler, C. J., Cartwright, G. E., and Wintrobe, M. M. (1957). *J. Biol. Chem.* **244**, 533.
176. Gubler, C. J., Lahey, M. E., Chase, M. S., Cartwright, G. E., and Wintrobe, M. M. (1952). *Blood* **7**, 1075.
177. Guggenheim, K. (1964). *Blood* **23**, 786.
178. Guggenheim, K., Tal, E., and Zor, V. (1964). *Br. J. Nutr.* **18**, 529.
179. Haavikko, K., Antilla, A., Helle, A., and Pesonen, E. (1985). *Acta Pediatr. Scand. (Suppl.)* **318**, 213.
180. Hall, G. A., and Howell, J. McC. (1969). *Br. J. Nutr.* **23**, 41.
181. Halls, D. J., Fell, G. S., and Dunbar, P. M. (1981). *Clin. Chim. Acta* **114**, 21.
182. Halsted, J. A., Hackley, B. B., and Smith, J. C., Jr. (1968). *Lancet* **2**, 278.
183. Hambidge, K. M. (1973). *Am. J. CLin. Nutr.* **26**, 1212.
184. Hambidge, K. M., and Droegemuller, W. (1974). *Obstet. Gynecol.* **44**, 666.
185. Hamilton, T. S., and Mitchell, H. H. (1949). *J. Biol. Chem.* **178**, 345.
186. Harris, D. I. M., and Sass-Kortsak, A. (1967). *J. Clin. Invest.* **46**, 659.
187. Harris, E. D. (1976). *Proc. Natl. Acad. Sci. U.S.A.* **73**, 371.
188. Harris, E. D. (1986). *J. Nutr.* **116**, 252.
189. Hart, E. B., Steenbock, H., Waddell, J., and Elvehjem, C. A. (1928). *J. Biol. Chem.* **77**, 797.
190. Hartmans, J., and Bosman, M. S. M. (1970). *In* "Trace Element Metabolism in Animals," (C. F. Mills, ed.), Vol. 1, p. 362. Livingstone, Edinburgh.
191. Hassel, C. A., Marchello, J. A., and Lei, K. Y. (1983). *J. Nutr.* **113**, 1081.
192. Hatano, S., Nishi, Y., and Usui, T. (1982). *Am. J. Clin. Nutr.* **35**, 120.
193. Hayter, S., Wiener, G., and Field, A. C. (1973). *Anim. Prod.* **16**, 261.

194. Haywood, S. (1985). *J. Pathol.* **145**, 149.
195. Healy, W. B., Bate, L. C., and Ludwig, T. G. (1964). *N.Z. J. Agric. Res.* **7**, 603.
196. Hemingway, R. G., MacPherson, A., and Ritchie, N. S. (1970). *In* "Trace Element Metabolism in Animals" (C. F. Mills, ed.), Vol. 1, p. 264. Livingstone, Edinburgh.
197. Henkin, R. I. (1974). *In* "Trace Element Metabolism in Animals" (W. G. Hoekstra, J. W. Suttie, H. E. Ganther, and W. Mertz, eds.), p. 647. Univ. Park Press, Baltimore.
198. Henkin, R. I., Schulman, J. D., Schulman, C. B., and Bronzert, D. A. (1973). *J. Pediatr.* **82**, 831.
199. Hennig, A., Anke, M., Groppel, B., and Lüdke, H. (1974). *In* "Trace Element Metabolism in Animals" (W. G. Hoekstra, J. W. Suttie, H. E. Ganther, and W. Mertz, eds.), p. 726. Univ. Park Press, Baltimore.
200. Hicks, J. J., Hernandez-Perez, O., Aznar, R., Mendez, D., and Rosado, A. (1975). *Am. J. Obstet. Gynecol.* **121**, 981.
201. Hill, C. H., and Matrone, G. (1970). *Fed. Proc., Fed. Am. Soc. Exp. Biol.* **29**, 1474.
202. Hill, C. H., and Starcher, B. (1965). *J. Nutr.* **85**, 271.
203. Hill, C. H., Matrone, G., Payne, W. L., and Barber, C. W. (1963). *J. Nutr.* **80**, 227.
204. Hill, C. H., Starcher, B., and Kim, C. (1968). *Fed. Proc., Fed. Am. Soc. Exp. Biol.* **26**, 129.
205. Hill, C. H., Starcher, B., and Matrone, G. (1964). *J. Nutr.* **83**, 107.
206. Hill, J. M., and Mann, P. G. (1962). *Biochem. J.* **85**, 198.
207. Hill, R. (1974). *In* "Trace Element Metabolism in Animals" (W. G. Hoekstra, J. W. Suttie, H. E. Ganther, and W. Mertz, eds.), p. 632. Univ. Park Press, Baltimore.
208. Holden, J. M., Wolf, W. R., and Mertz, W. (1979). *J. Am. Diet. Assoc.* **75**, 23.
209. Holtzman, N. A., Elliot, D. A., and Heller, R. II. (1966). *N. Engl. J. Med.* **275**, 347.
210. Hooper, P. L., Visconti, L., Garry, P. J., and Johnson, G. E. (1980). *J. Am. Med. Assoc.* **244**, 1960.
211. Howell, J. McC. (1970). *In* "Trace Element Metabolism in Animals" (C. F. Mills, ed.), Vol. 1, p. 103. Livingstone, Edinburgh.
212. Howell, J. McC., and Davidson, A. N. (1959). *Biochem. J.* **72**, 365.
213. Howell, J. McC., and Hall, G. A. (1969). *Br. J. Nutr.* **23**. 47.
214. Howell, J. McC., and Hall, G. A. (1970). *In* "Trace Element Metabolism in Animals" (C. F. Mills, ed.), Vol. 1, p. 106. Livingstone, Edinburgh.
215. Howell, J. McC., and Kumaratilake, J. (1985). *In* "Trace Elements in Man and Animals" (C. F. Mills, I. Bremner, and J. K. Chesters, eds.), p. 184. Commonwealth Agric. Bureaux, Farnham Royal, U.K.
216. Howell, J. McC., Blakemore, W. F., Gopinath, C., Hall, G. A., and Parker, J. H. (1974). *Acta Neuropathol.* **29**, 9.
217. Howell, J. McC., Davidson, A. N., and Oxberry, J. (1966). *Res. Vet. Sci.* **5**, 376.
218. Howell, J. McC., Edington, N., and Ewbank, R. (1968). *Res. Vet. Sci.* **9**, 160.
219. Hunt, C. E., and Carlton, W. W. (1965). *J. Nutr.* **87**, 385.
220. Hunt, C. E., Carlton, W. W., and Newberne, P. M. (1966). *Fed. Proc., Fed. Am. Soc. Exp. Biol.* **25**, 432.
221. Hunt, D. M. (1974). *Nature (London)* **249**, 852.
222. Hurley, L. S., and Bell, L. T. (1975). *Proc. Soc. Exp. Biol. Med.* **149**, 830.
223. Innes, J. R. M., and Shearer, G. D. (1940). *J. Comp. Pathol. Ther.* **53**, 1.
224. Ishmael, J., Gopinath, C., and Treeby, P. J. (1971). *J. Comp. Pathol.* **81**, 455.
225. Ishmael, J., Howell, J. McC., and Treeby, P. J. (1969). *Vet. Rec.* **85**, 205; *In* "Trace Element Metabolism in Animals" (C. F. Mills, ed.), Vol. 1, p. 268. Livingstone, Edinburgh.
226. Ivan, M., and Grieve, C. M. (1976). *J. Dairy Sci.* **59**, 1964.
227. Ivan, M., and Veira, D. M. (1985). *J. Dairy Sci.* **68**, 891.
228. Jameson, M. H., Sharma, H., Braganza, J. M., and Case, R. M. (1983). *Br. J. Nutr.* **50**, 113.

229. Jenkins, N. K., Morris, T. R., and Valamotis, D. (1970). *Br. Poult. Sci.* **11**, 241.
230. Jenkinson, S. G., Lawrence, R. A., Burk, R. F., and Williams, D. M. (1982). *J. Nutr.* **112**, 197.
231. Jenkinson, S. G., Lawrence, R. A., Grafton, W. D., Gregory, B. E., and McKinney, M. A. (1984). *Appl. Toxicol.* **4**, 170.
232. Jensen, L. S., and Maurice, D. V. (1979). *J. Nutr.* **109**, 91.
233. Jensen, L. S., Peterson, R. P., and Falen, L. (1964). *Poult. Sci.* **53**, 57.
234. Johnson, N. C., Kheim, T., and Kountz, W. B. (1959). *Proc. Soc. Exp. Biol. Med.* **102**, 98.
235. Joyce, J. M. (1955). *N.Z. Vet. J.* **3**, 157.
236. Kelleher, C. A. and Ivan, M. (1985). *In* "Trace Elements in Man and Animals" (C. F. Mills, I. Bemner, and J. K. Chesters, eds.), p. 364. Commonwealth Agric. Bureaux, Farnham Royal, U.K.
237. Kelsay, J. L., and Prather, E. S. (1983). *Am. J. Clin. Nutr.* **38**, 12.
238. Kidder, R. W. (1949). *J. Anim. Sci.* **8**, 623.
239. Kiilholma, P., Gronoos, M., Liukko, P., Pakarinen, P., Hyora, H., and Erkkola, R. (1984). *Gynecol. Obstet. Invest.* **8**, 212.
240. King, J. C., and Wright, A. L. (1985). *In* "Trace Elements in Man and Animals" (C. F. Mills, I. Bremner, and J. K. Chesters, eds.), p. 318. Commonwealth Agric. Bureaux, Farnham Royal, U.K.
241. King, J. C., Raynolds, W. L., and Margen, S. (1978). *Am. J. Clin. Nutr.* **31**, 1198.
242. King, J. O. L. (1972). *Br. Poult. Sci.* **13**, 61.
243. King. J. O. L. (1975). *Br. Poult. Sci.* **16**, 409.
244. King. R. L., and Williams, W. F. (1963). *J. Dairy Sci.* **46**, 11.
245. Kirchgessner, M., and Giessler, H. (1961). *Z. Tierphysiol. Tierernaehr. Futtermittelkd.* **16**, 297.
246. Kirchgessner, M. and Grassmann, E. (1970). *In* "Trace Element Metabolism in Animals" (C. F. Mills, ed.), p. 277. Livingstone, Edinburgh.
247. Kirchgessner, M., and Weser, U. (1965). *Z. Tierphysiol. Tierernaehr. Futtermittelkd.* **20**, 44.
248. Kirchgessner, M., Beyer, M. G., and Steinhart, H. (1976). *Br. J. Nutr.* **36**, p. 15.
249. Kirchgessner, M., Schwarz, F. J., Grassmann, E., and Steinhart, H. (1979). *In* "Copper in the Environment, Part 2" (J. O. Nriagu, ed.), p. 433. Wiley, New York.
250. Kirsten, G. F., Hesse, H. D., Watermeyer, S., Dempster, W. S., Pocock, F., and Varke-visser, H. S. (1985). *Afr. Med. J.* **68**, 402.
251. Klauder, D. S., Murthy, L., and Petering, H. G. (1972). *Trace Subst. Environ. Health* **6**, 131.
252. Klevay, L. M. (1970). *Am. J. Clin. Nutr.* **23**, 1194.
253. Klevay, L. M. (1974). *Nutr. Rep. Int.* **9**, 393.
254. Klevay, L. M. (1974). *Trace Subst. Environ. Health* **8**, 9.
255. Klevay, L. M. (1975). *Am. J. Clin. Nutr.* **28**, 764.
256. Klevay, L. M. (1975). *Nutr. Rep. Int.* **11**, 237.
257. Klevay, L. M. (1982). *Nutr. Rep. Int.* **26**, 329.
258. Klevay, L. M., Canfield, W. K., Gallagher, S. K., Henriksen, R. D., Lukaski, H. C., Bolonchuk, W., Johnson, L. K., Milne, D. B., and Sandstead, H. H. (1986). *Nutr. Rep. Int.* **33**, 371.
259. Klevay, L. M., Inman, L., Johnson, L. K., Lawler, M., Mahalko, J. R., Milne, D. B., Lukaski, H. C., Bolonchuk, W., and Sandstead, H. H. (1984). *Metabolism* **33**, 1112.
260. Klevay, L. M. (1984). *In* "Metabolism of Trace Metals in Man" (O. W. Rennert and W. Y. Chan, eds.), Vol. 1, p. 129. CRC Press, Boca Raton, Florida.
261. Klevay, L. M., Reck, S. J., and Barcome, D. F. (1977). *Fed. Proc., Fed. Am. Soc. Exp. Biol.* **36**, 1175.
262. Klevay, L. M., Reck, S. J., and Barcome, D. F. (1979). *J. Am. Med. Assoc.* **241**, 1916.

263. Klevay, L. M., Reck, S. J., Jacob, R. A., Logan, G. H., Munoz, J. M., and Sandstead, H. H. (1980). *Am. J. Clin. Nutr.* **33**, 45.
264. Kline, R. D., Corzo, M. A., Hays, V. W., and Cromwell, G. L. (1973). *J. Anim. Sci.* **37**, 936.
265. Kline, R. D., Hays, B. W., and Cromwell, G. L. (1971). *J. Anim. Sci.* **33**, 771.
266. Koivistoinen, P. (1980). *Acta Agric. Scand.* (Suppl.) **22.**
267. Kopp, S. J., Klevay, L. M., and Feliksik, J. M. (1983). *Am. J. Physiol.* **245H**, 855.
268. Koval'skiy, V. V., Yarovaya, G. A., and Shmavonyan, D. M. (1961). *Zh. Obsc. Biol.* **22**, 179.
269. Kowalczyk, T., Pope, A. L., and Sorensen, D. K. (1962). *J. Am. Vet. Med. Assoc.* **141**, 362.
270. Kowalczyk, T., Pope, A. L., Berger, K. C., and Muggenberg, B. A. J. (1964). *Am. Vet. Med. Assoc.* **145**, 352.
271. Krishnamachari, K. A. V. R. (1974). *Am. J. Clin. Nutr.* **27**, 108.
272. Krishnamachari, K. A. V. R., and Krishnaswamy, K. (1974). *Indian J. Med. Res.* **62**, 1415.
273. L'Abbé, M. R., and Fischer, P. W. F. (1984). *J. Nutr.* **114**, 813.
274. Lahey, M. E., Gubler, C. J., Chase, M. S., Cartwright, G. E., and Wintrobe, M. M. (1952). *Blood* **7**, 053.
275. Lal, S., and Sourkes, T. L. (1971). *Toxicol. Appl. Pharmacol.* **18**, 562.
276. "Latin American Table of Feed Composition." Univ. of Florida, Gainesville, 1974.
277. Lawrence, C. B., Davies, N. T., MIlls, C. F., and Nicol, F. (1985). *Biochim. Biophys. Acta* **809**, 351.
278. Lea, C. M., and Luttrell, V. A. S. (1965). *Nature (London)* **296**, 413.
279. Lee, H. J. (1956). *J. Agric. Sci.* **47**, 218.
279a. Lee, G. R., Cartwright, G. E., and Wintrobe, M. M. (1968). *Proc. Soc. Exp. Biol. Med.* **127**, 977.
280. Lehmann, K. B. (1891). *Munch, Med. Wochenschr.* **38**, 631.
281. Lei, K. Y. (1983). *J. Nutr.* **113**, 2178.
282. Leverton, R. M., and Binkley, E. S. (1944). *J. Nutr.* **27**, 43.
283. Lewis, G., Terlecki, S., and Allcroft, R. (1967). *Vet. Rec.* **81**, 415.
284. Lindow, C. W., Peterson, W. H., and Steenbock, H. (1929). *J. Biol. Chem.* **84**, 419.
285. Liu, C. F., and Medeiros, D. M. (1986). *Biol. Trace Elem. Res.* **9**, 15.
286. Lorenzen, E. J., and Smith, S. E. (1947). *J. Nutr.* **33**, 143.
287. Losee, F. L., Cutress, T. W., and Brown, R. (1974). *Caries Res.* **8**, 123.
288. Lukasewycz, O. A., Prohaska, J. R., Mayer, S. G., Schmidtke, J. R., Hatfield, S. M., and Marder, P. (1985). *Infect. Immun.* **48**, 644.
289. Lukasewycz, O. A., and Prohaska, J. R. (1982). *J. Natl. Cancer Inst.* **69**, 489.
290. Lyon, T. D. B., Smith, L. B., and Lenihan, J. M. A. (1985). *In* "Trace Elements in Man and Animals" (C. F. Mills, I. Bremner, and J. K. Chesters, eds.), p. 625. Commonwealth Agric. Bureaux, Farnham Royal, U.K.
291. Mahoney, J. P., Bush, J. A., Gubler, C. J., Moretz, W. H., Cartwright, G. E., and Wintrobe, M. M. (1955). *J. Lab. Clin. Med.* **46**, 702.
292. Manzler, A. D., and Schreiner, A. W. (1970). *Ann. Intern. Med.* **73**, 409.
293. Marshall, J., and Ottaway, J. M. (1983). *Talanta* **30**, 571.
294. Marston, H. R. (1946). *Proc. Symp. Fibrous Proteins, Soc. Dyers Color, Leeds.*
295. Marston, H. R., and Lee, H. J. (1948). *Aust. J. Sci. Res. Ser. B* **1**, 376.
296. Marston, H. R., and Lee, H. J. (1948). *J. Agric. Sci.* **38**, 229.
297. Marston, H. R., Lee, H. J., and McDonald, I. W. (1948). *J. Agric. Sci.* **38**, 216, 222.
298. Marston, H. R., Thomas, R. G., Murnane, D., Lines, E. W., McDonald, I. W., Moore, H. O., and Bull, L. B. (1938). *Aust. Commonw. Counc. Sci. Ind. Res. Bull.* **113.**
299. Mason, K. E. (1979). *J. Nutr.* **109**, 1079.

300. Masters, H. G., Smith, G. M., and Casey, R. H. (1985). *In* "Trace Elements in Man and Animals" (C. F. Mills, I. Bremner, and J. K. Chesters, eds.), p. 575. Commonwealth Agric. Bureaux, Farnham Royal, U.K.
301. Matsuda, I., Higashi, A., Ikeda, T., Uahara, I., and Kuroki, Y. (1984). *J. Pediatr. Gastroenterol. Nutr.* **3**, 421.
302. McCall, J. T., and Davis, G. K. (1961). *J. Nutr.* **74**, 45.
303. McCormick, L. D. (1985). *Biophys. J.* **47**, 381.
304. McCoskers, P. J. (1961). *Nature (London)* **190**, 887.
305. McCoskers, P. J. (1968). *Res. Vet. Sci.* **9**, 103.
306. McFarlane, W. D., and Milne, H. I. (1934). *J. Biol. Chem.* **107**, 309.
307. McGavin, M. D., Ranby, P. D., and Tammemagi, L. (1962). *Aust. Vet. J.* **38**, 8.
308. McHargue, J. S. (1925). *Am. J. Physiol.* **72**, 583.
309. McHargue, J. S. (1926). *Am. J. Physiol.* **77**, 245.
310. Medeiros, D. M., Lin, K. N., Liu, C. C. F., and Thorne, B. M. (1984). *Nutr. Rep. Int.* **30**, 559.
311. Menkes, J. H., Alter, M., Steigleder, G. K., Weakley, D. R., and Sung, J. H. (1962). *Pediatrics* **29**, 794.
312. Mertz, W. (1981). *Science* **213**, 1332.
313. Miller, E. J., Martin, E. R., Mecca, C. E., and Piez, K. A. (1965). *J. Biol. CHem.* **240**, 3623.
314. Mills, C. F. (1954). *Biochem. J.* **57**, 603.
315. Mills, C. F. (1955). *Br. J. Nutr.* **9**, 398.
316. Mills, C. F. (1956). *Biochem. J.* **63**, 187, 190.
317. Mills, C. F., and Murray, G. (1960). *J. Sci. Food Agric.* **9**, 547.
318. Mills, C. F., and Williams, R. B. (1962). *Biochem. J.* **85**, 629.
319. Milne, D. B., and Weswig, P. H. (1968). *J. Nutr.* **95**, 429.
320. Milne, D. B., Omaye, S. T., and Amos, W. H., Jr. (1981). *Am. J. Clin. Nutr.* **34**, 2389.
321. Miltimore, J. E., Kalin, C. M., and Clapp, J. B. (1978). *Can. J. Anim. Sci.* **58**, 525.
322. Miyata, S., Okuno, T., Nakamura, S., Nagata, H., and Kemeyama, M. (1985). *In* "Trace Elements in Man and Animals" (C. F. Mills, I. Bremner, and J. K. Chesters, eds.), p. 652. Commonwealth Agric. Bureaux, Franham Royal, U.K.
323. Morgan, K. T. (1983). *Res. Vet. Sci.* **15**, 88.
324. Morris, E. R., and Ellis, R. (1985). *In* "Trace Elements in Man and Animals" (C. F. Mills, I. Bremner, and J. K. Chesters, eds.), p. 443. Commonwealth Agric. Bureaux, Farnham Royal U.K.
325. Moule, G. R., Sutherland, A. K., and Harvey, J. M. (1959). *Queensl. J. Agric. Sci.* **18**, 93.
326. Mortzavi, S. H., Bani-hashemi, A., Mozafari, M., and Raffi, A. (1972). *Cancer* **29**, 1193.
327. Munch-Petersen, S. (1950). *Acta Pediatr.* **39**, 378.
328. Murthy, G. K., and Rhea, U. S. (1971). *J. Dairy Sci.* **54**, 1001.
329. National Research Council, "Copper." National Academy of Sciences, Washington, D.C., 1977.
330. National Research Council, "Mineral Tolerance of Domestic Animals." National Academy of Sciences, Washington, D.C., 1980.
331. National Research Council, "Recommended Dietary Allowances," 9th Ed. National Academy of Sciences, Washington, D.C., 1980.
332. National Research Council, "Underutilized Resources as Animal Feedstuffs," Table 2, p. 241. National Academy of Sciences, Washington, D.C., 1983.
333. Neal, W. M., Becker, R. B., and Shealy, A. L. (1931). *Science* **74**, 418.
334. Neumann, P. Z., and Sass-Kortsak, A. I. (1967). *Clin Invest.* **46**, 646.
335. Nielsen, A. L. (1944). *Acta Med. Scand.* **118**, 92.

336. Nriagu, J. O. (1979). "Copper in the Environment, Part I" (J. O. Nriagu, ed.). Wiley, New York.
337. O'Dell, B. L., Bird, D. W., Ruggles, D. F., and Savage, J. E. (1966). *J. Nutr.* **88**, 9.
338. O'Dell, B. L., Hardwick, B. C., Reynolds, G., and Savage, J. E. (1961). *Proc. Soc. Exp. Biol. Med.* **108**, 402.
339. Oestreicher, P., and Cousins, R. J. (1985). *J. Nutr.* **115**, 159.
340. Ogiso, T., Moriyama, K., Sasaki, S., Ishimura, Y., and Minato, A. (1974). *Chem. Pharm. Bull.* **22**, 55.
341. O'Hara, P. J., Newman, A. P., and Jackson, R. (1960). *Aust. Vet. J.* **36**, 225.
342. Olson, K. J., Fontenot, J. P., and Failla, M. L. (1984). *J. Anim. Sci.* **59**, 210.
343. Ono, K., Steele, N., Richards, M., Darcey, S., Fields, M., Scholfield, D., Smith, J. C., and Reiser, S. (1986). *Fed. Proc., Fed. Am. Soc. Exp. Biol.* **45**, 357.
344. Opsahl, W., Abbott, U., Kenney, C., and Rucker, R. (1984). *Science* **225**, 440.
345. Osaki, S., Johnson, D. A., and Frieden, E. (1966). *J. Biol. Chem.* **241**, 2746.
346. Osborn, G. R. (1968). *In* "Le Rôle de la paroi artérielle dans l'athérogenèse" (M. J. Lenegre, L. Scebat, and J. Renais, eds.). CNRS, Paris.
347. Owen, C. A. (1982). "Copper Proteins, Ceruloplasmin and Copper Protein Binding," p. 250. Noyes, Park Ridge, New Jersey.
348. Owen, C. A. (1981). "Copper Deficiency and Toxicity." Noyes, Park Ridge, New Jersey.
349. Owen, E. C., Proudfoot, R., Robertson, J. M., Barlow, R. M., Butler, E. J., and Smith, B. S. W. (1965). *J. Comp. Pathol.* **75**, 241.
350. Panić, B., Bezbradica, L. J., Nedeljkov, N., and Istwani, A. G. (1974). *In* "Trace Element Metabolism in Animals" (W. G. Hoekstra, J. W. Suttie, H. E. Ganther, and W. Mertz, eds.), p. 635. Univ. Park Press, Baltimore.
351. Parry, W. H., Jackson, P. G., Rao, S. R. R., and Cooke. B. C. (1985). *In* "Trace Elements in Man and Animals" (C. F. Mills, I. Bremner, and J. K. Chesters, eds.), p. 376. Commonwealth Agric. Bureaux, Farnham Royal, U.K.
352. Partridge, S. M., Elsden, D. F., Thomas, J., Dorfman, A., Telser, A., and Ho, P. L. (1964). *Biochem. J.* **93**, 30c.
353. Paynter, D. I., Moir, R. J., and Underwood, E. J. (1979). *J. Nutr.* **109**, 1570.
354. Pekarek, R. S., Burghen, G. A., Bartelloni, P. J., Calia, F. M., Bostian, K. A., and Beisel, W. R. (1970). *J. Lab. Clin. Med.* **76**, 293.
355. Pekarek, R. S., Kluge, R. M., Dupont, H. L., Wannemacher, R. W., Jr., Hornick, R. B., Bostian, K. A., and Beisel, W. R. (1975). *Clin. Chem.* **21**, 528.
356. Pennington, J. T., and Calloway, D. H. (1973). *J. Am. Diet. Assoc.* **63**, 143.
357. Picciano, M. F., and Guthrie, H. A. (1976). *Am. J. Clin. Nutr.* **29**, 242.
358. Picciano, M. F., and Guthrie, H. A. (1973). *Fed. Proc., Fed. Am. Soc. Exp. Biol.* **32** (Abstt.), 929.
359. Porter, H. (1974). *Biochem. Biophys. Res. Commun.* **56**, 661.
360. Porter, H., Johnston, J., and Porter, E. M. (1962). *Biochim. Biophys. Acta* **65**, 66.
361. Porter, H., Sweeney, M., and Porter, E. M. (1964). *Arch. Biochem. Biophys.* **104**, 97.
362. Porter, H., Wiener, W., and Barker, M. (1961). *Biochim. Biophys. Acta* **52**, 419.
363. Prasad, A. S. (1978). "Trace Elements and Iron in Human Metabolism." Plenum, New York.
364. Prasad, A. S., Brewer, G. J., Schoomaker, E. B., and Rabbani, P. (1978). *J. Am. Med. Assoc.* **240**, 2166.
365. Price, J., and Suttle, N. F. (1975). *Proc. Nutr. Soc.* **34**, 9A.
366. Prohaska, J. R. (1983). *J. Nutr.* **113**, 2048.
367. Prohaska, J. R. (1984). *J. Nutr.* **114**, 422.
368. Prohaska, J. R., and Cox, D. A. (1983). *J. Nutr.* **113**, 2623.

369. Prohaska, J. R., and Heller, L. (1982). *J. Nutr.* **112,** 2142.
370. Prohaska, J. R., and Lukasewycz, O. A. (1981). *Science,* **213,** 559.
371. Prohaska, J. R., Downing, S. W., and Lukasewycz, O. A. (1983). *J. Nutr.* **113,** 1583.
372. Prohaska, J. R., and Gutsch, D. E. (1983). *Biol. Trace Elem. Res.* **5,** 35.
373. Pryor, W. J. (1963). *Aust. J. Sci.* **25,** 498.
374. Punsar, S., Erametsa, O., Karvonen, M. J., Ryhanen, A., Hilska, P., Vornamo, H. (1975). *J. Chronic Dis.* **28,** 259.
375. Radhakrishnamurthy, B., Rucker, R., and Berenson, G. (1985). *Proc. Soc. Exp. Biol. Med.* **180,** 392.
376. Raper, H. S. (1928). *Physiol. Rev.* **8,** 245.
377. Recant, L., Voyles, N. R., Zalenski, C., Fields, M., and Bhathena, S. J. (1986). *Peptides,* **7,** 1061.
378. Reed, G. A., *Aust. Vet. J.* **51,** 107 (1975).
379. Reed, D. W., Passon, P. G., and Hultquist, D. E. (1970). *J. Biol. Chem.* **245,** 2954.
380. Reinhold, J. G., Kfoury, G. A., and Thomas, T. A. (1967). *J. Nutr.* **92,** 173.
381. Reiser, S., Ferretti, R. J., Fields, M., and Smith, J. C., Jr. (1983) *Am. J. Clin. Nutr.* **38,** 214.
382. Reiser, S., Smith, J. C., Mertz, W., Holbrook, J. T., Scholfield, D. J., Powell, A. S., Canfield, W. K., and Canary, J. J. (1985). *Am. J. Clin. Nutr.* **42,** 242.
383. Ritchie, H. D., Luecke, R. W., Balzer, B. V., Miller, E. R., Ullrey, D. E., and Hoefer, J. A. (1963). *J. Nutr.* **79,** 117.
384. Robinson, M. F., McKenzie, J. M., Thompson, C. D., and van Rij, A. L. (1973). *Br. J. Nutr.* **30,** 195.
385. Ross, D. B. (1964). *Vet. Rec.* **76,** 875.
386. Ross, D. B. (1966). *Br. Vet. J.* **122,** 279.
387. Rucker, R. B., Parker, H. E., and Rogler, J. C. (1969). *J. Nutr.* **98,** 57.
388. Rucker, R. B., Riggins, R. S., Laughlin, R., Chan, M. M., Chen, M., and Tom K. (1975). *J. Nutr.* **105,** 1062.
389. Rupp, H., and Weser, U. (1976). *Biochem. Biophys. Res. Commun.* **72,** 223.
390. Russanov, E. M., and Kassabova, T. A. (1982). *Int. J. Biochem.* **14,** 321.
391. Salmon, M. A., and Wright, T. (1971). *Arch. Dis. Child.* **46,** 108.
392. Saltman, P., Alex, T., and McCornack, B. (1959). *Arch. Biochem. Biophys.* **83,** 538.
393. Saylor, W. W., and Leach, R. M., Jr. (1980). *J. Nutr.* **110,** 448.
394. Savage, J. E. (1968). *Fed. Proc., Fed. Am. Soc. Exp. Biol.* **27,** 927.
395. Scheinberg, I. H., and Gitlin, D. (1952). *Science* **116,** 484.
396. Scheinberg, I. H., and Sternlieb, I. (1965). *Annu. Rev. Med.* **16,** 119.
397. Scheinberg, I. H., Cook, C. D., and Murphy, J. A. (1954). *J. Clin. Invest.* **33,** 963.
398. Scheinberg, I. H. (1984). *In* "Metalle in der Umwelt" (E. Merian, ed.), p. 451. Verlag Chemie, Weinheim.
399. Schelenz, R. F. W., and Harmuth-Hoene, A. E. (1985). *In* "Trace Elements in Man and Animals" (C. F. Mills, I. Bremner, and J. K. Chesters, eds.), p. 620. Commonwealth Agric. Bureaux, Farnham Royal, U.K.
400. Schroeder, H. A., Nason, A. P., Tipton, I. H., and Balassa, J. J. (1966). *J. Chronic Dis.* **19,** 1007.
401. Schultz, K. C. A., Van der Merwe, P. K., Van Rensburg, P. J., Swart, J. S. (1961). *Onderstepoort J. Vet. Res.* **25,** 35.
402. Schultze, M. O. (1940). *Physiol. Rev.* **20,** 37.
403. Schultze, M. O., Elvehjem, C. A., and Hart, E. B. (1936). *J. Biol. Chem.* **116,** 93.
404. Schultze, M. O., Elvehjem, C. A., and Hart, E. B. (1936). *J. Biol. Chem.* **115,** 453.
405. Schwarz, F. J., and Kirchgessner, M. (1974). *Int. J. Vitam. Nutr. Res.* **44,** 116, 258.
406. Shand, A., and Lewis, G. (1957). *Vet. Rec.* **69,** 618.

407. Shapcott, D., Vobecky, J. S., Vobecky, J., and Demers, P. P. (1985). *Nutr. Res. (Suppl.)* **1**, 289.
408. Shapiro, J., Morell, A. G., and Scheinberg, I. H. (1961). *J. Clin. Invest.* **40**, 1081.
409. Shearer, G. D., Innes, J. R. M., and McDougal, E. I. (1940). *Vet. J.* **96**, 309.
410. Sher, R., Shulman, G., Bailey, P., and Politzer, W. M. (1981). *Am. J. Clin. Nutr.* **34**, 1918.
411. Sherman, A. R., Helyar, L., and Wolinsky, I. (1985). *J. Nutr.* **115**, 607.
412. Shuler, T. R., and Nielsen, F. H. (1985). *In* "Trace Elements in Man and Animals" (C. F. Mills, I. Bremner, and J. K. Chesters, eds.), p. 382. Commonwealth Agric. Bureaux, Farnham Royal, U.K.
413. Sigel, H. (1981). *In* "Metal Ions in Biological Systems" (H. Sigel ed.), p. 13. Dekker, New York.
414. Simpson, C. F., and Harms, R. H. (1964). *Exp. Mol. Pathol.* **3**, 390.
415. Singh, N. P., and Medeiros, D. M. (1984). *Biol. Trace Elem. Res.* **6**, 423.
416. Sjollema, B. (1933). *Biochem. Z.* **267**, 151.
417. Sjollema, B. (1938). *Biochem. Z.* **295**, 372.
418. Smallwood, R. A., Williams, H. A., Rosenoer, V. M., and Sherlock, S. (1968). *Lancet* **2**, 1310.
419. Smeyers-Verbeke, J., May, C., Brochmans, P., and Massart, D. L. (1977). *Ann. Biochem.* **83**, 746.
420. Smith, H. W., and Jones, J. E. T. (1963). *J. Appl. Bacteriol.* **26**, 262.
421. Smith, M. S. (1969). *Br. Poult. Sci.* **10**, 97.
422. Smith, R. M. (1985). *In* "Trace Elements in Man and Animals" (C. F. Mills, I. Bremner, and J. K. Chesters, eds.), p. 567. Commonwealth Agric. Bureaux, Farnham Royal, U.K.
423. Smith, S. E., and Ellis, G. H. (1947). *Arch. Biochem.* **15**, 81.
424. Solomons, N. W. (1979). *Am. J. Clin. Nutr.* **32**, 856.
425. Sorenson, J. R. J. (ed.) (1982). "Inflammatory Diseases and Copper." Humana Press, Clifton, N.J.
426. Sorensen, J. R. J. (ed.) (1987). "Biology of Copper Complexes." Humana Press, Clifton, N.J., in press.
427. Sourkes, T. L., Lloyd, K., and Birnbaum, H. (1968). *Can. J. Biochem.* **46**, 267.
428. Spoerl, R., and Kirchgessner, M. (1976). *Arch. Tierernaehr.* **26**, 25.
429. Spray, C. M., and Widdowson, E. M. (1950). *Br. J. Nutr.* **4**, 332.
430. Starcher, B., and Hill, C. H. (1965). *Comp. Biochem. Physiol.* **15**, 429.
431. Starcher, B., Hill, C. H., and Matrone, G. (1964). *J. Nutr.* **82**, 318.
432. Steele, N., Richards, M., Darcey, S., Fields, M., Smith, J., and Reiser, S. (1986). *Fed. Proc., Fed. Am. Soc. Exp. Biol.* **45**, (Abstr.), 357.
433. Stemner, K. L., Petering, H. G., Murthy, L., Finelli, V. N., and Menden, E. E. (1985). *Ann. Nutr. Metab.* **29**, 332.
434. Stevenson, M. H., and Jackson, N. (1980). *Br. J. Nutr.* **43**, 205.
435. Stogdale, L. (1978). *Aust. Vet. J.* **554**, 139.
436. Strause, L. G., Hegenauer, J., Saltman, P., Cone, R., and Resnick, D. (1986). *J. Nutr.* **116**, 135.
437. Strickland, G. T., Beckner, W. M., and Leu, M. L. (1972). *Clin. Sci.* **43**, 617.
438. Sutherland, A. K., Moule, G. R., and Harvey, J. M. (1955). *Aust. Vet. J.* **31**, 141.
439. Suttle, N. F. (1973). *Proc. Nutr. Soc.* **32**, 24A.
440. Suttle, N. F. (1974). *Trace Subst. Environ. Health-7, Proc. Univ. Mo. Annu. Conf.*, p. 245.
441. Suttle, N. F. (1983). *J. Agric. Sci.* **100**, 651.
442. Suttle, N. F., and Field, A. C. (1968). *J. Comp. Pathol.* **78**, 351, 363.
443. Suttle, N. F., and Mills, C. F. (1966). *Br. J. Nutr.* **20**, 135, 149.
444. Suttle, N. F., Angus, K. W., Nisbet, D. I., and Field, A. C. (1972). *J. Comp. Pathol.* **82**, 93.
445. Suttle, N. F., Field, A. C., and Barlow, R. M. (1970). *J. Comp. Pathol.* **80**, 151.

446. Suveges, T., Ratz, F., and Salge, G. (1971). *Acta Vet.* **21**, 383.
447. Tanaka, Y., Atano, S., Michi, Y., and Usui, T. (1980). *J. Pediatr.* **96**, 255.
448. Tatum, H. J. (1973). *Am. J. Obstet. Gynecol.* **117**, 602.
449. Thomas, J., Elsden, D. F., and Partridge, S. (1963). *Nature (London)* **200**, 661.
450. Thompsett, S. L. (1940). *Biochem. J.* **34**, 961.
451. Thornton, E., Kershaw, G. F., and Davies, M. K. (1972). *J. Agric. Sci.* **78**, 157, 165.
452. Todd, J. R. (1969). *Proc. Nutr. Soc.* **28**, 189.
453. Todd, J. R. (1970). *In* "Trace Element Metabolism in Animals" (C. F. Mills, ed.), Vol. 1, p. 448. Livingstone, Edinburgh.
454. Todd, J. R., and Thompson, R. H. (1963). *Br. Vet. J.* **119**, 189.
455. Turnlund, J. R., Swanson, C. A., and King, J. C. (1983). *J. Nutr.* **113**, 2346.
456. Ullrey, D. E., Miller, E. R., Thompson, O. A., Zutaut, C. L., Schmidt, D. A., Ritchie, H. D., Hoefer, J. A., and Luecke, R. W. (1960). *J. Anim. Sci.* **19**, 1298.
457. Ulsen, F. W., Van (1972). *Tijdchr. Diergeneeskd.* **97**, 735.
458. Underwood, E. J. (1977). "Trace Elements in Human and Animal Nutrition," 4th Ed., p. 87. Academic Press, New York.
459. Van Campen, D. R. (1966). *J. Nutr.* **88**, 125.
460. Van Campen, D. R. (1971). *In* "Intestinal Absorption of Metal Ions, Trace Elements and Radionuclides" (S. Skoryna and D. Waldron-Edward, eds.), p. 211. Pergamon, Oxford.
461. Van Campen, D. R., and Gross, E. (1968). *J. Nutr.* **95**, 617.
462. Van Campen, D. R., and Mitchell, E. A. (1965). *J. Nutr.* **86**, 120.
463. Van den Hamer, C. J. A. (1975). *In* "Physiological and Biochemical Aspects of Heavy Elements in Our Environment" (J. B. W. Houtman and C.J. A. Van den Hamer, eds.), p. 63. Delft Univ. Press.
464. Van Koetsveld, E. E. (1958). *Tijdsehr. Diergeneeskd.* **83**, 229.
465. Van Ravesteyn, A. H. (1944). *Acta Med. Scand.* **118**, 163.
466. Van Wyk, J. J., Baxter, J. H., Akeroyd, J. H., and Motulsky, A. G. (1953). *Bull. Johns Hopkins Hosp.* **93**, 51.
467. Varo, P., Nuurtamo, M., Sarri, E., and Koivistoinen, P. (1980). *In* "Mineral Element Composition of Finnish Foods," p. 115. *Acta Agric. Scand. Suppl.* **22.**
468. Vilter, R. W., Bozian, R. C., Hess, E. V., Zellner, D. C., and Petering, H. G. (1974). *N. Engl. J. Med.* **291**, 188.
469. Vir, S. C., Love, A. H. G., and Thompson, W. (1981). *Am. J. Clin. Nutr.* **34**, 2382.
470. Vuori, E., Salmela, S., Akerboom, H. K., Viikarri, J., Uhari, M., Suoninen, P., Pietikainen, M., Pesonen, E., Lahde, P. L., and Dahl, M. (1985). *Acta Paediatr. (Stockholm) (Suppl.)* **318**, 205.
471. Wahle, K. W. J., and Davies, N. T. (1975). *Br. J. Nutr.* **34**, 105.
472. Walker-Smith, J., and Blomfield, J. (1973). *Arch. Dis. Child.* **48**, 476.
473. Wallace, H. D., McCall, J. T., Bass, B., and Combs, G. (1960). *J. Anim. Sci.* **19**, 1155.
474. Walravens, P. J. A., and Hambidge, K. M. (1976). *Am. J. Clin. Nutr.* **29**, 1114.
475. Wawschinek, O., and Höfler, H. (1979). *At. Absorpt. News Lett.* **18**, 97.
476. Weiner, A. L., and Cousins, R. J. (1980). *Biochim. Biophys. Acta* **629**, 113.
477. Weisenberg, E., Halbreich, A., and Mager, J. (1980). *Biochem. J.* **188**, 633.
478. Werther, C. A., Cloud, H., Ohtake, M., and Tamura, T. (1986). *Drug Nutr. Interact.* **4**, 256.
479. *W.H.O. Tech. Rep. Ser.* (532), 70 (1973).
480. Wiederanders, R. E. (1968). *Proc. Soc. Exp. Biol. Med.* **128**, 627.
481. Wiener, G., and Field, A. C. (1969). *J. Comp. Pathol.* **79**, 7.
482. Wiener, G., and Field, A. C. (1974). *J. Agric. Sci.* **83**, 403.
483. Wiener, G., Wilmut, I., Woolliams, C., Woolliams, J. A., and Field, A. C. (1984). *Anim. Prod.* **39**, 207.

484. Wilkie, W. J. (1959). *Aust. Vet. J.* **35,** 203.
485. Williams, D. M. (1981). *Am. J. Clin. Nutr.* **34,** 1694.
486. Williams, D. M., Burk, R. F., Jenkinson, S. G., and Lawrence, R. A. (1981). *J. Nutr.* **111,** 979.
487. Williams, D. M. (1982). *Curr. Top. Nutr. Dis.* **6,** 277.
488. Wintrobe, M. M., Cartwright, G. E., and Gubler, C. J. (1953). *J. Nutr.* **50,** 394.
489. "Wissenschaftliche Tabellen Geigy," 8th rev. Ed., p. 239. Ciby-Geigy, Basel, 1977.
490. Woolliams, J. A., Suttle, N. F., Wiener, G., Field, A. C., and Woolliams, C. (1983). *J. Agric. Sci.* **100,** 441.
491. Woolliams, J. A., Wiener, G., Suttle, N. F., and Field, A. C. (1983). *J. Agric. Sci.* **100,** 505.
492. Woolliams, J. A., Wiener, G., Suttle, N. F., and Field, A. C. (1985). *In* "Trace Elements in Man and Animals" (C. F. Mills, I. Bremner, and J. K. Chesters, eds.), p. 358. Commonwealth Agric. Bureaux, Farnham Royal, U.K.
493. Wu, B. N., Medeiros, D. M., Lin, K. N., and Thorne, B. M. (1984). *Nutr. Res.* **4,** 305.

11

Fluorine

K.A.V.R. KRISHNAMACHARI

National Institute of Nutrition
Tarnaka
Hyderabad, India

I. FLUORIDE TOXICITY IN HUMANS

A. Mechanism of Clinical Events

Ingested fluoride (F) is rapidly absorbed and transferred into the blood, where it reacts almost instantaneously with calcium to form calcium fluoride. In that form fluorine makes its way into the hard tissues, displaces surface anions, and, in due course, enters the bone crystal lattice. The large surface of amorphous bones available for ionic interaction, particularly in children, enables substantial amounts of fluorine to enter the crystal, resulting in increased crystal size. Due to the great affinity of fluorine toward calcium, serum calcium is gradually linked with fluoride, and there is a tendency for ionic serum calcium to decline when much fluorine is deposited in the bones. When the biologically available dietary calcium is low, as is commonly the situation in undernourished population groups living in fluorosis-endemic regions, the reduced serum ionic calcium levels may possibly stimulate the parathyroid, resulting in secondary hyperparathyroidism. The response of the body in secreting parathyroid hormone (PTH) and calcitonin into the circulation varies greatly between individuals. There is also evidence for the existence of a mechanism for a direct stimulation of the parathyroid by excess circulating fluorine. Depending on the body's response to the triggering mechanism, the relative amounts of circulating IPTH and calcitonin vary greatly. It is this relative proportion of the hormonal changes which determines the further course of the events. In the event of excess IPTH, bone resorption can be demonstrated both by kinetic studies and by radiological evidence. In due course, the clinical consequences of secondary hyperparathyroidism become obvious. The bone changes are attributed to the movement of fluorine, calcium, and phosphate into the bone structure, to stimulation of osteoblastic cells, and to metabolic changes in the matrix. Evidence for an effect of excess fluorine on collagen formation is not consistent. Studies in monkeys using [^{14}C]proline *in vivo* indicate that in this species, fluorine administration does not reduce the synthesis of bone collagen (238). In contrast, later studies in rabbits fed sodium fluoride, 50 mg/kg body weight per day using similar labeling techniques indicate a decrease in the collagen synthesis rate (244). Although the mechanism in humans is unknown, there may exist a difference in the matrix collagen synthesis between subjects who develop osteoporosis and osteomalacia and those who do not. The former group is the osteoporotic type, while the latter is the osteosclerotic type.

Crystallographic and X-ray diffraction studies have added to our understanding of the basic pathophysiology of fluorine entry into the crystal lattice of bone. The amorphous bones, because of their greater surface area, are more reactive than the more crystalline bones and permit entry of the fluoride ion by isoionic and heteroionic exchanges to a much greater degree. Thus, the spine and iliac crest tend to accumulate more fluorine per unit volume than the cortical limb bones. Of the fluorine that enters the bone structure, the major proportion is firmly retained while the remainder is loosely bound and exchangeable. The status of a bone at a given time depends on the balance between accretion and resorption. The former helps to increase the radiodensity of the bone, while the latter determines that fraction of fluoride which is available for excretion. Fluorine is primarily a bone-seeking mineral. About 60–80% or more of the ingested fluorine is retained by the skeletal tissue. The remaining fraction of fluorine is distributed in other tissues including teeth and soft tissues or is excreted.

Intact renal function is a prerequisite for renal elimination of fluoride ion. The fluorine excreted in urine comprises not only ionic fluoride but also complexes of fluorine with small molecules. Impaired renal clearance results in greater retention of the fluoride ion. Urinary hydroxyproline, an indicator of bone collagen breakdown, is often increased in fluorosis. Depending on the relative concentrations of hormones controlling bone metabolism, the net result may either be an increase in the quantity of urinary hydroxyproline or no change. In the former case there can be a specific increase of the fractions representing destruction of newly laid bone tissue and nascent collagen.

Deposition of excess fluorine renders bone more crystalline, more stable, and less reactive. In contrast, less crystalline bone yields to the influence of elevated circulating PTH, thereby causing urinary hydroxyproline to increase. In the classical osteosclerotic type of fluorosis, there is evidence for reutilization of the bone collagen breakdown products. Because of this metabolic event, urinary hydroxyproline remains unaltered in spite of rapid turnover. In the osteoporotic type of fluorosis, however, in addition to osteosclerosis of the axial skeleton, there is evidence of loss of bone elsewhere (at the epiphyseal ends of the long bones and the shafts of the short long bones), thereby causing an increase in total urinary hydroxyproline.

Hydrogen bonding of fluorine *in vivo* may explain the toxicity induced by chronic fluorine accumulation. Emsley (52) studied that aspect of fluorine metabolism and demonstrated the existence of strong bonding between fluoride ions and amides (RCONHR') and related molecules. Fluorine is a potent inhibitor of several enzyme systems. The N-H$\cdots$ F$^-$ hydrogen bond between fluorine and nitrogen–hydrogen linkage of amides is the strongest heteronuclear hydrogen bond known (35 kcal/mol). The deleterious effects of excessive fluoride administration on several enzyme systems have been well documented in experimental animals. Although most of these effects are inhibitory *in vivo* and *in vitro*, there are a few enzymes such as pyruvate kinase and glucose-6-phosphate dehydrogenase (G6PD) which are stimulated *in vivo* by fluoride.

The pathological changes brought about by fluorine depend on dosage and duration of exposure. The most characteristic feature of fluorine that determines its biological effects is the strong affinity of this element to calcium. The interaction between fluorine and calcium is discussed elsewhere (Section I,E). Apart from bone and dental structures which constitute the main targets of accumulation of ingested fluorine, some soft tissues undergo calcification under the influence of fluorine *in vivo* (26,69). Several workers described Monckeberg-type calcification of vessels (33,95,112,128,228,271). The demonstration of the presence of fluorine only in the calcium type of renal stones and its absence in the noncalcium stones suggests a possible etiological role for fluorine in certain types of renal stones (7).

B. Clinical Classification of Fluorosis

1. Hydric fluorosis
    a. Endemic dental fluorosis
    b. Endemic skeletal fluorosis
2. Industrial fluorosis
3. Foodborne skeletal fluorosis
4. Neighborhood fluorosis
5. Wine fluorosis
6. Drug-induced fluorosis

1. Hydric Fluorosis

a. Endemic Dental Fluorosis. This was graded by Dean (39) into five categories based on the macroscopic appearance of the teeth.

Classification	Score
Normal	0
Questionable	0.5
Very mild	1
Mild	2
Moderate	3
Severe	4

This classification is useful for quick field surveys in large population groups. Takamori and Kawahara (252) designed a more intricate system to grade each tooth separately taking into account streaks on the enamel surface, and marble appearance as well as brown staining of the enamel. Thylstrup *et al.*'s (264) classification is more contemporary and useful in practice. It takes into account the buccal, lingual, and occlusal surfaces, and uses nine scores. The dental changes are compared with histology and are correlated to the extent of subsurface porosity. Scores 1–5 indicate degrees of opacity; scores 6–9 indicate defective enamel.

b. Endemic Skeletal Fluorosis. By far, the most important aspect of chronic exposure to fluorine is the skeletal involvement. Fluorine being a cumulative, bone-seeking mineral, the resultant skeletal changes are progressive. According to the natural course of the disease, skeletal fluorosis may be classified into the following phases: preclinical, musculoskeletal, degenerative and destructive, crippling fluorosis, and complications.

The biological effects in humans due to chronic fluorine ingestion depend not only on the total dosage and duration of exposure, but also on associated factors such as nutritional status, functional status of the renal tissue, and interaction with other trace elements. Since the effect of fluorine is cumulative, the less serious consequences occur early in the natural course of the disease. Whatever may be the type of fluorine exposure, the clinical picture in chronic poisoning occurs in a phased manner.

Several reports of histological and histochemical changes in some soft tissues of experimental animals induced by fluorine suggest that such changes may also occur in humans. However, the dosage used in the animal studies was several fold higher than any expected human exposure (108). Zhavoronkov (283) described details of the involvement of several nonskeletal organs, including enzymatic, histological, and histochemical changes. Roholm (193) described a syndrome of nonskeletal fluorosis involving soft tissues.

The early clinical picture of fluoride toxicity is predominantly symptomatic and afebrile. Arthralgia, malaise, weakness, and joint stiffness—particularly in the lower limbs and the spine—are among the early features. In industrial exposure, ocular symptoms and headache also occur. In the established skeletal phase of the disease, symptoms and signs of bone and joint involvement are evident both clinically and radiologically. Pain is a cardinal feature due to arthritic lesions and to secondary peripheral nerve involvement. Jolly *et al.* (101) described paraplegia of fluorotic origin as a common complication.

On the other hand, typical radiological changes may precede any clinical signs or symptoms. Jolly *et al.* (101) described 309 symtom-free cases of established fluorosis, which constituted 25% of the samples studied.

2. Industrial Fluorosis

The following industries have been identified as posing a risk of fluorosis among their workers: (1) aluminum smelters, (2) phosphate fertilizer factories, (3) ceramics, (4) steel industry, and (5) glass industry.

Moller and Gudjonsson (164) were the first to describe skeletal and dental changes attributable to industrial exposure to fluoride. Roholm (193) gave a detailed description of the clinical picture and also the pathophysiology of cryolite workers. The number of countries with this problem and the variety of industries involved in fluoride release have since increased (Table I).

Bishop (21) reported a single case of a fertilizer factory worker who developed bone disease. Speder (230) described seven cases of fluorosis among phosphate miners in Morocco. Reports from the United Kingdom and France appeared in the early 1940s and 1960s, respectively. In 1960 reports emerged of a Norwegian aluminum smelter (155) and a Czecholslovak superphosphate factory (171a). De Vries and Lowenberg (41) reported industrial fluorosis from Holland. Around the same time Franke *et al.* (65) from the German Democratic Republic described three cases of industrial skeletal fluorosis. Schlegal (195a) described 61 cases from Switzerland among workers in aluminum smelters. In 1984 Tsunoda *et al.* (266) reported studies on fluorine excretion in industry attempting to correlate urinary fluorine with several parameters such as duration of exposure and type of job.

In a study of 43 aluminum workers, the fluoride content in ashed biopsy bone was reported to be high but did not bear any relation to the duration of exposure. There was an inverse relation, however, between the fluorine content and the time since the end of the exposure. It took 20 years for 50% of fluorine to disappear from the bone (12).

Discontinuation of Occupational Exposure to Fluorine. Czerwinski *et al.* (38) studied the effect of exposure cessation on clinical, radiological, and analytical aspects of bones in 60 former exployees of an aluminum plant. The duration

Table I
Industrial Fluorosis Cases Reported by Various Countries

Country	References	Mean bone fluorine (μg/g)	Urinary fluorine (mg/liter)
Japan	Tsunoda *et al.* (266)	—	3.68
German Democratic Republic	Dominok (45)	2500–12,900	—
India	Bhavsar *et al.* (19)	—	—
Switzerland	Baud *et al.* (11)		
Fluorosis		5616	—
Controls		1036	—
Poland	Rydzewska *et al.* (194)		
Fluorosis		—	0.63
Controls		—	0.50

of withdrawal ranged from 4 months to 7 years, and the subjects were divided into two groups, according to the time since withdrawal. Symptoms were less in the prolonged-withdrawal group, but urinary fluorine was elevated in both groups, with concentrations of 1.74 and 1.30 μg/ml, respectively, while serum biochemistry was perfectly normal. Roholm (193) observed high urinary fluorine levels for several years following withdrawal. Largent and Heyroth (131) reported negative fluorine balances, immediately after the end of exposure, followed by equilibrium, achieved within 3–4 years. It took nearly 8 years in these studies for 50% of the body's skeletal fluoride to be eliminated.

3. Foodborne Skeletal Fluorosis

Skeletal fluorosis exclusively attributable to dietary sources is extremely rare. Daijei (95) investigated 34 Chinese suffering from skeletal fluorosis who lived in an area with low levels of water fluoride but very high soil concentrations of up to 1000 μg/g. Each of the seven predominant and mainly vegetarian foods contained between 8.3 and 11.7 μg/g of fluorine. The affected persons developed typical radiological features of skeletal fluorosis and had osteoporosis and osteomalacia. Mean urinary fluorine was 6.9 μg/g. Dental fluorosis due to foodborne fluoride was reported from Thailand (76) and Vietnam (121). In a later study, Daijei (96) reported the fluorine content of rice, corn, and potatoes in four regions in Guizhou Province of China and the clinical profile of 396 patients with skeletal fluorosis. From the study, it is obvious that in spite of the very low content of fluorine in the drinking water ranging from 0.07 to 0.5 μg/g, severe forms of skeletal fluorosis are widely prevalent in China. Thus, primary foodborne skeletal fluorosis seems to be unique to China. The reason, according to the authors, is the high soil fluorine which is retained and available to the plants, as the acid nature of the clay does not permit percolation of the fluorine to deeper regions of the soil.

4. Neighborhood Fluorosis

Some industries discharge fluorine in gaseous or other form, polluting their respective neighborhoods with abnormally high amounts of fluorine, which may result in human and/or cattle disease. The clinical picture is similar to endemic skeletal fluorosis with dental and bone changes. While the human problem is attributed to either chronic inhalation of fluorine-bearing gaseous discharges or due to ingestion of fluorine via subsoil water, which acquires fluorine from the liquid effluents, the animals develop the disease due to consumption of fodder contaminated with droplets containing fluorine.

Waldbott (272) described human fluorosis in the neighborhood of an Ohio enamel factory affecting 23 persons and dogs in the neighborhood. The symptoms were indicative of the preskeletal phase. Four had "Chizzola maculae." Urinary fluoride excretion in 21 persons ranged from 0.4 to 2.4 mg/day.

Thiers (263) described follow-up human studies in the neighborhood of an enamel factory. Mean 24-hr urine excretion of the residents was 2 μg/g; the latter suffered from gastrointestinal disorders. Analysis of teeth showed fluorine values of 185 μg/g.

C. Radiological Changes

Radiological changes reflect the cumulative effects of chronic exposure. The response of the axial skeleton to chronic fluoride exposure differs from that of the limb bones and the short long bones. The radiological changes observed in established cases of fluoride toxicity include the following:

1. Osteosclerosis
   a. Thickening of trabeculation
   b. Calcification of interosseous membranes, ligaments, and tendons
   c. Thickening of cortical areas of long bones
   d. Osteophytosis and exostoses, involving vertebrae, flat bones, and ends and surfaces of long bones.
2. Osteoporosis involving ends of long bones and short long bones
3. Osteomalacia involving long bones
4. Periosteal new bone formation
5. Growth arrest lines and modeling defects
6. Changes indicative of secondary hyperparathyroidism

D. Tissue Changes

The ability of fluoride to enter the hard tissues has been ascribed to its capacity to replace other anions on the bone crystal lattice such as hydroxide (OH^-) and citrate ion. Considering that tissues other than bone and teeth cannot offer anions for exchange with fluoride ion, it is understandable that fluoride ion does not, in fact, accumulate in large quantities in organs like liver, spleen, or heart. Although there are a few reports on elevated fluorine contents in soft tissues of small laboratory animals, most of the available information indicates that in humans and in domestic animals, soft tissues accumulate very small quantities of fluorine both in normal situations and in response to fluorine exposure.

Fluorine is also secreted into biological fluids such as saliva, milk, and urine. Fluoride is predominantly excreted via the kidneys; thus, these organs are constantly exposed to high levels in chronic fluoride toxicity and can accumulate appreciable amounts. Nails, hair, calcified aorta, exostoses, renal calculi, and, to a lesser extent, other organs concentrate fluorine to various degrees.

Some studies suggest that reliable information on tissue fluorine content in biopsy and autopsy materials can be obtained in humans when the latest microanalytical techniques are applied. Some data on the fluorine content of biological materials from Algieria (50) and the Soviet Union (284) are summarized in Table II.

Analytical studies carried out in fluorosis areas of North Kazakhstan reported 0.62 μg/g fluorine in blood, 77.6 μg/g fluorine in teeth, 0.5 μg/g fluorine in milk, and 72.0 μg/g in hair (284). In this area, the fluoride content of water was 4 μg/g and the prevalence of dental fluorosis was 91%. Postmortem bone analyses from fluorotic industrial workers yielded a range of 3500–9910 μg/g fluorine in the whole skeleton and 2500–12,900 in the long bones (46). The fluorine content of tissues correlated with the clinical severity of osteofluorosis.

Although it is well established that the fluorine content of bone gradually decreases after withdrawal from the exposure (12), Dominok et al. (46), were able to demonstrate significantly elevated fluorine concentrations in bone as late as 21 years after cessation of exposure.

Rydzewska et al. (194) analyzed teeth in children residing in the vicinity of an aluminum smelter and found 610.5 μg/g fluorine in the enamel as compared to 450.8 μg/g in control children. Krishnamachari (123) estimated the fluorine content in ashed samples of biopsy bone material secured from subjects with genu valgum due to fluorosis and compared the values with that of bone tissues obtained from normal controls. In the former group, the mean fluorine content of the bone was 7283 μg/g while the control group had only 3190 μg/g fluorine.

Renal stones, reported to be common in endemic fluorosis areas (103), are capable of accumulating considerable amounts of fluorine. Dillon observed 2000 μg/g (42), Spira 1790 μg/g (235), Herman 1560 μg/g (85), and Auermann and

Table II

Fluorine Content of Various Tissues and Biological Fluids

Country	Area[a]	References	Sample size (n)	Tissue or biological fluid					
				Blood (mg/liter)	Teeth (μg/g)	Nails (μg/g)	Hair (μg/g)	Milk (mg/liter)	Water (mg/liter)
Algeria	N	Elsair et al. (50)	79	0.6	—	35.0	20	—	0.3
	A			3.6	—	17.0	30	—	1.2
	B		98	3.7	—	16.0	35	—	2.35
	C			4.0	—	320	40	—	4.5
Soviet Union	—	Zhavoronkov and Steochkova (284)		0.6	77.6	—	72	0.5	4.0

[a] N, Normal area; A, B, C, high-fluoride areas.

Kuhn 1130 μg/g (6). In independent studies, Zipkin (285) and Jolly (103) reported the fluoride content of renal stones from high-fluorine endemic areas and low-fluorine nonendemic areas. The values observed in high-fluoride areas were 3700 and 2200 μg/g. The corresponding values for fluorine in low-fluorine areas were 2600 and 200 μg/g, respectively. In the United States, Call et al. (26) and Geever et al. (69) reported the fluorine content for renal stones from Utah, New York, and Grand Rapids as 258, 2340, and 8400 μg/g, respectively. In an area with a water fluorine level of 1 μg/g, Opalko et al. (173) determined the fluorine content of saliva of people living in the vicinity of an industry and found it to be 0.138 μg/g, in contrast to 0.023 μg/g obtained in the saliva of the controls.

Hanhijarvi (78) studied the ionic plasma fluoride levels of 1600 subjects living in an endemic and a nonendemic area and found progressive increases in plasma ionic fluoride with age both in the controls and in the residents of endemic areas. The mean value for fluorine for each age in the latter group was always higher than that of the former at all ages studied. Taves et al. (255) summarized that of 13 tissues studied in nine normal men who died in accidents, only kidney, aorta, thymus, and lung had values for fluorine greater than the blood levels. Smith et al. (226) reported 2.0–2.5 μg/g of fluorine in liver and muscle on a dry-weight basis.

1. Fluorine in Human Milk

Fluorine is actively secreted into human breast milk. No diurnal variation in the fluorine content was seen by Spak et al. (229), who also reported no significant difference in fluorine content between fore and hind milk samples. Colostrum as well as mature milk contained fluorine at 0.36 μmol/liter in areas where fluorine content of water was 1 μg/g. The corresponding value for fluorine content of milk was 0.28 μmol/liter in areas with only 0.2 μg/g fluorine content in water. Administration of small supplements of fluoride (1.5 mg/day) to lactating mothers did not alter the breast milk fluorine content. However, chronically exposed mothers secreted more fluorine in breast milk in high fluorine areas. Hanhijarvi et al. (79) reported a mean value of 0.51, 0.59, and 0.61 μmol/liter, respectively, for breast milk ionic fluoride on the third, fifth, and seventh day of lactation among Finnish mothers who consumed water containing 0.2 μg/g fluorine.

2. Serum Changes

Chronic fluorine toxicity surprisingly brings about very little change, if any, in many biochemical parameters studied.

The results of investigations into the biochemical indices of serum in 31 severely affected skeletal fluorosis subjects in comparison with those in 10 normal subjects have been reported (236). In human subjects, there is no significant change induced by chronic fluorine toxicity in serum proteins, serum calcium, and inorganic phosphorus, and only alkaline phosphatase shows a significant rise in many cases. These observations were confirmed by investigators working independently in many endemic zones in India (99,187,236,259,260).

Singh *et al.* (220) did not observe any significant difference between the blood fluorine of patients with fluorosis with neurological changes and of those without neurological changes. In another study Jolly *et al.* (102) observed similar values for acid phosphatase, SGPT, SGOT, and cholinesterase between fluorosis subjects and controls.

By far, the most commonly observed change in serum in fluoride toxicity is an increase in alkaline phosphatase activity. However, in the osteoporotic type of fluorosis, this enzyme does not show any increase in serum (124).

Susheela *et al.* (245) reported a reduction in the serum levels of total protein-bound carbohydrates as well as seromucoid fraction of subjects with fluorosis residing in an area with 5.2 $\mu g/g$ fluorine in the drinking water. The exact mechanism responsible for the reduction of serum glycoproteins in fluorosis needs elucidation. Jolly *et al.* (101) investigated several subjects with fluorosis for serum biochemical changes but observed significant changes only in alkaline phosphatase.

Serum ionic fluoride was measured in a group representative of the Canadian population. The mean value for men <35 years of age was 0.88 ± 0.275 $\mu mol/liter$, while for men above that age it was 1.18 ± 0.350 $\mu mol/liter$. Women had values ranging from 0.22 to 0.32 $\mu mol/liter$ and showed a linear correlation of fluorine levels with age.

3. Plasma Fluoride versus Bone Fluorine

Parkins *et al.* (175) demonstrated a positive, linear correlation between plasma fluorine and iliac crest bone fluorine with age among populations residing in fluoridated communities in the United States. Tbe mean plasma fluoride of 19 men was 0.047 $\mu g/g$, while the corresponding mean value for iliac crest fluorine was 2824 $\mu g/g$. The fluorine contents of bone, blood, and urine from different endemic regions of India are presented in Tables III and IV.

E. Fluorine, Calcium, and Bone

Epidemiological studies in many developing countries strongly indicate that the more severe forms of skeletal fluorosis occur in populations subsisting on low intakes of dietary calcium (125,211). It is also observed that as the fluorine content of natural water supplies increases, the calcium content progressively declines (221). Fluorine is known to bind calcium in the body, causing ionic

Table III

Fluorine Content of Biological Materials in Fluorosis

Reference	Experimental group	Blood (mg/liter)	Urine (mg/liter)	Bone[a] (μg/g)	Teeth (μg/g)	Urinary calculi (μg/g)
Singh et al. (220)	Fluorosis	0.5–5.6	1.8–25.5	600–6800	—	—
	Controls	Traces	<1.0	<300	—	—
Jolly et al. (101)	Fluorosis	0.5–8.0	1.0–25.7	700–7000	0.5–6.9	8.5–41.5
	Controls	Traces	<1.9	1100	0.09	6.2
Bagga et al. (9)	Fluorosis	0.5–6.1	1.7–25.0	700–6800	—	—
Srikantia and Siddiqui (236)	Fluorosis	0.9–7.5	3.8–19.5	—	—	—

[a] Only seven subjects.

Table IV
Fluorine Content of Bone Samples in Fluorosis

Reference	Fluorine content of drinking water (mg/liter)	No. of samples analyzed	Mean fluorine content (μg/g)	Fluorine content of controls if any (μg/g)
Raman (183)	8.0–5.4	2	3800 and 4700	Nil
Singh and Jolly (219)	9.5	10	3430 (700–6800)	Nil
Krishnamachari (123)	8.0–14.0	37	7283	3190[a]
Jolly et al. (101)	5.0	20	3187 (700–7000)	1100 + 200
Singh et al. (220)	3.5–14.0	7	2610 (600–6800)	—
Baud et al. (10)[b]	—	2	2630 and 8560	280–790

[a] Mean of six values.
[b] Baud described occurrence of niflumic acid-induced fluorosis in subjects of arthritis who were treated with the drug for chronic arthritis (drug-induced fluorosis).

calcium to decrease; this, in turn, causes secondary hyperparathyroidism. Studies with ^{47}Ca in patients with skeletal fluorosis with and without genu valgum and in matched controls indicated that the total calcium turnover is significantly higher in chronic fluorosis, while the external turnover of calcium is less. Bone mineralization rate is higher in fluorosis. These parameters are significantly higher yet in those subjects who develop osteomalacia (169). The latter not only have increased bone mineralization but also increased bone resorption. Simultaneously, circulating IPTH levels are found to be elevated in fluorosis subjects (127).

Calcium-45 turnover studies in subjects with fluorosis kept on 800 mg of dietary calcium continuously indicated a high retention of dietary calcium and ^{45}Ca radioactivity. In addition, the bone formation rate and the exchangeable pool of calcium also showed an increase in response to the increased calcium intake (170). Chemical balance studies carried out simultaneously gave similar results.

1. Calcium Balance Studies

Indian investigators (170,212,258) working independently in different endemic zones investigated patients with established skeletal fluorosis for their calcium retention on a constant daily intake of 800 mg of calcium. In one of the studies (212) there was a normal control group on a similar intake of calcium. Skeletal fluorosis patients retained 53.2, 50.0, and 32% of ingested calcium, respectively, in the above three studies. In contrast, the control subjects retained only 17.5%. Fluorosis patients on low calcium intake (140 mg/day) had negative calcium balance of 30% of the intake. These studies indicate the high affinity of fluorine for calcium in chronic fluoride intoxication. Urinary calcium was significantly lower in all the fluorosis patients than in normals.

2. Fluoride Balance Studies

Few fluoride balance studies have been reported in fluorosis patients. Jolly (100) investigated fluorosis patients both on self-selected diets and on hospital rations and compared their fluorine balance with that of normals (Table V).

In this study, fluorine was predominantly derived from drinking water. In another study, fluorine balance in normal persons on two types of diet was reported. *Sorghum vulgare,* a minor millet which is consumed as a staple in rural parts of India, was associated with greater fluorine retention than a rice-based diet, providing the same amount of fluorine (129).

Spencer *et al.* (233) found positive balances in adult men, at a total fluorine intake of 4.5 mg/day from diet and drinking water. These balances were increased further when fluoride supplements were given. The amounts retained corresponded to about 40% of a fluoride intake of 4 or 14 mg/day and were about 30% of the highest intake of 44–48 mg/day (Table VI).

3. Bone Density in Fluorine Toxicity

In chronic fluorine intoxication due to prolonged environmental or industrial exposure, there is progressive deposition of fluorine in the bone tissue. In addition, there is new bone formation, and the mass of the bone tissue gradually increases. Spongy bone becomes more compact. Long before radiological evidence of new bone formation is apparent, increased bone density in different bones can be detected by radiodensitometry or photon absorptiometry. Early

Table V
Fluorine Balance Studies in Fluorosis[a]

Experimental group	n	Mean ingestion (mg)	Excretion (mg)	Balance (mg)
Controls	10	3.74	3.34	+0.41
Fluorosis subjects on home diet	10	9.88	7.93	+1.95
Fluorosis subjects on hospital diet	10	3.44	6.55	−3.21

[a] From Jolly (100).

Table VI

Fluorine Excretion and Balances in Adult Men Expressed as Percentage of
Fluoride Intake[a]

Fluoride intake (mg/day)	Fluorine excretion (%)		Fluorine balance (%)	Net absorption (%)
	Urine	Stool		
4	54	6	40	94
14	53	7	40	94
44–48	58	11	31	89
12[b]	50	12	38	88

[a] From Spencer *et al.* (233)
[b] Fluoride given as fish protein concentrate.

diagnosis is particularly useful in industrial exposure among workers who can be periodically investigated. Using morphometric measurements, Czerwinski (37) developed diagnostic criteria to detect the early phase of fluorosis. Li Yumin *et al.* (133), using density criteria, investigated aluminum plant workers in China. Apart from early diagnosis, bone density studies help in assessing clinical improvement following cessation of exposure.

4. Endemic Genu Valgum

A new manifestation of chronic fluorine toxicity, characterized by osteoporosis, osteomalacia, and osteosclerosis, occurring among adolescents and young adults exposed to high environmental fluorine was reported from India in 1973 (124). Osteoporosis of the peripheral bones associated with secondary hyperparathyroidism as a manifestation of chronic fluorine toxicity occurs among populations habitually consuming low dietary calcium. In this clinical situation there is increased calcium turnover, increased calcium accretion together with increased calcium resorption, elevated urinary hydroxyproline, radiological evidence of osteoporosis, decreased bone density, hormonal changes including increased circulating IPTH, decreased levels of thyrocalcitonin, and elevated levels of plasma growth hormone and normal serum alkaline phosphatase activity. A similar bone disease characterized by genu valgum, sabre tibia, and osteomalacia was reported among undernourished children living in endemic fluorosis areas of Kenhardt, South Africa (97). Later, endemic foodborne skeletal fluorosis characterized by osteomalacia and osteoporosis has been described in China (96). Osteomalacia was either peripheral or regional. Genu valgum and genu varum occur frequently in these populations. A similar disease has been reported also from the Arusha region in Tanzania, in which fluorosis is endemic. In addition to classical features, genu varum, genu valgum, osteomalacia, and osteoporosis have been observed among the younger generation (34). It is thus clear that the clinical picture of fluorosis includes softening of the bones and osteoporosis as well as secondary hyperparathyroidism on a global basis. It is not clear why in some endemic regions the osteosclerotic form of fluorosis prevails while in the other regions the osteoporotic form is prevalent.

5. Osteomalacia

Compston *et al.* (36) reported the occurrence of osteomalacia in a person who was treated for osteoporosis with sodium fluoride, 50 mg/day, and vitamin D. This occurred in spite of 800 mg of daily calcium intake.

The clinical pathology of fluorosis centers around bone tissue. Fluoroapatite is chemically different from normal bone, hydroxyapatite. Thus, osteomalacia is a basic change that occurs early in the process of fluorosis. Roholm (193) made a reference to the occurrence of osteomalacia with uncalcified osteoid in experimental animals receiving large doses of fluoride. Roholm fed cryolite and sodium fluoride to dogs and observed "a kind of osteomalacia." Endemic genu valgum, a condition often associated with endemic skeletal fluorosis in parts of India, is mainly characterized by osteomalacia (124). In addition to the contribution of the low calcium, discussed above, there is some evidence for a low copper status interacting with chronic fluoride intoxication in the genesis of endemic genu valgum (123). The associated secondary hyperparathyroidism further accentuates the bone loss and rarefaction. The study of Jackson (97) in South Africa and, later, reports from the Arusha region of Tanzania (34) and from the Guizhou Province in China (95,96), support the occurrence of osteomalacia as a component of fluorosis. Teotia *et al.* (262) described histological and radiological changes in bone which are in agreement with osteomalacic changes in fluorosis. These observations have a bearing on therapy.

6. Interactions of Fluorine with Other Trace Elements

The fluoride ion is highly reactive. *In vivo,* fluoride interacts with many organic and inorganic biological compounds, but has an especially high affinity for calcium and magnesium. While the calcium nutritional status modifies fluorine metabolism, long-term fluorine ingestion, in turn, influences calcium metabolism. Among the trace minerals, copper and molybdenum stand out in their interaction with fluorine. Singh and Kanwar (223) investigated the tissue contents of copper and iron in the liver, kidney, and bone of mice treated with graded doses of fluoride over a 16-week period. They reported copper depletion of liver and bone when treated with ≥ 50 µg/g of fluoride and also in kidney with 100 µg/g of fluoride. In contrast, the iron content of the liver and kidney increased as a result of this treatment. With higher doses of fluorine, bone also accumulated iron. In a study using ^{59}Fe Kahl *et al.* (106) demonstrated increased uptake of radioactive iron into liver and bone marrow following fluoride administration. Interestingly, they observed reduced incorporation of ^{59}Fe in the blood. Fluoride may inhibit ^{59}Fe incorporation into protoporphyrin.

Breast milk samples of mothers residing in endemic fluorosis areas and normal areas were analyzed for copper, zinc, and magnesium. The samples from the endemic areas tended to have low values for copper, that is, <100 μg/liter (126). The clinical significance of this finding is not known.

7. Fluorine and Osteoporosis

Osteoporosis is a condition associated with reduced bone mass per unit volume, and reduced bone density. The etiology of this debilitating condition which affects several millions of people all over the world is multifactorial. Age-related bone loss is by far the commonest cause of osteoporosis. In this condition, administration of fluorine has been shown to be beneficial both by reducing the severity of symptoms and by improving bone density. The ash content of bone obtained from iliac crest biopsy in cases of fluorosis has been stated as 49% compared with 37% in controls (64); hypermineralization has also been reported by others (193). Later studies, however, indicate that osteoporosis can also occur as a complication of fluorine toxicity in undernourished populations. Franke (64) treated 158 osteoporosis patients with sodium fluoride over 13 years, giving 80 mg/day sodium fluoride for 4 months, followed by 40 mg/day until the symptoms subsided. The success of the therapy was assessed histologically, radiologically, by densitometry, and by clinical picture, after years of treatment. Arthralgia and gastrointestinal disturbances, sometimes leading to gastric ulcers, were common side effects. The response of osteoporotic bone to fluoride administration depends on several factors including absorption capacity of the person for fluoride, nutritional status, renal threshold for fluoride, and dosage. If there is evidence of simultaneous osteomalacia, there is a need to give supportive vitamin D therapy as suggested by Jowsey et al. (105). Reutter et al. (191) and Olah et al. (172), however, contend that osteomalacia occurs transiently in the early phase of treatment with fluoride. It has been subsequently suggested to supply calcium simultaneously with fluoride, a combination that might obviate the need for vitamin D administration (104). The success of therapy is indicated by the achievement of a fasting fluorine level of 100–200 μg/liter in blood (253). Franke et al. (65) and Franke (64) did not observe any damage to liver and kidney during the treatment of osteoporosis with high doses of fluoride. The other complication of fluoride treatment is induction of secondary hyperparathyroidism as suggested by Faccini and Care (59) and Faccini (58). In skeletal fluorosis secondary hyperparathyroidism occurs in undernourished fluorotic subjects. The criteria for successful therapy of osteoporosis include an increase in bone volume and osteoid volume, increase in the thickness of trabeculae, increased bone formation surface, as well as bone formation rate (64). When treating osteoporosis with fluoride in undernourished populations, it is essential to assure an adequate calcium intake.

F. Metabolic and Endocrine Effects

1. Metabolic Effects

Geeraerts *et al.* (68) demonstrated a significant reduction in the metabolites of the serotonin pathway after a single oral dose of 10 mg sodium fluoride to rats, which returned to normal after 3 days. Zhavoronkov and Steochkova (284) demonstrated that sodium fluoride at 12 µg/kg body weight administered daily to rats depressed protein synthesis. Similar findings were reported in mice: fluoride inhibited DNA and protein synthesis (91).

Zhavoronkov and Steochkova (284) also cited reports from the literature of occurrence of oligospermia, azoospermia in men working in fluoride industries, and also of the occurrence of hypogonadism in fluorosis.

Transplacental Transfer of Fluorine. The opinions regrading transplacental passage of fluorine are divided. Experimental studies in pregnant ewes indicate that fluoride passes to the blood of the fetus as early as minutes following intravenous administration to the mother (140). Kauzal (109) believed that the death from bleeding gastric ulcers in five newborn children was caused by exposure of their mothers to industrial fumes containing fluorine. As early as 1955, Feltman and Kosel (61) demonstrated significantly higher fluorine content in the placentae of mothers who received a 2.2 mg/day sodium fluoride supplement for the entire duration of pregnancy. A similar situation was observed in women who consumed fluoridated water (109). Gedalia *et al.* (67), Held (84), and Brzezinsky *et al.* (24) described placental transfer of fluoride in animals. Studies with ^{18}F in ewes (13), in rabbits (56), and in pregnant women undergoing termination of pregnancy (56) indicated, however, that ^{18}F did not cross the placental barrier. The fluorine concentration was found to be no different between the maternal and fetal blood (3). Shen and Taves (202) had observed a linear relationship between maternal and fetal plasma fluorine concentration. In one study, plasma ionic fluoride in mothers' blood and in umbilical cord blood at the time of caesarean section in two groups of women, one from an endemic fluorosis area, the other from a control area, was measured. Plasma ionic fluoride was 8-fold higher in plasma as well as cord blood of those who resided in fluorosis areas. The values for fluorine in the maternal and cord blood were not different from each other in either group (257). This study confirms that a placental barrier does not exist for fluoride and that the latter passes through the placenta to the fetus in significant amounts. Many authors have reported progressive decline in the plasma ionic fluoride with length of gestation, with a concomitant reduction in urinary fluorine (67,79). This may mean that maternal fluorine is transferred to the fetus. The existence of such transplacental transfer acquires great significance in endemic fluorosis areas where the fetal tissues, including endocrine organs, are likely to be subjected to an exposure of large amounts of fluorine in intrauterine life. This may have relevance to the occurrence of dental and skeletal changes in very young children, as is occasionally reported from some of the endemic fluorosis areas.

In yet another study, pregnant ewes were administered 2 mg of sodium fluoride intravenously before parturition. Serial analysis over 90 min indicated nearly identical fluorine concentrations in paired samples of maternal and fetal plasma, thereby confirming that transplacental passage of fluorine (140) does occur. Weiss and de Carlini (277) demonstrated that fluorine from the anesthetic methoxyflurane crossed the placental barrier when administered to pregnant women. The transfer to the fetus was immediate. The plasma concentration of fluorine in the fetus, however, was only half that of the mother.

2.   Endocrine Effects

During the last decade, evidence has accumulated both in humans and in animals to suggest that fluorine toxicity is associated with altered endocrine function as well as with structural changes of the endocrine system.

The chief cells in the rat parathyroid gland undergo increased metabolic activity in response to fluorine administration (189). Fluorine and calcitonin have contrasting effects on osteoid turnover. The former stabilizes osteoid by stimulating osteoblasts while calcitonin exerts an inhibitory effect on osteoclasts (162). Evidence for the existence of functionally hyperactive parathyroid glands in patients with chronic fluorine toxicity is accumulating in the literature. Singh et al. (222) and Srikantia and Siddiqui (236) observed elevated serum alkaline phosphatase activity in fluorosis patients. Later, Teotia and Teotia (259) and Sivakumar and Krishnamachari (225) independently reported the presence of increased amount of IPTH in the circulation of fluorotic patients, indicative of hyperparathyroidism. The serum levels were twice normal in this condition. When genu valgum was also present, the hormone was further elevated to about four times the normal level (225).

Makhni et al. (144) observed marked increases in the size and weight of the parathyroid glands at autopsy in fluorosis patients. Histologically, hyperplasia of the glands, nuclear changes, and occasional degenerative changes were seen. This is in agreement with several other reports, both in humans and in sheep exposed to fluorine intoxication. Sheep maintained on drinking water contained 100 μg/g fluorine developed hyperplasia of parathyroid glands and had elevated levels of IPTH (57,58), but rats given 125 μg/g fluoride in water did not show such changes (75).

The role of the PTH–calcitonin axis in calcium homeostasis is well known. Although the IPTH levels of fluorosis patients are elevated, there is no corresponding elevation of thyrocalcitonin in all cases. The only data available are those published by Teotia *et al.* (261), who demonstrated a significant increase in the blood levels of this hormone, and of Sivakumar, who reported lowered plasma calcitonin values in patients with fluorosis (171). Circulating growth hormone is elevated in fluorosis (224). The exact mechanism for that elevation if unknown, as are the practical implications of this observation. A few reports cover vitamin D metabolism in humans exposed to fluorine toxicity. The concentrations of 25-hydroxycholecalciferol are not different between fluorosis patients and matched controls (261).

A relation between iodine metabolism and fluoride metabolism has been suggested. Many years ago it was thought that the excess fluorine in the environment may be in some way related to iodine deficiency goiter. Detailed studies in humans, however, ruled out any relation between thyroid function and fluorine toxicity.

G.   Kidney in Fluorine Toxicity

1.   Renal Function

The most important route of fluorine excretion is the kidney. The kidney is constantly exposed to high levels of fluorine that enter the renal parenchyma; it bears the brunt of prolonged exposure. There are several studies which indicate functional impairment of varying degrees depending on the severity of fluorotic bone changes. More severe skeletal lesions (grade III) are often associated with obvious renal functional defects. Elsair *et al.* (50) investigated 40 subjects with varying clinical severity (grades II and III) and identified mild tubular dysfunction followed by glomerular impairment.

Jolly *et al.* (103) described reduction in creatinine and fluorine clearance, indicative of glomerular dysfunction, in patients with chronic skeletal fluorosis, but no change of tubular function. Light microscope studies revealed no histological abnormalities in renal biopsy samples.

There is evidence to suggest that the fluorine content of renal stones is elevated in subjects living in high-fluorine areas (103). No correlation has been found between the fluorine and calcium contents of kidney stones, but there is a good correlation between urinary fluorine and the fluorine content of stones (103). Uric acid stones contain the lowest and calcium stones the highest concentration of fluorine (48). Herman (85) described the presence of large amounts of fluorine in the tissues of renal stone formers.

Renal stones were observed in children under 10 years of age living in endemic fluorosis areas in India (103). The weight of the calculi ranged from 0.5 to

13.7 g, and the fluorine content varied from 0.83 to 4.15 mg per stone. Fluorine was twice as high in the stones of endemic fluorosis subjects than in the stones of nonfluorosis patients. Calcium–fluorine ratios in the above two groups of subjects were 13 : 1 and 46 : 1, respectively, while the calcium–phosphorus ratios were 30 : 1 and 31 : 1. Urinary fluorine was significantly higher in the former group of subjects, but no difference was observed in urinary calcium and phosphorus between the two study groups. Although the exact genesis of renal stones in fluorine toxocity is not known, it is conjectured that insoluble calcium fluoride is deposited in the urinary tract as a nucleus around which other salts are deposited. Zipkin *et al.* (285) reported very high values for fluorine in urinary tract calculi. Evidence for higher incidence of renal calculi in endemic fluorosis areas is still lacking.

Aluminum hydroxide administered to subjects with renal disease and maintained on 3.5 mg fluoride daily rapidly brings about a negative fluorine balance (46% reduction), predominantly through increased gastrointestinal elimination of dietary fluorine. This effect of aluminum hydroxide is seen both in normal subjects and in those with kidney disease (232).

Taylor *et al.* (256) demonstrated that rats fed diets containing 20, 50, or 100 $\mu g/g$ of sodium fluoride for 6 months essentially had no changes in renal parenchyma. Guinea pigs subjected to hydrogen fluoride exposure continuously for 18 months did not show histological or functional impairment of renal tissue (192).

2. Urinary Excretion of Fluorine

Fluorine entering the body rapidly moves to the hard tissues. A fraction of the ingested fluorine is excreted daily. There is not only interindividual variation but also intraindividual variation in respect to fluorine excretion, which depends on three factors: (1) total fluorine intake; (2) duration of exposure to fluorine; and (3) normal kidney function.

Adult males excrete more fluorine than females (266), with daily excretion reported as 0.78 and 0.56 mg, respectively. The corresponding figures for boys and girls are 0.23 and 0.20 mg. Seasonal variation is demonstrable in urinary fluorine among normal subjects (266). Among industrial workers engaged in the aluminum industry, the average urinary fluorine is positively related to the duration of employment. These studies suggest that, on a community basis, urinary fluorine excretion may be a dependable indicator of community exposure. This is also true for population groups living in endemic fluorosis areas, where a positive correlation exists between urinary fluorine values and the fluorine content of drinking water. Autocorrelation coefficients for urinary fluorine indicate the existence of circadian rhythm for this parameter in normal subjects (93).

McClure and Kinser (149) as early as 1944 pointed out the role of urinary

fluoride as an indicator of intake in chronically exposed persons. Chandra *et al.*
(31a) later studied this problem in an endemic area in Rajasthan in India, where
urinary fluorine was positively related to the severity of dental fluorosis in
children aged 7–14 years (31a).

H. Miscellaneous Effects

1. Muscle and Nerve

Neurological complications are well known in skeletal fluorosis (101). Muscle
weakness is a common complaint. In order to find out whether muscle is directly
or indirectly affected by skeletal fluorosis, Reddy *et al.* (190) investigated 36
cases and observed abnormal electromyogram patterns in 19 patients suggesting
neurogenic atrophy rather than primary muscle involvement. The proximal muscle groups of the upper limbs were most affected. Rao *et al.* (186) studied
morphological changes in the sural nerve obtained at biopsy in 13 subjects with
well-established skeletal fluorosis. A reduction in the number of nerve fibers per
square millimeter was observed, especially of small fibers (<7 μg size), and
features suggestive of demyelination and remyelination as well as axonal damage
were seen under the light microscope. The same group of investigators observed
evidence for denervation followed by reinnervation in skeletal muscle biopsy
specimens obtained from 11 subjects with skeletal fluorosis. Primary muscle
damage was ruled out.

2. Hemodialysis with Fluoridated Water

It has been shown that during hemodialysis with fluoridated (1 μg/ml) dialysate, serum ionic fluoride increases (254). Arterial blood fluoride levels gradually rise to four times the predialysis value at the end of 3 hr of dialysis and
remain high (three times the basal value) 2 hr following completion of dialysis
(80). Among patients with considerable renal damage, such a reverse dialysis of
fluoride into the human body can be deleterious.

Speder (230) described arterial lesions in fluorosis. A Moenckeberg type of
arterial calcification was described by several authors (33,95,112,128,228,271).
Tuncel (267) reported a significantly higher incidence of Moenckeberg calcification among fluorosis subjects, particularly in those 60 years or older, compared
to normals. Whether this feature is related in any way to the secondary hyperparathyroidism so frequently seen in fluorosis needs to be studied.

3. Antidotes for Fluorine

Calcium given orally or intravenously is known to counteract the effects of
fluorine, particularly those of acute nature. Calcium helps to reduce fluorine

absorption by the intestines when given orally. Aluminum sulfate in combination with calcium carbonate has been used in cattle against fluorine toxicity with limited benefits (1). Magnesium metasilicate can bind enormous quantities of fluorine *in vitro* (186). Its role *in vivo* has been tested in chronic fluorosis patients, who responded with a partial clinical amelioration of their symptoms (188). Studies in rabbits suggest that boron given daily in a dose of one-third of the fluorine exposure dose (on a body weight basis) can reduce clinical and radiological manifestations of fluorine toxicity and correct hypocalcemia and hyperfluoremia (51). Boron is believed to induce the formation of a complex, BF_4, with fluorine, which is excreted in the urine. No drug or chemical has yet been shown to cure the chronic effects of fluorine toxicity.

II. FLUORIDE TOXICOSIS IN CATTLE

A. Clinical and Pathological Considerations

As early as in 1912, a disease resembling osteomalacia was described in cattle from Italy. The disease was prevalent in the neighborhood of a superphosphate factory. A disease characterized by bone deformities and bone thickening in horses consuming hay contaminated with fluoride in the neighborhood of aluminum industry was reported from Switzerland in the post-World War I period. These are examples of fluorosis in cattle caused by industrial discharge.

In contrast to these situations, joint affliction of cattle was reported in the endemic fluorosis regions in India as early as in 1934 (270). Occurrence of osteomalacic lesions in the bones of cattle exposed to fluorine was reported independently by Mahajan (141) and Majumdar *et al.* (142). Adult bulls administered 1 g of sodium fluoride daily developed skeletal lesions. Simultaneous administration of calcium reduced the clinical severity of the lesions and facilitated maintenance of constant body weight. When the fluoride dose was increased (3 mg/kg body weight), only aluminum phosphate supplementation could protect against the lesions.

Milk cattle, beef cattle, sheep, and dogs suffer from the effects of chronic exposure to fluoride in endemic fluorosis areas and in the neighborhood of fluorine industries. In the nonendemic areas livestock consume small amounts of fluorine which generally do not produce any health hazard. Comprehensive accounts of livestock fluorosis have been presented by Shupe and Alther (205) and Shupe and Olson (207).

Consumption of fluorine-contaminated forage close to aluminum factories, superphosphate factories, fluorine processing industries, and cryolite mining areas contributes to chronic fluoride intoxication in cattle. Severe forms of skeletal and dental fluorosis in cattle have been reported from western parts of India (19). Cattle feeds and supplements can contribute large amounts of fluorine. As in humans, fluorine toxicity in cattle is due to cumulative effects of fluorine from various dietary sources.

Tolerance levels for various domestic animals vary from 30 to 150 $\mu g/g$ fluorine in the diet.

The ingestion of quantities of fluorine that eventually prove to be toxic does not in the initial phases produce any observable ill effects. During that latent period, the animal is protected by two major physiological mechanisms, namely, (1) accumulation of fluorine by the bone tissue and (2) excretion of part of the ingested fluorine through kidney. Thus, urinary fluorine shows an increase, long before pathological and clinical signs appear. There are several variables, some biological and others physical, which determine the ultimate outcome of fluorine ingestion, that is, emergence of fluorine toxicosis. These factors include the dose of the fluorine ingested, the frequency of exposure and age of the animal, its nutritional and general health status, the animal species under consideration, and biological interactions between fluorine and other nutritional and dietary parameters. The skeleton acts as a buffer until 30–40 times the normal quantities of fluorine accumulate in bone. Then the excess fluorine overspills from the saturated bone into the circulation and, in turn, into the soft tissues. This is the stage of metabolic breakdown, and symptoms and clinical signs of disease begin to appear.

Excessive ingested fluorine can cause acute, subacute, or chronic fluorosis in animals. Shupe and Olson (207) prefer to use the term "fluoride toxicosis." There is then a voluntary refusal of food, so that typical starvation phenomena are superimposed on signs of fluoride toxicosis (180). At higher fluorine intake the effect on growth and appetite is almost immediate (154). The speed with which symptoms develop will depend among other factors on the dosage and duration of exposure. The fluorine content of blood increases, as does the urinary fluorine excretion. Clinical signs may appear long before the bone tissue is saturated with fluorine. Tooth formation in the young and appetite at all ages are sensitive to increased plasma fluorine concentrations (27,213).

Anemia has been reported as characteristic of high fluorine intakes (30,196), but cattle affected by the chronic form do not show evidence of changes in hemoglobin levels (87,92). Subacute and chronic types often overlap and may not be distinguishable (207). Both endemic-type and neighborhood fluoride toxicosis exhibit similar pictures.

Animals exposed to high amounts of fluorine prior to the eruption of the permanent teeth develop dental defects, described as fluorotic teeth, consequent to elevated plasma levels of fluorine. Shupe et al. (206) classified dental fluorosis into the following grades:

Grade	Criteria
0—Normal	
1—Questionable	Presence of aberrations—cause not known; no mottling
2—Slight	Mottled enamel; slight discoloration
3—Moderate	Generalized mottling; increased tooth wear
4—Marked	Definite mottling; enamel discolored, pitted, and eroded; tooth aberration definite
5—Severe	More severe than stage 4

Examination of the incisor teeth generally helps to diagnose dental fluorosis in cattle. For greater accuracy, teeth should be examined directly and with back lighting. As the disease progresses, teeth become modified in shape, size, color, orientation, and structure. Incisors become pitted and molars abraded. Pulp cavities may become exposed due to fracture or wear. Once the permanent teeth are fully formed and erupted, their structure is not affected by high fluorine intakes (66,269).

Bone Changes (Osteofluorosis)

As fluorine accumulates further in the body of the animal, osseous lesions begin to appear at different times during exposure. The bone changes can be correlated with the amount of fluorine accumulated in the bones. Young animals are more susceptible and accumulate more fluorine per unit volume of the bone than older animals. Initially, fluorine accumulates in the bone, without causing any radiological change. With chronic exposure to high doses, thickening of the bone and exostosis formation of jaw bone and long bones, leading to visible distortion occur eventually. Ribs, metacarpals, and metatarsals often show changes. Osteofluorotic lesions appear wherever pressure and tensile stress is physiologically operative; they look chalky white, rough, and porous compared to normal bones. The affected bones are larger in diameter and heavier than normal. As in humans, calcification of membranes, ligaments, and tendons occurs. Involvement of soft structures around the joints reduces mobility and may result in ankylosis. Walking at this stage becomes progressively difficult.

The bone changes may exhibit one or more of the histological features of osteoporosis, osteosclerosis, or osteomalacia. A combination of the above changes is often seen in different bones. Osteophyte formation and hyperostosis are the other changes. Periosteal hyperostosis is a typical feature. Based on studies in sheep and rabbit, Weatherall and Weidmann (274) indicated that bone resorption is a marked but not an invariable feature in osteofluorosis. The new bone formed in the exostoses has normal bone structure, but wide areas of uncalcified osteoid also occur. Collagen fibers are irregularly arranged, and increased osteoclastic resorption on the endosteal side is often seen. Overproduction of osteoid in fluorosis (35,176,274) and impaired blood supply to the bone due to the formation of severe periosteal hyperostosis has been reported (204). Belanger et al. (16) have demonstrated evidence for altered mucopolysaccharide metabolism in pigs exposed to fluorides. Later studies of Susheela et al. (245) in experimental rabbits also suggest that fluorine interferes with metabolism of glycoproteins.

In the early stages of fluorosis loss of appetite is not common, but in the chronically exposed states appetite and growth are adversely affected. There is evidence that fluorine toxicity adversely influences milk secretion in cows (250). Studies in dairy cows with ^{18}F indicate that the ionic fluoride concentration in milk and blood are equal, thereby indicating that the fluoride ion does pass into milk (177). A considerable proportion of fluorine in the milk is bound to casein, fat, and albumin (53). The reduced milk production in some cattle with fluorosis has been ascribed to the effect of general inanition. The tolerance limit is between 30 and 40 μg/g of fluorine in the ration. Cows fed 93 μg/g of sodium fluoride passed successfully through their first lactation, but continued feeding of fluoride at this level reduced the milk yield during the second and subsequent lactation periods (239). Smaller doses permit normal lactation for four lactation periods. Prolonged administration of fluoride does not influence the reproductive performance of the cows, but the poor nutritional status of the mother may adversely affect the outcome of pregnancy (53). The reproductive performance of sheep is relatively unaffected by fluorosis (250).

Stiffness and lameness, when they occur with other signs of fluoride toxicosis, can be considered part of the toxicity state. Severely diseased cattle may move around on their knees, due to spurring and bridging of the joints in the late stages. Also, calcification of periarticular ligaments and tendons may impede mobility (204,208). At this stage rigidity of spine and limbs is manifest.

Dental fluorosis in cattle grazing in the neighborhood of fluorine-related industries causes the following changes: (1) hypercementosis; (2) delayed eruption of permanent teeth; (3) alveolar bone necrosis; (4) hypoplasia of teeth; and (5) dental changes.

B. Fluorine in Tissues

The fluorine concentrations in the soft tissues of animals are normally low and do not increase with age. Fluorine does not concentrate in any tissues other than the bones and teeth, although the placenta and the aorta sometimes carry elevated levels of this element, probably as a result of calcification which secondarily accumulates fluorine (89). The kidneys are usually richer in fluorine than other

Table VII
Fluorine Levels in Soft Tissues of Animals[a]

Tissue	Sheep[b]		Cows[c]	
	Normal	10 mg/g liter fluorine in water for 2 years	Normal	50 μg/g fluorine in ration for 5.5 years
Liver	3.5	2.4	2.3	3.6
Kidney	4.2	20.0	3.5	19.3
Thyroid	3.0	7.6	2.1	7.3
Heart	3.0	2.3	2.3	4.6
Pancreas	—	3.2	2.8	4.2

[a] Expressed as micrograms per gram, dry basis.
[b] From Suttie and Kolstad (249).
[c] From Harvey and Queens (82).

organs, owing in part to retained urine. The thyroid gland has no special ability to concentrate fluorine, even when intakes are high (273).

The data presented in Table VII reveal only small increases in soft tissue fluorine concentrations in chronic fluorosis. Dairy cows receiving a ration supplemented with 50 μg/g fluoride as sodium fluoride for 5.5 years only increased the fluorine levels in heart, liver, thyroid, and pancreas 2- to 3-fold above the 2–3 μg/g found for these tissues in control cows (251). Even smaller increases were observed in the same tissues of sheep consuming water with 5 or 10 μg/ml fluoride for 2 years (53). Similar results were reported for rats receiving 2 mg/day fluoride for 62 days (268), while Schroeder et al. (197) were unable to detect any soft tissue fluorine accumulation in mice given 10 μg/ml fluoride in the drinking water for 2 years.

The tissues of newborn and suckling animals are usually lower in fluorine than those of their mothers. In one study, rats and rabbits were fed varying levels of added fluoride, up to 300 μg/g of the diet, for several weeks before and during pregnancy. The total fluorine in the fetuses at term was found to be negligible. Some increases occurred at the higher intakes, but the amounts were extremely small compared with those in the mothers (145). Rats and puppies subsisting on the milk of fluoride-fed mothers accumulated more fluorine in their bodies than similar animals consuming the milk of normally fed mothers (35,145). Some of this may have been acquired from the diet or water of the fluoride-fed mothers. Normal cow's milk contains only 1–2 μg/g fluoride (150).

Hillman et al. (86) observed signs of fluorosis in cattle which received mineral supplements containing 6300 μg/g fluoride or protein supplements containing 1088 μg/g. In a study of fluorosis in cattle from Michigan, the values in the

accompanying table were reported for fluoride content of various biological
materials (86).

	Bone (μg/g)	Milk (μg/g)	Urine (μg/g)
Fluorosis cattle	885–6918 (2406)	0.072–0.64	5.31
Controls	350–1000	—	—

Suttie *et al.* (250) reported fluorine values of 0.1–0.4 μg/g in milk of the
cattle. In another study Greenwood *et al.* (72) reported the fluorine content of
milk in cattle at varying fluoride intakes. The values were 0.06, 0.10, 0.14, and
0.20 μg/g at estimated dietary fluoride levels of 12, 27, 49, and 93 μg/g,
respectively.

Five-year-old grade Holstein cows were fed fluoride at 50, 100, 150, and 200
μg/g in the diet for 4 weeks at each level. At the highest fluoride intake the total
milk production decreased to 10–15% of the normal yield (246). Suttie and
Kolstad (249) reported plasma fluorine and urinary fluorine in cows fed fluoride
at various doses (see table).

Fluoride intake (μg/g)	Plasma fluorine (μg/g)	Urinary fluorine (μg/g)
Control	0.24	12.8
50	0.83	43.3
100	1.13	53.5
150	1.23	58.5
200	1.50	80.3

Miller *et al.* (156) demonstrated a reduction of citric acid concentration of
heart and kidney tissues of cows which developed skeletal fluorosis in the neighborhood of an industry. In contrast, aconitate hydratase activity showed an
increase by 25% in the heart tissue and 50% reduction in the kidney tissues.

Shupe *et al.* (208), in an in-depth study in cattle, showed no retention of
fluorine in the soft tissues when 110 μg/g of fluoride were ingested daily. Shupe
and Olson (207) reported values of >0.25 μg/g of fluorine in milk of fluoride-
intoxicated cattle.

Birds consuming high-fluoride diets can readily transfer fluorine to the egg,
especially into the yolk. The fluorine content of yolk of eggs from hens on a
normal low-fluoride diet was raised from 0.8–0.9 μg/g to as high as 3 μg/g by
supplementing the hen's diet with 2% rock phosphate (178).

C. Fluorine Tolerance

Tolerance to fluorine exposure in humans or animals is determined by several physiological and nonphysiological parameters including the age and species of the animals, the dosage, the form and the duration of exposure to fluorine, and the nutritional status. Although these variables are relatively unimportant in cases of overwhelming exposure, they are highly operative in chronic fluoride toxicosis where smaller amounts of daily intakes accumulate over a long period of time. Poultry is more resistant to the fluorine effect than livestock (70,77); chicks and turkeys can tolerate up to 200 $\mu g/g$ fluoride (2). The bigger animals, including cattle, sheep, and pigs, are tolerant only up to 80–100 $\mu g/g$ of fluoride in the dry ration (179). The margin between tolerance limit and the minimum toxicity limit is rather narrow for dairy cows, the two levels being 40 $\mu g/g$ fluoride and 50 $\mu g/g$ fluoride, respectively (248). Young calves are more susceptible than mature animals, and their tolerance for fluoride is limited to 30 $\mu g/g$ of the total dry diet. Sheep consuming water with 20 $\mu g/g$ fluoride for 2–3 years were not adversely affected with regard to food consumption, dental development, or wool production (181), but young sheep, <1 year of age, developed mottling of the teeth from drinking water with ≥5 $\mu g/g$ fluoride. Seasonal variations are known to affect fluorine consumption of humans and sheep; this is possibly related to water intake. During the periods of low fluoride intake the exchangeable skeletal stores of fluorine are depleted and excreted in the urine. Therefore, during the subsequent periods of high intake, more fluorine enters the skeleton.

Experiments in rats (132), sheep (53), and cows (27), comparing continuous and intermittent fluoride dosage, indicate that in both types of exposure the total yearly intake and skeletal storage of fluorine is similar (27). Short-term administration of high levels of fluorine has deleterious effects on experimental animals (29). Cattle are more resistant to fluoride from rock phosphate than from sodium fluoride. When administered in comparable doses, soluble and insoluble fluorine fed via contaminated hay showed similar toxicity (210). Similarly, for rats at an intake level of ≤14 $\mu g/g$, there is little difference in tolerance between soluble and insoluble forms of fluorine. When cryolite is administered in fine-powder form the attendant fluorine toxicity reaches the extent caused by comparable levels of sodium fluoride. Compared to fluorine administered through a dry diet, sodium fluoride and cryolite promote greater fluorine retention in rats (132,276). Suttie and Faltin (247a) showed that generalized undernutrition tends to increase the toxic effects of fluorine in cattle, as has also been observed in humans.

D. Enzyme Changes

Several enzyme systems are sensitive to relatively low concentrations of fluorine. The effect of fluorine on enzymes in higher plants is operative both *in vivo*

Table VIII
Effect of Fluorine on Enzymes in Higher Plants[a]

Type of effect	Enzymes and systems involved
Very sensitive *in vitro*	Glycolysis: phosphoglucomutase, enolase
	Respiration: mitochondrial ATPase
Inhibited *in vitro*	Glycolysis: UDPglucose fructose transglucosylase; hexokinase
	Respiration: peroxidase, polyphenoloxidase, succinic dehydrogenase
	Others: phytase, amylase, phosphatase, pyrophosphatase, esterase, urease, etc.
Enhanced *in vivo*	Pyruvate kinase, G6PD, pentose phosphate shunt, phosphoenolpyruvate carboxylase
Inhibited *in vivo*	Malic dehydrogenase, cytochrome oxidase, succinic dehydrogenase, peroxidase, catalase, ATPase

[a] We know that fluorine either activates or inhibits enzyme systems. This table summarizes data given by Miller *et al.* (157).

and *in vitro* (157). There are several enzymes which are specifically inhibited *in vivo* while the others are stimulated. A third group is inhibited *in vitro* only. A summary of these enzymes is given in Table VIII.

Fluoride has been shown to decrease the level of liver G6PD in rats and the rate of glycogen turnover (29,282). Whether this effect is directly attributable to fluorine, or is brought about by the reduced food consumption, is debatable. Rats consuming very high levels of fluoride in the order of 450 μg/g in the diet catabolize glucose at a normal rate but are unable to metabolize glycogen. This is due perhaps to some effect on the liver enzyme level (281). There is also another possibility that these alterations in carbohydrate metabolism may be related to the slower growth rate brought about by the fluorine.

Suttie *et al.* (247) observed fluorine-induced suppression of enolase activity. Alkaline phosphatase is one of the widely studied enzymes in the blood in human fluorosis and in experimental fluoride toxicosis. In a variety of animals the blood levels of glycogen phosphatase are elevated. Phillips (178) observed a marked stimulation of these enzymes in the blood of fluorotic cows, while Motzok and Branian (166) observed elevation of alkaline phosphatase in serum and bones of fluorotic chicks. Alkaline phosphatase activity in bone is more closely related to fluorine intakes than the activity of the enzyme in blood. Miller and Shupe (159) demonstrated a close correlation in cows and heifers between the amount of fluorine ingested and the bone alkaline phosphatase activity on one hand and the fluorine content of bone on the other. It appears that a significant increase in bone alkaline phosphatase occurs only when excessive doses of fluoride are administered, but not with low doses. In human subjects suffering from skeletal fluorosis the serum levels of this enzyme are not consistently elevated (124).

III. STUDIES ON EXPERIMENTAL ANIMALS

A.   Studies in Monkeys

In order to test the role of calcium and fluoride in the development of bone changes, monkeys were fed diets with various combinations of these two elements (73). After 5 years of observation by periodic X-ray examinations, ^{125}I photon absorptiometry, and bone histological findings, it was concluded that the low-calcium diet alone led to osteoporosis. The same diet with high fluoride caused the appearance of osteomalacia. Addition of high fluoride concentrations to a diet containing adequate calcium caused an increase in radiographic density.

Sriranga Reddy *et al.* (237) studied the effects of dietary protein, calcium, and vitamin C on fluoride-induced ^{45}Ca turnover in monkeys. Accretion rate, exchangeable calcium pool, and half-life were reduced by a low-protein diet, regardless of fluoride. Maximum accretion was observed when fluoride was administered along with a high-protein diet. ^{45}Ca-Specific activity was maximum in osteosclerotic bones.

B.   Studies in Rats

The mitigating effect of dietary calcium on fluorine toxicity in rats was suggested as early as 1941 (185), when it was demonstrated that the survival of rats administered toxic amounts of fluoride was prolonged by a ragi (*Eleusine coracana*) based diet, rich in calcium.

Sodium Fluoride and Microsomal Mixed-Function Oxidase

There is some evidence that drug-metabolizing enzymes are altered by sodium fluoride administration. Sodium fluoride administered to rats in an acute dose of 100 mg/kg body weight reduced renal and hepatic microsomal enzymes such as NADPH-linked aminopyrine, *N*-demethylase, and acetanilide hydroxylase. Cytochrome *P*-450 levels of kidney and liver were also marginally reduced under sodium fluoride treatment (227). The same authors also studied the *in vitro* effect of fluoride concentration ranging from zero to 100 m*M* on these enzymes. *In vitro* effects, for example, on acetanilide hydroxylase, were observed only with very high concentrations of fluoride (227).

Rats administered 10 or 25 μg/g fluoride for a period of 10 months show alterations in enzymes in liver and kidney to a varying extent (223). While alkaline phosphatase and acid phosphatase decreased in liver and kidney, ATPase was elevated in both organs. Lactic dehydrogenase was not affected, while glutamic oxaloacetic and pyruvic transaminases (GOT, GPT) showed a decrease in the kidney only. Thus, there is a variation in the response of various enzymes to fluoride administration (223). This variation is probably related to

Table IX
Fluorine Levels in Soft Tissues of Rats[a,b]

	Experimental group	
Tissue	Normal	2 mg fluorine daily for 76 days
Liver	0.21	0.28
Kidney	0.62	1.50
Thyroid	—	—
Heart	2.6	5.4
Pancreas	—	—
Muscle	0.53	1.60

[a] From Venkateswarlu and Narayana Rao (268).
[b] Expressed as micrograms per gram, dry basis.

the distribution pattern of fluorine within the organism, to the capacity of the hard tissues to accumulate excessive fluorine, and thus protect the circulatory system and the soft tissues.

Wallace-Durbin (273) and Hein et al. (83) demonstrated equal distribution of ^{18}F between blood and tissues in rats. This was further confirmed by Knaus et al. (116). In another study in rats Boros et al. (23) demonstrated a significant elevation of the fluorine content of femur (1300 $\mu g/g$) and incisor (300 $\mu g/g$), compared with 0.07 $\mu g/g$ in serum, after 4 weeks of 25 and 100 $\mu g/g$ of dietary fluoride. Venkateswarlu and Narayana Rao (268) administered 2 $\mu g/day$ fluoride for 76 days to rats; they found fluorine contents in the soft tissues of these animals that were only marginally higher than those of normal controls (Table IX).

C. Studies in Mice

Mice reared on a low-fluoride diet (0.1–0.3 $\mu g/g$) progressively increased their fluorine content of bone and teeth with age (111). In a three-generation study this age-related trend of bone fluorine accumulation weakened in the subsequent generations.

Chromosomal Changes

Between 50 and 200 $\mu g/ml$ sodium fluoride in the drinking water for 3 weeks brings about dose-dependent chromosomal aberrations in bone marrow cells. The abnormalities consist of translocations, dicentrics, ring chromosomes, and fragments, among others. Meiotic chromosomes from testes show aberrations when mice are fed fluoride at the 10 $\mu g/g$ level (161). Fluorine has been shown to inhibit nucleic acid synthesis in rats (18).

Mice administered sodium fluoride at 11 $\mu g/ml$ in drinking water for 12 weeks showed alterations in myeloid cells of the bone marrow as reflected by an increased number of blast forms. In addition, intracellular changes such as the presence of irregular, punched-out nuclei, mitotic abnormalities, and irregular DNA distribution are seen in the immature cells (71). Selvaraj and Sharra (199) showed increased intracellular respiration and glucose oxidation in leukocytes under the influence of fluoride.

D. Studies in Rabbits

Boron and Fluoride Toxicity

The effect of simultaneous administration of boron on fluoride-induced changes in rabbits were investigated by balance studies and X-ray examination (51). Fluoride (60 $\mu g/g$) and boron (23 $\mu g/g$) were given daily to rabbits for 2 months, during which period 4-day balance studies were carried out for fluorine, calcium, and phosphorus. Boron was shown to prevent the loss of calcium and phosphorus, induced by fluoride intoxication. It also improved the food intake of the animals, depressed by fluorine, and reversed hypocalcemia. Radiological changes were less pronounced in the boron-supplemented group. Similar protective effects of boron have been observed by Baer *et al.* (8). The mechanism of this protective effect is not well understood; complex formation between boron and fluoride may result in reduced toxicity.

Elsair *et al.* (51) studied the protective effect of boron given to rabbits at 15 mg/kg body weight simultaneously with fluoride at 40 mg/kg. At these levels, fluoride-induced coagulation defects were prevented by boron. The mode of action was not clear at that time. In another study by the same author (50), fluoride at 60 mg/kg body weight was given for 60 days followed by boron therapy with 23 mg/kg. The latter brought about a negative fluorine balance through hyperfluoruria and corrected the negative balance of calcium and phosphorus. Further studies with boron on fluoride-induced acute and subacute intoxication confirmed the beneficial effects of boron. The latter was also shown to correct secondary hyperparathyroidism due to fluoride intoxication.

The fluoride effects observed by different workers are variable. A majority of these studies used very high levels of fluoride, which are not normally encountered in the daily life of animals or humans.

Kathpalia and Susheela (108) investigated the effect of sodium fluoride (50 mg/kg body weight daily) on soft tissue proteins of rabbits. Protein estimation was carried out on days 158, 165, 225, and 265 in 12 soft tissues, all of which showed a reduction in protein content from 10 to 46%. Maximum reduction was observed in the stomach.

Suheela and Sharma (243) demonstrated a fall in serum total proteins and of albumin in rabbits given fluoride at 0.5–5.0 mg/kg body weight.

Sriranga Reddy and Narasiuga Rao (238) in their studies of *in vivo* biosynthesis of bone matrix used 165–250 mg sodium fluoride per kilogram diet.

They did not observe any significant difference in the mean collagen protein concentration between the groups even after 44 weeks of experimentation, although radiologically discernible new bone formation had occurred. In contrast, Susheela and Mukherjee (242) suggested that in rabbits, fluoride administration (50 mg/kg body weight per day) brought forth a significant reduction in collagen synthesis. They had also shown a reduction in the hydroxyproline content of cortical bone in rabbits given fluoride at 10 mg/kg body weight (240).

Makhni *et al.* (143) demonstrated osteoporosis, generalized reduction of bone density, honeycomb appearance of articular ends of cancellous bone, and thinning of the cortex in rabbits given increasing doses of 0.25–20 mg/kg body weight over 2–3 years.

Susheela and Mohan Jha (241) demonstrated a significant, 20% reduction in sialic acid content following a daily dose of 10 mg/kg body weight. This effect was observed in the cancellous, but not in cortical bones. The same investigators demonstrated a reduction in serum seromucoids by large doses of sodium fluoride. An increase in the sulfate content of cancellous and cortical bones was demonstrated without any change in the total hexosamines. Fluoride treatment brought about an increase in total glycosaminoglycan content in blood. Preliminary studies in rabbits suggest that fluoride treatment may bring about inadequate cross-linkages in collagen fibers in all tissues and particularly in tendon.

E. Studies in Guinea Pigs

Kour *et al.* (118) used very high doses of fluoride (500 and 1000 μg/g) to induce histological changes in the liver in guinea pigs within 3 months. With only 10 μg/g fluoride, no evidence of histological changes was observed. The pathology was characterized by central venidilation, focal necrosis, fatty change, and presence of calcified areas. The very high dose used in such studies may have no relevance to practical situations.

F. Studies in Dogs

Loeffler *et al.* (135,136) analyzed bone samples from 36 normal dogs and also samples of exostoses obtained at biopsy. The fluorine content was found to be higher in exostoses (990–3670 μg/g, versus 280–1700 μg/g in normal bone).

IV. FLUORIDE METABOLISM

A. Absorption

Soluble fluorides are rapidly and almost completely absorbed from the gastrointestinal tract, even at high intakes. Rats administered small oral doses of

^{18}NaF absorbed 75% in 1 hr (273) and 80–90% within 8 hr (53). When a dilute solution of sodium fluoride was given orally to rats and the animals killed immediately or 5, 15, 30, 60, and 90 min later, 12, 22, 36, 50, 72, and 86%, respectively, of the administered doses was absorbed (49). In humans given small amounts of soluble fluoride, maximum blood levels were observed within 1 hr, and 20–30% of fluoride appeared in the urine in the succeeding 3–4 hr (286).

The speed and extent of fluorine absorption varies with the physical and chemical form of the compound. In small amounts, the insoluble forms of fluorine are almost as well absorbed as the more soluble ones. Thus, the absorption of fluorine from calcium fluoride was 96% and that of cyrolite 93% in the rat when so administered (130). When added to the diet in solid form, the absorption of fluorine was only 60–70% from calcium fluoride and 60–77% from cryolite (138). Still poorer absorption, ranging from 37 to 54%, has been reported for the fluorine from bone meal. The fluorine in fish protein concentrate (FPC) has also been reported to be only 42–52% as available to young rats as sodium fluoride (287). In contrast, metabolic balance studies in adult men revealed a very high net fluorine absorption (88% for FPC and 94% from sodium fluoride) when administered to supply similar amounts of fluorine (234).

Fluorine absorption is known to be affected by several dietary factors. From the results of mineral balance studies in sheep it appears that aluminum salts exert a protective effect against high intakes of fluorine by reducing its absorption from the intestinal tract (14). This is consistent with human studies published in 1975 (188). Calcium salts function similarly. Thus, 3 mg fluorine was found in the whole carcass of rats at the end of a 2-week period when 1% calcium chloride was administered with sodium fluoride in the drinking water, compared with 6 and 11 mg, respectively, when this calcium salt was added at lower levels of 0.2 and 0.01% (276). When magnesium chloride and aluminum chloride were administered at the same levels, fluorine retention was similarly inhibited but to a smaller extent. High dietary calcium and phosphorus, independently and together, increase fecal fluorine excretion in humans, but the amounts so excreted are small so that the fluorine balance remains unaltered and plasma fluorine levels do not change (231).

The level of fat in the diet also influences fluorine toxicity, although the extent to which this is due to an effect on absorption is not clear. Raising the level of dietary fat from 5 to 15 or 20% enhances the growth-retarding effect of high fluorine intakes in rats (25,158), and chicks (22). The effect is unrelated to the chain length of the fat (158) and has been attributed, in part, to increased fluorine retention in the heart, kidneys, and skeletal tissues (22,25). Subsequently McGown and Suttie (151) have found that fluorine in the presence of fat causes delayed gastric emptying. They suggest that this effect probably accounts for the increased toxicity of fluorine to rats fed high-fat diets. However, Ericsson (54) reported no differences in femur accumulation of ^{18}F, 4 hr after administration of 4 μg/g fluoride with 28% olive oil, compared with controls which received only ^{18}F and water.

B. Fluorine in Blood

By the use of a suitable micromethod (215), Armstrong and co-workers (4) demonstrated significant increases in rat plasma fluorine values from levels of about 0.1 μg/g to as high as 1.0 μg/g when dietary fluorine was increased. Concentrations up to 3.3 μg/g fluorine occurred in rats receiving 600 μg/g fluoride as sodium fluoride (213). Singer and Armstrong (216) found that ionic rather than bound fluoride reflects increased total plasma fluorine values in the rat and rabbit, whereas in human and bovine plasma it is the ionic fluoride rather than the bound fluorine concentration that tends to remain low and uniform.

In a long-term study of dairy cows continuously or periodically exposed to high fluorine intakes, plasma fluorine concentrations were related to the current level of fluorine ingestion (27). The levels of control animals were consistently <0.1 μg/g, whereas those of the fluoride-treated animals were significantly higher. Levels of 1.0 μg/g in the plasma were observed only after extended periods of ingestion. A feature of these findings was the rapidity with which plasma fluorine levels changed in response to changes in fluoride intake.

Marked diurnal variation in plasma fluorine levels has been observed in sheep (213), rats (201), dogs, and humans (167) following the administration of a single oral dose of fluoride. Shearer and Suttie (201) have pointed out that plasma samples must be taken very soon after the actual ingestion of the fluoride if they are to reflect the total daily intakes. The loss of fluorine in the urine and to the skeleton is so rapid that control concentrations are approached within hours of the completion of intake. No diurnal variations in plasma fluorine are apparent in humans at lower continuous fluorine intakes (214,233). When plasma fluorine concentrations approach ≥0.5 μg/g, severe dental lesions appear. At levels >0.2 μg/g fluorine, less severe damage occurs, while at plasma fluorine levels <0.2 μg/g, no adverse effects are apparent (27). It seems, therefore, that 0.2 μg/g fluorine can be regarded as a critical plasma concentration in cattle.

C. Fluorine Distribution and Retention

Absorbed fluorine is distributed rapidly throughout the body, particularly in the hard tissues. It readily crosses cell membranes, including those of the erythrocytes. The speed with which fluorine leaves the blood is illustrated by several experiments. Perkinson et al. (177) computed that a removal rate of 40%/min for intravenously injected fluoride in lambs and 32%/min in cows, while Bell and co-workers (17) found that only 53% of the ^{18}F tracer dose administered to cattle remained in the blood after 2 min. The disappearance curve of ^{18}F from the blood is triphasic in character. In lambs, the first and most rapid phase has a half-time of only 3–4 min, the second about 1 hr, and the third about 3.3 hr (177). The first phase presumably represents mixing of the fluorine with body water, the second uptake by the skeleton, and the third excretion in the urine.

Metabolic studies with rats suggest that there is no level at which complete elimination occurs (273), and long-term studies with this species indicate continuous retention in the bones at the lowest levels of dietary fluorine employed (184,280). Machle and co-workers (139) found the input and output of fluorine in humans to be approximately equal, over many weeks, when the daily intake was as low as 0.5 mg, whereas definite retention occurred when the intake ranged from 3 to 36 mg daily.

Fluorine in Bones and Teeth

An extremely high proportion of the total-body fluorine is present in the skeleton. The higher the intake the higher the amount and the proportion of fluorine in bone. In normal adult farm animals not unduly exposed to fluorine, the concentrations of fluorine in whole dry, fat-free bones usually lie within the range 300–600 µg/g (53,176,251). The fluorine concentrations of normal teeth parallel those of the long bones but usually at lower levels. Thus, normal enamel is reported to contain 100–270 µg/g, normal dentine 240–625 µg/g, and normal molar teeth 200–537 µg/g fluorine on a dry, fat-free basis (32,148). These levels are dependent on the amount, chemical form, duration, and continuity of the fluorine intake and on the age of the animal.

Fluorine levels in bones and teeth many times higher than those cited above as "normal" can occur following prolonged high fluorine intakes, long before clinical signs and symptoms of toxicity appear. Swine bones appearing normal have been reported to contain upwards of 3000–4000 µg/g fluorine (113), and fluorine concentrations in compact bones from cattle below 4500 µg/g are considered to be innocuous (251). In dairy cattle fluorine toxicosis is associated with levels in excess of 5500 µg/g fluorine in compact and 7000 µg/g fluorine in cancellous bone, with a "saturation" point of the order of 15,000–20,000 µg/g (180,251). Concentrations between 4500 and 5500 µg/g fluorine indicate a marginal zone (251). The toxic thresholds for fluoride in the bones of sheep have been placed lower, 2000–3000 µg/g in bulk cortical and 4000–6000 µg/g in bulk cancellous bone (98).

D. Fluorine in Urine

A positive correlation between urinary fluorine excretion and intake of fluorine has been demonstrated in several species. The urinary fluorine concentration of sheep and cattle not exposed to excess fluorine rarely exceeds 10 µg/ml and is usually closer to 5 µ/ml. With elevated fluorine intakes the urinary fluorine levels rise quickly to 15–30 μg/ml and may reach upper limits of 70–80 μg/ml. Higher values are occasionally observed among animals consuming the same amounts of fluorine, as well as among samples from the same animal taken on different days or at different times on the same day (209,210,248). In the experiments of Suttie and co-workers (248), normal cows excreted urine containing <5 μg/ml fluorine. Cows that were on the borderline of fluorine toxicity, as judged by other criteria, excreted 20–30 μg/ml fluorine in the urine. Essentially similar results were obtained by Shupe *et al.* (209,210) with heifers fed different amounts and forms of fluoride for prolonged periods.

Urinary fluorine may come from the release of the element from the skeleton, as well as from the food and water supply. Mobilization and excretion of excess fluorine in the bones of animals previously fed a high-fluorine diet has been observed in rats (158,195), humans (134), and cattle (28). High urinary fluorine levels can therefore reflect either current ingestion or previous exposure. A scheme summarizing the many aspects of fluorine metabolism is presented in Fig. 1.

V. FLUORIDE AND DENTAL CARIES

A. Essentiality of Fluorine

The dominant health-related function of fluorine in humans and large animals is in the maintenance of the structure of teeth and protection against dental caries.

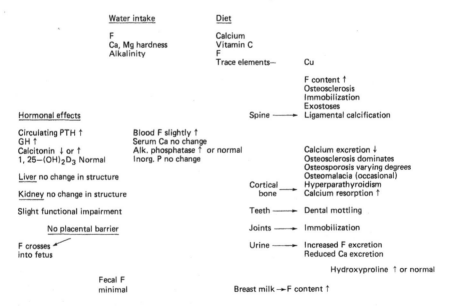

Fig. 1. Fluorine metabolism in humans—a schematic representation.

Several attempts in the past to demonstrate its essential role on the basis of other than dental criteria in rat experiments were not successful (44,147,200). Schroeder *et al.* (197) showed that mice fed 10 $\mu g/g$ fluoride in drinking water improved their body weight. Similarly, Schwarz and Milne (198) observed a 30% growth increment in rats fed fluoride in an ultraclean environment. Different fluorine compounds had different degrees of effect on the deposition of incisor pigments (160). In contrast to the different biological effect, all the fluorine compounds studied deposited nearly the same amounts of fluorine in rat femur. Messer and co-workers (152,154) demonstrated that mice receiving no fluorine in their supplies developed progressive infertility over two generations. They also observed anemia in fluorine-inadequate mice (152,153). On the other hand, Weber and Reid (275) reported that in a six-generation study in mice no definite differences were seen between the low- (0.2 $\mu g/g$) and the high-fluorine group (6 $\mu g/g$). In humans, however, information has been accumulating for several decades which strengthens the view that an optimal level of fluorine is required to reduce the prevalence of dental caries, and that fluorine should be considered an essential element on that basis.

B. Role of Fluorine in Dental Caries

The beneficial role of fluorine in reducing dental caries in children was recognized at the beginning of the twentieth century (47). Dean was the first to establish a correlation between the fluoride concentration of drinking water and community prevalence of dental caries (39) based on a study of 7257 children in 21 American cities. Hodge demonstrated that 1 $\mu g/ml$ fluoride is an optimum limit in drinking water (88). This has been verified subsequently by several workers who demonstrated higher community caries incidence when the fluoride level in water was either far above or far below this value. It is now known that it is the total amount of fluorine ingested from water and food that has relevance both in preventing caries and in the development of fluorine toxicity.

Based on the above studies, fluoridation of drinking water supplies has been implemented in many countries where the natural waters are low in fluorine. Follow-up studies have indicated that water fluoridation brought about a reduction of the caries incidence of children by 40–70% (43,55). As an alternative to fluoridation of community water supplies, the use of fluoride tablets and fortification of dietary items such as salt (146,278) and flour (49,146) have been investigated. Studies carried out in children, pregnant women, and lactating mothers indicate that fluoride tablets can reduce the incidence of caries (5,62,279). In addition, topical application of fluoride is beneficial in reducing caries incidence in children (94). Use of fluoride-containing toothpastes can be successful to some extent, (60,117,163), but the strongest, best documented effects are those of fluoride ingestion during the formative stages of dentition and of topical application of fluorine compounds.

C. Enamel Fluorosis

Fluoride tablets administered early in childhood prevent dental caries of the primary teeth. Such a procedure, however, causes dental fluorosis in the permanent teeth (264). Holm and Anderson (90) described enamel fluorosis in 45% of children over 6 years of age who received fluoride tablets early in childhood. The relative risk for this group of children was four to five times greater than that for children who did not use fluoride tablets. Studies by Forsman (63) had similar results. It is possible that protection against caries as well as against enamel fluorosis can be obtained with a yet to be defined, narrow range of fluoride supplementation.

D. Anticariogenic Mode of Action of Fluoride

The beneficial effect of fluoride against caries is due perhaps to several factors, one of which is the structural role of fluoride at the tooth surface while the others may involve an involvement in the structural strengthening of the whole tooth. The two primary actions of fluoride include (1) enhancing the resistance of the tooth by either altering its morphology or by fluoride-induced enamel effects and (2) affecting directly cariogenic bacteria of the oral cavity. The latter effect involves biochemical and microbiological changes: for example, fluoride may act as a bactericidal agent, it may modify the nature of surface proteins and affect their adsorption properties, or it may alter bacterial colonization. Above all, the antimetabolic action of fluoride may be important. In a 1982 study Morena et al. (165) demonstrated that the presence of fluoride increased the adsorbing capacity of the lattice toward four proteins normally present in saliva. Fluoride was found not to alter the colonization pattern of bacteria on the human enamel (114). Studies of the bactericidal properties of fluoride yielded varying results. Kilian et al. (114) found no change in total plaque bacteria when exposed to 21 µg/ml fluoride in water. Ostrom et al. (174) confirmed these observations even for 100 µg/ml of fluoride. A yet higher dose (113 µg/ml) of fluoride was, however, observed to reduce S. mutans in the plaque (15). Interestingly, 16% reduction in plaque index was brought about by high airborne fluorine (182). In contrast to these ambiguous results, many workers have confirmed that topical fluoride applied in high doses (900–5000 µg/g) either as gel or as dental floss, immediately and significantly reduced S. mutans by 23–75% (20,31,137,168).

De Paola and Kashket (40) have postulated that mineral solubility and bactericidal effects may predominate after high-level application of fluoride, whereas low-dose fluoride topicals and fluoridated water may produce mainly metabolic effects and remineralization. The mechanism of caries prevention by fluoride is obviously multifactorial.

Haavikko and Halle (74) studied the prevalence of varying grades of enamel

Table X
Fluorine Content versus Enamel Defects in Finnish Populations[a]

Enamel defect	Prevalence in low-fluoride areas[b]	Prevalence in high-fluoride area[c]
Mild mottling	0.8%	4.1%
Moderate mottling		11.4%
Children with enamel defect	41–74%	98.2%
Brown stains	—	10.5%
Pitting enamel	0.4%	2.3%

[a] Data from Haavikko and Helle (74).
[b] Low-fluoride areas: Lohja, 0.1 µg/ml fluoride in water; Espoo, 0.2–0.7 µg/ml; Hanko, 0.9 µg/ml.
[c] High-fluoride area: Elimaka, 1.6 µg/ml fluoride in water.

changes in relation to community water fluoride levels. Accordingly, three areas with fluoride content of 0.1, 0.2–0.7, and 0.9 µg/ml were selected to represent the low-fluoride areas, while one area with 1.6 µg/ml fluoride in the water served as the high-fluoride area. The prevalence figures are presented in Table X.

VI. NATURAL DISTRIBUTION OF FLUORIDES

Fluorides are widely distributed in nature. A variety of rocks such as gray granite gneisses (fluoride content 0.06%), gray and pink granites (fluoride content 0.025%), and pink granite (fluoride content 0.06%) bear fluoride (107). These rocks constitute the primary source of environmental fluoride which leaches out from time to time into the soil and thence into subsoil water. The fluorine present in the water may be in the form of one of the salts of sodium or potassium. In addition, organic fluorine also occurs in nature. The fluorine-bearing minerals of importance include fluorspars (49%), cryolite (54%), apatite (3.4%), micas (8%), pigmatites including topaz (19%), and tourmaline (5%). Volcanic eruptions and geothermal waters contain large quantities of fluorine. Numerous industries emit fluoride in gas and effluents. Total fluorine deposits in the world are estimated to exceed 85 million tons. Subsoil water contributes small to very large quantities of soluble fluorine to the intake of humans wherever it is the chief source of drinking water. The biological availability of water fluorine is influenced by total alkalinity, the amount of calcium and magnesium salts present, and total hardness. In general, however, fluorine in gaseous, particulate, or dissolved form is readily available to humans and animals. In spite of this, there are several regions in the world where the environmental concentration of fluorine is so low as to necessitate supplementary intake of fluorine either through fluoridation of water supplies or by one of several other public health measures.

A. Fluorine in Surface Waters

Several studies indicate that surface waters contain little fluorine even in endemic fluorosis areas. A systematic analysis of paired water samples from villages, one from a subsoil source and the other from flowing surface water, has shown that the fluorine content of subsoil water in the endemic areas was at least four to eight times higher than that of surface water (122). This has practical relevance, since many endemic fluorosis regions in the world are located in the vicinity of flowing streams. Protected potable water from such sources may be beneficial in the prevention of dental and skeletal changes.

B. Fluorine Intake of Humans

Data of fluorine intakes by human populations on a global basis are inadequate. The computation of fluorine intake depends on accepting several presumptions and using several tables. Intake estimates based on analysis of duplicate diets are probably the most accurate, but difficult to obtain. Singer and Ophaug (217,218) summarized daily fluorine intake values for infants, toddlers, and adults in fluoridated communities and for adults in nonfluoridated communities in the United States. They also summarized tabulated data on the fluorine content of foods obtained by different analytical methods. In the past, it was believed that dietary intake of fluorine did not contribute much to the etiology of fluorosis, because much greater amounts of fluorine enter the human body through water, particularly in the endemic regions of the world. Much of the information on dietary fluorine originated in the 1950s after some of the methodological difficulties had been resolved. Tsunoda and Kunita (265) estimated the mean daily fluorine intakes in Japanese men and women at 1.34 and 1.12 mg, respectively. Kessabi and Amouzigh (110) reported fluorine values in straw, barley, and water in relation to soil fluorine content in Morocco (Table XI).

In endemic fluorosis areas, the fluorine content of self-selected local foods varies widely. Tea infusion has been confirmed to contain large amounts of fluorine. In addition, the quantity of water consumed locally can have a pronounced influence on the daily intake; it varies with climate, season, and locality. In some endemic regions, water contains as much as 10–25 μg/ml fluorine and consumption of 1 or 2 liters, as is common in these areas, will contribute 10–50 mg/day. In very small infants, the fluorine intake is negligible, <0.2 μg/day, increasing to 0.5–1.0 mg at 6 months of age. Singer and Ophaug (218) compiled published data for the intake of toddlers which ranges from 0.3 to 0.8

Table XI
Fluorine Content of Soil and Water in Morocco[a]

Item	Fluorine values (μg/g)	
	Range	Mean
Soil	410–12630	2785
Straw	52–147	104
Barley	10–43	22
Bone	1115–8350	3260
Water	0.42–1.90	1.17

[a] Data from Kessabi and Amouzigh (110).

μg. Adults living in nonfluoridated areas consume <1.0 mg/day, while the daily intake in fluoridated areas varies from 2 to 4.5 mg. Among American foods, beverages contribute significant amounts of fluorine. Hodge and Smith estimated that 0.5–1.5 mg/day fluorine is consumed by American adults (89). Kramer *et al.* (119) estimated the mean daily intake of adults in four nonfluoridated cities at 0.78–1.03 mg and at 1.73–3.44 mg in 12 fluoridated cities. Krepkogorsky (120) reported 2.5 mg as the average intake figure for England, 3.3 mg/day in the Ukraine, and 2.1 mg/day in other parts of the Soviet Union. The staple food of New Foundlanders provided 2.74 mg/day. Natural salt is another good source of fluorine. Cooking and processing modify the fluorine content of foods. Addition of water during cooking can enhance the total fluorine intake, particularly in endemic areas. Reconstituted artificial milk can have as much as 53 times the fluorine content of breast milk, which is 0.025 μg/g, depending on local water quality.

REFERENCES

1. Allcroft, R., and Burns, K. N. (1969). *Fluoride* **2,** 55–59.
2. Anderson, J. O., Horst, J. S., Strong, D. C., Nielsen, H., Greenwood, D. A., Robinson, W., Shupe, J. L., Binn, W., Bagley, R. A., and Draper, C. I. (1955). *Poult. Sci.* **34,** 1147.
3. Armstrong, W. D., Singer, L., and Makowski, E. L. (1970). *Am. J. Obstet. Gynecol.* **107,** 432–434.
4. Armstrong, W. D., Singer, L., and Vogel, J. J. (1966). *Fed. Proc., Fed. Am. Soc. Exp. Biol.* **25,** 696.
5. Arnold, F. A., McClure, F. J., and White, C. L. (1960). *Dent. Prog.* **1,** 8.
6. Auermann, E., and Kuhn, H. (1980). Editorial. *Fluoride* **13,** 1–3.
7. Auermann, E., and Kuhn, H. (1981). Editorial. *Fluoride* **14,** 101.
8. Baer, H. P., Bech, R., Franke, J., Grunewald, A., Kochmann, W., Melson, F., Runge, H., and Weidner, W. (1980). *Fluoride* **13,** 96–98.

9. Bagga, O. P., Mehta, S. P., Prakash, V., Raizada, A., Singh, K. N., Sankhyan, K. A., Gupta, R., and Sood, B. (1979). *Fluoride* 12, 72.
10. Baud, C. A., Boivin, G., and Demeurisse, C. (1982). *Fluoride* 15, 54–56.
11. Baud, C. A., Boivin, G., and Lagier, R. (1979). *Fluoride* 12, 103–104 (Abstr.).
12. Baud, C. A., Lagier, R., Boivin, G., and Boillat, M. A. (1978). *Vorchows. Arch. A. Pathol. Anat. Histol.* 280, 283–297.
13. Bawden, J. W., Wolkoff, A. S., and Flowers, C. E. (1964). *J. Dent. Res.* 43, 678–683.
14. Becker, D. E., Griffith, J. M., Hobbs, C. S., and McIntyre, W. H. (1950). *J. Anim. Sci.* 9, 647.
15. Beighton, D., and McDougall, W. A. (1977). *J. Dent. Res.* 56, 1185–1191.
16. Belanger, L. F., Visek, W. J., Lotz, W. E., and Comar, C. L. (1958). *Am. J. Pathol.* 34, 25.
17. Bell, M. C., Merriman, G. M., and Greenwood, O. A. (1961). *J. Nutr.* 73, 379.
18. Bempong, M. A., and Tower, E. C. (1970). *J. Hered.* 64, 324.
19. Bhavsar, B. S., Desai, V. K., Mehta, N. R., and Krishnamachari, K. A. V. R. (1985). *Fluoride* 18.
20. Birkeland, J. M. (1972). *J. Dent. Res.* 80, 82–84.
21. Bishop, P. A. (1936). *Am. J. Roentgenol. Radium Ther.* 35, 577–585.
22. Bixler, D., and Muhler, J. C. (1960). *J. Nutr.* 70, 26.
23. Boros, I., Vegh, A., Schaper, R., Keszler, and Rittlop, B. (1984). *Fluoride* 17, 183–192.
24. Brzezinsky, A., Bercovici, B., and Gedalia, I. (1960). *Obstet. Gynecol.* 15, 329–331.
25. Buttner, W., and Muhler, J. C. (1958). *J. Nutr.* 65, 259.
26. Call, R. A., Greenwood, D. A., and Lecheminant, W. H. (1965). *Public Health Rep.* 80, 529–538.
27. Carlson, J. R. (1966). Doctoral thesis, University of Wisconsin, Madison.
28. Carlson, C. H., Armstrong, W. D., and Singer, L. (1960). *Am. J. Physiol.* 187, 199.
29. Carlson, J. R., and Suttie, J. W. (1966). *Am. J. Physiol.* 210, 79.
30. Cass, J. S. (1961). *J.O.M.* 3, 471–527.
31. Caulfield, P. W., Navia, J. M., Rogers, A. M., and Alvarez, C. (1981). *J. Dent. Res.* 60, 927–932.
32. Chang, C. Y., Phillips, P. H., and Hart, E. B. (1934). *J. Dairy Sci.* 17, 695.
33. Chawla, S., Kanwar, K., Bagga, P., and Anand, D. (1964). *J. Assoc. Physicians, India* 12, 221.
34. Christie, D. P. (1984). *Fluoride* 17, 55–56.
35. Comer, C. L., Visek, W. J., Lotz, W. E., and Rust, J. H. (1953). *Am. J. Anat.* 93, 361.
36. Compston, J. E., Chadha, S., and Merrett, A. L. (1982). *Br. Med. J.* 281, 910–911.
37. Czerwinski, E. (1978). *Fluoride* 11, 51–54.
38. Czerwinski, E., Pospulka, W., Nowacki, G., and Skolarczyk, K. (1981). *Fluoride* 14, 61–68.
39. Dean, H. T. (1942). *In* "Fluorine and Dental Health" (R. Moutton, ed.), pp. 6–11. American Association of Advanced Science, Washington, D.C.
40. DePaola, O. F., and Kashket, S. (1983). *In* "Fluorides" (J. L. Shupe, H. J. Peterson, and N. C. Leone, eds.). Paragon Press, Salt Lake City, Utah.
41. DeVries, K., and Lowenberg, H. E. V. (1974). *Pneumonologie* 150, 149–154.
42. Dillon, C. (1954). *Dent. Pract.* 4, 181.
43. Dirks, O. B. (1974). *Caries Res. (Suppl.)* 8, 1–2.
44. Doberenz, A. R., Kurnich, A. A., Kurtz, E. B., Kemmerer, A. R., and Reid, B. L. (1964). *Proc. Soc. Exp. Biol. Med.* 117, 689.
45. Dominok, B. (1984). *Fluoride* 17, 193–195.
46. Dominok, G., Siefert, K., Frege, J., and Dominok, B. (1984). *Fluoride* 17, 23–26.
47. Eager, J. M. (1901). *Public Health Rep.* 16, 2576.

48. Editorial (1980). *Fluoride* **13**, 1–2.
49. Ege, R. (1961). *Tandlaegebladet* **65**, 445, quoted by Ericson.
50. Elsair, J., Merad, R., Denine, R., Khelfat, K., Tabet Aoul, M., Assum Kumar, B., Reggabi, M., Azzouz, M., Hamrour, S., Alamir, B., Biebie, M., Naceur, J., and Benali, S. (1982). *Fluoride* **15**, 43–47.
51. Elsair, J., Merad, R., Denine, R., Reggabi, M., Alamir, B., Benali, S., Azzouz, M., and Khelafat, K. (1980). *Fluoride* **13**, 129–138.
52. Emsley, J. (1981). *J. Am. Chem. Soc.* **103**, 24–28.
53. Ericsson, Y. (1958). *Acta Odontol. Scand.* **16**, 51 & 127.
54. Ericsson, Y. (1968). *J. Nutr.* **96**, 60.
55. Ericsson, Y. (1974). *Caries Res.* **8** (Suppl. 1), 16.
56. Ericsson, Y., and Malmas, C. (1962). *Acta Obstet. Gynecol. Scand.* **41**, 144.
57. Faccini, J. M. (1967). *Nature (London)* **214**, 1269–1271.
58. Faccini, J. M. (1969). *Calcif. Tissue Res.* **3**, 1–16.
59. Faccini, J. M., and Care, A. D. (1965). *Nature (London)* **207**, 1399–1401.
60. Fanning, E. A., Gotjamanos, T., Vowles, N. J., Cellier, K. M., and Simmons, D. W. (1967). *Med. J. Aust.* **1**, 383.
61. Feltman, R., and Kosel, G. (1955). *Science* **122**, 560–561.
62. Feltman, R., and Kosel, G. (1961). *J. Dent. Med.* **16**, 190.
63. Forsman, B. (1977). *Scand. J. Dent. Res.* **85**, 22–30.
64. Franke, J. (1976). As quoted by Franke, J., and Barthold, L. *In* "Fluorides" (J. L. Shupe, H. B. Peterson, and N. E. Leone, eds.). Paragaon Press, Salt Lake City, Utah.
65. Franke, J., Rath, F., Runge, H., Fengler, F., Auermann, E., and Lennart, G. (1975). *Fluoride* **8**, 61–63.
66. Garlick, N. L. (1955). *Am. J. Vet. Res.* **16**, 38.
67. Gedalia, I., Brzezinski, A., Portugese, N., and Bercovici, B. (1964). *Arch. Oral. Biol.* **9**, 331–340.
68. Geeraerts, F., Schimpfessel, L., and Crokaert, R. (1981). *Fluoride* **14**, 155–160.
69. Geever, E. F., McCann, H. C., and Frank, M. S. (1971). *HSMHA Health Rep.* **86**, 820–828.
70. Gerry, R. W., Carrick, C. W., Roberts, R. E., and Hauge, S. M. (1949). *Poult. Sci.* **28**, 19.
71. Greenberg, S. R. (1980). *Anat. Rec.* **196.**
72. Greenwood, D. A., Shupe, J. L., Stoddard, G. E., Harris, L. E., Neilson, H. M., and Olson, L. E. (1964). *Utah Agric. Exp. Stn. Spec. Rep.* **17**, 36.
73. Griffiths, H. J., Hunt, R. D., Zimmerman, T., Firberg, H., and Cuttino, J. (1975). *Invest. Radiol.* **10**, 263–268.
74. Haavikko, K., and Helle, A. (1975). *Fluoride* **8**, 243–244.
75. Hac, L. R., Freeman, S., and Nock, W. B. (1967). *Am. J. Physiol.* **212**, 213–216.
76. Hadjmarkos, D. M., and Leatherwood, E. C. (1966). *Am. J. Public Health* **56**, 391–393.
77. Halpin, J. G., and Lamb, A. R. (1932). *Poult. Sci.* **11**, 5.
78. Hanhijarvi, H. (1975). *Fluoride* **4**, 198–207.
79. Hanhijarvi, H., Erkkola, R., and Kanto, J. (1977). *Fluoride* **10**, 169–173.
80. Hanhijarvi, H., Lampainen, E., and Pesonen, A. (1982). *Fluoride* **15**, 35–43.
81. Harinarayana Rao, S., Krishnamurthy, D., Sesikeran, B., and Raja Reddy, D. (1981). *Fluoride* **14**, 94.
82. Harvey, J. M., and Queens (1953). *J. Agric. Sci.* **46**, 720–726.
83. Hein, J. W., Bonner, J. F., Brudevold, F., Smith, F. A., and Hodge, H. C. (1956). *Nature (London)* **178**, 1295–1296.
84. Held, H. R. (1975). *Fluoride* **8**, 178–181.
85. Herman, J. R. (1956). *Proc. Soc. Exp. Biol. Med.* **91**, 189–191.
86. Hillman, D., Bolenbaugh, D., and Convey, E. M. (1979). *Fluoride* **12**, 100–102 (Abstr.).

87. Hobbs, C. S., and Merriman, G. M., (1962). *Agric. Exp. Stn. Bull.* **351.**
88. Hodge, H. C. (1950). *J. Am. Dent. Assoc.* **40,** 436.
89. Hodge, H. C., and Smith, F. A. (1965). *In* "Fluorine Chemistry" (J. H. Simon, ed.). Academic Press, New York.
90. Holm, A. K., and Anderson, R. (1984). *Fluoride* **17,** 58–59.
91. Holland, R. I. (1979). *Cell Biol. Int. Rep.* **3,** 701–705.
92. Hoogstratein, B., Leone, N. C., Shupe, J. L., Greenwood, D. A., and Lieberman, J. (1965). *J. Am. Med. Assoc.* **192,** 26.
93. Horiuchi, I., Nasu, and Morimoto, M. (1984). *Fluoride* **17,** 173–177.
94. Hownik, B., Dirks, O. B., and Kwant, G. W. (1974). *Caries Res.* **8,** 27.
95. Huo Daijei (1981). *Fluoride* **14,** 51–55.
96. Huo Daijei (1984). *Fluoride* **17,** 9–14.
97. Jackson, W. P. U. (1962). *S. Afr. Med. J.* **36,** 932.
98. Jackson, D., and Weidmann, S. M. (1958). *J. Pathol. Bacteriol.* **76,** 451.
99. Jolly, S. S. (1974). *In* "Proceedings of Symposium on Fluorosis". Indian Academy of Geosciences, Hyderabad.
100. Jolly, S. S. (1976). *Fluoride* **9,** 138–147.
101. Jolly, S. S., Prasad, S., Sharma, R., and Rai, B. (1971). *Fluoride* **4,** 64–79.
102. Jolly, S. S., Prasad, S., and Sharma, R. (1964). *J. Assoc. Physicians India* **18,** 491.
103. Jolly, S. S., Sharma, O. P., Garg, G., and Sharma, R. (1980). *Fluoride* **13,** 10–16.
104. Jowsey, J. (1977). *Geriatrics* **32,** 41–50.
105. Jowsey, J., Riggs, J. B. L., Kelly, P. J., and Hoffman, D. C. (1972). *Am. J. Med.* **53,** 43–49.
106. Kahl, S., Wojcik, K., and Ewy, Z. (1973). *Bull. Acad. Pol. Sci. Ser. Sci. Biol.* **21,** 389–393.
107. Karunakaran, C. (1977). *In* "Proceedings of Symposium on Fluorosis," pp. 3–18. Indian Academy of Geo-Sciences, Hyderabad.
108. Kathpalia, A., and Susheela, A. K. (1978). *Fluoride* **11,** 125–129.
109. Kauzal, G. (1975). *Fluoride* **8,** 178–181.
110. Kessabi, M., and Amouzigh, M. (1981). *Fluoride* **14,** 169–171.
111. Khalawan, S. A. (1981). *Fluoride* **14,** 56–61.
112. Khan, Y. M., and Wig, K. L. (1945). *Indian Med. Gaz.* **80,** 429–438.
113. Kick, C. H., Bethke, R. M., Edginton, B. H., Wilder, O. H. M., Record, R., Wilder, W., Hill, T. J., and Chase, S. M. (1935). *Agric. Exp. Stn. Bull.* **558.**
114. Kilian, M., Thylstrup, A., and Fejerskov (1979). *Caries Res.* **13,** 330–343.
115. Kleerekoper, M., and Parfit, A. M. (1983). *In* "Fluorides" (S. L. Shupe, H. B. Peterson, and N. C. Leone, eds.). Paragaon Press, Salt Lake City, Utah.
116. Knaus, R. M., Dost, F. N., Johnson, D. E., and Wang, C. H. (1976). *Toxicol. Appl. Pharmacol.* **38,** 335–343.
117. Knutson, J. W., and Armstrong, W. D. (1947). *Public Health Rep.* **62,** 425.
118. Kour, K., Koul, M. L., and Koul, R. L. (1981). *Fluoride* **14,** 119–123.
119. Kramer, L., Osis, D., Wiatrowski, E., and Spencer, A. (1974). *Am. J. Clin. Nutr.* **27,** 590.
120. Krepkogorski, L. N. (1963). *Gig. Sanit.* **28,** 30.
121. Krepkogorski, L. V. (1963). *Gig. Sanit.* **12,** 30–35.
122. Krishnamachari, K. A. V. R. (1975). *Indian J. Med. Res.* **63,** 475.
123. Krishnamachari, K. A. V. R. (1982). *Fluoride* **15,** 25–30.
124. Krishnamachari, K. A. V. R., and Krishnaswamy, K. (1973). *Lancet* **2,** 877.
125. Krishnamachari, K. A. V. R., and Krishnaswamy, K. (1974). *Indian J. Med. Res.* **62,** 1415.
126. Krishnamachari, K. A. V. R., Rajalakshmi, K., and Radhiah, G. (1982). *Fluoride* **15,** 81–87.

127. Krishnamachari, K. A. V. R., and Sivakumar, B. (1976). *Fluoride* **9**, 185–200.
128. Kumar, S. P., and Kemp Harper, R. A. K. (1963). *Br. J. Radiol.* **36**, 497.
129. Lakshmaiah, N., and Srikantia, S. G. (1977). *Indian J. Med. Res.* **65**, 543.
130. Largent, E. J. (1959). *In* "Fluoride and Dental Health" (J. C. Muhler and M. K. Hine, eds.), p. 132. Indiana Univ. Press, Bloomington.
131. Largent, E. J., and Heyroth, F. F. (1949). *J. Ind. Hyg. Toxicol.* **31**, 134.
132. Lawrenz, M., Mitchell, H. H., and Ruth, W. A. (1940). *J. Nutr.* **20**, 383.
133. Li Yumin and We Keqin (1984). *Fluoride* **17**, 148–154.
134. Likins, R. C., McClure, F. J., and Steere, A. C. (1956). *Public Health Rep.* **71**, 217.
135. Loeffler, K., Brosi, C., Oelschlager, W., and Feyler, L. (1981). *Fluoride* **14**, 192.
136. Loeffler, K., Brehm, H., Oelschlager, W., Schenkel, H., and Freyler, L. (1981). *Fluoride* **14**, 192.
137. Loesche, W. J., Murray, R. J., and Mellberg, J. R. (1973). *Caries Res.* **7**, 282–296.
138. Machle, W., and Largent, E. J. (1943). *J. Ind. Hyg. Toxicol.* **25**, 112.
139. Machle, W., Scott, E. W., and Largent, E. J. (1942). *J. Ind. Hyg. Toxicol.* **24**, 199.
140. Maduska, A. L., Ahokas, R. A., Anderson, G. D., Lipahitz, J., and Morrison, J. C. (1980). *Am. J. Obstet. Gynecol.* **136**, 84–86.
141. Mahajan, M. R. (1934). Annual Report of Imperial Council of Agricultural Research, New Delhi.
142. Majumdar, B. N., Ray, S. N., and Sen, K. C. (1943). *Indian J. Vet. Res.* **13**, 95.
143. Makhni, S. S., Singh, P., and Thapar, S. P. (1977). *Fluoride* **10**, 82–86.
144. Makhni, S. S., Sidhu, S. S., Singh, P., and Grover, A. S. (1979). *Fluoride* **12**, 124.
145. Maplesdon, D. C., Motzok, I., Oliver, W. T., and Branion, H. D. (1960). *J. Nutr.* **71**, 70.
146. Marthaler, M., and Schweiz (1962). *Bull. Eidg. Gesundheitsamt.* Part B. No.2.
147. Maurer, R. L., and Day, H. G. (1957). *J. Nutr.* **62**, 561.
148. McClure, F. J., (1949). *Public Health Rep.* **64**, 1061.
149. McClure, F. J., and Kinser, C. A. (1944). *Public Health Rep.* **59**, 1575.
150. McClure, F. J., and Likins, R. C. (1951). *J. Dent. Res.* **30**, 172.
151. McGown, E. L., and Suttie, J. W. (1974). *J. Nutr.* **104**, 909.
152. Messer, H. H., Armstrong, W. D., and Singer, L. (1974). *In* "Trace Element Metabolism in Animals—2," p. 425. Univ. Park Press, Baltimore.
153. Messer, H. H., Armstrong, W. D., and Singer, L. (1972). *Science* **177**, 893.
154. Messer, H. H., Armstrong, W. D., and Singer, L. (1973). *J. Nutr.* **103**, 1319.
155. Messer, H. H., Wong, K., Wegner, M. E., Singer, L., and Armstrong, W. D. (1972). *Nature (London) New Biol.* **240**, 218.
155a. Midttun, O. (1960). *Acta Allergy* **15**, 208–221.
156. Miller, G. W., Egyed, M. N., and Shupe, J. L. (1978). *Fluoride* **11**, 14–17.
157. Miller, G. W., Mingho Yu, and Pushnik, J. C. (1983). *In* "Fluorides" (J. L. Shupe, H. B. Peterson, and N. C. Leone, eds.). Paragaon Press, Salt Lake City, Utah.
158. Miller, R. F., and Phillips, P. H. (1956). *J. Nutr.* **59**, 425.
159. Miller, G. W., and Shupe, J. L. (1962). *Am. J. Vet. Res.* **23**, 24.
160. Milne, D. B., and Schwarz, K. (1974). *In* "Trace Element Metabolism in Animals—2" (W. G. Hoekstra *et al.*, eds.), p. 170. Univ. Park Press, Baltimore.
161. Mohammed, A. H., and Chandler, M. E. (1982). *Fluoride* **15**, 110–118.
162. Mohammedally, S. M., Phil, M., and Wix, P. (1982). *Fluoride* **15**, 137–143.
163. Moller, I. J. (1973). *Int. J. Environ. Stud.* **5**, 87.
164. Moller, P. F., and Gudjonsson, S. V. (1932). *Acta Radiol.* **13**, 269–294.
165. Moreno, E. C., Kresak, M., and Hay, D. I. (1982). *J. Biol. Chem.* **257**, 2981–2989.
166. Motzok, I., and Branion, H. D. (1958). *Poult. Sci.* **37**, 1469.

167. Muhler, J. C., Stookery, G. R., Spear, L. B., and Bixler, D. (1966). *J. Oral. Ther. Pharmacol.* **2**, 241.
168. Myers, M., and Handelman, S. L. (1971). *J. Dent. Res.* **50**, 597–599.
169. Narasinga Rao, B. S., Krishnamachari, K. A. V. R., and Vijayasarathy, C. (1979). *Br. J. Nutr.* **41**, 7–14.
170. Narasinga Rao, B. S., Siddiqui, A. H., and Srikantia, S. G. (1968). *Metabolism* **17**, 366.
171. National Institute of Nutrition, Hyderabad (1976). Annual Report, 166.
171a. Novratil, J., Barborik, M., and Haushan, L. (1960). *Cesk. Otolaryngol.* **9**, 199–201.
172. Olah, A. J., Reutter, F. W., and Dambaches, M. A. (1978). "Fluoride and Bone," pp. 242–248.
173. Opalko, K., Lisiecka, K., Mitrega, J., and Mietkiewska (1982). *Fluoride* **15**, 78–81.
174. Ostrom, C. A., Koulourides, T., Hickman, F., and Phantumvanit, P. (1977). *J. Dent. Res.* **56**, 212–221.
175. Parkins, F. M., Tinanoff, N., Moutinho, M., Anstey, M. B., and Waziri, M. H. (1974). *Calcif. Tissue Res.* **16**, 335–338.
176. Pendborg, J. J., and Plum, C. M. (1946). *Acta Pharmacol. Toxicol.* **2**, 294.
177. Perkinson, J. D., Whitney, I. B., Monroe, R. A., Lotz, W. E., and Comar, C. L. (1955). *Am. J. Physiol.* **182**, 383.
178. Phillips, P. H., Halpin, J. G., and Hart, E. B. (1935). *J. Nutr.* **10**, 93.
179. Phillips, P. H., Hart, E. B., and Bohstedt, G. (1934). *Wisc. Agric. Exp. Stn. Bull.* **123**.
180. Phillips, P. H., and Suttie, J. W. (1960). *Arch. Ind. Health* **21**, 343.
181. Pierce, A. W. (1959). *Aust. J. Agric. Res.* **10**, 186.
182. Poulsen, S., and Moller, I. J. (1974). *Arch. Oral Biol.* **19**, 951–954.
183. Raman, T. K. (1970). Quoted in *Fluoride* **3**, 105.
184. Ramseyer, W. F., Smith, C., and McCay, C. M. (1957). *J. Gerontol.* **12**, 14.
185. Ranganathan, S. (1941). *Indian J. Med. Res.* **29**, 693.
186. Rao, K. V., Purushottam, D., Khandekar, A. K., Vaidyanathan, D., and Francis, P. G. (1972). *Curr. Sci.* **41**, 841.
187. Rao, S. R., Murthy, K. J. R., Murthy, T. V. S. D., and Reddy, S. S. (1974). *Proc. Symp. Fluorosis,* pp. 441–448.
188. Rao, S. R., Murthy, K. J. R., Murthy, T. V. S. D., Reddy, S. S., and Saxena, M. K. (1975). *Fluoride* **8**, 144–153.
189. Ream, L. J., and Principato, R. (1981). *Am. J. Anat.* **162**, 233–241.
190. Reddy, M. V. R., Raja Reddy, D., Ramulu, S. B., and Mani, D. S. (1978). *Fluoride* **11**, 33.
191. Reutter, F. W., Schenk, R. K., and Merz, W. A. (1970). *Fluoride* **3**, 209.
192. Rioufol, C., Bourbon, P., and Philibert, C. (1982). *Fluoride* **15**, 157–161.
193. Roholm, K. (1937). "Fluorine Intoxication—A Clinical Hygienic Study." NYT. Nordisk Forlog, Arnold Busck. Copenhagen.
194. Rydzewska, A., Kamienski, A., Fiejrieroicz, Z., Chmielnik, M., and Cyplik, F. (1982). *Fluoride* **15**, 21–25.
195. Savchuck, W. B., and Armstrong, W. D. (1951). *J. Biol. Chem.* **193**, 575.
195a. Schlegal, H. H. (1974). *Sozial Praventivmed.* **19**, 269–274.
196. Schmidt, H. J., and Rand, W. E. (1952). *Am. J. Vet. Res.* **13**, 38.
197. Schroeder, H. A., Mitchener, M., Balassa, J. J., Kanisawa, M., and Nason, A. P. (1968). *J. Nutr.* **95**, 95.
198. Schwarz, K., and Milne, D. B. (1972). *Biol. Inorg. Chem.* **1**, 331.
199. Selvaraj, R. J., and Sharra, A. J. (1966). *Nature (London)* **211**, 1272–1276.
200. Sharpless, G. R., and McCollum, E. V. (1933). *J. Nutr.* **6**, 163.
201. Shearer, T. R., and Suttie, J. W. (1967). *Am. J. Physiol.* **212**, 1165.
202. Shen, Y. W., and Taves, D. R. (1974). *Am. J. Obstet. Gynecol.* **119**, 205.

203. Shivchandra, and Thergaonkar, V. P. (1984). *Fluoride* **17**, 155–159.
204. Shupe, J. L. (1967). Diagnosis of fluorosis in cattle. *In* "International meeting of the World Association for Buiatrics." Publ. No. 4. pp. 15–30. Zurich.
205. Shupe, J. L., and Alther, E. W., (1966). *Handb. Exp. Pharmacol.* **20**, Pt. 1, 307–354.
206. Shupe, J. L., Haries, L. E., Greenwood, D. A., Butcher, J. E., and Nielson, H. M. (1963). *Am. J. Vet. Res.* **24**, 300.
207. Shupe, J. L., and Olson, A. E. (1983). *In* "Fluorides" (J. L. Shupe, J. P. Peterson, and N. C. Leone, eds.), pp. 319–338. Paragaon Press, Salt Lake City, Utah.
208. Shupe, J. L., Mines, M. L., and Greenwood, D. A. (1964). Clinical and pathological aspects of fluorine toxicosis in cattle. Reprinted from *Ann. N.Y. Acad. Sci.* **3**, 618–637.
209. Shupe, J. L., Mines, M. C., Greenwood, D. A., Harris, L. E., and Stoddard, G. E. (1973). *Am. J. Vet. Res.* **24**, 964–984.
210. Shupe, J. L., Mines, M. L., Harris, L. E., and Greenwood, D. A. (1962). *Am. J. Vet. Res.* **23**, 777.
211. Siddiqui, A. H. (1955). *Br. Med. J.* **2**, 1408.
212. Siddiqui, A. H. (1968). *Fluoride* **1**, 76–85.
213. Simon, G., and Suttie, J. W. (1968). *J. Nutr.* **96**, 152.
214. Singer, L., and Armstrong, W. D. (1960). *J. Appl. Physiol.* **15**, 508.
215. Singer, L., and Armstrong, W. D., (1965). *Anal. Biochem.* **10**, 495.
216. Singer, L., and Armstrong, W. D. (1974). *In* "Trace Element Metabolism in Animals—2" (W. G. Hoekstra *et al.*, eds.), p. 698. Univ. Park Press, Baltimore.
217. Singer, L., and Ophaug, H. (1979). *In* "National Symposium on Dental Nutrition" (S. H. Y. Wei, ed.), pp. 47–62. Univ. of Iowa Press, Ames.
218. Singer, L., and Ophaug, R. H. (1983). *In* "Fluorides" (J. L. Shupe, H. B. Peterson, and N. C. Leone, eds.), pp. 157–166. Paragaon Press, Salt Lake City, Utah.
219. Singh, A., and Jolly, S. S. (1961). *Q.J. Med.* **30**, 357.
220. Singh, A., Jolly, S. S., and Bansal, B. C. (1961). *Lancet,* **1**, 197.
221. Singh, A., Jolly, S. S., Bansal, B. C., and Mathur, O. C. (1963). *Medicine* **42**, 229.
222. Singh, A., Singh, B. M., Singh, I. D., Jolly, S. S., and Malhotra, J. C. (1966). *Indian J. Med. Res.* **54**, 591–597.
223. Singh, M., and Kanwar, K. C. (1981). *Fluoride* **14**, 107–112.
224. Sivakumar, B. (1977). *Horm. Metab. Res.* **9**, 436–437.
225. Sivakumar, B., and Krishnamachari, K. A. V. R. (1976). *Horm. Metab. Res.* **8**, 317.
226. Smith, F. A., Gardener, D. E., Leone, H. C., and Hodge, H. C. (1960). *Am. Med. Assoc. Arch. Ind. Health* **21**, 330–332.
227. Soni, M. G., Kachole, M. S., and Pawar, S. S. (1982). *Fluoride* **15**, 132–136.
228. Soriano, M. (1968). *Fluoride* **1**, 56–64.
229. Spak, C. J., Hardell, L. I., and De Chateau, P. (1983). *Acta Paediatr. Scand.* **72**, 699–701.
230. Speder, E. (1936). *J. Radiol. Electrol.* **20**, 1–11.
231. Spencer, H. (1974). *In* "Trace Element Metabolism in Animals—2" (W. G. Hoekstra *et al.*, eds.), p. 696. Univ. Park Press, Baltimore.
232. Spencer, H., Kramer, L., Gatza, C., Norris, C., Wiatrowski, E., and Gandhi, V. C. (1980). *Arch. Intern. Med.* **140**, 1331–1335.
233. Spencer, H., Osis, D., and Wiatrowski, E. (1974). *Trace Subst. Environ. Health-7. Proc. Univ. Mo. Annu. Conf.* 289.
234. Spencer, H., Osis, D., Wiatrowski, E., and Samachson, J. (1970). *J. Nutr.* **100**, 1415.
235. Spira, L. (1956). *Exp. Med. Surg.* **14**, 72–78.
236. Srikantia, S. G., and Siddiqui, A. H. (1965). *Clin. Sci.* **28**, 477.
237. Sriranga Reddy, G., and Narasinga Rao, B. S. (1971). *Metabolism* **20**, 650.
238. Sriranga Reddy, G., and Narasinga Rao, B. S. (1972). *Calcif. Tissue Res.* **10**, 207–215.

239. Stoddard, G. E., Baterman, G. O., Harris, L. E., Shupe, J. L., and Greenwood, D. A. (1963). *J. Dairy Sci.* **46**, 720–726.
240. Susheela, A. K., and Mohan Jha (1981). *Experentia* **37**.
241. Susheela, A. K., and Mohan Jha (1981). *I.R.C.S. Med. Sci.* **9**, 898.
242. Susheela, A. K., and Mukherjee, D. (1981). *Toxicol. Eur. Res.* **3**.
243. Susheela, A. K., and Sharma, Y. D. (1980). *Fluoride* **13**, 151–159.
244. Susheela, A. K., and Sharma, Y. D. (1981). *I.R.C.S. Med. Sci.* **9**, 862.
245. Susheela, A. K., Sharma, Y. D., Jha, M., Rajalakshmi, K., and Ram Mohan Rao, N. V. (1981). *Fluoride* **14**, 150.
246. Suttie, (1977). *J. Occup. Med.* **19**, 40–48.
247. Suttie, J. W., Drescher, M. P., Quisell, D. O., and Young, K. L. (1974). *In* "Trace Element Metabolism in Animals—2" (G. W. Koekstra *et al.*, eds.), p. 327. Univ. Park Press, Baltimore.
247a. Suttie, J. W., and Faltin, E. C. (1973). *Am. J. Vet. Res.* **34**, 479.
248. Suttie, J. W., Gesteland, R., and Phillips, P. H. (1961). *J. Dairy Sci.* 44, 2250.
249. Suttie, J. W., and Kolstad, D. L. (1977). *J. Dairy Sci.* **60**, 1568–1573.
250. Suttie, J. W., Miller, R. F., and Philips, P. H. (1957). *J. Nutr.* **63**, 211–224.
251. Suttie, J. W., Phillips, P. H., and Miller, R. F. (1958). *J. Nutr.* **65**, 293.
252. Takamori, T., and Kawahara (1971). *Fluoride* **4**, 154–165.
253. Taves, D. R. (1970). *Fed. Proc., Fed. Am. Soc. Exp. Biol.* **29**, 1185–1187.
254. Taves, D. R., Freeman, R. B., and Kamm, D. E. (1968). *Trans. Am. Soc. Artif. Intern. Organs* **14**, 412.
255. Taves, D. R., Forbes, N., Silverman, D., and Hicks, D. (1983). *In* "Fluorides" (J. L. Shupe, H. B. Peterson, and N. C. Leone, eds.), pp. 189–194. Paragaon Press, Salt Lake City, Utah.
256. Taylor, J. M., Scott, J. K., and Maynard, E. A. (1960). *Toxicol. Appl. Pharmacol.* **3**, 278–289.
257. Teotia, M., Teotia, S. P. S., and Teotia, N. P. S. (1972). *Fluoride* **5**, 125.
258. Teotia, S. P. S., Kanwar, K. C., and Teotia, M. (1969). *Fluoride* **1**, 76–85.
259. Teotia, S. P. S., and Teotia, M. (1973). *Br. Med. J.* **1**, 637.
260. Teotia, S. P. S., Teotia, M., and Singh, R. K. (1976). *Fluoride* **9**, 68.
261. Teotia, S. P. S., Teotia, M., Singh, R. K., Teotia, N. P. S., Daver, D. R., Heela, S., and D'Hello, U. P. (1978). *Fluoride* **11**, 115.
262. Teotia, S. P. S., Teotia, M., and Teotia, N. P. S. (1976). *Fluoride* **9**, 91–97.
263. Thiers, G. (1979). *Fluoride* **12**, 109–110.
264. Thystrup, O., Fejerskov, C., Brunn, and Kann, J. (1984). *Fluoride* **17**, 57–58.
265. Tsunoda, H., and Kunita, H. (1973). *J. Environ. Pollut. Control* **9**, 613–619.
266. Tsunoda, H., Sakurai, S., Itai, K., Sato, T., Nakaya, S., Mita, M., and Tatsumi, M. (1984). *Fluoride* **17**, 159–167.
267. Tuncel, E. (1984). *Fluoride* **17**, 4–8.
268. Venkateswarlu, P., and Narayana Rao, D. (1957). *Indian J. Med. Res.* **45**, 387.
269. Venkataraman, K., and Krishnaswamy, N. (1949). *Indian J. Med. Res.* **37**, 277.
270. Viswanathan, G. R. (1946). *Indian Vet. J.* **23**, 4.
271. Waldbott, G. L. (1976). *Fluoride* **9**, 24–28.
272. Waldbott, G. L. (1979). *Vet. Hum. Toxicol.* **21**, 4–8
273. Wallace-Durbin, P., (1954). *J. Dent. Res.* **33**, 789.
274. Weatherall, J. A., and Weidmann, S. M. (1959). *J. Pathol. Bacteriol.* **78**, 233.
275. Weber, C. W., and Reid, B. L. (1974). *In* "Trace Element Metabolism in Animals—2" (G. W. Hoekstra *et al.*, eds.), p. 707. Univ. Park Press, Baltimore.
276. Weddle, D. A., and Muhler, J. C. (1954). *J. Nutr.* **54**, 437.

277. Weiss, V., and de Carlini, Ch. (1975). *Experimentia* **31,** 339–341.
278. Wespi, H. J., and Schweiz (1962). *Bull. Eidg. Gesundheitsamt* Part B. No.2.
279. Wrodek, G. (1959). Quoted by Hodge and Smith *Zahnaerzl. Mitt.* **47,** 258.
280. Wuthier, R. E., and Phillips, P. H. (1959). *J. Nutr.* **67,** 581.
281. Zebrowski, E. J., and Suttie, J. W. (1966). *J. Nutr.* **88,** 267.
282. Zebrowski, E. J., Suttie, J. W., and Phillips, P. H. (1964). *Fed. Proc., Fed. Am. Soc. Exp. Biol.* **23,** 184.
283. Zhavoronkov, A. A. (1977). *Arkh. Pathol.* **39,** 83–91.
284. Zhavoronkov, A. A., and Steochkova, L. S. (1981). *Fluoride* **14,** 182–191.
285. Zipkin, I., Lee, W. A., and Leone, N. C., (1962). *P.H.S. Publ.* **825,** 435–437.
286. Zipkin, I., and Likins, R. C. (1957). *Am. J. Physiol.* **191,** 549.
287. Zipkin, I., Zucas, S. M., and Stillings, B. R. (1969). *J. Nutr.* **100,** 293.

12

Mercury

THOMAS W. CLARKSON

Environmental Health Sciences Center
University of Rochester
School of Medicine
Rochester, New York

I. INTRODUCTION

Mercury (Hg) occurs widely in the biosphere and has long been known as a toxic element presenting occupational hazards associated with both ingestion and inhalation. No vital function for the element in living organisms has yet been found. The toxic properties of mercury have evoked increasing concern lately due to the extent of its use in industry and agriculture, and the recognition that alkyl derivatives of the element are more toxic than most other chemical forms and can enter the food chain through the activity of microorganisms with the ability to methylate mercury in sediments in bodies of fresh and ocean waters.

II. MERCURY IN ANIMAL TISSUES AND FLUIDS

Mercury was detected in all the tissues of human accident victims, with no known abnormal exposure to mercury other than dental repair, examined by a neutron activation technique (25,36). The mean concentrations mostly fell between 0.5 and 2.5 ppm mercury (dry basis), or about 0.1 to 0.5 ppm fresh weight. The highest mercury levels were present in the skin, nails, and hair. Among the internal organs the kidneys generally carry the highest mercury concentrations. For example, Joselow *et al.* (31) reported a mean of 2.7 ppm mercury, wet weight (range 0-26) for kidney, compared with means ranging from 0.05 to 0.30 ppm for 11 other tissues, including the liver and lungs. Kosta and co-workers (34) also obtained appreciably higher mean mercury levels in the human kidney than in the liver, brain, thyroid, and pituitary gland of postmortem samples from individuals not exposed to abnormal amounts of mercury. In comparable postmortem samples from mercury mine workers and from the population in which the mine was situated, kidney and brain mercury levels were substantially higher and thyroid and pituitary mercury levels remarkably higher than in the nonexposed group. In the miners the mercury accumulation and increases in the thyroid and pituitary glands were of the order of 1000-fold and in the town population 10-fold compared with the controls (Table I). In the organs displaying these mercury accumulations selenium was also accumulated, giving an approximately 1 : 1 molar ratio between mercury and selenium. Mercury levels in brain in otherwise nonexposed people are slightly increased according to the number of mercury amalgam tooth fillings (19).

Goldwater (22) reported that 74% of a normal population drawn from 16 countries had blood mercury concentrations <5 μg/liter and 98% had <50 μg/liter. In a 1975 study of 679 residents of Saskatchewan, blood mercury levels were found to range from 1 to 42 μg/liter, with a mean of 6.7 and a median of 5 μg/liter (17).

Dependence of Blood Levels on Fish Consumption

Mercury in the edible portion of fish is mainly present as methylmercury. This distributes preferentially to the red blood cells (cell–plasma ratio of about 20 : 1 (32,61)). A Swedish Expert Group (65) noted that people who never eat fish as a rule have very low levels in the blood cells (2–5 μg/kg). Persons with moderate fish consumption probably have mean mercury levels about 10 μg/kg. High fish consumers, especially of large predatory fish such as shark, tuna, and swordfish, have values as high as 400 ng/g (70). Cord blood levels in fish eaters correlate closely with maternal blood levels and are usually about 20% higher (61).

After inhalation of mercury vapor, levels in red cells are usually somewhat higher than those in plasma (11), but the difference is not as great as in the case of methylmercury.

Goldwater (23) reported that 79% of a normal population drawn from 15 countries had urine mercury concentrations <0.5 μg/liter. Occupationally exposed individuals may have urinary concentrations several orders of magnitude above levels found in the normal population.

Concentrations of mercury are influenced by diet and by airborne mercury. In individuals not exposed to airborne mercury, the mercury levels in hair relate to fish consumption. In a comprehensive worldwide evaluation, Airey (1) noted

Table I

Mean Mercury Content of Human Organs[a]

Group	Mercury content (ppm fresh weight)				
	Thyroid	Pituitary	Kidney	Liver	Brain
Mercury mine workers	35.2 ± 28.5 (8)[b]	27.1 ± 14.9 (7)	8.4 ± 4.9 (8)	0.26 ± 0.25 (8)	0.70 ± 0.64 (6)
Idrija population	0.70 ± 0.45 (10)	0.46 ± 0.54 (11)	0.66 ± 1.13 (11)	0.107 ± 0.059 (11)	0.038 ± 0.045 (9)
Nonexposed controls	0.03 ± 0.037 (16)	0.04 ± 0.026 (6)	0.14 ± 0.16 (7)	0.03 ± 0.017 (8)	0.0042 ± 0.0026 (5)

[a] From Kosta et al. (34).

[b] All figures in parentheses refer to the number of subjects analyzed.

that the average hair mercury concentration was 1.4 μg/g for groups consuming fish once a month or less; for groups consuming fish once every 2 weeks it was 1.9 μg/g, and for once or more a day, it was 11.6 μg/g. Certain very heavy fish consumers, particularly consumers of large predatory fish, may have hair mercury values up to 50 μg/g (70). Levels in occupationally exposed individuals have been reported to cover a range of 1–49 μg/g (29). This author also reviewed evidence that metal concentrations vary along the length of the hair according to sex and age, hair treatment such as shampoos, and the location of hair on the body. Few data have been published on nails or teeth.

The forms in which mercury occurs in the tissues are considered in the next section.

III. MERCURY METABOLISM

The metabolic behavior of mercury varies greatly with the chemical form in which it is presented to the animal, the extent to which other elements with which it interacts are present in the diet, and apparently also with genetic differences. Marked differences in metabolic behavior exist between inorganic and aryl mercury compounds and the alkyl mercury. Inorganic mercury compounds are absorbed to the extent of about 5 to 15% of an oral dose in both humans (56) and animals (12), whereas about 90% of phenyl and methylmercury compounds are absorbed (14,45). Mercury vapor is retained to an extent of 80% of the inhaled amount (26). Following absorption the inorganic, aryl, and methoxyalkyl mercury compounds behave similarly, due to the rapid degradation of the two latter forms to inorganic mercury (66). Simple alkyl forms of mercury are not only better absorbed, they are much better retained and induce higher brain mercury contents than aryl mercury compounds (67). The biological half-time of methylmercury in the whole body in humans is approximately 70 days as compared to 40 days for inorganic mercury after oral doses (45,56). Following inhalation of mercury vapor, the whole-body half-time in humans is about 50 days. Considerable species differences exist in excretion rates, (for review see Ref. 13).

Metabolic differences between simple alkyl mercury compounds and other mercurials extend further to their pattern of excretion. Methylmercury is secreted in bile as a methylmercury–glutathione complex, most of which is subsequently reabsorbed from the intestinal tract and redistributed to the tissues (16), thus contributing to the high retention of methylmercury. A fraction of the methylmercury in the intestinal tract is broken down to inorganic mercury by intestinal microflora; the latter, being poorly reabsorbed, is excreted in the feces (59). Biliary secretion (4) and demethylation of microflora (59) are greatly diminished in suckling animals, thus explaining the slow rate of excretion before weaning.

Diet affects the biological half-time of methylmercury, probably by influencing the demethylating activity of microflora (59).

The metabolic differences between methylmercury and inorganic mercury are shown also by studies of placental transfer. The placenta presents an effective barrier against the transfer of inorganic mercury in rats (21). Transfer of methylmercury to the fetus greatly exceeds that of either mercuric chloride or phenylmercuric acetate in mice (64). The higher mercury content of cord blood than of maternal blood in women exposed to methylmercury from fish (17) was detailed in Section I. The ready placental transfer of methylmercury to the fetus no doubt explains the occurrence of "congenital" Minamata disease in infants, as discussed in Section IV, although ingestion of methylmercury from the mother's milk may be a contributing factor, since mammary transfer of methylmercury has been demonstrated in humans (2).

Much remains to be learned of the chemical forms and intracellular distribution of mercury. Methyl forms do not occur in animal cells and tissues in significant amounts, unless ingested or injected as such. In other words, the animal body has an extremely limited capacity to convert inorganic and various organic forms of mercury into the more toxic methyl forms (74). This ability to transform mercury is principally confined to microorganisms, and it is their activity that can introduce dangerous methylated mercury compounds into the food chain.

A nonhistone protein component into which mercury is rapidly incorporated has been isolated from rat kidney nuclei (9) while the main mercury-binding protein in rat kidney cytosol appears to be a stable metallothioneinlike protein (76), together with other more labile mercury-binding fractions (28). The mercury-binding capacity of the thioneins is extremely high (54), and zinc and cadmium have been completely displaced from metallothionein from rat liver *in vitro,* yielding a purified protein containing 8 gram atoms of Hg^{2+} (63). However, the role of metallothionein in detoxifying inorganic mercury has been questioned (38).

The metabolic antagonism between mercury and selenium manifested in the protection against selenium toxicity displayed by mercury, as discussed in the previous section, is paralleled by a comparable protection against mercury toxicity by selenium. A mutual metabolic antagonism between the two elements thus exists. The first evidence for this came from the studies by Parizek and co-workers (51), who showed that selenite protected against the renal necrosis and mortality in rats caused by injected mercuric chloride and counteracted placental mercury transfer. A protective action of dietary selenium against methylmercury toxicity has also been demonstrated in rats and Japanese quail (20). Mercury selenium interactions have been extensively reviewed by Magos and Webb (40). Most experimental work has been conducted with sodium selenite salts. Magos *et al.* (39) have shown that forms of selenium more likely to occur in nature are less effective in interacting with and neutralizing the toxic effects of mercury. The role of dietary selenium in modifying the potential toxicity of dietary methylmercury in humans is as yet unknown (62). There is evidence that methylmercury forms a bismethylmercury–selenide complex (CH_3Hg-Se-Hg-CH_3), and inorganic mercury forms a selenide complex (HgSe) in tissues of animals dosed with selenite salts (27,39,41,49).

The biochemical mechanisms involved in the prevention or amelioration of mercury toxicity by selenium are little understood. It does not appear to be due to increased mercury excretion. In fact there is some evidence of an increase in mercury retention and definite evidence of a changed distribution of retained mercury, with increased levels in the liver and spleen and decreased levels in kidneys (53). Dietary selenite does not increase the amount of mercury in thionein when this mercury is fed as methylmercury, and methylmercury combines poorly with thionein *in vitro* (10). This suggests that selenium does not ameliorate or prevent methylmercury toxicity by increasing the rate of methylmercury breakdown.

The biochemical effects of mercury compounds may be many and varied depending on the severity of tissue injury. Few biochemical effects have been identified preceding gross tissue injury. Protein synthesis in brain cells precedes signs of intoxication by methylmercury in experimental animals. Binding to the protein tubulin and subsequent destruction of microtubules may be an important precursor to damage to the developing central nervous system (CNS) (for a review, see Ref. 15). Mild enzymuria and proteinuria precede more severe damage to the kidney by inorganic mercury (8,24).

IV. SOURCES OF MERCURY

Mercury and methylmercury are naturally occurring substances to which all living organisms have been exposed in varying degrees depending on natural biological, chemical, and physical processes. Modern technological developments involving the use of mercury compounds are responsible for the discharge of large and variable amounts of the element into the environment. The main industrial source is the chloralkali industry (48). Other major industrial uses of mercury are in the manufacture of electrical apparatus, paint, dental preparations, and pharmaceuticals, and in paper and pulp making as slimicides and algicides. The mercury used in agriculture for seed treatment can be a particularly hazardous source if methylmercury compounds were used (3). Mercury uses and their possible hazards as sources of mercury contamination in Canada (18) and in the United States (71) have been comprehensively reviewed.

It is apparent that mercury can enter the biosphere from a variety of man-made sources, as just described, and also from the burning of fossil fuels (71). Mercury is transported in the atmosphere mainly as mercury vapor over a global scale. It is converted to an as yet unidentified soluble form with a turnover time of a few weeks (71). Methylated mercury compounds enter the food chain mainly through the activity of microorganisms that have the ability to methylate the mercury present in industrial wastes. The mercury in fish from polluted waters occurs almost entirely as methylmercury (73). Methylated mercury compounds may also enter the food chain through their use as a disinfectant of grain. The decrease in the methylmercury content of foods of animal origin in Sweden after methoxyethyl mercury was substituted for methylmercury as a seed disinfectant has been documented by Westoo (75).

Dietary intake of mercury has been reviewed by the U.S. Environmental Protection agency (71). Fish and fish products are the dominant source. Average daily intake is about 4 μg, mostly in the form of methylmercury. However, in certain fish-eating communities intakes may be higher by an order of magnitude or more (52). In an extensive study of Swedish diets, a very wide range of 1.0–30.6 μg/day mercury was found (50). The importance of the fish content is evident from the fact that 90 of the diets containing no fish averaged only 3.5 μg/day mercury, whereas the 55 diets that included fish of any kind averaged 9.0 μg/day. The importance of fish is further emphasized by the fact that most of the mercury in fish is present as highly toxic methylmercury compounds. In foods of plant origin little or none of the mercury is normally present in this form, while in meat and dairy products the low levels of mercury can include a small proportion of methylmercury, presumably from residues in feeds containing fish meal or treated cereal grains. In an investigation of Canadian foods, concentrations ranging from 0.005 to 0.075 ppm mercury were reported for a variety of food items not abnormally exposed to mercury at any time, compared with levels of ≥ 1 ppm mercury in specimens from mercury-contaminated areas (30).

Although man-made mercury pollution of freshwater rivers and lakes can greatly increase the mercury levels in freshwater fish, it should be appreciated that the mercury levels are high in wide-ranging ocean fish where no such pollution could have occurred. Thus Miller *et al.* (46) found the mercury levels of museum specimens of seven tuna caught 62–93 years ago to range from 0.26 to 0.64 ppm (wet weight) and those of five specimens caught recently from 0.13 to 0.48 ppm (wet weight). Much higher mercury levels have been observed in marine mammals and birds (33). The range of mercury concentration in the livers of 22 marine mammals was extremely wide (0.37–326 ppm mercury wet weight). In four marine birds the mercury levels ranged more narrowly from 1.8 to 2.4 ppm in the liver and from 0.35 to 0.58 ppm in the brain. In the marine mammals the mercury was not present largely as methylmercury as it is in fish, and was highly correlated with the selenium levels. In fact, in marine mammals a 1 : 1 mercury–selenium molecular increment ratio and an almost perfect linear correlation between mercury and selenium were found. It was suggested that marine mammals are able to detoxify methylmercury by a specific chemical mechanism in which selenium is involved.

The contribution of mercury inhaled from the air is negligible compared with intake from the food, except where there is an occupational exposure (71). Mercury volatilizes so readily that it can constitute a health hazard to laboratory workers and others handling the metal for long periods, unless proper precautions are taken. The water supply is also a relatively insignificant source of mercury, except when contaminated from industrial or geological sources. The tentative upper limit for mercury in drinking water has been placed at 1 μg/liter (77), a figure related to levels found in natural waters for drinking purposes and to the high ingestion rate of 2.5 liters of water per day. At these upper limits 2.5 μg mercury per person per day would be provided, mostly in inorganic form.

Little is known of normal mercury intakes by grazing or even by stall-fed farm animals. Lunde (37) reported a mean mercury concentration of 0.18 ppm (range 0.03–0.40) for 12 commercial fish meals from different sources. This is well above the mercury levels in ordinary cereal grains or protein supplements of plant origin.

V. MERCURY TOXICITY

Mercury poisoning from inhalation of mercury vapor has been prominent at times among goldsmiths and mirror makers, and the term "mad as a hatter" derives from the symptoms shown by workers in the treatment of furs with mercuric nitrate. The manifestations of such subacute mercury poisoning are primarily neurological, with tremors, vertigo, irritability, moodiness, and depression, associated with salivation, stomatitis, and diarrhea. In poisoning from the ingestion of inorganic mercury salts, the liver and kidneys are the tissues most affected and there may also be proteinuria, necrosis of the intestinal tract, and diarrhea. When mercury is ingested as the more toxic alkyl derivative, the symptoms include progressive incoordination, loss of vision and hearing, and mental deterioration arising from a toxic neuroencephalopathy in which the neuronal cells of the cerebral and cerebellar cortex are selectively involved. These clinical and pathological changes, with particular impairment of scotopic vision, have been demonstrated in the squirrel monkey, a species reported to be a valuable model for the study of methylmercury poisoning in humans (7).

The changes just described were evident in victims of a tragic occurrence of methylmercury poisoning in Japan (Minamata disease) following the dumping into the Minamata Bay of mercury-containing factory wastes and the consumption of fish caught in the bay (35,68). The outbreak was characterized by a high incidence of "congenital" cases in infants. The mental retardation, cerebral palsy, and mortality that were evident resulted from the ease with which methyl-mercury passes the placental barrier and concentrates preferentially in fetal tissues, particularly the brain. The sensitivity of the fetus to methylmercury is apparent from these observations. Follow-up studies of the massive outbreak of methylmercury poisoning in Iraq (3) further documented the high susceptibility of the developing CNS (42). Native Indian communities in Quebec, Canada have been reported to exhibit mild effects in both adults and prenatally exposed infants as a result of consumption of fish with elevated levels of methylmercury (43,44). The finding in Canada that males are more susceptible to prenatal damage by methylmercury were confirmed in animal studies by Sager *et al.* (60).

Outbreaks of methylmercury poisoning have also occurred from the accidental consumption of bread made from seed grain treated with mercurial fungicides (3) and meat from animals fed on such grain.

The minimum safe levels of dietary mercury, or the maximum intakes compatible with the long-term health of humans, depend on pregnancy and age, as just mentioned, and on the form or forms in which the mercury is ingested. The joint FAO/WHO expert committee on food additives (78) established a provisional tolerable weekly intake of 0.3 mg mercury per person of which no more than 0.2 mg mercury should be present as the methylmercury ion CH_3Hg^+. These amounts are equivalent to 0.5 μg and 0.33 μg, respectively, per kilogram body weight, or assuming an average dry matter intake of 400 g/day, a total mercury content of the dry diet close to 0.08 and 0.06 ppm, respectively. It was stated further that "where the total Hg in the diet is found to exceed 0.3 mg/week the level of methylmercury compounds should also be investigated. If the excessive intake is attributable entirely to inorganic Hg, the above provisional limit for total Hg no longer applies and will need to be reappraised in the light of all the prevailing circumstances."

It is not known to what extent selenium actually protects against mercury poisoning in humans. In the case of inhaled mercury vapor, Kosta and co-workers (34) noted, in a study of mercury mine workers, that mercury and selenium were present in a 1 : 1 molar ratio in a number of organs and tissues. In these circumstances the abnormally high mercury levels found in the tissues were apparently without deleterious effects in the individuals, several of whom had been 10–16 years in retirement following long periods of mercury exposure in the mines. The selenium intakes from the diet were not reported but were said not to be abnormally high, suggesting that the coaccumulation with mercury is a natural or autoprotective effect. Where the main source of dietary mercury is fish or marine mammals, the diet is naturally enriched in selenium relative to mercury, and might provide some protection. It seems reasonable to suggest, further, that in areas naturally low in selenium, individuals would be at greater risk from mercury poisoning than those in areas of high selenium status.

In addition to dietary mercury determinations, measurements of the mercury levels in hair, urine, blood, and saliva have been given consideration as diagnostic procedures in the prediction of incipient mercury poisoning. An expert group that made an evaluation of the risks from methylmercury in fish contends that the best available index of the degree of exposure is the level of methylmercury or mercury in the red blood cells, and that the level in whole blood or hair is also valuable (65). Clinically manifest poisoning of adults sensitive to methylmercury, it is claimed, "may occur at a level in whole blood down to 0.2 μg Hg/g, which level seems to be reached on exposure to about 0.3 mg Hg as methylmercury/day, or about 4 μg Hg/kg body weight." If, as the group believed, a factor of 10 gives a sufficient margin of safety, the acceptable level in whole blood would be 0.02 μg/g, corresponding to about 0.04 μg/g in the red cells and about 6 μg/g in the hair. These conclusions have been supported in general by subsequent expert reviews (71,79,80). In the case of inhalation of mercury vapor, the critical urinary concentration above which mercury poisoning may appear in susceptible individuals has been estimated to be in the range 150–300 μg/liter. Certain biochemical effects have been detected at lower levels (71).

REFERENCES

1. Airey, D. (1983). *Sci. Total Environ.* **31,** 157–180.
2. Amin-Zaki, L., Elhassani, S., Majeed, M. A., Clarkson, T. W., Doherty, R. A., and Greenwood, M. R. (1976). *Am. J. Dis. Child* **130,** 1070–1076.
3. Bakir, F., Damluji, S. F., Amin-Zaki, L., Mutadha, M., Khalidi, A., Al-Rawi, N. Y., Tikriti, S., Dhahir, H. I., Clarkson, T. W., Smith, J. C., and Doherty, R. A. (1973). *Science* **181,** 230–241.
4. Ballatori, N., and Clarkson, T. W. (1982). *Science* **216,** 61–63.
5. Ballatori, N., and Clarkson, T. W. (1984). *Biochem. Pharmacol.* **33,** 1087–1092.
6. Ballatori, N., and Clarkson, T. W. (1985). *Fundam. Appl. Toxicol.* **5,** 816–831.
7. Berlin, M., Grant, C. A., Hellstrom, J., and Schutz, A. (1975). *Arch. Environ. Health* **30,** 340–348.
8. Buchet, J. P., Roels, B., and Lauwerys, R. (1980). *J. Occup. Med.* **22,** 741–750.
9. Chanda, S. K., and Cherian, M. G. (1973). *Biochem. Biophys. Res. Commun.* **50,** 1013–1019.
10. Chen, R. W., Ganther, H. E., and Hoekstra, W. G. (1973). *Biochem. Biophys. Res. Commun.* **51,** 383–390.
11. Cherian, M. G., Hursh, J. B., Clarkson, T. W., and Allen, J. (1978). *Arch. Environ. Health* **33,** 109–114.
12. Clarkson, T. W. (1971). *Toxicology* **9,** 229–243.
13. Clarkson, T. W. (1972). *CRC Crit. Rev. Toxicol.* **II,** 203–234.
14. Clarkson, T. W. (1973). *In* "Mercury in the Western Environment" (D. R. Buhler, ed.), pp. 332–360. Department of Printing, Oregon State University, Portland.
15. Clarkson, T. W. (1983). *In* "Biological Aspects of Metals and Metal-Related Diseases" (B. Sarker, ed.), pp. 183–197. Raven, New York.
16. Clarkson, T. W., Ballatori, N., and Cernichiari, E. (1983). *Bol. Natl. Acad. Med. Buenos Aires* **VI,** 163–174.

17. Dennis, C. A. R., and Fehr, F. (1975). *Sci. Total Environ.* **3**, 267 and 275–277.
18. Fimreite, N. (1970). *Environ. Pollut.* **6**, 119.
19. Friberg, L., Kullman, L., Lind, B., and Nylander, M. (1986). *Lakartidningen* **83**, 519–522.
20. Ganther, H. E., Gondie, C., Sunde, M. L., Wagner, P., Hoh, S., and Hoekstra, W. G. (1972). *Science* **175**, 1122–1124.
21. Garrett, N. E., Garrett, R. J. B., and Archdeacon, J. W. (1972). *Toxicol. Appl. Pharmacol.* **22**, 649–654.
22. Goldwater, L. J. (1964). *R. Inst. Public Health Hyg. J.* **27**, 279–301.
23. Goldwater, L. J. (1972). *In* "Mercury, A History of Quicksilver" (L. J. Goldwater, ed.), pp. 135–149. York Press, Baltimore.
24. Gotelli, C. A., Astolfi, E., Cox, C., Cernichiari, E., and Clarkson, T. W. (1985). *Science* **227**, 638–640.
25. Howie, R. A., and Smith, H. (1967). *J. Forensic Sci. Soc.* **7**, 90.
26. Hursh, J. B., Cherian, M. G., Clarkson, T. W., Vostal, J. J., and Vander Mallie, R. (1976). *Arch. Environ. Health* **31**, 302–309.
27. Iwata, H., Masukawa, T., Kito, H., and Hayashi, M. (1981). *Biochem. Pharmacol.* **30**, 3159–3163.
28. Jakubowski, M., Piotrowski, J., and Trojanowska, B. (1970). *Toxicol. Appl. Pharmacol.* **16**, 743–753.
29. Jenkins, D. W. (1979). Report No. EPA-600/4-79-049. U.S. Environmental Protection Agency, Las Vegas, Nevada.
30. Jervis, R. E., Deburn, D., Le Page, W., and Teifenbach, B. (1970). Report Dep. Chem. Eng. & Appl. Chem., University of Toronto.
31. Joselow, M. M., Goldwater, L. J., and Weinberg, S. B. (1967). *Arch. Environ. Health* **15**, 64–66.
32. Kershaw, T. G., Clarkson, T. W., and Dhahir, P. H. (1980). *Arch. Environ. Health* **35**, 28–36.
33. Koeman, J. H., van de Ven, W. S. M., de Goeij, J. J. M., Tjioe, P. S., and van Haaften, J. L. (1975). *Sci. Total Environ.* **3**, 279–287.
34. Kosta, L., Byrne, A. R., and Zelenko, V. (1975). *Nature (London)* **254**, 238–239.
35. Kurland, L. T., Faro, S. N., and Siedler, H. S. (1960). *World Neurol.* **1**, 370–395.
36. Lenihan, J. M., and Smith, H. (1967). *In* "Nuclear Activation Techniques in the Life Sciences." IAEA, Vienna.
37. Lunde, G. (1968). *J. Sci. Food Agric.* **19**, 432.
38. Magos, L. (1982). *In* "Nephrotoxicity, Assessment and Pathogenesis" (P. H. Bach, F. W. Bonner, J. W. Bridges, and E. A. Lock, eds.), pp. 325–337. Wiley, New York.
39. Magos, L., Clarkson, T. W., and Hudson, A. R. (1984). *J. Pharmacol. Exp. Ther.* **228**, 478–483.
40. Magos, L., and Webb, M. (1980). *CRC Crit. Rev. Toxicol.* **4**, 1–42.
41. Magos, L., Webb, M., and Hudson, A. R. (1979). *Chem Biol. Interact.* **28**, 359–362.
42. Marsh, D. O., Myers, G. J., Clarkson, T. W., Amin-Zaki, L, Tikriti, S., and Majeed, M. W. (1981). *Clin. Toxicol.* **18**, 1311–1318.
43. McKeown-Eyssen, G. E., and Reudy, J. (1983). *Am. J. Epidemiol.* **118**, 461–469.
44. McKeown-Eyssen, G. E., Reudy, J., and Neims, A. (1983). *Am. J. Epidemiol.* **118**, 470–479.
45. Miettinen, J. K. (1973). *In* "Mercury, Mercurials and Mercaptans" (M. W. Miller and T. W. Clarkson, eds.), pp. 233–246. Thomas, Springfield, Illinois.
46. Miller, G. E., Grant, P. M., Kishore, R., Steinkruger, F. J., Rowland, F. S., and Guinn, V. P. (1972). *Science* **175**, 1121–1122.
47. Mulder, K., and Kostyniak, P. J. (1985). *Toxicologist* **5**, 52.
48. Murozumi, M. (1967). *Electrochem. Technol.* **5**, 236.

49. Nasukawa, T., Kito, H., Hayashi, M., and Iwata, H. (1982). *Biochem. Pharmacol.* **31**, 75–78.
50. Norden, A., Dencker, I., and Schutz, A. (1970). *Naeringsforskning* **14**, 40.
51. Parizek, J., Benes, I., Ostadalova, I., Babicky, A., Benes, J., and Pitha, J. (1969). *In* "Mineral Metabolism in Pediatrics" (D. Barltrop and W. J. Burland, eds.), pp. 117–134. Blackwell, Oxford.
52. Piotrowski, J. K., And Inskip, M. J. (1981). Health Effects of Methylmercury. *MARC Report No. 24.* Monitoring and Assessment Research Center, Chelsea College, University of London.
53. Potter, S., and Matrone, G. J. (1974). *J. Nutr.* **104**, 638–647.
54. Pulido, P., Kagi, J. H. R., and Vallee, B. I. (1966). *Biochemistry* **5**, 1768–1777.
55. Rabenstein, D. L., and Isab, A. A. (1982). *Biochim. Biophys. Acta.* **721**, 374–384.
56. Rahola, T., Hattula, T., Karolainen, A., and Miettinen, J. K. (1973). *Ann. Clin. Res.* **5**, 214–219.
57. Refvsik, T., and Norseth, T. (1975). *Acta Pharmacol. Toxicol.* **36**, 67–78.
58. Richardson, R. T., and Murphy, S. D. (1975). *Toxicol Appl. Pharmacol.* **31**, 505–519.
59. Rowland, I. R., Robinson, R. D., and Doherty, R. A. (1983). *In* "Chemical Toxicology and Clinical Chemistry of Metals" (S. S. Brown and J. Savory, eds.), pp. 381–384. Academic Press, New York.
60. Sager, P. R., Aschner, M., and Rodier, P. M. (1984). *Dev. Brain Res.* **12**, 1–11.
61. Skerfving, S. (1974). *Toxicology* **2**, 3–23.
62. Skerfving, S. (1978). *Environ. Health Perspect.* **25**, 57–66.
63. Sokolowski, G., Pilz, W., and Wester, U. (1974). *FEBS Lett.* **48**, 222–225.
64. Suzuki, T., Miyama, T., and Katsunuma, H. (1967). *Ind. Health* **5**, 290–292.
65. Swedish Expert Group (1971). *Nord. Hyg. Tidsk.* (Suppl. 4).
66. Swensson, A., and Ulfvarson, U. (1968). *Acta Pharmacol. Toxicol.* **26**, 259–272.
67. Takeda, Y., Konugi, T., Hoshino, I. O., and Ukita, T. (1968). *Toxicol. Appl. Pharmacol.* **13**, 156–164.
68. Takeuchi, T. (1972). *In* "Environmental Mercury Contamination" (R. Hartung and B. D. Dinman, eds.), pp. 247–289. Science Publ. Ann Arbor, Michigan.
69. Thomas, D. J., and Smith, J. C. (1979). *Toxicol. Appl. Pharmacol.* **47**, 547–556.
70. Turner, M. D., Marsh, D. O., Smith, J. C., Rubio, E. C., Chiriboga, J., Chiriboga, C. C., Clarkson, T. W., and Inglis, J. B. (1980). *Arch. Environ. Health* **35**, 367–378.
71. U.S., E.P.A. (1984). Mercury Health Effects Update. Report No. EPA-600/8-84-019F. Office of Health and Environmental Assessment, Washington, D.C.
72. Webb, M. (1982). *In* "Nephrotoxicity Assessment and Pathogenesis" (P. B. Bach, F. W. Bonner, J. W. Bridges, and E. A. Lock, eds.), pp. 296–309. Wiley, New York.
73. Westoo, G. (1966). *Acta Chem. Scand.* **20**, 2131–2137.
74. Westoo, G. (1968). *Acta Chem. Scand.* **22**, 2277–2280.
75. Westoo, G. (1969). *In* "Chemical Fallout" (M. W. Miller and G. G. Berg, eds.), pp. 75–93. Thomas, Springfield, Illinois.
76. Wisniewska, J. M., Trojanowska, B., Piotrowski, J., and Jakubowski, M. (1970). *Toxicol. Appl. Pharmacol.* **16**, 754–763.
77. World Health Organization (1971). "International Standards for Drinking Water." World Health Organization, Geneva.
78. World Health Organization (1972). *W.H.O. Tech. Rep. Ser.* **505**
79. World Health Organization (1976). "Environmental Health Criteria. 1. Mercury." World Health Organization, Geneva.
80. World Health Organization (1980). "Health Risk Evaluation for Methylmercury—An Interim Review." EHE/EHC/80. 22. World Health Organization, Geneva.

13

Molybdenum

COLIN F. MILLS

The Rowett Research Institute
Bucksburn, Aberdeen, Scotland

GEORGE K. DAVIS[1]

University of Florida
Gainesville, Florida

I. MOLYBDENUM IN ANIMAL TISSUES

A. General Considerations

Molybdenum is widely distributed in soils, plants, and animal tissues. The average abundance of molybdenum in the earth's crust has been placed at 1 mg/kg, but individual rock types may range from near zero to as high as 3000 mg/kg (43,44). Molybdenum is essential for plants and is found at concentrations ranging from <0.5 to >100 mg/kg dry matter in plants grown on soils containing high levels of molybdenum, either naturally or as a result of contamination (69,163).

The molybdenum content of forages important for grazing animals can vary greatly in response to soil molybdenum content, soil pH, and the season of the year (150). The addition of fertilizers containing molybdenum (13,39,158), or of lime, (which increases soil pH, and thus molybdenum uptake), and contamination of soils by industrial (31,182) and mine wastes (184), may all increase the molybdenum content of forages. Forages containing molybdenum up to 26

---

[1]Present address: 2903 S.W. Second Court, Gainesville, Florida 32601.

429

Table I

Normal Molybdenum Concentrations in Animal Organs

Species	Molybdenum concentration (mg/kg, dry matter basis)						
	Liver	Kidney	Spleen	Lung	Brain	Muscle	Reference
Adult human	3.2	1.6	0.20	0.15	0.14	0.14	171
Adult rat	1.8	1.0	0.52	0.37	0.24	0.06	152
Chicken	3.6	4.4	—	—	—	0.14	152

mg/kg have been found in Manitoba, and sweet clover *(Melilotus officinalis)* found in the Great Plains area of the United States can contain up to 200 mg/kg dry matter (163).

Molybdenum in animal tissues also reflects molybdenum intake, but many factors influence the resulting tissue concentration. Some representative values are presented in Table I. In ruminants, particularly, the interrelationship between tissue molybdenum and intakes of copper and sulfur has been extensively investigated. The depressive effect of dietary sulfur on tissue molybdenum is illustrated in Table II. In addition, dietary protein, iron, zinc, lead, tungstate, ascorbic acid, and α-tocopherol have all been reported to influence the level of molybdenum in tissues (126).

B. Molybdenum in the Liver

Molybdenum concentration in the liver of mammals varies and depends on the amount of molybdenum in the diet, and especially the amount of copper, sulfate, and protein that are also being consumed at the same time. Liver values ranging from <1 to >30 mg/kg have been reported as reflecting the molybdenum in the diet of sheep when copper and sulfate have been held constant (38,180). A mean molybdenum concentration of 4.2 (range 3.9–13.7) mg/kg dry matter was reported from a study of 62 bovine liver samples in the Netherlands (176). Tipton and Cook (171) reported an average value of 3.2 mg/kg dry matter for adult humans, which is within the range found for most mammalian species consuming their usual diets with average molybdenum content. Mean values for human liver molybdenum reported in a compilation of multinational data (92) ranged from 0.36 to 0.80 mg/kg fresh tissue.

Judging from studies with rats, increases in dietary molybdenum have less effect on liver molybdenum content in nonruminants than in ruminants. Furthermore, in contrast to ruminants, the depressive effect of dietary sulfur sources on liver molybdenum accumulation is more specifically related to increases in sulfate intake.

Liver molybdenum is usually low in neonates, but molybdenum can cross the placental barrier with the result that high levels of the element in the diet of the dam increase hepatic molybdenum in the newborn (38). The high levels of molybdenum in the liver that can result from consumption of high-molybdenum diets can decline rapidly if molybdenum intake falls and copper and sulfate in the diet are increased.

Unless such dietary factors that can influence the concentration of molybdenum in the liver are evaluated simultaneously, liver molybdenum analysis is not a very good indication of the dietary intake.

Table II

Influence of Dietary Molybdenum and Inorganic Sulfate Intake on the Molybdenum Content of the Tissues of Sheep[a]

Tissue	Molybdenum content (mg)			
	Molybdenum intake 0.3 mg/day		Molybdenum intake 20.9 mg/day	
	Sulfate intake 0.9 g/day	Sulfate intake 6.3 g/day	Sulfate intake 0.9 g/day	Sulfate intake 6.3 g/day
Liver	1.58	0.48	5.79	1.93
Kidney	0.17	0.02	1.17	1.32
Spleen	0.52	0.02	0.57	0.14
Heart	0.18	0.01	0.94	0.04
Lung	0.65	0.09	3.96	0.42
Muscle	5.84	0.08	28.6	1.92
Brain	0.01	0.01	0.09	0.01
Skin	6.62	1.50	58.9	3.44
Wool	15.2	0.99	26.9	1.14
Small intestine	0.26	0.03	0.88	0.79
Cecum	0.24	0.01	1.64	0.58
Colon	0.63	0.04	4.21	0.62
Skeleton	61.0	13.0	164.0	16.0
Total body	92.9	16.8	297.7	28.4

[a] From Dick (53).

C. Molybdenum in Blood

Blood molybdenum concentrations tend to reflect dietary molybdenum intakes (52,107). Cunningham (38) noted that, normally, the blood plasma of sheep contained 0.01–0.02 μg molybdenum per milliliter, rising to 0.2 μg/ml when a molybdenum supplement was given. A further study (52) showed that molybdenum in the blood of sheep rose from 0.02 to 4.9 μg/ml when dietary molybdenum increased from 0.4 to 96 mg/day. With a low molybdenum and low sulfate intake, 70% of the molybdenum was in the erythrocytes. At high molybdenum and sulfate intakes a high proportion of the molybdenum was in the plasma in the form of "copper–molybdenum protein" compounds insoluble in trichloroacetic acid (127,159,160). As indicated later, this probably reflects the high affinity for plasma albumin of thio- and oxythiomolybdates absorbed into the bloodstream after their synthesis intraruminally when dietary intakes of molybdenum and sulfate are high.

Blood plasma molybdenum rose from ~0.05 μg/ml in cattle receiving an alfalfa diet containing 0.6 mg molybdenum/kg, to 3.0 μg/ml when the diet was supplemented with molybdenum to contain 101 mg/kg dry matter (106). Correspondingly large changes with increasing dietary molybdenum have been found in other studies with cattle (35,175) and guinea pigs (10). Simultaneous increases in dietary sources of sulfur reduce these increments of blood molybdenum in ruminants. In contrast, this ameliorative effect of dietary sulfur is only pronounced in nonruminants if inorganic sulfate supplements are given.

Reported concentrations of molybdenum in normal human whole blood differ very widely. Typical of the mean values and ranges (μg/ml) quoted are as follows: <0.005 to 0.41 (11), 0.003 (28), 0.01 ± 0.001 (14), 0.07 ± 0.003. Blood molybdenum increases in several human disorders (181). Whereas, normally, blood molybdenum in healthy humans is distributed almost equally between erythrocytes and plasma (14), in leukemia erythrocyte molybdenum rises. Anemic subjects were found to have lower contents of molybdenum in erythrocytes and plasma.

D. Molybdenum in Milk

The molybdenum content of cow, ewe, and goat milk reflects dietary molybdenum concentrations. Koivistoinen (97) has reported that whole milk and skim milk analyzed in Finland contained molybdenum at 0.05 mg/kg (wet weight), and milk powder, 0.4 mg/kg. Anke et al. (6) found that goat milk had 116 μg molybdenum per liter under usual feeding practices. However, when molybdenum in the feed was high or when molybdenum intake was increased by supplementation of the diet or by injection of molybdenum into the animal, milk molybdenum increased dramatically. Archibald (8) reported a range of from 18 to 120 μg molybdenum per liter in "normal" cow's milk, increasing to 371 μg/liter after feeding 500 mg of molybdenum as ammonium molybdate daily. Somewhat smaller additions of molybdenum, 50–100 mg/cow/day, increased milk molybdenum content by 25–40 μg/liter above the range of 20–30 μg/liter found when molybdenum intake was only 5 mg/day (81). Similar changes were found in goat's milk when molybdenum intake increased from 1 to 13 mg/day.

The milk of ewes also reflects the molybdenum in their feed. Hogan and Hutchinson (84) noted that milk of ewes on pasture with a low content of molybdenum had <10 μg/liter, while that from ewes on high-molybdenum pasture (13 mg/kg) had 980 μg molybdenum per liter. Sulfate in the diet has a marked effect on milk molybdenum in ewes. Thus, when 23 g of sulfate ion were fed daily to animals on a diet containing 25 mg molybdenum per kilogram, milk molybdenum content declined from 1043 μg/liter to 137 μg/liter in 3 days. A high proportion of the molybdenum in ewes' milk was found to be associated with the aqueous phase after removal of fat, casein, and albumin (84).

In a summary of published data from a variety of countries, Iyengar (92) reported mean molybdenum concentrations in human milk ranging from 0.001 to 0.004 μg/ml. Higher mean values were reported in samples from India and the Phillipines, 0.009 and 0.02, respectively.

Hart et al. (81) noted that although the molybdenum in cow's milk is almost all associated with the molybdenoenzyme, xanthine oxidase, when animals receive a low-molybdenum diet, increases in molybdenum intake do not produce a proportional increase in xanthine oxidase activity. On the other hand, addition of the molybdate antagonist, tungstate, to the diet reduced xanthine oxidase activity in goat's milk (132).

Nutritional rather than genetic factors appear to determine the composition of the molybdenum compounds in milk. It has been claimed that the proportion of active and inactivated forms of xanthine oxidase reflect the amounts of molybdenum and copper in the diet (82,101).

E. Molybdenum in Bones and Teeth

In contrast to early claims, there is no consistent evidence that molybdenum has an anticariogenic effect (78), but teeth do accumulate molybdenum if dietary molybdenum increases. Losee et al. (108) noted that human dental enamel is relatively rich in molybdenum. Mean molydenum levels were 5.5 ± 0.71 but ranged from 0.7 to 39 μg/g dry weight. Shirley and Easley (155) studied molybdenum retention in the femur and in tooth enamel and dentine in rats. Omission of vitamin E from the diet markedly increased molybdenum uptake into the enamel of molar and incisor teeth. Females tended to accumulate more molybdenum in their femurs than males when diets provided <1 mg/kg molybdenum. When given diets ranging in molybdenum content from <1 to 100 mg/kg, the content of the element in molar enamel ranged from 1 to 81 mg/kg and, in dentine, from 6 to 10 mg/kg. Incisors had about one-third the amount of molybdenum found in the molars.

Caudal bone tissue in cattle contained ~4 mg/kg molybdenum, increasing to mean concentrations of 8 or 15 mg/kg when molybdenum at 100 mg/kg was added to diets of grass or alfalfa. Many animals receiving such high-molybdenum diets developed lameness (106).

Changes in dietary molybdenum intakes are frequently reflected by changes in the level of this element in the skeleton. For example, Davis (42) increased the levels in the bones of rats from 0.2 to 9–12 mg/kg dry matter and in the livers from 1–2 to 11–12 mg/kg by raising the molybdenum content of the diet from <1 to 30 mg/kg. Molybdenum in the tibia of chicks was increased 100-fold when a diet low in molybdenum was supplemented with 2000 mg/kg molybdenum as molybdate (41). Even higher concentrations were observed in the bones and soft tissues of rabbits and guinea pigs fed highly toxic doses of calcium molybdate or molybdenum trioxide (62).

II. MOLYBDENUM IN FOODS AND FEEDS

A. Molybdenum in Human Foods and Diets

In an extensive compilation of inorganic elements in Finnish foods, Koivistoinen (97) reported that cereals such as wheat, rye, barley, and oats had molybdenum contents ranging from 0.2 to 0.4 mg/kg. Soy meal was higher in molybdenum content, with values between 9 and 10 mg/kg. Many meat products contained <0.1 mg molybdenum per kilogram (wet weight), but bovine liver contained 1.7, pork liver 2.0, sheep liver 0.7, beef kidneys 0.6, and pig kidneys 0.7 mg/kg (wet weight).

The mean molybdenum content of common vegetables ranged from 0.1 mg/kg in potatoes to 0.7 mg/kg in dried peas. Sweet red pepper contained 0.7 mg/kg, whole milk and skim milk averaged 0.05 mg/kg (wet weight), milk powder had 0.4 mg/kg, and cheese, 0.6 mg/kg. Green pea soup contained between 0.1 and 2.3 mg/kg (wet weight).

Warren et al. (186) reported 500-fold differences in the molybdenum contents of some vegetables sampled from different areas, thus illustrating the marked influence of environmental conditions on plant uptake of this element. Thus the molybdenum content of the average adult diet can vary considerably depending on both the pattern of foods consumed and the source of the vegetable constituents. Vegetarian diets could be rich in molybdenum if the vegetables were grown on neutral or alkaline soils with a high content of available molybdenum, or be very low in molybdenum if these items came from areas with acid soils low in molybdenum. Since animal products, with the exception of liver, tend to be quite low in molybdenum, these items in the diet would not be expected to contribute much to the molybdenum intake. The following ranking of food groups in decreasing order of their contribution to the molybdenum contained in "Western" diets is typical: beans > peas > cereals > milk and milk products > nonleguminous vegetables (152,153,173,174,190).

Estimates of molybdenum intake from typical U.S. diets range from 44 to 460 μg/day (22,172,174). One recent study (174) has pointed out that molybdenum intake per unit body weight declines markedly during life, falling from 8 to ~2 μg/kg between infancy and adulthood. Intake of adults in New Zealand was 48–96 μg/day (148), and in England, 128 ± 34 μg/day (80). Individuals consuming a high-sorghum diet might ingest as much as 1.5 mg of molybdenum daily (45).

Although the molybdenum-containing enzymes of plant and animal tissues have been studied extensively, and the conclusion has been reached that most contain a common molybdenum-containing "cofactor" (see later), the proportion of the element present in such organically bound form(s) in feeds derived from plant and animal sources is not known. Much of the molybdenum present in fresh herbage is water soluble but, for unknown reasons, its solubility declines markedly when herbage is dried or becomes frosted (65). The proportion of plant tissue molybdenum that is in the form of inorganic molybdates is not known.

B. Molybdenum in Animal Feeds and Forages

Legumes and their seeds have the highest and most variable content of molybdenum among animal feeds. The molybdenum content of mixed pastures varies according to the proportion of legumes (e.g., clovers) and, for both grasses and leguminous species, the uptake of the element increases with soil pH, particularly when soil molybdenum and moisture content are high (150). Furthermore, the content of molybdenum in many forages increases steadily throughout the growing season (150,185). Such factors make it difficult to estimate molybdenum intake from tabulated data on the content of the element in feeds. Although the following values for molybdenum (in milligrams per kilogram dry matter) may be regarded as representative, many exceptions reflecting local variations in soil conditions that affect molybdenum uptake into crops are likely to be found: cereal grains and straws 0.2–0.5; grasses 0.2–0.8; clovers and others legumes 0.5–1.5; vegetable protein concentrates 0.5–2.0. The following reports of anomalously high molybdenum contents of feeds (in milligrams per kilogram dry matter) are typical of data obtained during investigations of the significance of high molybdenum intakes as a cause of livestock disorders:

Mixed forages:	Manitoba, 1.7–26 (21,107); Southwestern England, 18–108 (65)
Wheat grass:	Great Plains United States, 1–44 (163)
Sweet clover:	Great Plains United States, 1–203 (163)
Barley grass:	Texas (on oil shale residues), 21 (185)
Oat grass:	Norway, 1.5–7.2 (68)

The marked influence of differences in soil pH on the molybdenum content of edible grasses and clovers is indicated in Table III, derived from data obtained in a Scottish study (150). It is evident that marked increases in molybdenum content occur well before the soil could be regarded as "overlimed." Since no corresponding change in copper content occurs, it is also evident that increases in soil pH, by widening the molybdenum–copper ratio of herbage, enhance the risks of molybdenum-induced copper deficiency in ruminants (see Section V,G).

III. MOLYBDENUM METABOLISM

Molybdenum is readily and rapidly absorbed from most diets and many inorganic forms of the element. The hexavalent water-soluble forms, sodium and ammonium molybdate (18,114), and the molybdenum of high-molybdenum herbage (65) are particularly well absorbed by ruminants. Even such insoluble compounds as molybdenum trioxide and calcium molybdate, but not molybdenum sulfate, are well absorbed by rabbits and guinea pigs when fed in large doses (126). Yearling cattle have been shown to absorb oral doses of molybdenum-99 (^{99}Mo) much less rapidly than growing pigs. The urine is a major route of excretion of molybdenum in pigs (17,156), rats (128), and humans (148,149,172), but apparently not in cattle (5) or sheep on low-sulfate intakes (49,52).

Studies of the forms of molybdenum–sulfur compounds probably involved as

Table III

Influence of Changes in Soil pH on the Molybdenum Content of Mixed Grasses and Clovers Grown on a Single Soil Type[a]

Plant	Molybdenum content at different soil pH values (mg/kg dry matter)					
	5.0	5.5	6.0	6.5	7.0	7.5
Grasses	1.1	1.6	2.7	4.0	4.3	5.2
Clovers	0.9	1.3	2.7	3.9	5.7	5.9

[a] The copper content of grasses (range 2.9–3.1 mg/kg dry matter) and clovers (6.9–7.2 mg/kg dry matter) was not affected significantly by changes in soil pH. From Ref. 150.

antagonists of copper utilization by ruminants have shown that molybdenum derived from water-soluble thiomolybdate anions (MoS_4^{2-}, $MoOS_3^{2-}$, and $MoO_2S_2^{2-}$) is readily absorbed by rats, sheep, and cattle (26,90,113,121,168).

The absorption, metabolism, and physiological effects of molybdenum are influenced very markedly by interactions involving dietary sources of sulfur and their metabolites. Such interrelationships govern not only the retention of molybdenum in tissues, but also the tolerance of potentially toxic dietary concentrations of the element and its effects on the utilization of copper. The powerful depressive effect of increases in dietary sulfur intake on the tissue retention of molybdenum by sheep is illustrated by data presented in Table II.

Absorption of the molybdate anion from the small intestine of rats and sheep is inhibited competitively by the sulfate anion, probably by mutual competition for a common carrier system (32,111). Understandably, therefore, both the tissue accumulation of molybdenum and the tolerance of potentially toxic intakes of molybdenum by rats and chicks have been found to be inversely related to dietary concentrations of inorganic sulfate (62,116,178).

However, metabolic interactions between molybdenum sources and dietary or endogenous sulfate are not confined to absorptive processes. Thus Dick (47,48) showed that, in sheep given 10 mg molybdenum daily, increases in dietary sulfur, whether achieved by substituting alfalfa hay for oat hay or by adding inorganic sulfate to the diet (Fig. 1), increased total molybdenum excretion, markedly increased the proportion of the dose excreted in urine, and strongly reduced molybdenum retention in a wide range of tissues (Table II) (53). Such observations have been amply confirmed in sheep (201), cattle (40) and marsupials (16).

Effects of sulfate on the urinary output, tissue retention, and the toxicity of molybdenum are highly specific to this anion and are not reproduced in rats by increasing intakes of citrate, tartrate, acetate, bromide, chloride, or nitrate (178). However, sulfate of endogenous origin may also have similar effects, since such effects are reproduced in rats by increasing the intake of methionine, cystine, or thiosulfate (74,178), and in sheep by dietary supplements of methionine, thiosulfate, or increases in protein intake (52,151). While the concept first advanced by Dick (52), that inorganic sulfate interferes with the transport of molybdenum across membranes, is supported by experimental evidence of such antagonism during intestinal absorption (32,111), extension of this postulate to events such as inhibition of kidney tubular reabsorption of molybdenum secreted in the glomerular filtrate (52) still awaits verification. Furthermore, as indicated in more detail when considering aspects of the metabolic antagonism between molybdenum and copper, upper limits exist both in ruminants and nonruminants to the extent to which sulfur sources ameliorate the adverse effects of increased intakes of molybdenum. In ruminants, virtually all dietary sources of sulfur degradable in the rumen to sulfide, exacerbate rather than ameliorate such ef-

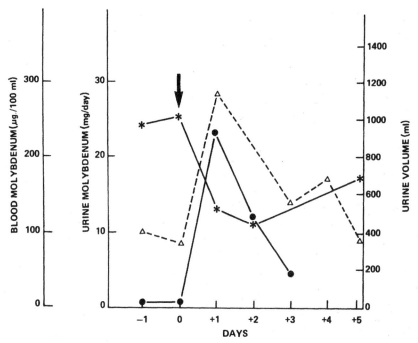

Fig. 1.   Effect on blood (*) and urine (●) molybdenum of 11 g potassium sulfate given by mouth, at the time indicated by the arrow, to a sheep on a diet of chaffed hay and a molybdenum intake of 10 mg/day. It should be noted that induction of a similar diuresis by administration of potassium chloride did not increase urinary molybdenum output. △, Urine volume. From Dick (48).

fects. In rats excessive dietary intakes of sulfate (75) or cystine (79) act similarly, presumably from the endogenous generation of sulfide in the lower gut.

IV. FUNCTIONS OF MOLYBDENUM AND EFFECTS OF DEFICIENCY

A. General Aspects

The established biochemical roles of molybdenum all center around the redox functions of the element, exploiting its capacity to attain valence states of $+4$, $+5$, or $+6$ during the catalytic cycles of enzymes of which it is a constituent. Molybdenum-dependent enzymes have been identified in all living systems. In mammals, they are involved in the metabolism of purines, pyrimidines, pteridines, and aldehydes, and in the oxidation of sulfite. There are good grounds for the belief that these enzymes have a common essential cofactor, tentatively designated molybdopterin—an alkylphosphate-substituted pterin to which molybdenum is coordinated through two sulfur atoms (94,144)—the suggested

Structure I. Suggested structure of molybdopterin (see text).

structure of which is indicated above. Detailed accounts of the structure and properties of these enzymes are published elsewhere (36,143).

B. Xanthine Oxidase/Dehydrogenase

The massive biochemical literature on xanthine oxidase reflects not only the fact that it was the first molybdenoenzyme to be identified but also the frequency with which it has formed the basis for modeling of the role of molybdenum in the active centers of other enzymes (for review see Ref. 37). Unfortunately, this intense interest is not accompanied by a corresponding weight of evidence that variations in xanthine oxidase activity induced by changes in molybdenum supply have great pathological relevance in the context of nutritional deficiencies of molybdenum.

The enzyme catalyzes a range of oxidation reactions typical of which is

$$\text{Xanthine} + H_2O + O_2 \rightarrow \text{uric acid} + H_2O_2$$

In addition to its molybdopterin cofactor, the site at which oxidation of purines, pyrimidines, pterins, pyrimidines, and some aldehydes occurs (143), the enzyme also contains one flavin adenine dinucleotide (FAD) and two iron–sulfur clusters in each of two subunits. When acting as an oxidase (i.e., using oxygen as electron acceptor), significant yields of superoxide (O_2^-) are often noted (67). While it has been suspected that such superoxide generation, if excessive, may account for some of the features of molybdenum toxicity, it is now evident that, within tissues the enzyme functions as a dehydrogenase using nicotinamide adenine dinucleotide (NAD^+) or other electron carriers rather than oxygen as electron acceptor. If, as suspected, conversion to the oxidase form occurs principally during enzyme isolation (188,189), the pathological significance of the enzyme's capacity to generate O_2^- would be reduced, substantially.

The finding that low-molybdenum diets reduced intestinal and liver xanthine "oxidase" activity was made almost simultaneously by DeRenzo (46) and Westerfeld and Richert (192). Evidence of a positive linear relationship between rat intestinal xanthine oxidase activity and dietary molybdenum concentrations within the range 0–0.1 mg/kg diet was exploited as a technique for assay of "available molybdenum" in diets (191). Inclusion of 45 or 94 mg tungsten as tungstate per kilogram of the diet as a molybdenum antagonist produced much greater (95–97%) losses of rat intestine and liver xanthine oxidase than resulted from dietary "deficiency" of molybdenum (83). Although such treatment increased urinary molybdenum losses and depleted tissue molybdenum, no clinical effects were observed and the pattern of purine metabolites in urine did not change (83). Possibly because of their greater dependence on the metabolism of purines, through xanthine to uric acid as the primary pathway of nitrogen excretion, chicks appear more sensitive to indirect induction of molybdenum deficiency by dietary tungstate (83). Thus the above dietary concentrations of tungsten decreased uric acid excretion, increased fecal xanthine and hypoxanthine output, reduced growth, and increased mortality in chicks.

Further evidence of the tolerance of a low xanthine oxidase activity in mammalian tissues is provided by the disorder, heritable xanthinuria, in humans. Clinical manifestations of xanthine oxidase "deficiency" develop only if excessive urinary output of xanthine leads to the development of kidney calculi (37,154,200) or deposition of xanthine or hypoxanthine in muscle induces a mild myopathy. It has been suggested, tentatively, that an increased incidence of xanthine calculi in sheep in some areas of New Zealand may be associated with the consumption of molybdenum-deficient forage (12). Other suggested relationships between changes in tissue xanthine oxidase activity and pathological changes include claims that the high incidence of human gout in some regions of Armenia is associated with high-molybdenum geochemical anomalies and high dietary intakes of molybdenum (10–15 mg/day) (99,100). Though not yet independently verified, studies (77) in rats showing that high intakes of molybdenum hyperactivate xanthine oxidase activity and increase blood and urinary concentrations of its primary product, uric acid, have led to the postulation of a direct causal relationship between molybdenum intake and the incidence of this human disorder.

C. Aldehyde Oxidase

While aldehyde oxidase is distinguishable from xanthine oxidase on the basis of its sensitivity to a range of enzyme inhibitors, the two enzymes share many common substrates and exhibit a similar distribution between tissues (141). Although the substantially higher tissue aldehyde oxidase activity of herbivores than carnivores suggests a form of nutritional adaptation (103,104), the primary substrate(s) of this enzyme *in vivo* are not known nor is it clear whether activity is influenced significantly by changes in molybdenum intake. Its involvement in nicotinic acid metabolism has been suggested.

D. Sulfite Oxidase

With its high specificity for the oxidation of sulfite to sulfate, this mitochondrial molybdenoenzyme is involved in the final step of metabolism of sulfur derived from methionine, cysteine, and related compounds. Isolated now from chicken, rat, bovine and human liver, kidney, and heart and, at lower yield, from many other tissues (187) the enzyme has a molecular weight range of 110,000–120,000, and contains two molybdopterin- and two cytochrome b_5-type heme residues (141,197). Detection of this biological role of molybdenum (95,143) followed detailed investigation of neurological abnormalities in children associated with abnormally high urinary outputs of sulfite, thiosulfate, and S-sulfo-L-cysteine induced by a genetic deficiency of sulfite oxidase (96). Confirmation of the functional role of molybdenum was provided by evidence that its activity declined when tungsten replaced molybdenum in its active center after rats were offered tungstate-supplemented diets (76,95).

Strong presumptive evidence of a nutritional deficiency of molybdenum of sufficient severity to induce pathological changes attributable to a loss of sulfite oxidase activity has now been obtained from a human patient undergoing prolonged total parenteral nutrition (1). Clinical symptoms, totally eliminated by supplementation of the infusate to provide 300 μg ammonium molybdate per day, included irritability leading to coma, tachycardia, tachypnea, and night blindness. Symptoms were alleviated by reducing the total protein or sulfur-amino acid contents of infusates and aggravated by sulfite infusion. Biochemical indications of a low tissue sulfite oxidase activity included a 25-fold increase in thiosulfate excretion, a 70% decline in sulfate output, and a marked increase in plasma methionine. Whether aspects of this syndrome also reflect a loss of tissue xanthine oxidase activity for which there was additional biochemical evidence is not known.

E.   Additional Reported Responses to Molybdenum "Deficiency"

From their early studies of the effects of dietary molybdenum intake on tissue xanthine oxidase activity, Higgins et al. (83) concluded that the molybdenum requirements of chicks and rats were probably <0.02 mg/kg diet, unless tungsten or other factors reducing the efficiency of molybdenum utilization are present. Other studies support this conclusion (105,147). While few, if any, animal diets in commercial use are likely to be so deficient in molybdenum, the question of whether unidentified dietary constituents sufficiently reduce molybdenum availability to induce deficiency remains open to debate.

Thus, suggestions that factors other than dietary tungstate may influence molybdenum availability, arise from evidence that, under incompletely defined circumstances, the growth of chicks or turkey poults given diets based on soya or soya protein isolates containing 0.9–1.6 mg/kg molybdenum was enhanced 10–20% by molybdenum supplements providing only an additional 0.013 (146) or 0.025–0.5 mg/kg diet (145). That the molybdenum of some soya protein preparations may be relatively unavailable was suggested by earlier data (105,147,191).

Increases in chick growth rate have also been observed when diets based on dried skim milk and glucose were supplemented with 5 mg/kg molybdenum (5).

Other responses to increases in dietary molybdenum have also been described. These include improved egg hatchability and improved viability of chicks exhibiting characteristic feather abnormalities when given a single dose of 40 μg molybdenum (as molybdate) (133). Femoral abnormalities, "scabby" hips and feathering defects in commercial broilers were eliminated either by injecting molybdates (0.2–0.5 μg molybdenum im) or by fortifying the commercial diet with 0.2–2.5 mg/kg molybdenum (135). Additional studies with broilers suggested that when such syndromes were provoked by injection of copper sulfate (2 mg copper), they were ameliorated by subsequent injection of 0.5 mg molybdenum (134). Attempts to reproduce these syndromes by offering to chicks and broilers diets with molybdenum contents from 0.3 to 0.7 mg/kg failed, even when transfer of the element to the developing egg yolk was restricted by increasing dietary sulfate from 0.02 to 0.32%, or when chicks from dams given tungstate (1 g/kg diet) were reared on diets very low in molybdenum (0.03 mg/kg) (129,130).

Such conflicting reports may well reflect the complexities of interactions between sulfur, the transition elements, and molybdenum which, as indicated in the following section, can modify the solubility of the latter in biological fluids, and perhaps thus influence its availability as an essential nutrient.

The possibility that prolonged exposure to a molybdenum-poor regime may be required to deplete tissue molybdenum sufficiently to induce a clinical response is suggested by studies with goats (7). Diets providing <0.06 mg molybdenum per kilogram decreased conception rate, parity, and fetal survival, and the offspring of dams so treated initially developed normally but later declined in growth rate and suffered a high mortality. Such effects were not observed if the diets of dams and their kids contained 1 mg/kg molybdenum.

Evidence that dietary molybdenum intake influences the cellulolytic activity of rumen microorganisms in sheep is conflicting. Increases in the rates of gain of sheep given semipurified rations were attributed to a 24% enhancement of cellulose degradation when dietary molybdenum was 2.4 rather than 0.4 mg/kg in one study (61). Supplements of molybdate also accelerated cellulose degradation during *in vitro* incubations of rumen contents (61). However, variables as yet unidentified can apparently eliminate this response (60).

Much more consistent is the evidence that variations in either the intake of molybdenum or of its potential antagonist, tungstate, modify the activity of the molybdenoenzyme nitrate reductase in rumen microorganisms. Thus, low-molybdenum diets decrease rumenal nitrate reductase activity (29,170), increase rumen nitrate levels, and decrease blood nitrite in sheep (170). Generation of nitrite from ingested nitrate within the rumen and its subsequent absorption is a prelude to the development of "nitrite" poisoning in ruminants, and it has been verified experimentally that tolerance of dietary nitrate is increased substantially if rumen nitrate reductase activity is inhibited by increasing tungstate intake (72,98).

V. RESPONSES TO ELEVATED INTAKES OF MOLYBDENUM

A. General Considerations

The effect on the animal of high dietary molybdenum intakes depends on its species and age, the amount and chemical form of the ingested molybdenum, the copper status and copper intake of the animal, and the dietary content of inorganic sulfate and of other sulfur sources such as proteins or sulfur-amino acids. The complexity of these interactions with other nutrients is such that it is often difficult to identify the effects of molybdenum toxicity per se, to predict pathological responses to changes in molybdenum intake, and—particularly in ruminants—to distinguish the direct adverse effects of molybdenum exposure from those caused by the secondary copper deficiency that frequently develops. The latter topic is considered in detail in the next section.

The literature on molybdenum "toxicity" is extensive. Some reports, particularly those written before the major influence of the above variables was fully appreciated, are difficult to interpret except in the light of more recent, if incomplete, understanding of the mode of action of such factors. The following generalizations based on evidence considered later influence the interpretation and the general applicability of many "case reports" of the effects of diets high in molybdenum.

Variations in sulfur intake, or more precisely, the concentrations and forms in which sulfur-containing metabolites are presented to tissues accumulating molybdenum, have a marked influence on molybdenum tolerance. With exceptions to be discussed later, the sulfate ion, whether originating from the diet or from sulfur-amino acid breakdown, restricts molybdenum uptake, enhances its excretion, and thus increases tolerance. In very marked contrast, generation of acid-labile sulfides within the gut and, probably, within tissues, potentiates molybdenum toxicity. There are indications that restricted reoxidation of sulfides may be a fundamental lesion of molybdenum intoxication per se (112,123) that may indirectly be the basis of (1) the antagonistic effect of molybdenum on the utilization of copper and (2) the partial or sometimes complete protection against molybdenum intoxication afforded by increases in copper status (12,65,79). Such considerations—determining the response of the molybdenum-loaded animal to the intake and metabolism fate of its dietary sources of sulfur and to changes in copper intake—influence both the pathological consequences of molybdenum exposure and the even wider importance of molybdenum as a cause of copper deficiency in ruminants.

It is clear, however, that not all manifestations of molybdenum intoxication are attributable to an inhibition of copper-dependent processes. Thus, reductions in the rate of growth of rats, guinea pigs, and chicks given high-molybdenum diets are only partially eliminated by increases in copper intake (10,11,74). Furthermore, the histopathological features of molybdenum intoxication in rat liver and kidney, and concurrent declines in the activities of enzymes such as glucose-6-phosphatase, succinic dehydrogenase, and acetylcholinesterase in kidney, liver, and brain tissue (15) are not typical of copper deficiency. Other features probably characteristic of molybdenosis per se are the development of mandibular exostoses (131), testicular damage in rats (93) and cattle (169), and disorders in phosphorus metabolism in cattle that result in skeletal and joint abnormalities, osteoporosis, and high serum phosphatase activity (41). The biochemical basis of these effects is not understood. Molybdenum is known to be a potent antagonist of the ATP-dependent system responsible for the synthesis of "active" sulfate (phosphoadenosine phosphosulfate), an essential substrate for many tissue reactions leading to the formation of sulfate conjugates (196). It also readily forms stable complexes with o-diphenols, some of which (e.g., epinephrine and the catecholestrogens) have important biological roles. It is not yet clear whether such properties are relevant to any of the effects of molybdenum excess *in vivo*.

Species differences in tolerance to molybdenum are substantial (126). Cattle are by far the least tolerant, followed by sheep, while horses and pigs are the most tolerant of farm livestock. The high tolerance of horses is apparent from their failure to show signs of toxicity on "teart" pastures (i.e., causing molybdenosis; see Section V,B) that severely affect cattle. Foals, however, have been reported to develop rickets when their dams have been maintained on high-molybdenum pastures. The high tolerance of pigs is evident from the report of Davis (42) that consumption of diets containing 1000 mg/kg molybdenum for 3 months induced no ill effects. This is 10–20 times the levels that result in drastic scouring in cattle. These species differences are not fully understood. The high tolerance of pigs cannot be due to poor absorption, judging by the results of experiments with ^{99}Mo which indicate rapid absorption and rapid excretion of the metal in this species (17,156). Rats, rabbits, guinea pigs, and poultry are not so tolerant of molybdenum as pigs but are much more tolerant than cattle. Diets providing 2000 mg/kg molybdenum induce severe growth depression in chicks, accompanied by anemia when the dietary level is raised to 4000 mg/kg (11,41,102). At 200 mg molybdenum per kilogram, some inhibition of chick growth occurs (74), and at 3000 mg/kg diet turkey poult growth is depressed (102).

The clinical manifestations of molybdenum toxicity also vary among different species. Growth retardation or loss of body weight is an invariable result of high molybdenum intakes, but diarrhea is a conspicuous feature only in cattle. Diarrhea has not been reported in rabbits and guinea pigs ingesting sufficient molybdenum to produce marked loss of weight and death (62). In young rabbits the molybdenosis syndrome is characterized further by alopecia, dermatosis, and severe anemia, with a deformity in the front legs of some animals but no achromotrichia or diarrhea (9). In molybdenotic rats, deficient lactation and male sterility associated with some testicular degeneration (93) and connective tissue changes resulting in mandibular or maxillary exostoses (178) have been observed. Connective tissue changes, associated mainly with the humerus, have been reported in sheep grazing high-molybdenum pastures (5–20 mg/kg dry matter) (85). The lesions were characterized by lifting and hemorrhage of the periosteum and the muscle insertions and occasional spontaneous fractures. The condition was largely prevented by supplementary copper. Body weight gains, wool weights, and blood hemoglobin levels were also significantly improved when sheep grazing high-molybdenum pastures received supplementary copper. Although supplementary copper similarly alleviates the growth depression brought about by high- molybdenum intakes in rabbits (2), chicks (11), and rats (75), its protective effect is less than that afforded by an increase in dietary sulfate (75,178,193). Provided an adequate copper status is maintained, supplementary methionine and cystine can alleviate molybdenum toxicity in rats (56,74,75,177,179,194) and sheep (151), and thiosulfate administration can be equally effective in sheep. As these substances are all potentially oxidizable to sulfate in tissues, they probably act in the same manner as inorganic sulfate, by promoting its excretion in urine and reducing molybdenum retention in tissues as discussed previously. High-protein diets—also potential sources of endogenous sulfate—have been claimed to alleviate molybdenum toxicity in rats (15,115), and this effect was associated with a decline in blood and liver molybdenum levels. In the chick, neither methionine nor sulfate appeared to reduce molybdenum retention, although the latter was highly effective in preventing the growth depression induced by high-molybdenum diets (41).

Despite the above convincing evidence of the protective effects of many dietary sources of sulfur, it is now clear that circumstances also exist in which supplements of sulfate, methionine, or cystine strongly exacerbate rather than alleviate molybdenum intoxication. Thus, when rats marginally deficient in copper were given diets supplemented with high levels of L-cystine (9.4 g/kg) and molybdenum (800 mg/kg as molybdate), the addition of this sulfur-amino acid produced a profound anemia and diarrhea and increased mortality (79). Excessive fortification of rat diets with sulfate (6.4 or 27 g/kg) or methionine (10 g/kg) alleviated the toxic effects of 800 mg molybdenum per kilogram diet in rats of normal copper status but strongly enhanced such effects if dietary copper was reduced from ≥ 3 mg/kg to <1 mg/kg (75). In the light of later evidence that as little as 35 mg molybdenum (as molybdate) per kilogram of diet is extremely toxic to rats if traces of sulfide are present in the diet (120), it appears probable that endogenous generation of sulfide within the gut in response to a dietary excess of sulfur-sources could account for the above striking reversal of effects on susceptibility to molybdenum toxicity. Also relevant to this phenomenon of sulfide intolerance is direct and indirect evidence that the capacity of the liver to oxidize sulfide is reduced markedly in rats offered high-molybdenum diets (112,123), unless such diets are fortified with copper (15 mg Cu/kg) (157).

Several other metabolic responses to molybdenum intoxication have been noted. Molybdenosis, severe enough to kill some adult rabbits in 3–4 weeks, was accompanied by falls in plasma thyroxine (T_4) to 2.31 ± 0.34 μg/100 ml, compared with control values of 4.4 ± 0.34 μg/100 ml. The thyroid showed increased colloidal storage, epithelial inactivity and a decrease of epithelial cells from 25.8% in the controls to 12.8% in the molybdenum-treated rabbits. No such glandular changes were observed in rabbits given a diet restricted to the intakes of the molybdenum-fed rabbits but without the molybdenum (122). Dinu and co-workers (56) demonstrated a significant fall in liver glucose-6-phosphatase activity in rats given molybdenum, 1 and 10 mg/kg, but the activities of acid and alkaline phosphatase, xanthine oxidase, aldolase, and succinic oxidase and dehydrogenase were unaffected by these relatively low rates of molybdenum supplementation. In rats suffering from molybdenum toxicity, alkaline phosphatase activity was decreased in the liver and increased in the kidney and intestine (177,194). Alkaline phosphatase in the blood serum of young sheep grazing moderately high- to high-molybdenum pastures was also increased except when additional copper was given (85).

The anorexia of molybdenosis results from a voluntary rejection of diets as a consequence of the rats developing an ability to recognize the presence of molybdenum in the diet (124). The ability to reject high-molybdenum diets through sensory, probably olfactory, recognition of the molybdate requires a learning or conditioning period. Since sulfate alleviates the diarrhea that develops on the high-molybdenum diet, Monty and Click (124) suggested that rats learn to associate a gastrointestinal disturbance with a sensory attribute of diets containing toxic levels of molybdenum.

Responses to cold stress are increased by elevated dietary levels of molybdenum. Winston and co-workers (199) studied rats given Na_2MoO_4 in the drinking water at 10 and 0 mg molybdenum per liter from birth to 6–11 months of age and kept at 2°–3°C for 4 days, or at room temperature for this period. The higher intake of molybdenum had a significant effect on the response to cold, judging by the increased body weight loss, even for animals acclimated to the metal for some time. The basis for this effect has not yet been determined.

In humans, a high incidence of gout has been associated with abnormally high molybdenum concentrations in the soils and plants in some parts of the Soviet Union (100). Humans and livestock exposed to these high molybdenum intakes (10–15 mg/day in humans) displayed abnormally high serum uric acid levels and tissue xanthine oxidase activities. Increases in uric acid excretion were noted in some individuals but not when diets supplied less molybdenum (<1.5 mg/day).

Both these elevated levels of molybdenum intake enhanced urinary copper excretion (45).

B. Molybdenosis in Cattle

Severe molybdenosis in cattle occurs under natural grazing conditions in many parts of the world. In England, where the condition was first studied, it is known as teart and in New Zealand as "peat scours". All cattle are susceptible to molybdenosis, with milking cows and young stock suffering most. Sheep are much less affected, and horses are not affected at all in teart areas. The characteristic scouring varies from a mild form to a debilitating condition so severe that cattle can suffer permanent injury or death. Within a few days of being placed on some teart pastures, cattle begin to scour profusely and develop harsh, staring discolored coats. They usually recover rapidly when transferred to normal pastures lower in molybdenum. Typical teart pastures contain from 20 mg to as high as 100 mg molybdenum per kilogram dry matter, compared with 3–5 mg/kg or less in nearby areas in which this disorder does not occur (65). An apparently identical disorder was induced by iv injection of molybdate over a 2- or 3-week period (3).

Following a report from the Netherlands of a scouring condition responsive to copper therapy in cattle (27), teart was successfully treated with copper sulfate (64). Treatment at the very high rate of 2 g/day for cows and 1 g/day for young stock, or the iv injection of 200–300 mg copper daily, effectively controlled the molybdenum-induced scouring and provided a practical means of field control. If the high molybdenum intakes are prolonged and supplementary copper is not given, a depletion of tissue copper to deficiency levels occurs with associated hypocupremia, but the scouring and loss of condition of severe molybdenosis can occur without a concomitant hypocuprosis. Dick and co-workers (55) have suggested the possibility that the scouring that occurs in cattle, but not in horses, on teart pastures of high molybdenum content may be the result from the formation of thiomolybdates in the rumen, and that the curative effect of copper sulfate may reflect the precipitation within the gut of insoluble copper thiomolybdates (see next section) as a consequence of such treatment.

The possibility that the sulfate radical may also be of some significance in the treatment of molybdenum toxicity with copper sulfate has been suggested by Dick (51). He showed that molybdenum-induced scouring in a calf could be corrected within 4 days by daily drenching with 2 g copper sulfate and within 6 days by drenching with an equivalent amount of sulfate as potassium sulfate.

Scouring and weight loss are such dominant manifestations of molybdenum toxicity in cattle at very high molybdenum intakes that other disorders tend to be obscured. At moderately high levels of molybdenum in the herbage of some areas, a disturbance of phosphorus metabolism occurs, giving rise to lameness, joint abnormalities, osteoporosis, and high serum phosphatase levels (42). Connective tissue changes and some spontaneous bone fractures have been observed in sheep at comparable molybdenum intakes in the absence of supplementary copper (85). Under such conditions, cows may conceive with difficulty and young male bovines exhibit a complete lack of libido. The testes reveal marked damage to interstitial cells and germinal epithelium with little spermatogenesis (169). These changes are comparable with those reported in molybdenotic rats (93,178).

C.   Molybdenum and Bovine Reproduction

Several studies (136,137) indicate that relatively modest increases in dietary molybdenum intake can impair the reproductive performance of female cattle. Increasing the molybdenum content (from 0.1 to 5.1 mg/kg) of a basal ration providing 4 mg copper and 2.8 g total sulfur per kilogram delayed puberty and markedly reduced conception rate in Friesian heifers. In two replicates of the experiment, the conception rate of molybdenum-treated animals was 30% and 17% compared with 72% and 66% for untreated control heifers. Molybdenum supplementation reduced the frequency and size of plasma "pulses" of luteinizing hormone suggesting an impairment of pituitary function.

Although both control and molybdenum-supplemented animals were declining in copper status during these studies, it was clear that the above effects were not attributable to a copper deficiency induced by increased dietary molybdenum; thus they were not evident in animals in which an equal degree of copper depletion was induced by high dietary iron (600 mg/kg).

From this study and other evidence that high dietary molybdenum induces irregularities in the estrous cycle of rats (93,198), it appears probable that molybdenum can exert a direct adverse effect on events controlling estrus. If so, this may explain conflicting evidence on the significance of copper deficiency as a cause of bovine infertility (138). It is notable that many of the investigations from which such a relationship has been claimed were conducted on herds grazing pastures with high molybdenum contents (20,66,125).

D.   Molybdenum–Copper Interrelations

There is now abundant evidence that interactions between molybdenum and copper influence the health and productivity of ruminants under circumstances much less extreme than those that produce the dramatic clinical manipulations of acute debilitation and scouring considered above. Thus, in 1945 (54) it was shown that pastures very low in molybdenum (<0.1 mg/kg dry matter) permitted a rapid accumulation of copper and increased the risks of copper toxicity in sheep, whereas, in contrast, clinical signs of copper deficiency in sheep were shown in 1956 to be attributable to the presence of as little as 5 mg molybdenum per kilogram of diet (201).

The complexities of interactions involving molybdenum and copper that influence the physiological availability of the latter first began to emerge from the pioneering studies of Dick (48,50) and of Wynne and McClymont (201). These workers showed that the potency of molybdenum as an antagonist of cooper in sheep was not constant but was influenced markedly by the nature of the diet and potentiated by increased intakes of inorganic sulfate. That this important finding was nevertheless an oversimplification of events governing the quantitative effects of molybdenum as a copper antagonist will be considered later in this section, after outlining current understanding of the mechanisms involved.

E. Mechanism of Interactions Involving Molybdenum, Sulfur, and Copper

The paradoxical finding that while in ruminants dietary sulfate potentiated the "toxic" effects of molybdenum acting as a copper antagonist, in nonruminants sulfate frequently alleviated molybdenum toxicity, was difficult to resolve. Concepts initially advanced to account for the molybdenum–copper antagonism were that copper transport was inhibited if membranes were presented simultaneously with molybdate and sulfate ions (52), or that formation *in vivo* of an insoluble cupric molybdate complex restricted absorption and utilization of copper (57,58,86). However, neither accounted for marked species differences in the response to diets supplemented simultaneously with molybdate and sulfate.

Subsequent progress has centered around consideration of the major species differences between ruminants and nonruminants in the nature of metabolites of dietary sulfur sources within the digestive tract and their possible relevance to the action of molybdenum. Evidence that sulfide (S^{2-}) or hydrosulfide (HS^-) were important metabolites in the rumen but rarely appeared in the gastrointestinal tract of nonruminants led to investigation of the effects of traces of dietary sulfide on the response of rats given molybdate (117,120). Such treatment led to a marked inhibition of growth accompanied by achromotrichia and frequent diarrhea, all ameliorated by provision of extra dietary copper.

The concept that, in the ruminant, sulfide generated either by rumen bacteria reduction of dietary or salivary sulfate or the degradation of ingested sulfur-amino acids may be involved received further support from several lines of evidence. Thus, in sheep given supplementary molybdenum, rumen concentrations of sulfide increased (117), the biological turnover of sulfide decreased (71) and other potential sources of rumen sulfide such as dietary methionine or cystine were also found to potentiate the adverse effects of molybdenum on copper utilization (165).

From evidence that increases in rumenal sulfide decrease the flow of soluble copper from the rumen (19,117) and insoluble copper sulfide(s) are physiologically unavailable, it can be argued that the basis for the molybdenum–copper antagonism in ruminants is merely the promotion of conditions favoring the formation of copper sulfide. It was clear, however, that such a process restricting the absorption of copper could not account for all the effects observed when molybdenum induces a physiological deficiency of copper in ruminants. Thus, both in sheep (53) and cattle (175), it has been noted that as dietary molybdenum concentrations increase (typically to ≥ 10 mg/kg), clinical signs of copper deficiency such as fleece lesions, hair depigmentation, and retarded growth can still develop, but with little or no decline in liver copper and with blood copper normal or occasionally supranormal. In such an apparently anomalous situation, Smith and Wright (159,161) found that much of the molybdenum and copper of sheep plasma were in forms which, in contrast to normal, were not released into the supernatant when plasma proteins were precipitated with 5% trichloroacetic acid. Fractionation of such plasma revealed, first, that formation of this component was contingent upon a simultaneous increase in the dietary intakes of both molybdenum and sulfur (25), and, second, that radioactive copper incorporated into this fraction was not readily taken up by the liver.

Investigating reasons for the marked intolerance of molybdenum exposed animals to sulfide, Van Reen in 1959 (178) demonstrated the extreme toxicity of tetrathiomolybdate to rats. The subsequent suggestion (164) that thiomolybdates might be involved in the molybdenum–copper antagonism was first given credence by experiments of Dick *et al.* (55), who found not only that addition of di- or tetrathiomolybdate reproduced, *in vitro*, the effects of high dietary molybdenum and sulfur on plasma copper described above, but also that incubation of rumen microorganisms with molybdate and sulfate generated optical spectra characteristic of the thiomolybdates. Aspects of the "thiomolybdate hypothesis" suggesting that initial events involved in antagonistic interactions between molybdenum and copper are the intrarumenal generation of thiomolybdates have been reviewed extensively elsewhere (4,33,70,109,110,119,166).

The principal features of the thiomolybdate hypothesis invoked to explain the effects of dietary molybdenum and sulfur, first, on the absorption of copper by ruminants, and second, their additional effects in restricting the utilization of tissue copper, are as follows:

1. Sulfide (S^{2-}) or hydrosulfide (HS^-) ions are generated within the rumen both by microbial reduction of ingested sulfate and partial degradation of sulfur amino acids derived from dietary or rumen bacterial proteins.
2. This reactive sulfide progressively displaces oxygen from ingested molybdate ions to yield oxythio- or tetrathiomolybdate(s), as follows:

$$MoO_4^{2-} \rightarrow MoO_3S^{2-} \rightarrow MoO_2S_2^{2-} \rightarrow MoOS_3^{2-} \rightarrow MoS_4^{2-}$$

3. Subsequent reaction of the thio- or oxythiomolybdates with copper yields products from which copper is physiologically unavailable.

Variables influencing the production, absorption, and rumen accumulation of sulfide and particularly of the HS^- ion, probably the principal sulfur donor for thiomolybdate synthesis, have been described in detail elsewhere (24). Variations in rumen pH, in the intake of dietary sulfur both from inorganic and organic (e.g., sulfur-amino acid) sources and in energy and nitrogen intake all influence HS^- concentrations.

Although the extent to which rumen conditions favor formation either of oxythiomolybdates or of the fully sulfur-substituted tetrathiomolybdate (MoS_4^{2-}) has been a matter of controversy (23,34,59,110), much evidence strongly suggests that the principal initial products are tri- and tetrathiomolybdates retained primarily upon particulate or bacterial fractions of rumen digesta (40). Those fractions of sheep rumen digesta from which it was subsequently shown that such tri- or tetrathiomolybdate accounted for more than half their total molybdenum were found to contain copper in a form virtually unavailable to rats (139). The utilization of copper from corresponding fractions of intestinal digesta was inhibited similarly. Furthermore, supplementation of the diets of sheep with tetrathiomolybdate, before or after weaning, restricted both the absorption of dietary copper and its hepatic storage (168).

Indirect evidence suggests that the yield of higher thiomolybdates (e.g., MoS_4^{2-} and $MoOS_3^{2-}$) is related, directly, to the intake of dietary sulfur sources potentially degradable to sulfide in the rumen. The extent to which molybdenum derived from inclusion of tetrathiomolybdate in the diet appears in blood plasma is related inversely to dietary copper content (122,168). Thus there are grounds to suspect that this effect and the reduced availability of dietary copper both arise from the formation within the digestive tract of insoluble copper- and molybdenum-containing products whose precise nature has yet to be determined. While evidence of late is consistent with the possibility that Cu(I) derivatives of tetrathiomolybdate account for a considerable proportion of these (140), it is suspected that insoluble copper–molybdenum–sulfur complexes of substantially higher molecular weight may also be formed. Whether these are polymeric thiomolybdo complexes with copper remains open to speculation (118).

Essentially these concepts of the molybdenum–copper antagonism in ruminants envisage that molybdenum acts, not by a direct interaction with copper as suggested previously (57,58), but as a secondary consequence of its affinity for sulfide generated within the rumen. Sulfide itself has a marked affinity for copper, but although their mutual reaction is known to reduce copper solubility within the gut (19), the opportunities for this are normally limited by rapid rumen absorption of sulfide and its reoxidation to sulfate within tissues (71). The intervention of molybdenum in this process appears to sequester potentially reactive sulfido groups, delay their removal from the rumen (33) or proximal small intestine, and thus increase opportunities for their reaction with copper to yield products from which the latter element is not readily utilized.

The suggestion that the basis for these interactions is the rumenal synthesis of thiomolybdates may prove to be an oversimplification, but other evidence indicates that some features of the response of ruminants to high dietary intakes of molybdenum are consistent with absorption of monomeric thiomolybdates from the gut to restrict utilization of tissue copper. Such "systemic" effects on copper metabolism, demonstrated initially in rats (121) and then in ruminants (109,168) given tetrathiomolybdate either orally or parenterally, induce achromotrichia skeletal lesions, anemia, and growth failure.

Rats given ammonium tetrathiomolybdate rapidly developed exostoses of the femur, tibia, fibula, and mandibles, and "beading" of the ribs. Long bones exhibited increased cellular activity but with cartilaginous dysplasia and defective endochondral ossification in their growth plates. Other skeletal lesions included disorganized subperiostial proliferation of bone, resorption of trabecular bone, and interference with fibrogenesis at the points of ligamentous attachments to bone (122,162). The similarity of many of these lesions to those reported in lambs (85) and calves (91) grazing pastures high in molybdenum content has been commented on (162). All are preventable if the copper content of the diet is increased. While extensive mitochondrial lesions found in the gut of tetrathiomolybdate-treated rats may also reflect the secondary induction of a copper deficiency, other features of gut disorder suggested the possibility of more direct toxic effects (63).

Some characteristic responses of the rat to supplementation of its diet with tetrathiomolybdate are illustrated in Fig 2. After administration of tetra- or trithiomolybdate, ceruloplasmin activity declines abruptly and is followed by a reduced activity of other copper-dependent enzymes such as cytochrome oxidase and (copper,zinc) superoxide dismutase. Despite such effects and the protection afforded by an increased supply of copper (121,122), it is noteworthy that they are often initiated before a major depletion of tissue copper has occurred. In such circumstances the proportion of liver or kidney copper present in soluble forms declines significantly, and marked increases occur in both total plasma copper and the proportion of plasma copper and molybdenum associated with plasma albumin and not released therefrom when plasma proteins are precipitated with (5%) trichloroacetic acid (for reviews see Refs. 109,119).

Unequivocal evidence of the presence of di- and trithiomolybdates in the plasma of sheep given diets containing as little as 6 mg molybdenum per kilogram has now been obtained (140). From this and other studies with cattle (90) (see Fig. 3) it appears likely that, ruminant diets low in copper but high in molybdenum and sulfur which optimize conditions for absorption may also limit the systemic utilization of tissue copper in addition to their more frequently observed effect in restricting copper absorption. If so, this may well explain the apparently higher susceptibility to clinical manifestations of copper deficiency of cattle in which a low copper status has been induced by elevated intakes of

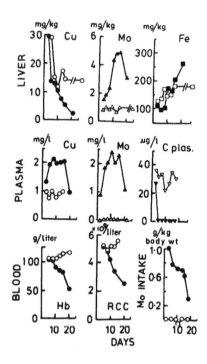

Fig. 2. Typical responses of weanling rats given a semisynthetic diet containing copper (3 mg/kg) either unsupplemented (open symbols) or supplemented with molybdenum (6 mg/kg) as ammonium tetrathiomolybdate (solid symbols). Notable features are the effects of MoS_4^{2-} in depleting plasma ceruloplasmin (C plas.) activity (despite an initial rise in plasma copper), the continual decline in liver copper and sustained increase in liver iron, and the rapid induction of anemia (Hb, hemoglobin; RCC, red blood cell count). From Mills *et al.* (121).

molybdenum rather than by low-copper diets or diets with an excessive content of iron (88).

F. Control of Copper "Toxicity" Syndromes with Tetrathiomolybdate

Evidence of the powerful potency of tetrathiomolybdate as a copper antagonist emerging from studies with rats led Howell and his colleagues (73) to investigate its value for the control of copper intoxication in sheep. Intravenous administration of ammonium tetrathiomolybdate prevented the onset of the hemolytic crisis. These encouraging results have been confirmed in an extensive series of field studies of the prophylactic and therapeutic value of thiomolybdate injection previous to or during the chronic and acute phases of the copper toxicity syndrome in sheep (87).

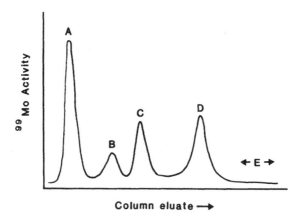

Fig. 3. Characteristic profile of [99]Mo fractions in plasma derived from cattle infused intraruminally with [[99]Mo]molybdate while receiving a hay–concentrate diet supplemented to provide 15 g sulfur daily. Before application to a Sephadex G25 column, plasma samples were treated with (nonradioactive) tetrathiomolybdate to liberate protein-bound [99]Mo-containing constituents. Peaks: A, residual protein-bound [99]Mo; B, [[99]Mo]molybdate; C, [[99]Mo]dithiomolybdate; D, [[99]Mo]trithiomolybdate; E, anticipated position of [[99]Mo]tetrathiomolybdate if present. Derived from the data of Hynes *et al.* (90). Similar evidence of the presence of di- and trithiomolybdates, but not of tetrathiomolybdate, in the plasma of sheep given 5 mg molybdenum as molybdate per kilogram of diet has been obtained (140).

Similar considerations led Walshe (183) to investigate the therapeutic value of ammonium tetrathiomolybdate in cases of Wilson's disease in human subjects. Initial trials indicate that oral tetrathiomolybdate promotes remission of clinical symptoms and offers an alternative therapeutic approach in subjects intolerant of the penicillamine therapy previously used to restrict copper accumulation in the liver and central nervous system.

G. Quantitative Aspects of the Molybdenum–Copper Antagonism

In view of the above evidence that responses to increased intakes of molybdenum are influenced both by dietary composition and by factors such as differences in the extent of microbial activity within the gut, it is now understandable why past views on the tolerable levels of dietary molybdenum have differed so markedly.

Evidence that the tolerance of dietary molybdenum by nonruminants is determined primarily by its adverse effects on copper utilization is mostly confined to species such as guinea pigs or rabbits in which cecal microbial activity is an important component of digestive processes (9,10). For other nonruminant species, the molybdenum–copper antagonism appears to influence the tolerance of dietary molybdenum only under conditions in which excessive intakes of sulfate (75) or sulfur-amino acids (79) probably promote sulfide generation by microbial activity within the lower gut (89).

In marked contrast, the tolerance of molybdenum by ruminants is very frequently governed by its adverse effects on copper utilization and thus by the extent to which dietary molybdenum reduces the supply of dietary copper that is physiologically available. In response to this situation, attempts were initially made to interpret data for the molybdenum content of ruminant diets on the basis of tolerable copper–molybdenum ratios derived from consideration of evidence obtained during "field" investigations of copper-responsive disorders associated with elevated intakes of molybdenum. Reasons for the poor predictive value of this approach emerged, first, from evidence that dietary sulfate potentiates the action of molybdenum as a copper antagonist (52), and the later finding that this property is shared by dietary sources of the sulfur-amino acids (165). Thus, efforts to predict the effects of molybdenum in ruminant diets must, at least, include consideration of the relationship between total copper supply and that of molybdenum and of dietary sources of sulfur, both organic and inorganic, potentially metabolizable to sulfide by rumen microorganisms.

Progress toward prediction of the effect of these variables on the physiological availability of dietary copper has been made in a systematic series of studies with sheep by Suttle and his colleagues (167). In this approach, data on the plasma copper response of sheep, previously depleted of copper, was monitored when diets differing in copper, molybdenum, and total sulfur content were used to derive a series of equations describing the influence of these variables on the percentage "absorbability" ($A_{Cu\%}$) of dietary copper. Typical equations for predicting the influence of molybdenum and total sulfur contents of hay on copper absorbability from representative feeds for sheep are

For summer herbage: $A_{Cu\%} = 5.72 - 1.297\ S - 2.785\ \log_e Mo$
$$+ 0.227\ (Mo \times S)$$

For hay: $A_{Cu\%} = 8.9 - 0.70\ \log_e Mo - 2.61\ \log_e S$

Units for Mo, mg/kg; for S, g/kg, dietary dry matter. Figure 4 illustrates these relationships for summer herbage.

Such estimates of the absorbability of dietary copper, when used in conjunction with data on total dietary copper intake and estimates of copper requirement, are proving of value for identification of situations under which dietary molybdenum and sulfur enhance risks of copper deficiency. That these studies are revealing the need for differing predictive equations for herbage at different stages of growth, for conserved forages, and for semisynthetic diets is fully understandable in view of their probable influence on events in the rumen which modify the potency of molybdenum as a copper antagonist.

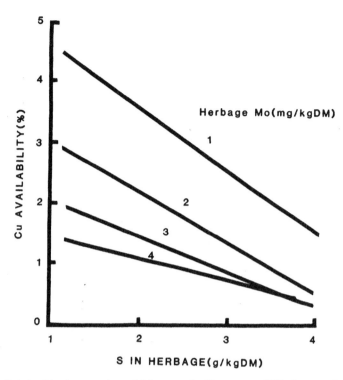

Fig. 4. Relationships between the molybdenum and sulfur contents (DM, dry matter) of summer herbage and the availability (%) of dietary copper derived from studies with sheep. Reproduced with permission from Suttle (167).

REFERENCES

1. Abumrad, N., Schnieder, A. J., Steel, D., and Rogers, L. S. (1981). *Am. J. Clin. Nutr.* **34**, 2551.
2. Allaway, W. H., Kubota, J., Losee, F., and Roth, M. (1968). *Arch. Environ. Health* **16**, 342.
3. Allcroft, R., and Lewis, G. (1956). *Lantsouwk. Tijdschr.* **68**, 711.
4. Allen, J. D., and Gawthorne, J. M. (1985). *In* "Trace Elements in Man and Animals—TEMA 5" (C. F. Mills, I. Bremner, and J. K. Chesters, eds.), p. 361. Commonwealth Agric. Bureaux, Farnham Royal, U.K.
5. Anders, E., and Hill, C. (1970). *Fed. Proc., Fed. Am. Soc. Exp. Biol.* **29**, 766.
6. Anke, M., Groppel, B., and Grün, M. (1985). *In* "Trace Elements in Man and Animals" (C. F. Mills, I. Bremner, and J. K. Chesters, eds.), p. 154. Commonwealth Agric. Bureaux, Farnham Royal, U.K.
7. Anke, M., Grün, M., Partschefeld, M., and Groppel, B. (1978). *In* "Trace Element Metabolism in Man and Animals—TEMA 3" (M. Kirchgessner, ed.), p. 230. Institut für Ernährungsphysiologie, Technische Universität, München, Freising, Weihenstephan.
8. Archibald, J. G. (1951). *J. Dairy Sci.* **34**, 1026–1029.

9. Arrington, L. R., and Davis, G. K. (1953). *J. Nutr.* **51**, 295–304.
10. Arthur, D. (1965). *J. Nutr.* 87, 69.
11. Arthur, D., Motzok, I., and Brannion, H. D. (1958). *Poult. Sci.* **37**, 1181.
12. Askew, H. O. (1958). *N.Z. J. Agric. Res.* **1**, 447.
13. Askew, H. O., Chittenden, E., Watson, J., and Waters, A. J. (1958). *N.Z. J. Agric. Res.* **1**, 874.
14. Bala, Y. M., and Lifshits, V. M. (1966). *Fed. Proc., Fed. Am. Soc. Exp. Biol.* **25**, 370–372.
15. Bandyopadhyay, S. K., Chatterjee, K., Tiwari, R. K., Mitra, A., Bannerjee, A., Ghosh, K. K., and Chatterjee, G. C. (1981). *Int. J. Vitam. Nutr. Res.* **51**, 401.
16. Barker, S. (1960). *Nature (London)* **185**, 41.
17. Bell, M. C., Higgs, B. G., Lowrey, R. S., and Wright, P. L. (1964). *Fed. Proc., Fed. Am. Soc. Exp. Biol.* **23**, 873 (Abstr.).
18. Bell, M. C., Higgs, B. G., Lowrey, R. S., and Wright, P. L. (1964). *J. Nutr.* **84**, 367.
19. Bird, P. R. (1970). *Proc. Aust. Soc. Anim. Prod.* **8**, 212.
20. Blakemore, F., and Venn, J. A. J. (1950). *Vet. Res.* **62**, 756.
21. Boila, R. J., Devlin, T. J., Drysdale, R. A., and Lillie, L. E. (1984). *Can. J. Anim. Sci.* **64**, 919–936.
22. Bostrom, H., and Wester, P. O. (1968). *Acta Med. Scand.* **190**, 155.
23. Bray, A. C., Suttle, N. F., and Field, A. C. (1982). *Proc. Nutr. Soc.* **41**, 67 A.
24. Bray, A. C., and Till, A. R. (1975). *In* "Digestion and Metabolism in the Ruminant" (I. W. McDonald and A. C. I. Warner, eds.), p. 241. Univ. of New Engl. Press, Henniker, New Hampshire.
25. Bremner, I. (1976). *Proc. Nutr. Soc.* **35**, 21A.
26. Bremner, I., Mills, C. F., and Young, B. W. (1982). *J. Inorg. Biochem.* **16**, 109.
27. Bronwer, F., Frens, A. M., Reitsma, P., and Kalesvaart, C. (1938). *Versl. Landbouwkde. Onderz.* **C44**, 267.
28. Brune, D., Samsahl, K., and Wester, P. O. (1966). *Clin. Chim. Acta* **13**, 285.
29. Buchman, D. T. (1966). "Effect of Molybdenum and Copper in Forage on Nitrate Reduction in Ruminants." Ph.D. thesis, University of Florida.
30. Butt, E. M. (1964). *Arch. Environ. Health* **8**, 52.
31. Buxton, J. C., and Allcroft, R. (1955). *Vet. Rec.* **67**, 273.
32. Cardin, C. J., and Mason, J. (1975). *Biochim. Biophys. Acta* **394**, 46; (1976). **455**, 937.
33. Chesters, J. K., Mills, C. F., and Price, J. (1985). *In* "Trace Elements in Man and Animals—TEMA 5" (C. F. Mills, I. Bremner, and J. K. Chesters, eds.), p. 351. Commonwealth Agric. Bureaux, Farnham Royal, U.K.
34. Clarke, N. J., and Laurie, S. H. (1979). *J. Inorg. Biochem.* **12**, 37.
35. Clawson, W. J., Lesperance, A. L., Bohman, V. R., and Layhee, D. C. (1972). *J. Anim. Sci.* **34**, 516–520.
36. Coughlan, M. P. (1980). *In* "Molybdenum and Molybdenum-containing Enzymes" (M. P. Coughlan, ed.), p. 119. Pergamon, New York.
37. Coughlan, M. P. (1983). *J. Inherited Metab. Dis.* **6** (Suppl. 1), 70.
38. Cunningham, I. J. (1950). *In* "A Symposium on Copper Metabolism" (W. D. McElroy and B. Glass, eds.), pp. 246–273. Johns Hopkins Press, Baltimore.
39. Cunningham, I. J., and Hogan, K. G. (1956). *N.Z. J. Sci. Technol. (A)* **38**, 248.
40. Cunningham, I. J., Hogan, K. G., and Lawson, B. M. (1959). *N.Z. J. Agric. Res.* **2**, 145.
41. Davies, R. E., Reed, B. L., Kurnick, A. A., and Couch, J. R. (1960). *J. Nutr.* **70**, 193.
42. Davis, G. K. (1950). *In* "A Symposium on Copper Metabolism" (W. D. McElroy and B. Glass, eds.), pp. 216–228. Johns Hopkins Press, Baltimore.
43. Davis, G. K. (1984). *In* "Metalle in der Umwelt" (E. Merian, ed.), pp. 479–485. Verlag Chemie, Weinheim.

44. Davis, G. K., Jorden, R., Kubota, H., Laitenen, A., Matrone, G., Newberne, P. M., O'Dell, B. L., and Webb, J. S. (1974). *In* "Geochemistry and the Environment," Vol. 1, Chap. IX, pp. 68–79. National Academy of Sciences, Washington, D.C.
45. Deosthale, Y. G., and Gopalan, C. (1974). *Br. J. Nutr.* **31**, 351–355.
46. De Renzo, E. C. (1954). *Ann. N.Y. Acad. Sci.* **57**, 905.
47. Dick, A. T. (1952). *Aust. Vet. J.* **28**, 30.
48. Dick, A. T. (1953). *Aust. Vet. J.* **29**, 18, 233.
49. Dick, A. T. (1954). *Aust. Vet. J.* **30**, 196.
50. Dick, A. T. (1954). *Aust. J. Agric. Res.* **5**, 511.
51. Dick, A. T. (1954). Ph.D. thesis, University of Melbourne.
52. Dick, A. T. (1956). *In* "A Symposium on Inorganic Nitrogen Metabolism" (W. D. McElroy and B. Glass, eds.), pp. 445–473. Johns Hopkins Press, Baltimore.
53. Dick, A. T. (1956). *Soil Sci.* **81**, 229–236.
54. Dick, A. T., and Bull, L. B. (1945). *Aust. Vet. J.* **21**, 70–72.
55. Dick, A. T., Dewey, D. W., and Gawthorne, J. M. (1975). *J. Agric. Sci.* **85**, 567–568.
56. Dinu, I., Boghianu, L., and Sporn, A. (1972). *Rev. Roum. Biochim.* **9**, 215.
57. Dowdy, R. P., and Matrone, G. (1968). *J. Nutr.* **95**, 191.
58. Dowdy, R. P., and Matrone, G. (1968). *J. Nutr.* **95**, 197.
59. El-Gallad, T. T., Mills, C. F., Bremner, I., and Summers, R. (1983). *J. Inorg. Biochem.* **18**, 323.
60. Ellis, W. C., and Pfander, W. H. (1960). *J. Anim. Sci.* **19**, 1260.
61. Ellis, W. C., Pfander, W. H., Muhrer, M. E., and Pickett, E. E. (1958). *J. Anim. Sci.* **17**, 1980.
62. Fairhall, L. T., Dunn, R. D., Sharpless, N. E., and Pritchard, E. A. (1945). *U.S. Public Health Serv. Bull.* **293**.
63. Fell, B. F., Dinsdale, D., and El-Gallad, T. T. (1979). *J. Comp. Pathol.* **89**, 495.
64. Ferguson, W., Lewis, A. H., and Watson, S. J. (1938). *Nature (London)* **141**, 553.
65. Ferguson, W. S., Lewis, A. H., and Watson, S. J. (1943). *J. Agric. Sci.* **33**, 44–51.
66. Field, H. I. (1957). *Vet. Rec.* **69**, 788, 832.
66a. Frieder, E. (1978). *In* "Trace Element Metabolism in Man and Animals—TEMA 3" (M. Kirchgessner, ed.). Institut für Ernährungsphysiologie, Technische Universität, München, Freising, Weihenstephan.
67. Fridovich, I. (1970). *J. Biol. Chem.* **245**, 4053.
68. Froslie, A., Havre, G. N., and Norheim, G. (1985). *In* "Trace Elements in Man and Animals—TEMA 5" (C. F. Mills, I. Bremner, and J. K. Chesters, eds.), p. 855. Commonwealth Agric. Bureaux, Farnham Royal, U.K.
69. Gardner, A. W., and Hallpatch, P. K. (1962). *Vet. Rec.* **74**, 113–116.
70. Gawthorne, J. M., Allen, J. D., and Nader, C. J. (1985). *In* "Trace Elements in Man and Animals—TEMA 5" (C. F. Mills, I. Bremner, and J. K. Chesters, eds.), p. 346. Commonwealth Agric. Bureaux, Farnham Royal, U.K.
71. Gawthorne, J. M., and Nader, C. J. (1976). *Br. J. Nutr.* **35**, 11.
72. Geurink, J. H., Malestein, A., Kemp, A., Korzeniowski, A., and Vant'Klooster, A. T. (1982). *Neth. J. Agric. Sci.* **30**, 105.
73. Gooneratne, S. R., Howell, J. M., and Gawthorne, J. M. (1981). *Br. J. Nutr.* **46**, 457.
74. Gray, L. F., and Daniel L. J. (1954). *J. Nutr.* **53**, 43–51.
75. Gray, L. F., and Daniel, L. J. (1964). *J. Nutr.* **84**, 31–37.
76. Gunnison, A. F., Farraggella, T. J., Chiang, G., Dulak, L., Zaccardi, J., and Birkner, J. (1981). *Food Cosmet. Toxicol.* **10**, 209.
77. Gusev, E. V. (1969). *Gig. Sanit.* **34**, 63.

Colin F. Mills and George K. Davis

78. Hadjimarkos, D. M. (1973). *In* "Trace Substances in Environmental Health—VII" (D. D. Hemphill, ed.), pp. 25–29. Univ. of Missouri, Columbia.
79. Halverson, A. W., Phifer, J. H., and Monty, K. J. (1960). *J. Nutr.* **71,** 95.
80. Hamilton, E. I., and Minski, M. J. (1972–3). *J. Sci. Total Environ.* **1,** 341.
81. Hart, L. I., Owen, E. C., and Proudfoot, R. (1967). *Br. J. Nutr.* **21,** 617–630.
82. Hart, L. I., McGartoll, M. A., Chapman, H. R., and Bray, R. C. (1970). *Biochem. J.* **116,** 851–864.
83. Higgins, E. S., Richert, D. A., and Westerfield, W. W. (1956). *J. Nutr.* **59,** 539.
84. Hogan, K. G., and Hutchinson, A. J. (1965). *N.Z. J. Agric. Res.* **8,** 625–629.
85. Hogan, K. G., Money, D. F. L., White, D. A., and Walker, R. (1971). *N.Z. J. Agric. Res.* **14,** 687.
86. Huisingh, J., and Matrone, G. (1976). *In* "Molybdenum in the Environment" (W. R. Chappell and K. K. Petersen, eds.), p. 125. Dekker, New York.
87. Humphries, W. R., Mills, C. F., Greig, A., Roberts, L., Inglis, D., and Halliday, G. J. (1986). *Vet. Rec.* **119,** 596.
88. Humphries, W. R., Phillippo, M., Young, B. W., and Bremner, I. (1983). *Br. J. Nutr.* **49,** 77.
89. Huovinen, J. A., and Gustafson, B. E. (1967). *Biochim. Biophys. Acta* **136,** 441.
90. Hynes, M., Woods, M., Poole, D. B. R., Rogers, P., and Mason, J. (1985). *J. Inorg. Biochem.* **24,** 279.
91. Irwin, M. R., Poulos, P. W., Smith, B. P., and Fisher, G. L. (1974). *J. Comp. Pathol.* **84,** 611.
92. Iyengar, G. V. (1985). "Concentrations of 15 Trace Elements in Some Selected Adult Human Tissues and Body Fluids." Rept. 1974, Inst. Medicine, Jülich Nuclear Res. Center, Jülich, F.R.G.
93. Jeter, M. A., and Davis, G. K. (1954). *J. Nutr.* **54,** 215–220.
94. Johnson, J. L., and Rajagopalan, K. V. (1982). *Proc. Natl. Acad. Sci. U.S.A.* **79,** 6856.
95. Johnson, J. L., Rajagopalan, K. V., and Cohen, H. J. (1974). *J. Biol. Chem.* **249,** 859.
96. Johnson, J. L., Waud, W. R., Rajagopalan, K. V., Duran, M., Beemer, F. A., and Wadman, S. K. (1980). *Proc. Natl. Acad. Sci. U.S.A.* **77,** 3715.
97. Koivistoinen, P. (1980). *Acta Agric. Scand. (Suppl.)* **22.**
98. Korzeniowski, A., Geurink, J. H., and Kemp, A. (1981). *Neth. J. Agric. Sci.* **29,** 37.
99. Kovalskij, V. V., and Vorotnitskaya, I. E. (1970). *In* "Trace Element Metabolism in Animals—TEMA 1" (C. F. Mills, ed.), p. 176. Livingstone, Edinburgh.
100. Kovalskij, V. V., Yarovaya, G. A., and Shmavonyan, D. M. (1961). *Z. Obsc. Biol.* **22,** 179.
101. Kovalskij, V. V., Vorotnitskaya, I. E., and Tsoi, G. G. (1974). *In* "Trace Element Metabolism in Animals" (W. G. Hoekstra, J. W. Suttie, H. E. Ganther, and W. Mertz, eds.), pp. 161–170. Univ. Park Press, Baltimore.
102. Kratzer, F. H. (1952). *Proc. Soc. Exp. Biol. Med.* **80,** 483.
103. Krenitsky, T. A., Neil, S. M., Elion, G. B., and Hitchings, G. H. (1972). *Arch. Biochem. Biophys.* **150,** 585.
104. Krenitsky, T. A., Tuttle, J. V., Cattan, E. L., and Wang, P. (1974). *Comp. Biochem. Physiol.* **49B,** 687.
105. Leach, R. M., Turle, D. E., Zeigler, T. R., and Norris, L. C. (1961). *Poult. Sci.* **41,** 300.
106. Lesperance, A. L., and Bohman, V. R. (1963). *J. Anim. Sci.* **22,** 686–694.
107. Lesperance, A. L., Bohman, V. R., and Oldfield, J. E. (1985). *J. Anim. Sci.* **60,** 791–802.
108. Losee, F., Cutress, T. W., and Brown, R. (1973). *In* "Trace Substances in Environmental Health VII" (D. D. Hemphill, ed.), pp. 19–24. Univ. of Missouri, Columbia.
109. Mason, J. (1982). *Ir. Vet. J.* **36,** 164.
110. Mason, J. (1986). *Toxicology* **42,** 99.
111. Mason, J., and Cardin, C. J. (1977). *Res. Vet. Sci.* **22,** 313.

112. Mason, J., Cardin, C. J., and Dennehy, A. (1978). *Res. Vet. Sci.* **24,** 104.
113. Mason, J., Lamand, M., and Kellerher, C. A. (1982). *J. Comp. Pathol.* **92,** 509.
114. Mason, J., Lamand, M., Tressol, J. C., and Lab, C. (1978). *Ann. Rech. Vet.* **9,** 577.
115. Miller, R. F., and Engel, R. W. (1960). *Fed. Proc., Fed. Am. Soc. Exp. Biol.* **19,** 666.
116. Miller, R. F., Price, N. O., and Engel, R. W. (1956). *J. Nutr.* **60,** 539.
117. Mills, C. F. (1960). *Proc. Nutr. Soc.* **19,** 162.
118. Mills, C. F. (1982). *In* "The Chemistry and Uses of Molybdenum," 4th Int. Conf. (H. F. Barry and P. H. C. Mitchell, eds.), p. 134. Climax Molybdenum, Ann Arbor.
119. Mills, C. F., and Bremner, I. (1980). *In* "Molybdenum and Molybdenum-Containing Enzymes" (M. Coughlan, ed.), p. 517. Plenum, New York.
120. Mills, C. F., and Mitchell, R. L. (1971). *Br. J. Nutr.* **26,** 117.
121. Mills, C. F., El-Gallad, T. T., and Bremner, I. (1981). *J. Inorg. Biochem.* **14,** 189.
122. Mills, C. F., El-Gallad, T. T., Bremner, I., and Wenham, G. (1981). *J. Inorg. Biochem.* **14,** 163.
123. Mills, C. F., Monty, K. J., Ichihara, A., and Pearson, P. B. (1958). *J. Nutr.* **65,** 129.
124. Monty, K. J., and Click, E. M. (1961). *J. Nutr.* **75,** 303–308.
125. Munro, I. B. (1957). *Vet. Rec.* **69,** 125.
126. National Research Council (1980). "Mineral Tolerance of Domestic Animals." National Academy of Sciences, Washington, D.C.
127. Nederbragt, H. (1982). *Br. J. Nutr.* **48,** 353.
128. Neilands, J. B., Strong, F. M., and Elvehjem, C. A. (1948). *J. Biol. Chem.* **172,** 431.
129. Nell, J. A., and Annison, E. F. (1980). *Br. Poult. Sci.* **21,** 183.
130. Nell, J. A., Annison, E. F., and Balnave, D. (1980). *Br. Poult. Sci.* **21,** 193.
131. Ostrom, C. A., Van Reen, R., and Miller, C. W. (1961). *J. Dent. Res.* **40,** 520.
132. Owen, E. C., and Proudfood, R. (1968). *Br. J. Nutr.* **22,** 331–340.
133. Payne, C. G. (1977). *Br. Poult. Sci.* **18,** 427.
134. Payne, C. G. (1978). *In* "Trace Element Metabolism in Man and Animals" (M. Kirchgessner, ed.), p. 515. Institut für Ernährungphysiologie, Technische Universität, München, Freising-Weihenstephan.
135. Payne, C. G., and Bains, B. S. (1975). *Vet. Rec.* **97,** 436.
136. Phillippo, M., Humphries, W. R., and Atkinson, T. (1985). *Proc. Nutr. Soc.* **44,** 82A.
137. Phillippo, M., Humphries, W. R., Bremner, I., Atkinson, T., and Henderson, G. (1985). *In* "Trace Elements in Man and Animals—TEMA 5" (C. F. Mills, I. Bremner, and J. K. Chesters, eds.), pp. 176–180. Commonwealth Agric. Bureaux, Farnham Royal, U.K.
138. Phillippo, M., Humphries, W. R., Lawrence, B. L., and Price, J. (1982). *J. Agric. Sci.* **99,** 359.
139. Price, J., and Chesters, J. K. (1985). *Br. J. Nutr.* **53,** 323.
140. Price, J., Will, A. M., Paschaleris, G., and Chesters, J. K. (1987). *Br. J. Nutr.* In press.
141. Rajagopalan, K. V. (1980). *In* "Molybdenum and Molybdenum-Containing Enzymes" (M. P. Coughlan, ed.), p. 241. Academic Press, New York.
142. Rajagopalan, K. V. (1980). *In* "Enzymatic Basis of Detoxification" (W. B. Jacoby, ed.), Chap. 14. Academic Press, New York.
143. Rajagopalan, K. V. (1984). *In* "Biochemistry of the Essential Ultratrace Elements" (E. Frieden, ed.), p. 149. Plenum, New York.
144. Rajagopalan, K. V., Johnson, J. L., and Hainline, B. E. (1982). *Fed. Proc., Fed. Am. Soc. Exp. Biol.* **41,** 2608.
145. Reid, B. L., Kurnick, A. A., Burroughs, R. N., Svacha, R. L., and Couch, J. R. (1957). *Proc. Soc. Exp. Biol. Med.* **94,** 737.
146. Reid, B. L., Kurnick, A. A., Svacha, R. L., and Couch, J. R. (1956). *Proc. Soc. Exp. Biol. Med.* **93,** 245.
147. Richert, D. A., and Westerfield, W. W. (1953). *J. Biol. Chem.* **203,** 915.

148. Robinson, M. F., McKenzie, J. M., Thomson, C. D., and Van Rij, A. L. (1973). *Br. J. Nutr.* **30**, 195–205.
149. Rosoff, B., and Spencer, H. (1964). *Nature (London)* **202**, 410.
150. SAC/SARI (1982). "Trace Element Deficiency in Ruminants." Rept. Study G., Scot. Agr. Coll/Res. Insts., Edinburgh.
151. Scaife, J. F. (1963). *N.Z. J. Sci. Technol. Sect. A* **38**, 285, 293.
152. Schroeder, H. A., Balassa, J. J., and Tipton, I. H. (1970). *J. Chronic Dis.* **23**, 481–499.
153. Schurz, H., Kloos, G., and Senser, F. (1986). "Food Consumption and Nutrition Tables" 1986–1987," 3rd Ed. Wissenschaftlicher Verlag, Stuttgart.
154. Seegmiller, J. E. (1980). *In* "Metabolic Control and Disease" (P. K. Bondy and L. E. Rosenberg, eds.), p. 777. Saunders, Philadelphia.
155. Shirley, R. L., and Easley, J. F. (1976). *In* "Molybdenum in the Environment" (W. R. Chappell and K. K. Petersen, eds.), pp. 221–228. Dekker, New York.
156. Shirley, R. L., Jeter, M. A., Feaster, J. P., McCall, J. T., Outler, J. C., and Davis, G. K. (1954). *J. Nutr.* **54**, 59–64.
157. Siegel, L. M., and Monty, K. J. (1961). *J. Nutr.* **74**, 167.
158. Smith, B., and Coup, M. R. (1973). *N.Z. Vet. J.* **21**, 252.
159. Smith, B. S. W., and Wright, H. (1975). *Clin. Chim. Acta* **62**, 55.
160. Smith, B. S. W., and Wright, H. (1975). *J. Comp. Pathol.* **85**, 299.
161. Smith, B. S. W., and Wright, H. (1976). *Int. Symp. Nuclear Tech. Anim. Prod., IAEA/FAO, Vienna.* p. 109.
162. Spence, J. A., Suttle, N. F., Wenham, G., El-Gallad, T. T., and Bremner, I. (1980). *J. Comp. Pathol.* **90**, 139.
163. Stone, L. R., Erdman, J. A., Fedder, G. L., and Holland, H. D. (1983). *J. Range Manage.* **36**, 280–285.
164. Suttle, N. F. (1974). *Proc. Nutr. Soc.* **33**, 299.
165. Suttle, N. F. (1975). *Br. J. Nutr.* **34**, 411.
166. Suttle, N. F. (1980). *Ann. N.Y. Acad. Sci.* **355**, 195.
167. Suttle, N. F. (1983). *In* "Feed Information and Animal Production" (G. E. Robards and R. G. Packham, eds.), p. 211. Commonwealth Agric. Bureaux, Farnham Royal, U.K.
168. Suttle, N. F., and Field, A. C. (1983). *J. Comp. Pathol.* **93**, 379.
169. Thomas, J. W., and Moss. (1951). *J. Dairy Sci.* **34**, 929.
170. Tillman, A. D., Sheria, G. M., and Sirny, R. J. (1965). *J. Anim. Sci.* **24**, 1140.
171. Tipton, I. H., and Cook, M. J. (1963). *Health Phys.* **9**, 103–145.
172. Tipton, I. H., Stewart, P. L., and Martin, P. G. (1966). *Health Phys.* **12**, 1683.
173. Tretkova, I. N. (1969). *Hyg. Sanit.* **34**, 115.
174. Tsongas, T. A., Meglen, R. R., Walravens, P. A., and Chappell, W. R. (1980). *Am. J. Clin. Nutr.* **33**, 1103.
175. Vanderveen, J. E., and Keener, H. A. (1964). *J. Dairy Sci.* **47**, 1224–1230.
176. Van Esch, G. J., and Hart, P. C. (1953). *Landbouwkd. Tijdschr.* **65**, 195.
177. Van Reen, R. (1954). *Arch. Biochem. Biophys.* **53**, 77.
178. Van Reen, R. (1959). *J. Nutr.* **68**, 243–250.
179. Van Reen, R., and Williams, M. A. (1956). *Arch. Biochem. Biophys.* **63**, 1.
180. Van Ryssen, J. B. J., and Stielau, W. J. (1981). *Br. J. Nutr.* **45**, 203–210.
181. Versieck, J., Hoste, J., Barbier, F., Vallenberghe, L., DeRudder, J., and Cornelis, R. (1981). *In* "Trace Element Metabolism in Man and Animals—TEMA 4" (J. M. McHowell, J. W. Gawthorne, and C. L. White, eds.), p. 534. Australian Academy of Science, Canberra.
182. Verwey, J. H. P. (1968). Proc. 1st European Congress, "Influence of Air Pollution on Plants and Animals," p. 269. Centre for Agric. Publ. Document., Wageningen.

183. Walshe, J. M. (1986). *In* "Orphan Diseases and Orphan Drugs" (I. H. Scheinberg and J. M. Walshe, eds.), p. 76. Manchester University Press, Manchester.
184. Ward, G. M. (1978). *J. Anim. Sci.* **46**, 1078.
185. Ward, G. M., and Nagy, J. G. (1976). *In* "Molybdenum in the Environment" (W. R. Chappell and K. K. Petersen, eds.), pp. 97–113. Dekker, New York.
186. Warren, H. V., Delavault, R. E., Fletcher, K., and Wilks, E. (1971). *In* "Trace Substances in Environmental Health" (D. D. Hemphill, ed.), pp. 94–103. Univ. of Missouri, Columbia.
187. Wattiaux-DeConinck, S., and Wattiaux, R. (1971). *Eur. J. Biochem.* **19**, 552.
188. Waud, W. R., and Rajagopalan, K. V. (1976). *Arch. Biochem. Biophys.* **172**, 365.
189. Waud, W. R., and Rajagopalan, K. V. (1976). *Arch. Biochem. Biophys.* **172**, 354.
190. Wester, P. O. (1976). *Atherosclerosis* **20**, 207.
191. Westerfeld, W. W., and Richert, D. A. (1953). *J. Nutr.* **51**, 85.
192. Westerfeld, W. W., and Richert, D. A. (1954). *Ann. N.Y. Acad. Sci.* **57**, 896.
193. Widjajaksuma, M. C. R., Basrur, P. K., and Robinson, G. A. (1973). *J. Endocrinol.* **57**, 419.
194. Williams, M. A., and Van Reen, R. (1956). *Proc. Soc. Exp. Biol. Med.* **91**, 638.
195. Williams, R. J. P. (1978). "The Biological Role of Molybdenum." Climax Molybdenum, London.
196. Wilson, L. G., and Bandurski, R. S. (1958). *J. Biol. Chem.* **223**, 975.
197. Winge, D. R., Southerland, W. M., and Rajagopalan, K. V. (1978). *Biochemistry* **17**, 1846.
198. Winston, P. W. (1981). *In* "Disorders of Mineral Metabolism" (F. Bronner and J. W. Coburn, eds.), Vol. 1, p. 295. Academic Press, New York.
199. Winston, P. W., Hoffman, L., and Smith, W. (1973). *In* "Trace Substances in Environmental Health—7" (D. Hemphill, ed.), Univ. of Missouri, Columbia.
200. Wyngaarden, J. B. (1978). *In* "Metabolic Basis of Inherited Disease" (J. B. Stanbury, J. B. Wyngaarden, and D. S. Frederickson, eds.), p. 1037. McGraw-Hill, New York.
201. Wynne, K. N., and McClymont, G. L. (1956). *Aust. J. Agric. Res.* **7**, 45.

Index

A

AAS, *see* Atomic absorption spectrometry

Adenosylcobalamin
-dependent reactions, mechanism of, 175, 176
methylmalonyl coenzyme A mutase dependence on, 165
in vitamin B_{12} deficiency, sheep, 165
structure, 146

Adenylate cyclase, activation by vanadium, 284

Albumin, nickel binding in serum, 249–250

Aldehyde oxidase, molybdenum intake and, 441

Alkaline phosphatase, in bone, fluorine intake and, 394

Amine oxidase, chicken
in bone, reduction in copper deficiency, 328–329

Amino acids
metabolism in chromium deficiency, 236–237
nonheme iron absorption stimulation, 94

Ammonium tetrathiomolybdate, Wilson's disease treatment, 455

Anemia, in copper deficiency
ceruloplasmin role, 327–328
iron metabolism impairment, 327

Anorexia, in molybdenum toxicity, rat, 447

Arginase, manganese-containing, diet effect, 196

Arteriosclerosis, chromium in hair decrease, 229

Ascidian worm, vanadium in blood, 275

Ascorbic acid
copper absorption inhibition, 320, 322
nickel toxicity alleviation, 266
nonheme iron absorption stimulation, 92–93, 121–123
vanadium toxicity reduction, 293

Ataxia
congenital, in manganese deficiency, mouse, 201
neonatal
in copper deficiency, lamb
cytochrome oxidase depression, 331–332
myelopathy, 331–332
neuron degeneration in brain stem and spinal cord, 331
phospholipid synthesis impairment, 332
uncoordinated movements, 329–330
in manganese deficiency, guinea pig, rat, 200–201

Atomic absorption spectrometry (AAS)
description, 28–29, 67
flame atomization, 67–69
graphite furnace atomization, 68–69
(Na,K)-ATPase, inhibition by vanadium, 282–283
manic-depressive illness, and, 283
reversal by catecholamines, 287

B

Blood
chromium in serum, human, 43, 226–227
after glucose load, 44
increase by stress, 238

copper in serum and plasma
  ceruloplasmin and, 45, 311, 312
  diet effects, 315–316
  disease effects, chicken, human, sheep,
    316–317
  normal level, 44–45, 312–313
  during pregnancy and neonatal growth,
    313–314
  SOD in erythrocytes and, 45, 311–312
fluorine in plasma, intake effect, 392, 400
iron content
  ferritin, 84–86
    -hemosiderin ratio, 87–89
  hemoglobin, 81–82
  hemopexin and haptoglobin, 86
  transferrin, 82
manganese content
  in animals, 189
  in chicken plasma, egg-laying and, 189
  in whole blood and erythrocytes, human,
    188–189
    disease and diet effects, 189
    sex differences, 190
  menstrual, iron excretion, 105
mercury content, poisoning manifestation
    and, human, 426
molybdenum content, intake effects, 433
  copper-molybdenum protein and, 433
  inhibition by dietary sulfate, 439
nickel in serum
  albumin-bound, 249–250
  animal species, 247
  decrease in patients with hepatic cirrhosis
    and psoriasis, 248
  increase in patients with
    acute myocardial infarction, 247–248
    acute stroke and severe burns, 248
  toxicity test, 266
vanadium–transferrin complex in plasma,
    278, 279
zinc in serum and plasma, human, 47–48
  metallothionein and, 49
Bone
  copper deficiency-induced disorders
    amine and lysyl oxidases reduction, 328–
      329
    osteoblastic activity depression, 328
  fluorine content, 392
    intake effect, 401
  in fluorosis, human
    density increase, 378–379

fluorine entry in crystal lattice, 366, 372
    radiological changes, 371
  manganese
    accumulation, dietary supply effect, 186
    deficiency-induced abnormalities, 198–
      199
      mucopolysaccharide synthesis impair-
        ment and, 199–200
  mineralization, promotion by vanadium, 289
Bone marrow
  ferritin–hemosiderin ratio, 88
  plasma iron flow to, 105–106
Boron, fluorosis symptom amelioration, 387,
    397
Brain, manganese accumulation and function,
    197

C

Cadmium
  copper absorption inhibition, 315, 323
  iron absorption inhibition, 97
Calcineurin, nickel-activated in vitro, 257
Calcium
  balance in fluorosis, human, 377–378
  copper absorption and, 323, 324
  cytoplasmic, increase by vanadium, 288,
    290
  -fluorine interaction, human, 365, 367, 375,
    377, 386–387
  manganese absorption inhibition, poultry,
    192
Carcinogenesis, blocked by vanadium, rat, 291
Cardiovascular system
  in copper deficiency, disorders in
    animals, 334–335
    humans, 343–344
      zinc–copper ratio effect, 346
  in fluorosis, arterial lesions, 386
Catecholamines
  reversal of (Na,K)-ATPase inhibition by
    vanadium, 287
Central nervous system (CNS)
  changes in Phalaris staggers, ruminants,
    170–171
Certified reference material (CRM), analytical
    methods validation and, 72–73
Ceruloplasmin
  anemia in copper deficiency and, 327–328
  iron mobilization from stores and, 106–107
  in serum, 45, 311, 312

disease effects, 316–317
transport from liver to target organ, 325
Chicken
  copper
    deficiency, 329, 334–335
    dietary requirements, 340
  iron requirements, 118–119
  manganese requirement during development,
    209–210
  nickel deficiency signs, 252
Chlorophyll synthesis
  vanadium-stimulated, *Chlorella fusca,* toma-
    to, 286
Cholesterol, serum
  decrease in manganese deficiency, 204
  increase in deficiency of
    chromium, 235–236
    copper, animals, 337; human, 343
  vanadium dose-dependent effect, 288–289
Chondroitin sulfate, manganese deficiency
  and, 199
Chromium
  absorption
    accumulation in nuclear fraction, 231, 237
    dietary intake and, human, 229
    hexavalent
      high absorption and retention, 230–231
      transport to erythrocytes, 231
    inhibition by
      iron, vanadium, zinc, 230
      phytate, 230
    stimulation by oxalate, 230
    trivalent
      low absorption and retention, 230–231
      transferrin-bound, transport to tissues,
        231
  content in human organs and fluids
    decline with age, 225
    exposure to chromium and, 226
    geographic region and, 225–226
    hair, age and disease effects, 44, 228–
      229
    milk, 229
    serum, 43–44, 226–227
      increase by stress, 238
  dietary
    errors in analysis, 239–240
    intakes in various countries, 238–240
    longevity increase, male mouse, 238
    toxicity, higher of hexavalent than of triv-
      alent, 240

interaction with nuclear RNA in liver, 237
in poultry eggs, 229
urinary excretion, human, 43, 227–228
  increase by stress, 238
vanadium toxicity reduction, 293
deficiency
  impairment of
    amino acid metabolism, 236–237
    glucose tolerance, 232–235
    serum cholesterol increase, 235–236
    signs and syptoms of, 232 (table)
CNS, *see* Central nervous system
Cobalt, *see also* Vitamin B$_{12}$
  absorption
    from diet, 150–151
    enhancement by iron deficiency, 151–
      152
    poor in ruminants, 152
  content in animals and humans
    human blood, variation, 150
    increase by cobalt injection, *animals,*
      149
    tissue distributions and, 152
    in milk, 150
    organs, mainly liver, 148–149
  excretion
    fecal, 152
    with sweat and hair, 152
    urinary
      intestinal absorption and, 151, 152
      iron status and, 151
  in human nutrition
    content in diet, 172
    erythropoiesis stimulation, 172–173
    supply only in form of vitamin B$_{12}$, 172
    thyroid gland function and, 173
  iron absorption inhibition, 97–98
  in nonruminant animal nutrition
    low requirements, 173
    vitamin B$_{12}$ supplement, 174
  toxicity
    cardiac failure in heavy beer drinkers,
      174–175
    cardiomyopathy, dog, 175
    low order of, 174
Cobalt, in ruminant nutrition
  deficiency
    "bush sickness", cattle, New Zealand,
      22, 153
    "coast disease", sheep, South Australia,
      22, 153–154

manifestation
  content decline in liver and kidney,
      149, 155
  symptoms, 154–155
  Phalaris staggers, chronic, 169–172
  prevention
    direct administration to staff-fed ani-
        mals, 159–160
    pasture top dressing with cobalt-con-
        taining salts, 159
    vitamin B_{12} subcutaneous injection, 159
  treatment
    cobalt pellet therapy, 161
    injection ineffectiveness, 160–161
    vitamin B_{12} depletion, 155–158, 161–162
    white liver disease, lamb, 168–169
  requirements, 156–157
Cobamides
  biological activities, 146 (table)
  production by some microorganisms, 143–
      148
    evolution and, 146–147
    methanogenesis and, 147–148
  structure, 145
  synthesis by rumen organisms, 162–163
    cobalt deficiency and, 162
Coffee, nonheme iron absorption inhibition, 98
Complement metabolism, C3 convertase, nick-
    el-activated, 258
Conalbumin, iron-bound in egg white, chick-
    en, 90–91
Copper
  absorption
    age and diet effects, 320
    calcium effects, 323, 324
    inhibition by
      ascorbic acid, 320, 322
      cadmium, 315, 323
      iron, 315
      molybdenum, 315, 320, 322–323; *see*
          also Molybdenum toxicity
      zinc, 315, 321, 323, 324
    mechanisms of, 321
    metallothionein role, 321, 324
  content in animals and humans
    blood, 44–45, 311–317
    distribution in tissues, 304, 307–308
    eggs, hen, 319–320
    fluorine-induced decrease, human, 380
    hair, 45, 318–319

liver, 308–311
    milk, 317–318
      decrease in fluorosis, human, 381
    wool, 319
  excretion
    fecal, 325
    urinary, 46, 325–326
  importance of, early history, 301–303
  iron absorption inhibition, 97
  nickel toxicity affected by, 264–265
  transport through liver, 324–325
Copper deficiency, animals
  anemia, iron metabolism and, 327–328
  bone disorders, species differences, 328–
      329
  cardiovascular disorders, 334–335
  enzymes in development of, 326
  glucose tolerance reduction, 337
  immune system impairment, 335–336
  infertility, 333–334
  keratinization impairment in fur and wool,
      332–333
  lipid metabolism changes, 336–337
  neonatal ataxia, 329–332
  neuropeptides and, 326–327
  pigmentation failure in fur and wool, 332
  scouring, cattle, 329
Copper, dietary
  in animal feed, 303–304, 306
  deficiency in human nutrition, 342–344
    cardiovascular disturbances, 343–344
    SOD activity decrease, 343
  in human foods, 304–306
  requirements
    chicken, 340
    human, 344–346
    rodents, 338
    ruminants, 341–342
  toxicity, animals
    pig, 348
    ruminants, 348–350
    species differences, 347–348
  -zinc ratio, coronary heart disease and, 346
Copper sulfate, scouring therapy in molyb-
    denum toxicity, cattle, 448
Copper sulfide, synthesis in rumen
    molybdenum–copper antagonism and, 450–
      451
Creatine phosphokinase, serum, increase in
    copper toxicosis, 349

CRM, *see* Certified reference material
Cytochrome oxidase, depression in copper deficiency, 331–332

D

Dental caries, *see* Teeth, caries
Desferrioxamine
 iron urinary excretion stimulation, 104
 nonheme iron absorption inhibition, 94–95
Diabetes
 chromium content in hair decrease, 229
 effect of chromium supplementation, 234–235
 -like glucose tolerance curves in manganese deficiency, rat, 204–205
Dog, fluorine toxicity, 398

E

EDTA
 nonheme iron absorption inhibition, 95
 vanadium toxicity reduction, 293
Eggs, hen
 chromium content, higher in yolk, 229
 copper content, 319–320
 iron
  conalbumin-bound in white, 90–91
  phosvitin-bound in yolk, 90
 manganese content
  deficient, shell poor formation, 200
  higher in yolk, 190
Elastin, aortic, decrease in copper deficiency, 334–335
Emission spectrography, 28
Epilepsy, seisure increase in manganese deficiency, rat, 197
Erythrocytes
 copper bound to SOD, 311–312
 hexavalent chromium transport to, 231
 manganese level, 189
 mercury content, fish consumption and, human, 418
Escherichia coli (E. coli)
 anemic piglet infection, 115–116
 SOD, manganese-containing, 196
Ethanol, dietary, manganese absorption stimulation, 192

F

Fatty acids, composition in copper deficiency, 336
Feathers, manganese content, 188
Feces, excretion of
 cobalt, 152
 copper, 325
 iron, 103
 nickel, 248–249
 trace elements, metabolic balance technique, human, 38–39
 vanadium, 277, 278
Ferritin
 content in human serum, 84–86
  during development, 84–85
  inflammation and, 86
  liver disease and, 85–86
  sex differences, 84
 histochemistry, 87
 iron storage and, 87–89
 in liver and spleen cells, 106–107
 mucosal iron absorption and, 100
 properties, 87
 -vanadium complexes in plasma, 279
Ferritin.-hemosiderin ratio
 in bone marrow and muscles, 88–89
 human diseases and, 88
 iron storage and, 88
 in liver and spleen, 87
Fish, methylmercury high content, 423–424
Fluorine
 absorption
  dietary factor effects
   inhibitory by calcium and aluminum, 399
   stimulatory by fat, 399–400
  high speed of, 398–399
 animal tolerance, species differences, 393
 content in animals and humans, intake effect
  blood, 400
  bones and teeth, 401
  distribution throughout body, 400–401
  urine, 401–402
 dental caries reduction, human, 403
  bactericidal action and, 404
  fluorine content in water and, 405
 effects on enzymes
  in animals, 394, 395
  in plants, 393–394

intake with water and diet, human, 406–407
metabolism, human, scheme, 402
natural distribution, 405–406
in surface waters, 406
Fluorine toxicity, animals
cattle
content
in bone and fluids, 392
in soft tissues, 390–391
symptoms, 387–390
dental fluorosis, 388–389
osteofluorosis, 389–390
tolerance, 390, 393
dog, 398
guinea pig, 398
monkey
bone changes, 395
calcium turnover and, 395
mouse, chromosomal aberrations in bone
marrow cells, 396–397
rabbit
bone composition and, 398
boron protective effect, 397
osteoporosis, 398
protein in serum and tissues decrease,
397
rat, microsomal enzyme reduction in liver
and kidney, 395
Fluorosis, human
amelioration with calcium, magnesium metasilicate, boron, 386–387
arterial lesions, 386
bone
density increase, 378–379
radiological changes, 371
calcium balance and, 377–378
clinical events
calcium fluoride formation, 365, 367,
375, 377
enzyme inhibition, 367
fluorine entry in bone crystal lattice, 366,
372
hydroxyproline increase in urine, 366–
367
hyperparathyroidism, 365–366, 380, 383–
384
enamel, induction by fluoride tablets, 404
endemic genu valgum, low dietary calcium
and, 379–380
endocrine effects, 383–384

fluorine metabolism
balance in body, 378–379
aluminum hydroxide effect, 385
content increase in tissues and fluids,
372–376
milk, 374
renal stones, 372, 374, 384–385
transplacental transfer and, 382–383
urinary excretion increases, 385–386
foodborne, 370–371
during hemodialysis with fluoridated water, 386
hydric
dental, endemic 368
skeletal, endemic 368–369
industrial, 369–370
kidney functional impairment, 384–385
metabolic effects, 382
neighborhood, fluorine discharge by industry
and, 371
neurogenic atrophy, 386
muscle weakness and, 386
osteomalacia, 379–380
osteoporosis
fluoride treatment causing complications,
381–382
in undernourished populations, 381
serum composition and, 374–375
trace element content and, 380–381
Folates, in ruminant liver, vitamin B_{12} deficiency and, 167

G

Gastric secretion, iron absorption and, 102
Genu valgam, endemic, in human fluorosis,
379–380
Gluconeogenesis, manganese deficiency and,
205–206
Glucose-6-phosphatase
inhibition by vanadium, 288
molybdenum toxicity and, 445, 447
Glucose tolerance
diabeticlike in manganese deficiency, rat,
204–205
improvement by
chromium supplementation, 232–235
in diabetics, 234–235
in hypoglycemics, 235
insulin-potentiating chromium complexes and, 233

in patients with total parenteral nutri-
    tion, 233–234
in protein-calorie malnutrition, 234
cytoplasmic Ca^{2+} increase and, 288
glucose-6-phosphates inhibition and,
    288
insulin receptor phosphorylation and,
    287–288
reduction in copper deficiency, 337
Glutamic oxaloacetic transaminase, serum
    increase in copper toxicosis, sheep, 349
Glutathione peroxidase
    depression in copper deficiency, 336–337
    in platelets, selenium nutriture in humans
        and, 46
Glycoprotein, nickel-binding
    histidine-rich in serum, 250
    in kidney and urine, 251
Glycosyltransferases, specific activation by
    manganese, 195
Gout, in molybdenum toxicity, human, 447
Growth hormone, increase in fluorosis,
    human, 384
Guinea pig
    fluorine toxicity, 398
    neonatal ataxia in manganese deficiency,
        200

H

Hair, human, content of
    chromium
        advantage of determination, 228
        decrease connected with
            age, 228
            arteriosclerosis, 229
            diabetes, 229
        cobalt, 152
        copper, individual variation, 45, 318–319
        iron, 91, 104
        manganese, dietary supply and, 187–188
        mercury, fish consumption and, 418, 420
        vanadium
            decrease in multiple sclerosis, 276
            increase in chronic renal failure, 276
Haploglobin, hemoglobin return to liver and,
    86
Heliotropium europaeum, in pasture
    liver damage, copper toxicity and, 348–349

Hemoglobin
    content in blood, 81–82
    metabolic cycle, iron turnover and, 105–106
    return to liver by hemopexin and
        haploglobin, 86
Hemopexin, hemoglobin return to liver and, 86
Hemosiderin
    formation from ferritin, liver and spleen
        cells, 107
    histochemistry, 87
    iron storage and, 87–89; *see also* Ferritin–
        hemosiderin ratio
    properties, 87
Hydroxyproline, urinary
    increase in fluorosis, human, 366–367
Hyperparathyroidism, human
    in fluorosis, 365–366, 377, 380, 383–384
    in osteoporosis treatment with fluoride,
        381–382
Hypoglycemy, improvement by chromium,
    235

I

Immune system
    in copper deficiency, 335–336
    in managanese deficiency, 206–207
Infertility, *see* Reproduction impairment
Insulin
    activity potentiation by chromium
        in amino acid metabolism, 237
        in glucose tolerance, 233
    synthesis, impairment in manganese defi-
        ciency, 205
Iodine deficiency, goiter and, 21–22
Ionophores, nickel toxicity alleviation, 266
Iron
    chromium absorption inhibition, 230
    content in animals and humans
        blood hemoglobin, 81–82
        fluorine-induced increase, 380–381
        hair and wool, 91
        hen's eggs, 90–91
        milk, 89–90
        organs, 80
        plasma hemopexin and haptoglobin, 86
        serum
            in diseases, human, 83–84
            ferritin, 84–89
            species differences, 83

TIBC and, 82–84
    transferrin, 82
    total in body, 79–80
copper absorption inhibition, 315
excretion
    fecal, 103
    menstrual blood losses by women, 105
    with sweat, hair, nails, 104
    total quantity, 104–105
    urinary, 103–104
ferrous, vanadium toxicity reduction, 293
metabolism
    ceruloplasmin role, 106–107
        copper deficiency and, 327–328
    incorporation into
        ferritin in liver and spleen cells, 106–
            107
        reticulocyte hemoglobin, 106
    placental transport, unidirectional, 107–
        108
    plasma flow to bone marrow, 105–106
    transferrin receptor and, 107
nickel toxicity affected by, 264–265
storage compounds, 87–89; see also Fer-
    ritin, Hemosiderin
Iron deficiency
human
    adolescent females, menstrual blood loss
        and, 112–113
    adults
        in developing countries, 113
        during pregnancy, supplementary iron
            and, 113–114
    infants, iron supplement role, 111–112
    preadolescents, 112
    immunity and, 110
    manifestation
        blood iron and storage compound de-
            crease, 108
        folic acid decrease in serum, 109
        gastrointestinal tract abnormalities,
            109
        heme enzyme inhibition in blood and
            tissues, 109–110
        lipid abnormalities, 108–109
        muscle myoglobin reduction, 109
        TIBC increase, 109
    pigs on high copper diet
        older pigs, amelioration by ascorbic acid,
            115

piglets
    E. coli infection and, 115–116
    iron complex injection and, 115
    symptoms, 114
ruminants, young, 116
vanadium effect, 286
Iron, dietary
in animal feed
    content, 124–125
    manganese absorption inhibition, 191–192
    toxicity, 126
in human foods
    content, 119–120
    efficacy, bioavailability factors, 120–123
    fortification, 123–124
    toxicity, 125–126
requirements
    for cattle and sheep, 119
    for humans
        high during growth and pregnancy,
            116–117
        in vegetal diet higher than in animal,
            117–118
    for poultry, 118–119
    for pigs, 118
Iron, nonheme, absorption, 90–103
    from diets of animal and plant origin, 98–99
    during diseases and trace element deficien-
        cies, 99
    inhibition by
        cadmium, copper, manganese, 97
        cobalt, 97–98, 151–152
        coffee and tea, 98
        desferrioxamine, 94–95
        EDTA, 95
        nickel, rat, 254–255, 264–265
        phosphate high level, 97
        sodium phytate, 95–96
        soy products, 96–97
        zinc, 97–98
    mechanism and regulation
        ferritin role, 100–101
        gastric secretion and, 102
        mucosal block theory, animal model, 99–100
        pancreatic secretion and, 102–103
        transferrin role, 100–101
    radiolabeled extrinsic tag technique, 92
    species differences, 101–102
    stimulation by
        amino acids, 94

ascorbic acid, 92–93, 121–123
carbohydrates, 94
"meat factor", 93–94, 121
organic acids, 93

K

Kidney
chronic failure, vanadium increase in hair,
276
in fluorosis, human
fluorine-containing stones, 372, 374,
384–385
tubular and glomerular dysfunction, 384
nickel
glycoprotein-bound, 251
retention, 250–251

L

Lactoferrin, in milk, 90
Liver
cirrhosis, serum nickel increase, 248
cobalt content, 148–149
vitamin B_{12} and, 149–150
copper content, effects of
age, 308–309
diet, 309–311
diseases, 311
species, 308
ferritin–hemosiderin ratio, 87
hemoglobin return to, hemopexin and
haploglobin role, 86
hepatocytes, copper uptake *in vitro,* 324–
325
manganese content, 185–186
dietary supply and, 187
molybdenum content, age and intake effects,
431
Lysyl oxidase, chicken, depression in copper
deficiency
in aortic tissues, 335
in lung, 329

M

Magnesium metasilicate, fluorosis symptom
amelioration, 387

Man
chromium in organs and fluids, 225–229
cobalt in nutrition, 172–173
copper dietary requirements, 344–346
fluorosis, 365–381
iron deficiency, 110–114
trace element studies, 37–49
Manganese
absorption
age-dependent decrease, rat, 193–194
assay with ^{54}Mn, 191
dietary factor effects
calcium excess, inhibitory, 192
ethanol, stimulatory, 192
iron, competitive inhibition, 191–192
manganese, 191, 195
pregnancy and, 192
bioavailability from diets, 192–193
content in animals and humans
blood, 188–190
concentrated in mitochondria, 185
hair, dietary supply and, 187–188
hen's eggs, higher in yolk, 190
milk, 190
organs, dietary supply effects, 186–187
glucocorticoid-induced redistribution,
mouse, 192
total in body, human, 185
wool and feathers, 188
enzyme activation
nonspecific, 195
specific, 195–196
excretion
effect on homeostasis, 194–195
fecal, 194
urinary, very low, 194
in metalloenzymes, 196
transport
clearance from blood stream, 193
hepatic release, glucocorticoids and, 193
retention higher in premature infants, 193
Manganese deficiency
ataxia, prevention by manganese supply during pregnancy
congenital, mouse, 201
neonatal, guinea pig, rat, 200
vestibular function impairment, 200–201
carbohydrate metabolism and, 205–206
immune function and, 206–207
iron absorption inhibiton, 97, 191

lipid metabolism and, 202–204
melanin abnormal formation, pallid mouse,
 201
mitochondria
 SOD activity decrease, 202–203, 204
 ultrastructural abnormalities, 202–203
pancreatic islets
 insulin synthesis impairment, 205
 number of islet decrease, 204
 SOD activity decrease, 205
reproductive function impairment
 female mammals, 201, 202
 hen, 201
 male mammals, 201–202
seizures in epileptics and, 197
skeletal abnormalities and, 198–199
 mucopolysaccharide synthesis impairment,
  199–200
 glycosyltransferase activities and, 199
Manganese, dietary
 in animal feed and forages, 212–213
 in human foods and beverages, 211–212
 requirements during development
  humans, 210–211
  pigs, 208
  poultry, 209–210
  rodents, 207–208
  ruminants, 208–209
 toxicity
  for animals, 213–214
  for humans, 214–215
Manic-depressive illness
 (Na,K)-ATPase inhibition by vanadium and,
  283
"Meat factor", nonheme iron absorption stimulation, 93–94, 121
Menkes' disease, male infants
 copper absorption and utilization impairment, 324, 342
 serum copper and ceruloplasmin decrease,
  316
Mercury
 content in human tissues and fluids, 417–
  419
  accumulation in thyroid and pituitary
   glands, exposed workers, 418–419
  erythrocytes, fish consumption and, 418
  hair, fish consumption and, 418, 420
 metabolism
  antagonism with selenium, 421–422

binding by proteins, 421
inorganic and methylmercury, comparison
 absorption, 420
 excretion, 420–421
 placental transfer, 421
sources
 air and water, 424
 dietary methylmercury, mainly from fish,
  423–424
 man-made, 422–423
toxicity
 manifestation, blood level and, 426
 methylmercury poisoning, 424–425
 neurological symptoms, 424–425
 protection by selenium, 421–422, 425
 safe dietary level, 425
Metalloenzymes, nickel-containing, in plants
 and microorganisms, 255–257
Metallothionein
 in copper metabolism, 321, 324, 325
 zinc nutriture in humans and, 49
Methanosarcina barkeri, cobamines in methane synthesis, 147–148
Methylcobalamin
 intermediate in synthesis of methane and
  methionine, 175
 methyltransferase and, 175
 structure, 147
Methylmalonic acid (MMA), in vitamin B_{12}
 deficiency, ruminants, 165–166
Methylmalonyl coenzyme A mutase
 adenosylcobalamin requirement for activity,
  165
 in propionate metabolism, ruminants, 164–
  165
 vitamin B_{12} deficiency and, 165, 167
5-Methyltetrahydrofolate: homocysteine methyltransferase, ruminant liver
 methylcobalamin and, 175
 vitamin B_{12} deficiency and, 167
 white liver disease in lambs and, 169
Microorganisms
 cobalt-dependent, cobamide production,
  143–148; *see also* Cobamides
 nickel-containing metalloenzymes, 255–257
 vitamin B_{12} synthesis in rumen, 143, 162–
  163
Milk, content of
 chromium, 229
 cobalt, 150

copper, 317–318
  decrease in fluorosis, human, 381
  fluorine, intake effect, cattle, 392
  iron
    association with casein, lactoferrin, trans-
      ferrin, 90
    iron status and, 89
    species differences, 89
  manganese
    accumulation in
      casein fraction, bovine, 193
      whey, human, 193
    dietary supply effect, 190
  molybdenum, intake effect, 433–434
    xanthine oxidase and, 434
  nickel, 248
MMA, see Methylmalonic acid
Molybdenum
  in animal feed and forages, 436, 437
    mixed grasses and clovers, soil pH effect,
      437 (table)
  content in animal organs and fluids
    blood, 433
    bone, 435
    dietary sulfur effect, sheep, 431, 432
    liver, 431
    milk, 433–434
    species differences, 430 (table)
    teeth, 434–435
  copper absorption inhibition, 315, 322–323
    see also Molybdenum toxicity, cattle
  deficiency
    conflicting results, 442–443
    nitrate reductase decrease in rumen, 443–
      444
  functions
    aldehyde oxidase and, 441
    sulfite oxidase and, 441–442
    xanthine oxidase and, 434, 440–441, 447
  in human foods and diets, 435–436
  metabolism
    absorption, inhibition by sulfate, 437–438
    -tungsten interaction, 440, 442, 443
    urinary excretion, increase by sulfate,
      438–439
  natural sources, 429, 431
Molybdenum toxicity
  cattle
    copper antagonism, 438, 449–457
      sulfate effects, 438, 439, 444, 450–452

      thiomolybdate synthesis and, 451–453,
        455
      tolerable dietary copper–molybdenum
        ratio, estimation, 456–457
    reproduction impairment, 449
    scouring, treatment with copper sulfate,
      448
  clinical manifestation, 445–447
    gout, high incidence, human, 447
    metabolic changes, 445, 447
    species differences, 445
    sulfide-potentiated, 444, 446–447
Molybdopterin
  as cofactor for
    sulfite oxidase, 442
    xanthine oxidase, 440
  structure, 439–440
Monkey, fluorine toxicity, 395
Mouse
  chromium in diet, longevity increase in
    males, 238
  fluorine toxicity, chromosomal aberration in
    bone marrow cells, 396–397
  manganese deficiency
    congenital ataxia and, 201
    pallid form, melanin abnormalities, 201
Mucopolysaccharides, in manganese deficiency
  egg shell poor formation, hen, 200
  skeletal abnormalities, 198–200
Multiple sclerosis, vanadium decrease in hair,
  276
Muscles
  cardiac, vanadium effect on contractile
    force, 290–291
  myoglobin reduction in iron deficiency, 109
  skeletal, ferritin–hemosiderin ratio, 88–89
Myocardial infarction, acute, serum nickel increase, 247–248

N

NAA, see Neutron activation analysis
NADH, oxidation by vanadate, 284–285
Neuropeptides, reduction in copper deficiency,
  326–327
Neutron activation analysis (NAA), 29–29, 69
Nickel
  content in animals and humans
    animal tissues, species differences, 246–
      247

human tissues, 245–246
milk, human, 248
serum
   response to pathological conditions,
      human, 247–248
   species differences, 247
sweat, human, 248, 249
functions
   enzyme activation *in vitro*, 257–258
   iron absorption enhancement, 258
   in metalloenzymes, plants, micro-
      organisms, 255–257
   placenta separation and, 259
   prolactin secretion and, 259
metabolism
   absorption, very low, 248
   complexes with serum proteins, 249–250
   excretion
     fecal, 248–249
     urinary, 249
   retention in kidney, 250–251
     bound to glycoprotein and peptide, 251
   transport across mucosal epithelium, 249
parenteral nutrition, toxicity signs, 263–264
Nickel carbonyl, carcinogenic, toxicity by inhalation, 266
Nickel, dietary
in animal feed and pastures, 261–262
deficiency signs, 251–255
   cattle, 252
   chicken, 252
   goat, 252–253
   pig, 253
   rat, iron status and, 253–255
   sheep, 255
in human foods, 260–261
requirements for animals, 259–260
toxicity
   animals, 262
   clinical tests, content in urine and serum,
     266
   effects of
     ascorbic acid, protein, ionophores, 266
     copper, antagonistic interaction, 265
     iron, competitive interactions, 264–265
     zinc status, 265–266
   human, 263
   poultry, 262–263
Nickeloplasmin, in human and rabbit sera,
   250

Nitrate reductase, in molybdenum deficiency,
   ruminants, 443–444

O

Oral contraceptives, copper increase in blood
   and, 313
Organic acids, nonheme iron absorption stimulation, 93
Osteofluorosis, cattle, 389–390
Osteomalacia, human
   in fluorosis, 379–380
   in osteoporosis treatment with fluoride, 380,
     381
Osteoporosis, human
   as fluorosis complication in undernourished
     populations, 381
   treatment with fluoride, 381–382

P

Pancreas
   insulin synthesis in islets, manganese defi-
     ciency and, 204–205
   secretion, iron absorption and, 102–103
Peroxidase, vanadium-dependent, *Ascophyllum
   nodosum,* 285–286
Phalaris species, in grazing pastures, effects
   of sheeps
   chronic *Phalaris* staggers, 170–172
   peracute disease, heart failure, 169–170
Phalaris staggers, chronic
   CNS changes in ruminants
     myelin degeneration in spinal cord, 170–
      171
     pigment in nerve cells, 170
   prevention by cobalt salt oral dosing, 170
Phosphate, high level, iron absorption inhibition, 97
Phosphoenolpyruvate carboxykinase
   in manganese deficiency, 205–206
   specific activation by manganese, 196
Phospholipid synthesis
   in liver microsomal fraction, inhibition by
     vanadium, 289
Phosphoryl transfer enzymes, inhibition by
   vanadium, 283–284
Phosvitin, iron-bound in egg yolk, hen, 90
Phytate, chromium absorption inhibition, 230

Pig
  copper, dietary
    requirements, 338–340
    toxicity, 348
    very high, iron deficiency induction
      ascorbic acid treatment, 115
      E. coli infection and, 115–116
      iron complex injection and, 115
  manganese requirement during development, 208
  nickel deficiency signs, 253
Pituitary gland, mercury accumulation in exposed workers, 418–420
Placenta, human
  maternal fluorine transfer to fetus, 382–383
  separation, nickel role, 259
Progesterone receptor activation, inhibition by vanadium, 286–287
Prolactin secretion, nickel effect, 259
Propionate metabolism, ruminants
  glucose synthesis in liver and, 165
  impairment in vitamin B_{12} deficiency, 165
    clearance from blood and, 165–167
    MMA accumulation and excretion, 165–166
    methylmalonyl coenzyme A mutase and, 164–165
Protein, nickel toxicity alleviation, 266
Protein-calorie malnutrition, improvement by chromium, 234
Protein kinase, tyrosine-specific, activation by vanadate, 285
Psoriasis, serum nickel decrease, 248
Pyruvate carboxylase
  manganese-containing, diet effect, 196
  in manganese deficiency, 205–206

R

Rabbit, fluorine toxicity, 397–398
Rat
  copper
    dietary requirements, 338
    -molybdenum antagonism, 453–454
  fluorine toxicity, 395–396
  manganese requirement during development, 207–208
  nickel deficiency signs, 253–255
    iron status and, 253–255

Reproduction, impairment
  in copper deficiency, cattle, hen, rodents, 333–334
  in manganese deficiency
    hen, 201
    rodents, 201–202
    ruminants, 202
  in molybdenum toxicity, 449
Rheumatoid arthritis, SOD decrease in neutrophils, 203
RNA, nuclear in liver, chromium effects
  protection against heat denaturation, 237
  synthesis stimulation, 237
Ruminants
  cobalt in nutrition, *see* Cobalt, in ruminant nutrition
  copper
    deficiency, 22, 329–332
    dietary requirements, 341–342
    toxicity
      creatine phosphokinase increase, 349
      glutamic oxaloacetic transaminase increase, 349
      hemolytic crises, 349–350
      liver damage by *Heliotropium europaeum* and, 348–349
  fluorine toxicity, 387–394
  manganese requirement during development, 208–209
    deficiency, nitrate reductase decrease in rumen, 443–444
  molybdenum
    dietary, content in pastures, 436–437
    in organs, dietary sulfate effect, 431–432
    toxicity, 448–449
      copper and sulfate interaction, 438–439, 446, 450–453, 455
  nickel deficiency signs, 252–253, 255
  vitamin B_{12} synthesis by microorganisms in rumen, 143, 162–163

S

Scouring, cattle
  in copper deficiency, 329
  in molybdenum toxicity, copper sulfate therapy, 448
Selenium
  deficiency causing diseases
    in humans (China), 23

in livestock, 22
nutritional status assessment, human
    concentration in blood and hair, 46
    glutathione peroxidase activity in platelets
        and, 46
    protection against mercury toxicity, 421–
        422, 425
    toxicity, protection by mercury, 421
SOD, *see* Superoxidse dismutase
Sodium phytate, nonheme iron absorption inhi-
    bition, 95–96
Soy products, iron absorption and, 96–97, 123
Stress, chromium metabolism and, 238
Stroke, acute, serum nickel increase, 248
Sulfate, effect on molybdenum metabolism
    absorption and retention decrease, 432,
        437–439
    copper interaction, 438–439, 450–453, 455
    toxicity potentiation, 444, 446
    urinary excretion increase, 438–439
Sulfite oxidase, molybdenum intake and, 441–
    442
Superoxide dismutase (SOD)
    copper binding in erythrocytes, 311–312
    in copper deficiency, decrease
        animals, 336
        human, 343
    from *E. coli* and chicken, comparison, 196
    in erythrocytes, copper nutriture, human,
        and, 45
    manganese deficiency-induced decrease, 196
    in mitochondria, 202–203, 204
    in pancreatic islets, 205
    in neutrophils, decrease in rheumatoid ar-
        thritis, 203
Sweat
    cobalt loss, 152
    iron excretion, 104
    nickel content, 248, 249

T

Tea, nonheme iron absorption inhibition, 98
Teeth
    caries, human
        induction by fluoride tablets taken in early
            childhood, 404
        reduction by fluorine in water, 403
        vanadium effects, 289–290
    defects in fluorine toxicosis, cattle, 388–389
    fluorine content, intake effect, 401

molybdenum content, intake effect, 435
Thiomolybdates
    in copper-containing diet, effect on copper
        and iron metabolism, rat, 453–454
    synthesis in rumen, molybdenum–copper
        antagonism and, 451–453, 455
Thyroid gland
    mercury accumulation in exposed workers,
        418–419
    in molybdenum toxicity, 447
Total iron binding capacity (TIBC)
    species differences, 82–83
    variation in diseases, human, 83–94
Total parenteral nutrition, diabeticlike symp-
    toms, 233
    improvement by chromium, 234–235
Trace elements, *see also specific elements*
    analytical methods
        accuracy-based measurement, 59–62
        calibration standards, 70–71
        criteria for routine procedures, 69
        data handling and evaluation, 74–76
        general laboratory practices, 64–66
        history, 26–27
        quality
            assurance systems, 57–58
            control, 64, 74–75
        sample digestion, 69–70
            dry ashing, 70
            wet, with strong acids and oxidizing
                agents, 70
        sampling, 62–64
        signal measurement error, sources of, 71–
            72
        techniques available
            AAS, 28–29, 66–69
            NAA, 28–29, 66
        validation
            common exchange sample and, 73–74
            CRM and, 72–73
        valid data generation, 64
    animal model, future research, 49–51
    classification based on essentiality, 3–5
    deficiencies
        in animals, induction by
            caging, all-plastic, 36–37
            diets, 23–24, 30–33
            drinking water, ultrapure, 34
            environment, ultraclean, all-plastic, 34–
                36
        criteria of, 14–15

nature's experiments, 21–23
phases of
    clinical (IV), 17
    compensated metabolic (II), 16
    decompensated metabolic (III), 16–17
    initial depletion (I), 15–16
discovery
    detection in living organisms, 5–6
    diseases induced by deficient or excessive
        intakes, 6–7
    new trace elements, 26
    nutritional significance, 8
    overexposure risks, 9
dose–response curves, 2–3
human studies
    metabolic balance technique, 37–40
        feces collection, 38–39
        food, duplicate samples, 38
        limitations, 39–40
        urine collection, 38
    nutritional status assessment, 42–49
    radioisotopes, 41
    stable isotopes, 41
    subject selection, 41–42
mode of action, 9–11
    enzyme system catalysis, 10
    homeostatic control of concentration, 10
needs and tolerances
    absolute and dietary requirements, 11–12
    criteria for, 13–14
    dietary recommendations, 12–13
    safe dietary levels, 13
Transferrin
    milk, 90
    mucosal iron absorption and, 100–101
    reticulocytes, binding and transportation,
        106
    serum
        defence mechanism and, 82
        diurnal rhythm, 82
        iron transport, 82
        structure, 82
        trivalent chromium binding, 231
    -vanadium complexes in plasma, 278, 279
Tungsten, interaction with molybdenum, 440,
    442, 443

U

Uric acid, serum, in molybdenum toxicity, 447
Urine, excretion of
    chromium, 227–228
        increase by stress, 238
    cobalt, 151, 152
    copper, 325–326
        plasma cortisone effect, 326
    fluorine
        increase in fluorosis, human, 385–386
        intake effect, 392, 401–402
    iron, 103–104
    nickel, 249
        toxicity test, 266
    trace elements, metabolic balance technique,
        human, 38
    vanadium excretion, 277, 278

V

Vanadium
    content
        in animal tissues, diet and age effects,
            267
        in ascidian worm blood, 275
        in human tissues and fluids
            blood, transferrin-bound in plasma,
                278, 279
            data variating with methods, 276
            hair, disease effects, 276
    deficiency
        contradictory findings, 279–280
        diet composition effects, 281–282
    functions
        adenylate cyclase activation, 284
        anticarcinogenic action, rat, 291
        (Na,K)-ATPase inhibition, 282–283
            manic-depressive illness and, 283
            reversal by catecholamines, 287
        calcium metabolism and, 290
        cardiac muscle contractile force and, 290–
            291
        chlorophyll synthesis stimulation, Chlo-
            rella fusca, tomato, 286
        chromium absorption inhibition, 230
        dental caries and, 289–290
        glucose tolerance improvement, 287–288
        hormone interactions with, 286–287
        iron metabolism and, 286
        lipid metabolism dose-dependent changes,
            288–289
        mineralization promotion, of bones and
            teeth, 289
        NADH oxidation by vanadate, 284–285

peroxidase cofactor, *Ascophyllum
nodosum,* 285–286
phosphoryl transfer enzyme inhibition,
283–284
tyrosine-specific protein kinase stimulation, 285
metabolism
absorption, species differences, 277–278
complexes with transferrin and ferritin,
278, 279
excretion
fecal, 277, 278
urinary, 277, 278
retention in kidney, 278
increase by endocrine deficiency, 286
Vanadium, dietary
in animal feed, 292
in human foods, 291–292
toxicity
animals and poultry
age-dependent, 293
dosage, 292–293
reduction by EDTA and other substances, 293
species differences, 293
humans, 294
manifestation, 294
Vitamin B_{12}, *see also* Adenosylcobalamin,
Methylcobalamin
in cobalt-deficient ruminants
absorption in small intestine and, 164
cobamide activity and, 162–163
content decrease, sheep
liver, 149–150, 155, 158, 161
serum, 158
rumen, 155, 161
whole blood, 155, 161
folate store depletion in liver, 167
propionate metabolism impairment, 164–
167
methylmalonyl coenzyme A mutase
and, 165, 167
methyltransferase and, 167
requirement for parenteral and oral administration, 157
subcutaneous injection for prophylaxis,
159
synthesis by rumen organisms and, 162–
163

in human diet, 172
in nonruminant animal nutrition, 173–174
structure, 145
white liver disease treatment, lamb, 168

W

White liver disease, in cobalt deficiency, lamb
liver pathology, 169
neurobiological symptoms, 168–169
photosensitivity and, 168–169
reversal by cobalt or vitamin B_{12}, 168
Wilson's disease, human
ammonium tetrathiomolybdate therapy, 455
serum copper and ceruloplasmin decrease,
316
Wool
copper content, variable, 319
in copper deficiency, impairment of
keratinization, 332–333
pigmentation, 332
iron content, 91
manganese content, 188

X

Xanthine oxidase, molybdenum intake effects
intestine, 440–441
milk, 434
tissues, 447
Xylosyltransferase, specific activation by manganese, 196

Z

Zinc
blood, human, 47–48
metallothionein and, 49
chromium absorption inhibition, 230
copper absorption inhibition, 315, 321, 323,
324
-copper ratio in diet, coronary heart disease
risk and, 346
deficiency causing dwarfism, human, 22–23
iron absorption inhibition, 97–98
nickel toxicity affected by, 265–266

Printed and bound by CPI Group (UK) Ltd, Croydon, CR0 4YY

03/10/2024

01040428-0004